D0465181

OTHER TITLES OF RELATED INTEREST

Sociology in Our Times, by Diana Kendall: An introductory sociology textbook with special emphasis on the intersection of race, class, and ethnicity

Americans All: Race and Ethnic Relations in Historical, Structural, and Comparative Perspectives, by Peter Kivisto

Race, Class, and Gender: An Anthology, 2d edition, by Margaret L. Andersen and Patricia Hill Collins

Origins and Destinies: Immigration, Race, and Ethnicity in America, by Silvia Pedraza and Ruben G. Rumbaut

The American Class Structure: A New Synthesis, 4th edition, by Dennis Gilbert and Joseph A. Kahl

The Black Family: Essays and Studies, 5th edition, by Robert Staples

RACE AND ETHNIC RELATIONS
American and Global Perspectives

FOURTH EDITION

Martin N. Marger
Michigan State University

Wadsworth Publishing Company
I(T)P® An International Thomson Publishing Company

Belmont, CA • Albany, NY • Bonn • Boston • Cincinnati • Detroit • Johannesburg •
London • Madrid • Melbourne • Mexico City • New York • Paris •
San Francisco • Singapore • Tokyo • Toronto • Washington

Sociology Editor: Eve Howard
Editorial Assistant: Deirdre McGill
Marketing Manager: Michael Dew
Project Editor: Jerilyn Emori
Managing Designer: Andrew Ogus
Print Buyer: Karen Hunt
Permissions Editor: Jeanne Bosschart
Advertising Project Manager: Joseph Jodar
Copy Editor: Adrienne Armstrong
Illustrator: Craig Hanson
Compositor: ColorType
Printer: Quebecor Printing/Fairfield

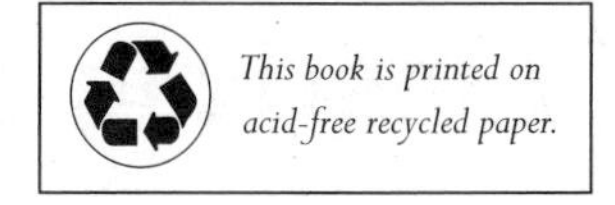

Printed in the United States of America
1 2 3 4 5 6 7 8 9 10

For more information, contact Wadsworth Publishing Company, 10 Davis Drive, Belmont, CA 94002, or electronically at http://www.thomson.com/wadsworth.html

International Thomson Publishing Europe
Berkshire House 168-173
High Holborn
London, WC1V 7AA, England

Thomas Nelson Australia
102 Dodds Street
South Melbourne 3205
Victoria, Australia

Nelson Canada
1120 Birchmount Road
Scarborough, Ontario
Canada M1K 5G4

International Thomson Publishing GmbH
Königswinterer Strasse 418
53227 Bonn, Germany

International Thomson Editores
Campos Eliseos 385, Piso 7
Col. Polanco
11560 México D.F. México

International Thomson Publishing Asia
221 Henderson Road
#05-10 Henderson Building
Singapore 0315

International Thomson Publishing Japan
Hirakawacho Kyowa Building, 3F
2-2-1 Hirakawacho
Chiyoda-ku, Tokyo 102, Japan

International Thomson Publishing Southern Africa
Building 18, Constantia Park
240 Old Pretoria Road
Halfway House, 1685 South Africa

Library of Congress Cataloging-in-Publication Data
Marger, Martin.
Race and ethnic relations : American and global perspectives / Martin N. Marger. — 4th ed.
p. cm.
Includes bibliographical references and index.
ISBN 0-534-50563-5
1. Race relations. 2. Ethnic relations. 3. United States—Ethnic relations. 4. United States—Race relations. 5. Ethnic relations—Case studies. I. Title.
HT1521.M275 1996
305.8—dc20 96-18143

CONTENTS

PART III: ETHNIC RELATIONS IN COMPARATIVE PERSPECTIVE

PREFACE

IN RECENT YEARS Americans have become progressively aware of the worldwide scope of race and ethnic relations. In addition to events in their own society, through the electronic and print media they have been exposed to ethnic conflict in societies as distant and exotic as Sri Lanka and Rwanda as well as those closer, geographically and culturally, like Canada and Northern Ireland. Issues of race and ethnicity abound on every continent, and it is evident that these issues increasingly defy national boundaries.

Curiously, American social scientists have not always kept pace in adapting to the global context of race and ethnic relations. Some continue to focus almost exclusively on the United States, paying only incidental attention to ethnic patterns and events in other societies. In line with this view, texts in the field of race and ethnic relations have ordinarily provided no more than cursory coverage to ethnic issues outside the American sphere, if at all. Students, therefore, often continue to think of racial and ethnic, or minority, issues as uniquely American phenomena.

Many other social scientists, however, have come to see the utility and relevance of a more cross-national, or comparative, approach to the study of race and ethnicity. Such an approach set off *Race and Ethnic Relations* from other texts when it was first published in 1985. Its objective was to provide readers with a comparative perspective without sacrificing a strong American component. That objective was retained in subsequent editions and remains unchanged in this, the fourth edition. The book's overriding theme is the global nature of ethnicity and the prevalence of ethnic conflict in the modern world.

At the same time that an international perspective on race and ethnicity seems more compelling than ever, a close and careful analysis of American ethnic relations is surely imperative. For better or worse, the United States, the most diverse of multiethnic societies, more often than not is a global pacesetter in ethnic relations. More important, most readers of *Race and Ethnic Relations* continue to be American students, who require a solid understanding of their own society that subsequently can be used as a comparative frame of reference. *Race and Ethnic Relations,* therefore, provides thorough coverage of America's major ethnic groups and issues. Indeed, as in previous editions, the United States remains the *essential* focus of the fourth edition.

The theoretical and conceptual thrust of this edition will be familiar to past readers: a power-conflict perspective, emphasizing the power dynamics among ethnic groups. Race and ethnic relations are seen as manifestations of stratification and of the competition and conflict that develop over societal rewards—power, wealth, and prestige. In accord with this perspective, I have emphasized the structural, or macrolevel, patterns of race and ethnic relations rather than the social-psychological patterns, though the latter are interspersed throughout.

In the relatively brief span of time since the publication of the third edition of *Race and Ethnic Relations,* much has changed in the United States and in other parts of the world that bears directly on the study of race and ethnicity. In response to those changes, the fourth edition has been revised with the latest data and the most relevant examples from a variety of societies and includes references to important new studies. In fact, virtually every chapter has been revised substantively.

The basic organization of the book has not been altered. The intent of the chapters that make up Part I is to introduce the principal terms, concepts, and theories of the field of race and ethnic relations. Although illustrations of ideas and concepts and applications of theory are related primarily to American society, whenever appropriate they are presented in an international context. New discussions have been incorporated into these chapters where they seemed warranted.

Part II, which focuses on American society, has been kept intact organizationally, though timely changes in content have been made throughout. Chapter 5 has been revamped to better explain the formation of the American ethnic hierarchy and has been expanded to describe more thoroughly the place of Native Americans in the ethnic hierarchy. Chapters 6, 7, 8, 9, and 10, covering the other major U.S. ethnic groups and categories, have been brought up to date with the most current statistical data and topical illustrations. Chapter 11 is intended to present what I believe are the major ongoing issues of race and ethnic relations in the United States: the persistent gap between Euro-Americans and racial-ethnic groups; policies designed to address that gap, such as affirmative action; and large-scale immigration. Discussion of these issues has been updated to reflect pertinent societal events and policy changes that have occurred in the last few years.

Much of Part III has been substantially revised. Chapter 12, dealing with South Africa, has been almost completely rewritten in light of the political transformation that has taken place in that society. Canada and Northern Ireland have also experienced important events in recent years that have had a profound impact on the nature of their ethnic relations; Chapters 14 and 15 have been revised accordingly.

Chapter 16 is entirely new. With accelerating global migration, political turmoil, and changing national boundaries, ethnic conflict has seemed to gain momentum in almost every world region. Often this has been the result of immigration and expanding ethnic heterogeneity, as in Germany, France, and other western European countries. In other cases, ethnic nationalism has been the driving force of conflict, demonstrated most vividly in recent years in Africa and eastern Europe. After a discussion of the impact of immigration and increasing ethnic diversity in western Europe, the remainder of this chapter is devoted to the conflict in the former Yugoslavia, a tragic affair that in the 1990s gripped the American media, policymakers, and public.

With the addition of Chapter 16, an entire range of ethnically diverse societies is presented in Part III, each of which continues to be a case of major significance in the study of race and ethnic relations in the contemporary world. Moreover, each stands as an intriguing comparative case vis-à–vis the United States.

The number of American college and university courses with ethnic content has grown enormously in recent years. This, I believe, is a reflection of the pressing problems and commanding issues of race and ethnicity in America and the growing awareness of ethnic divisions and inequalities in an increasingly diverse society. The content of *Race and Ethnic Relations* is comprehensive and thus appropriate for a variety of courses that may be differently titled and structured (for example, "race and ethnicity," "minority relations," "ethnic stratification," "multiculturalism") but which all deal in some fashion with ethnic issues.

ACKNOWLEDGMENTS

As with past editions of *Race and Ethnic Relations,* I wish to acknowledge those at Wadsworth who helped prepare the fourth: Eve Howard, Sociology Editor, and Jerilyn Emori, Project Editor. Adrienne Armstrong did a meticulous job of copyediting. Serina Beauparlant, former sociology editor at Wadsworth, continued to exert a guiding influence and I express my gratitude to her. Also to be acknowledged are the reviewers for this edition who provided helpful insights and suggestions: Andrew Barlow, Diablo Valley College; Margaret Brooks-Terry, Baldwin-Wallace College; and William S. Packard, Tacoma Community College. A writing project such as this can be a demanding and stressful, at times even painful, process not only for the author but for those closest to him. In this case, my wife, Connie, never flinched. Always she remained a fount of comfort, joy, and support. For her, much love and thanks.

ABOUT THE AUTHOR

Martin N. Marger received his bachelor's degree from the University of Miami, his master's from Florida State University, and his doctorate from Michigan State University. In addition to his research and writing in the field of race and ethnic relations, his work includes studies in political sociology and social inequality. He is the author of *Elites and Masses: An Introduction to Political Sociology*, 2nd ed. (Wadsworth, 1987) and coeditor, with the late Marvin Olsen, of *Power in Modern Societies* (Westview, 1993). His articles have appeared in many sociological journals including *Social Problems, Ethnic and Racial Studies,* and the *International Journal of Comparative Sociology.* Currently he teaches in the Center for Integrative Studies in Social Science at Michigan State, where he is also associate director of the Canadian Studies Center.

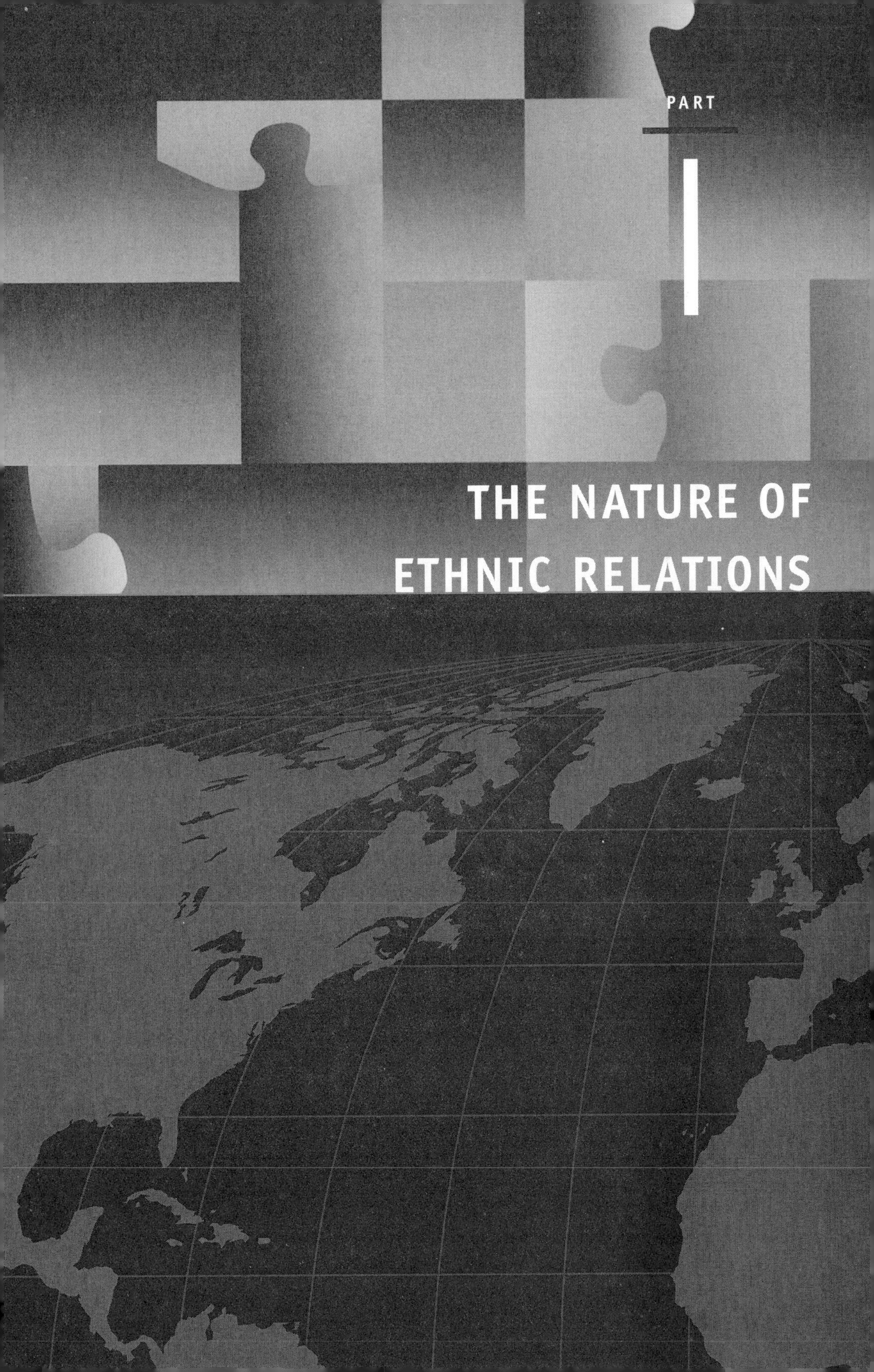

PART

I

THE NATURE OF ETHNIC RELATIONS

THE FOLLOWING four chapters erect a framework for analyzing ethnic relations in the United States (in Part II) and in five other societies (in Part III). Each chapter in Part I deals with a particular dimension of interethnic relations. Chapter 1 introduces some basic concepts and terminology of the field, in particular *ethnicity* and *race,* the latter an especially misunderstood term in everyday usage. Chapter 2 concerns the structure of inequality, which seems to be an inevitable outcome when diverse ethnic groups exist under a common political and economic system. Chapter 3 explains the techniques by which this system of inequality is maintained. Finally, Chapter 4 introduces the notion of conflict and explains it as an inherent characteristic of interethnic relations. However, there are differing degrees of group inequality and differing levels of group conflict, and these variations create different types of multiethnic societies; several of these types are described.

One objective of Part I is to explain the sociological approach to race and ethnic relations. This approach is fundamentally different from the manner in which relations among racial and ethnic groups are commonly viewed and interpreted. Sociologists see everyday social occurrences differently from laypersons, and they describe them differently as well. They go beneath the superficial to uncover the unseen and often unwitting workings of society, frequently exposing the erroneousness of much of what is considered well-established knowledge. One sociologist, Peter Berger, has put it well: "It can be said that the first wisdom of sociology is this — things are not what they seem" (1963:23). This is particularly so in the study of race and ethnic relations.

As an example, most people, if asked, could attempt an explanation of why black–white relations in the United States have been customarily discordant, and they might even venture to explain why conflict is also so commonplace among ethnic groups in other parts of the world. They would probably explain that humans are belligerent "by nature" or that there are "inherent" differences among groups, creating unavoidable fear and distrust. Although these explanations are direct and apparently simple to comprehend, they do not necessarily stand up when subjected to sociological analysis. Groups with different cultural origins and physical traits may indeed clash quite commonly, but, as we will see, there are social factors more significant than any "innate" tendencies that account for that discord.

The subject matter of sociology—or any of the other social sciences—is not the abstruse world of physics or chemistry but the everyday life of people. Because the objects of their study are so much a part of common human experience, sociologists often seem to make unnecessarily complex what appears to be quite simple. But the application of rigorous theory and methods and the use of precise terminology are the chief distinguishing features of the sociological approach, in contrast to the more unencumbered ways of problem solving that are employed by most people. In short, sociologists apply a scientific approach to analyzing human relations. And in doing so, they find that much of what is taken for granted as commonsensical is not so simple or common and perhaps not at all "sensical." In studying race and ethnic relations, therefore, it is necessary to establish more precise terms for various racial and ethnic phenomena and to become aware of the major theories and research findings that underlie the sociological approach to this field.

Race and ethnic relations—of whatever sort—cannot be explained with a single analytic tool or with one general theory. Thus, we will find that sociologists are not in agreement on all issues in the field. At times, in fact, they may be diametrically opposed. But it is necessary to understand that the very nature of scientific inquiry makes such a lack of consensus an almost foregone conclusion. Scientists, whether physical or social, recognize no absolute explanations or unchanging theories. Like any science, sociology poses more questions than it answers. In the following chapters, therefore, we will find no explanations that have not been questioned, tested, and retested. Though we can obtain no final answers to many puzzling questions using the sociological approach, we can sharpen immeasurably our insight into the whys and wherefores of relations among different racial and ethnic groups. But we must be prepared to accept the frustration that often accompanies new ideas about what is old and familiar and new methods of observing what has customarily been seen uncritically.

CHAPTER

1

INTRODUCTION: SOME BASIC CONCEPTS

ON THE AFTERNOON of October 3, 1995, most Americans paused momentarily to listen to the verdict in the trial of O. J. Simpson, in what had become, arguably, the most celebrated judicial case in U.S. history. The defendant had been charged with the murder of his former wife and her acquaintance. The trial and the jury's verdict, however, went well beyond the arguments presented by the prosecution and the defense, and captivated the nation for much of an entire year. Most important, they touched upon the society's most basic ethnic cleavage, with public opinion polarized along ethnic lines. The trial and the verdict were interpreted differently by blacks and whites. Most blacks saw the judicial process in the case as tainted and suspect; most whites saw it as fair. Much of blacks' dubiousness stemmed from their perception of the criminal justice system, particularly the police, as incapable of treating blacks with the same probity as whites. As to the "not guilty" verdict, most blacks saw it as just, while most whites saw it as a perversion of justice. Nothing could better demonstrate the divergent perspectives than televised pictures of blacks spiritedly celebrating the verdict's announcement while whites stared glumly in disbelief. The trial and its aftermath dramatized the depth of division between black and white segments of multiethnic America and, in the view of many, reinforced and even sharpened that division.

Ironically, only three years earlier in Los Angeles, another case with massive ethnic implications had gripped the nation. A black motorist, Rodney King, was stopped by police after a high-speed car chase. At the scene, King was subsequently beaten by four white Los Angeles police officers. Because the beating had been videotaped by a passing motorist, national television showed it to the entire country. One year later, the four officers were acquitted, by a mostly white jury, of all charges stemming from the incident. Following the verdict, the worst race-related riot in American history erupted in the South Central district of Los Angeles, an area populated primarily by blacks and Latinos. Looting, arson, and shootings were rampant. Stores owned by Koreans seemed to be special targets of the arsonists and looters. During the riots, 51 people were killed, 2,000 were injured, and more than 1,000 buildings were destroyed or damaged.

These two events, however, represented only the most sensational and well-publicized occurrences of ethnic drama in the early 1990s. Consider that during these few years, violent neo-Nazi youths called "skinheads" were operating in numerous cities, attacking and vandalizing blacks, Jews, Asians, and Latinos; for four days in 1991, the Brooklyn neighborhood of Crown Heights was consumed by violence between black and Hasidic Jewish residents, culminating in the stabbing death of a Hasidic student; tensions between Korean shopowners and black customers in several large cities turned to violence; turf wars between Cambodians and Latinos in Long Beach, California, resulted in the shooting deaths and wounding of dozens; a white supremacist and former Ku Klux Klan leader almost won the governorship of Louisiana; and dozens of racially or ethnically motivated incidents occurred on a variety of university and college campuses. The list could be extended almost without end. Moreover, the divisions among racial and ethnic groups were sharpened by the public airing of emotional issues of affirmative action, multiculturalism, and immigration. Indeed, by the latter years of the decade, issues of race and ethnicity were causing great strains in the society's social fabric and were threatening to divide the country more sharply than at any time in the post–World War II era.

Most Americans today acknowledge the seriousness of division and conflict between people of various racial and ethnic groups in the United States. Indeed, many would readily assert that problems stemming from race and ethnicity are the most severe, persistent, and irresolvable facing the society. In recognizing the critical nature of these problems, however, many forget or fail to observe that they are not unique to the United States. Consider a few prominent news stories of the 1990s.

In late 1992, in Mölln, Germany, a 51-year-old woman and two girls, 10 and 14, all of Turkish nationality, were burned to death after firebombs were thrown into their home. The bombing was carried out by neo-Nazis, who had been leading a rising wave of violence against immigrants and foreign workers in Germany. Just two months before the Mölln attack, buildings housing refugee Romanian gypsies and Vietnamese guestworkers were firebombed. During 1992, right-wing extremists made more than 2,000 ethnically motivated attacks, killing at least 16 people. The attacks brought visions of a Germany reverting to the violent racism of the Nazi period.

In Yugoslavia, the ruling Communist party, which had been the country's principal binding force, collapsed in 1990. The subsequent revival of historic rivalries among three of the country's major ethnic groups—Serbs, Croats, and Muslims—reached epic proportions. The country was split into several independent states, but Serbian leaders refused to relinquish control of lands where Serbs lived alongside other ethnic groups inside the borders of the newly created states. Parts of Croatia and Bosnia were overrun by Serbs using savage military tactics, including the bombardment, internment, torture, and mass murder of men, women, and children. By the time peace efforts had begun in late 1995, at least 200,000 had been brutally killed, hundreds of thousands more had been tortured or raped, and two million persons had been displaced, forced to flee their homes. The Serbian efforts were referred to as "ethnic cleansing," literally a purging of non-Serbs in these territories. Croats and Muslims engaged in counteractions with similar intent, if not as widespread. The savagery of the conflict was the most extraordinary in Europe since World War II.

In 1994, in the tiny African nation of Rwanda, a conflict of almost unimaginable proportions ensued between the country's two major ethnic groups, Hutus and Tutsis. Armed mostly with machetes and clubs, Hutu militias, whipped into a frenzy by political propaganda, carried out massacres of Tutsis, the country's minority ethnic group. Not satisfied merely with killing, bodies were often taken apart in the most ghastly fashion. In little more than a month, perhaps half a million—no one can be certain of the exact number—had been slaughtered and a million more had fled their villages. The United Nations referred to the situation as "genocide." Despite past periods of interethnic violence, the two groups, ironically, had shared an essentially common culture and language and, after many generations, the physical differences that had been said to separate them had diminished to the point where it was no longer possible to distinguish one from the other.

These events aroused shock and astonishment and were given extensive coverage by the American media. But simultaneously, serious conflicts with ethnic or racial dimensions, marked by periodic outbursts of violence, were occurring in Sri Lanka, India, Burundi, Sudan, Nigeria, Lebanon, Israel, Spain, Indonesia, Russia, and several of the former republics of the Soviet Union. Other ethnic-related conflicts, less volatile or sizable, were occurring in many other countries as well. As one looks at the contemporary world, it becomes evident that ethnic conflict in some degree has been a basic feature of almost all modern societies with diverse populations.[1]

Societies comprising numerous racial, religious, and cultural groups can be described as *multiethnic.* It is mainly these societies with which we will be concerned. Although racial and ethnic diversity in the modern world is most apparent in those societies in which ethnic conflict has been continual for many generations—such as the United States, Northern Ireland, or South Africa—or in which it has recently emerged with great intensity—such as Germany—ethnic heterogeneity in contemporary societies is commonplace, not exceptional. In fact, few nations today are not multiethnic. Only a handful of the more than 180 member countries of the United Nations are ethnically homogeneous. "Multiethnicity," notes sociologist Robin Williams, "is the rule" (1994:50). Moreover, the extent of diversity within these societies is very great. In most, the largest group is not even half the society's population (Connor, 1972).

Table 1.1 indicates the relative ethnic diversity of some current societies. Ethnic diversity includes differences among groups in language, religion, race, and culture. As can be seen, ethnic heterogeneity characterizes societies on every continent and in various stages of development.

Social scientists had maintained for many years that industrialization and the forces of modernization would diminish the significance of race and ethnicity in heterogeneous societies (Deutsch, 1966). They felt that with the breakdown of small, particularistic social units and the emergence of large, impersonal bureaucratic institutions, people's loyalty and identity would be directed primarily to the national state rather than to internal racial and ethnic communities. The opposite trend, however, seems to have characterized the contemporary world.[2]

During the past several decades in the industrialized nations, ethnic groups thought to be well absorbed into the national society have reemphasized their cultural identity, and new groups have demanded political recognition. In the United States, the emergence of black ethnic consciousness in the 1960s stimulated ongoing social movements among other ethnic groups, including Hispanics, Asians, American Indians, and even those of European origin. A new wave of immigration begun in the 1970s has further heightened American ethnic issues. In Western Europe, ethnically based political movements have been evident in

---

[1]The seriousness of ethnic conflict in the modern world is highlighted by the fact that, since 1945, perhaps as many as twenty million people have died as a result of ethnic violence (Williams, 1994).

[2]Connor (1972), in fact, contends that there is much evidence to support the thesis that modernization results in increasing demands for ethnic separatism. Blumer (1965) has also shown that industrialization does not necessarily lead to more benign ethnic relations or to displacement of the established ethnic order.

Table 1.1 Ethnic Diversity of Selected Nation-States

HIGH	MEDIUM	LOW
United States	United Kingdom	Sweden
Canada	Australia	Irish Republic
Russia	Israel	Egypt
India	Argentina	Colombia
Nigeria	Malaysia	Japan

Britain, France, Spain, and Belgium. Throughout Eastern Europe, the late 1980s brought massive economic and political change, rekindling ethnic loyalties that had been suppressed for several decades. And in Canada, the traditional schism dividing English- and French-speaking groups has in recent years widened, threatening to break up the Canadian nation.

In the developing, or Third World, nations, too, the ethnic factor has emerged with great strength. World War II marked the end of several centuries of imperialist domination of non-Western peoples by European powers, and a great number of new nations were created, the political boundaries of which were often carved out of the administrative districts of the old colonial states. In many cases, these artificial boundaries were drawn up with little consideration of the areas' ethnic composition. As a result, the new nation-states found themselves faced with the problem of integrating diverse cultural groups, speaking different languages and even maintaining different belief systems, into a single national society. The upshot has been numerous and constant ethnic conflicts in which one group refuses to recognize the political authority of the central government, which is usually dominated by members of a rival group. In 1947, for example, following independence from Britain, the Indian subcontinent was thrown into violent discord between Hindus and Muslims, the area's two major religious groups. This conflict was resolved only by the eventual partitioning of the region into two autonomous nation-states, India and Pakistan. Among developing societies, another ethnic conflict of great magnitude occurred in 1969 in Nigeria, an African nation of considerable ethnic diversity. One group, the Ibos, sought independence, leading to a bitter civil war in which thousands were killed. The already-mentioned violence of Rwanda in 1994 was perhaps the most frightful.

It must be remembered, however, that these conflicts represent only the most severe and violent. Others, more moderate and less violent, typify ethnic relations in multiethnic societies. In short, racial and ethnic forces, though varying in scope and intensity, are important bases of both group solidarity and cleavage in the modern world. Moreover, their impact is not likely to diminish in the foreseeable future.

RACE AND ETHNIC RELATIONS

The study of race and ethnic relations is concerned generally with the ways in which the various groups of a multiethnic society come together and interact over extended periods. As we proceed in our investigation, we will be looking specifically for answers to four key questions.

Basic Questions

1. *What is the nature of intergroup relations in multiethnic societies?* As we will see, ethnic relations commonly take the form of conflict and competition. Indeed, we can easily observe this by following the popular media accounts of ethnic relations in the United States and other nations. As our opening illustrations reveal, these are usually descriptions of hostility and violence. However, intergroup relations are never totally conflictual. If this were not so, ethnic groups would exist in a perpetual whirlwind of discord and strife. Cooperation and accommodation also characterize ethnic relations. Just as we will be concerned with understanding why conflict and competition are so common among diverse groups, it will also be our concern to investigate harmonious conditions and the social factors that contribute to them.

2. *How are the various ethnic groups ranked, and what are the consequences of that ranking system?* In all multiethnic societies, members of various groups are treated differently and receive unequal amounts of the society's valued resources—wealth, prestige, and power. In short, some get more than others and are treated more favorably. Moreover, this inequality is not random but is well established and persistent for many generations. A structure of inequality emerges in which one or a few ethnic groups, called the *dominant* group or groups, are automatically favored by the society's institutions, particularly the state and the economy, whereas other ethnic groups remain in lower positions. These subordinate groups are called ethnic *minorities.* We will be concerned with describing this hierarchy and determining how such systems of ethnic inequality come about.

3. *How does the dominant ethnic group in a multiethnic society maintain its place at the top of the ethnic hierarchy, and what attempts are made by subordinate groups to change their positions?* The dominant group employs a number of direct and

indirect methods (various forms of prejudice and discrimination) in protecting its power and privilege. This does not mean, however, that subordinate groups do not attempt to change this arrangement from time to time. In fact, organized movements may arise, such as the black civil rights movement of the 1960s in the United States, in which concerted efforts are made by subordinate groups to change their position in the ethnic hierarchy. One of our chief concerns, then, will be the ways in which systems of ethnic inequality are maintained and how they change.

4. *What are the long-range outcomes of ethnic interrelations?* When ethnic groups exist side by side in the same society for long periods, either they move toward some form of unification or they maintain or even intensify their differences. These various forms of integration and separation are called *assimilation* and *pluralism*. Numerous outcomes are possible, extending from complete assimilation, involving the cultural and physical integration of the various groups, to extreme pluralism, including even expulsion or annihilation of groups. Usually, less extreme patterns are evident, and groups may display both integration and separation in different spheres of social life. Again, our concern is not only with discerning these outcomes but also with explaining the social forces that favor one or the other.

A Comparative Approach

The study of race and ethnic relations has a long tradition in American sociology, beginning in the 1920s with the research of Robert Park, Everett Hughes, and Louis Wirth. These scholars were among the first to focus attention on the relations among ethnic groups, particularly within the ethnic mélange of large American cities like Chicago and New York. The sociology of race and ethnic relations has progressed enormously since that time so that it now constitutes one of the chief subareas of the sociological discipline.

With few exceptions, however, American sociologists have continued to concentrate mainly on American groups and relations, often neglecting the analysis of similarities and differences between the United States and other heterogeneous societies. But if we are to try to understand the general nature of race and ethnic relations, it is necessary to go beyond the United States—or any particular society—and place our analyses into a comparative, or cross-societal, framework. As we have seen, ethnic diversity, conflict, and accommodation are worldwide phenomena, not unique to American society. However, because most research in race and ethnic relations has been the product of American sociologists dealing with the American experience, we are often led to assume that patterns evident in the United States are the same in other societies.

Our purpose in this book is to investigate race and ethnic relations using a comparative perspective in which the United States will be seen as one among many contemporary multiethnic societies. This requires paying close attention to patterns of race and ethnicity in non-American societies. Because we are likely to be most familiar with the United States, however, we will most often focus on American groups and relations. Even in those chapters in which cases outside the United States are the major topic (Part III), some room will be allotted for comparisons with the United States.

A comparative approach will not only enable us to learn about race and ethnicity in other societies but also provide us with a sharper insight into race and ethnicity in the United States. It has often been observed that we cannot begin to truly understand our own society without some knowledge of others. Moreover, in addition to the differences revealed among societies, similarities may also become apparent. As sociologists Tamotsu Shibutani and Kian Kwan point out, comparing American ethnic relations with those of other societies "reveals that patterns of human experience, though infinitely varied, repeat themselves over and over in diverse cultural contexts" (1965:21). Discovering such generalizable patterns of the human experience is the ultimate aim of all sociological efforts.

ETHNIC GROUPS

The most fundamental concept with which we will deal is *ethnic group*. The ideas *ethnic group* and *ethnicity* are relatively new. As sociologists Nathan Glazer and Daniel P. Moynihan (1975) note, these terms did not even appear in standard English dictionaries until the 1960s. Groups generally referred to today as "ethnic" were previously thought of as races or nations, but these are terms clearly different in meaning. What are the chief characteristics of ethnic groups?

Characteristics of Ethnic Groups

Unique Cultural Traits Basically, ethnic groups are groups within a larger society that display a unique set of cultural traits. As sociologist Melvin Tumin defined it, an ethnic group is "a social group which, within a larger cultural and social system, claims or is accorded special status in terms of a complex of traits (ethnic traits) which it exhibits or is believed to exhibit" (1964:243). Ethnic groups, then, are subcultures, maintaining certain behavioral characteristics that, in some degree, set them off from the society's mainstream, or modal, culture. Such unique cultural traits are not trivial but are fundamental features of social life such as language and religion.

Unique cultural traits, however, are not sufficient alone to define ethnic groups. Can we speak of physicians as an ethnic group? or truck drivers? or college students? Obviously, we would consider none of these categories "ethnic" even though they are groups that exhibit some common behavioral traits setting them off from the larger society. They too, in a sense, are subcultures. Clearly, we need further qualifications for distinguishing ethnic groups.

Sense of Community In addition to a common set of cultural traits, ethnic groups display a sense of community among members, that is, a consciousness of kind or an awareness of close association. In simple terms, there exists a "we" feeling among members. Sociologist Milton Gordon (1964) suggests that the ethnic group serves above all as a social-psychological referent in creating a "sense of peoplehood." This sense of community, or oneness, derives from an understanding of a shared ancestry, or heritage. Ethnic group members view themselves as having common roots, as it were. When people share what they believe to be common origins and experiences, "they feel an affinity for one another," writes sociologist Bob Blauner, "a 'comfort zone' that leads to congregating together, even when this is not forced by exclusionary barriers" (1992:61).

Such common ancestry, however, need not be real. As long as people regard themselves as alike by virtue of their perceived heritage, and as long as others in the society so regard them, they constitute an ethnic group, whether such a common background is genuine or fictitious (Shibutani and Kwan, 1965). Sociologists Everett and Helen Hughes have perceptively observed that "an ethnic group is not one because of the degree of measurable or observable difference from other groups; it is an ethnic group, on the contrary, because the people in it and the people out of it know that it is one; because both the *ins* and *outs* talk, feel, and act as if it were a separate group" (1952:156). Ethnic groups, then, are social creations wherein ethnic differences are basically a matter of group perception. Groups may be objectively quite similar but perceive themselves as very different, and the converse is equally true.

In recent years, sociologists have debated the relative significance of the cultural element and the sense of community as most critical to the formation of an ethnic group (Dorman, 1980). The argument boils down to a question of whether ethnic groups are objective social units that can be identified by their unique culture or merely collectivities that people themselves define as ethnic groups. Whereas some view the cultural features of the group as its key distinctive element, others argue that stressing its unique culture minimizes the importance of the subjective boundaries of the group that people themselves draw (Barth, 1969). Most simply, the latter maintain that if people define themselves and are defined by others as an ethnic group, they *are* an ethnic group, whether or not they display unique cultural patterns. If this is the case, the cultural stuff of which the ethnic group is composed is unimportant.

Although this may seem like a relatively minor theoretical point, it is of importance when ethnic groups in a society begin to blend into the dominant cultural system. Sociologists have traditionally assumed that as groups integrate into the mainstream society, the basis of retention of ethnicity diminishes. But whether people continue to practice ethnic ways may matter little as long as they continue to define themselves and are defined by others in ethnic terms. Many Americans continue to think of themselves as ethnics even though they exhibit little or no understanding of or interest in their ethnic culture. Do third- or fourth-generation Irish Americans, for example, really share a common culture with their first-generation ancestors? Wearing a button proclaiming "I'm proud to be Irish" is hardly a display of the traits of one's Irish-American forebears. Yet an ethnic identity may remain intact for such persons, and they may continue to recognize their uniqueness within the larger society. Thus, despite the lack of a strong cultural factor, the sense of Irish-American identity may be sufficient to sustain an Irish-American ethnic group.

We thus have two views of the ethnic group: (1) it is an objective unit that can be identified by a people's distinct cultural traits, or (2) it is merely the product of people's thinking of and proclaiming it as an ethnic group. To avoid the extreme of either of these views, sociologist Pierre van den Berghe defines the ethnic group as both an objective and a subjective unit: "An ethnic group is one that shares a cultural tradition *and* has some degree of consciousness of being different from other such groups" (1976:242). As he points out, it is foolish to think that ethnic groups simply arise when people so will it. Fans of a particular football team may feel a sense of commonality and even community, but they surely do not compose an ethnic group. In short, there must be some common cultural basis and sense of ancestry to which ethnic group members can relate. As van den Berghe notes, "There can be no ethnicity (or race) without some conception and consciousness of a distinction between 'them' and 'us.' But these subjective perceptions do not develop at random; they crystallize around clusters of objective characteristics that become badges of inclusion or exclusion" (1978:xvii). Although ethnic boundaries are very flexible, they are always founded on a cultural basis. At the same time, however, an ethnic group cannot exist in an objective sense independent of what its members think and believe. There must be a sense of commonality, and such a feeling of oneness arises generally through the perception of a unique cultural heritage.

Ethnocentrism The "we" feeling of ethnic groups ordinarily leads naturally to *ethnocentrism,* the tendency to judge other groups by the standards and values of one's own. Inevitably, this produces a view of one's own group as superior to others. The ways of one's own group (*in-group*) become "correct" and "natural," and the ways of other groups (*out-groups*) are seen as "odd," "immoral," or "unnatural." Sociologists and anthropologists have found this inclination to judge other groups by the standards of one's own and to view out-groups as inferior or deficient to be a universal practice.

In multiethnic societies, such feelings of group superiority become a basis for group solidarity. In addition to fostering cohesiveness within one group, however, ethnocentrism also serves as the basis of conflict between different groups. As Bonacich and Modell have explained, "Ethnicity is a communalistic form of social affiliation, depending, first, upon an assumption of a special bond among people of like origins, and, second, upon the obverse, a disdain for people of dissimilar origins" (1980:1). Here we can begin to understand why conflict among ethnic groups is so pervasive and intractable.

Ascribed Membership Ethnic group membership is ordinarily *ascribed.* This means that one's ethnicity is a characteristic acquired at birth and not subject to basic change. Being born a member of an ethnic group, one does not leave it except in unusual circumstances. One might change ethnic affiliation by "passing"—that is, by changing one's name or other outward signs of ethnicity—or by denying group membership. But it is extremely difficult to divest oneself completely of one's ethnic heritage. Through the socialization process, individuals come to learn their group membership early and effectively and to understand the differences between themselves and members of other groups. So well internalized is this group identification that one comes to accept it almost as naturally as accepting one's gender. As Hughes and Hughes have suggested, "If it is easy to resign from the group, it is not truly an ethnic group" (1952:156). This understanding of ethnic descent is what creates such sharp and, at times, deadly divisions among ethnic groups (Williams, 1994).

Those who attempt to shed their ethnic identity find that the society will rarely permit this fully. Gordon notes in this regard that in American society a person who attempts to relinquish his or her ethnic identity finds "that the institutional structure of the society and the set of built-in social and psychological categories with which most Americans are equipped to place him—to give him a 'name'—are loaded against him" (1964:29). Once the ethnic categories in a society are set, explains Gordon, placing people into them is almost automatic and is by no means subject entirely to people's volition.

In multiethnic societies where ethnic boundaries are not rigid and where there is much marriage across ethnic lines, like the United States and Canada, the voluntary nature of ethnicity becomes more salient. The ethnic origins of third- or fourth-generation Euro-Americans, for example, may be quite varied. Individuals therefore make decisions about "who they are" ethnically, some in a contrived fashion and others almost unconsciously. A person whose family origins may contain Italian, Polish, and Irish elements might emphasize the Italian part and identify himself as "Italian American," disregarding his other ancestral links. In such cases, the volitional component of ethnic identity is strong. For those whose ethnic identity is based also on physical, or racial, characteristics, however, the capacity to choose becomes more limited. For such people, ascription is paramount.

Territoriality Ethnic groups often occupy a distinct territory within the larger society. Most of the multiethnic societies of Europe consist of groups that are regionally concentrated. Basques and Catalans in Spain, Welsh and Scots in Britain, and Flemings and Walloons in Belgium are groups that maintain a definable territory within the greater society. Such multiethnic societies are quite different from the United States or Australia, where ethnic groups have for the most part immigrated voluntarily and, though sometimes concentrated in particular areas, are not regionally confined.

When ethnic groups occupy a definable territory, they also maintain or aspire to some degree of political autonomy. They are, in a sense, "nations within nations." In some societies, the political status of ethnic groups is formally recognized. Each group's cultural integrity is acknowledged, and provision is made for its political representation in central governmental bodies. Such societies are best referred to not only as multiethnic but as multinational (van den Berghe, 1981).

In other societies, where such multinationality is not formally recognized, certain ethnic groups may aspire to some degree of political autonomy or perhaps even full independence from the national state, usually dominated by other ethnic groups. The Spanish Basques represent such a case. With a culture and language distinct from other groups in Spain, the Basques have traditionally seen themselves as a separate nation and have negotiated with the Spanish government at various times to promote their sovereignty. A Basque nationalist movement has been evident since the late nineteenth century, but in the past two decades it has taken on a particularly virulent and often violent form (Clark, 1980; Ramirez and Sullivan, 1987). Similar nationalist movements, perhaps not so hostile, have, during most of the post–World War II era, typified many multiethnic societies. We will look at one such movement, among French-speaking Canadians in Quebec, in Chapter 14.

Where ethnic groups do not continue to maintain significant aspects of culture (like language) for many generations and where they are geographically dispersed rather than concentrated, such nationalist movements do not ordinarily arise. This is the case in the United States, where ethnic groups are scattered throughout the society, have generally taken on the major cultural ways of the dominant white Anglo-Saxon Protestant, or "WASP," group after a generation or two, and seek greater power within the prevailing political system. Nonetheless, even where ethnic groups are no longer territorially based, they usually retain sentimental ties to their society of origin.

Ethnicity as a Variable

Each of these characteristics—unique culture, sense of community, ethnocentrism, ascribed membership, and territoriality—will be displayed to a different degree by different ethnic groups. These traits are variables and will not only differ from group to group but also change at various historical times within any single ethnic group. Thus, we should not expect to find all ethnic groups in a society equally unique in cultural ways, strongly self-conscious and recognized by out-groups, or even ethnocentric. The extent to which ethnic groups are noticeable and maintain a strong consciousness among members depends on both in-group and out-group responses. Some ethnic groups seek rapid assimilation and are accepted into the society's mainstream relatively quickly. Others, however, may retain their group identity for many generations because of rejection by the dominant group, their own desire to maintain the ethnic community, or a combination of these two. Jews, for example, have maintained a strong group consciousness in most societies, largely as a result of the historically consistent hostility to which they have been subjected but also because they have consciously sought to preserve their group identity.

Here we might consider sociologist Robin Williams's distinction between what he calls ethnic "collectivities" and ethnic "categories" (1964, 1979, 1994). The former are made up of people who have a recognizable culture, are aware of themselves as a unit, are recognized as a unit by outsiders, interact with other group members, and feel a sense of obligation to support and defend the group. Ethnic categories, in contrast, are merely groupings of people with one or a number of similar characteristics but with little sense of membership and little interaction among them. At different times and in different social contexts, ethnic groups may be both collectivities and categories. No single Italian American, for example, can interact with every other Italian American. At the level of interaction, then, there are really many smaller Italian-American ethnic groups or communities. All Italian Americans, however, can recognize similarities of culture and behavior among themselves, can identify themselves as Italian Americans, and can maintain a sense of group solidarity. Members of out-groups may also recognize these similarities and identify an Italian-American ethnic group. There is, then, in addition to numerous ethnic communities, one larger, more abstract collectivity called Italian Americans. Going one step further, journalists, pollsters, or even social scientists may enumerate a group referred to as Italian Americans, regardless of the degree of ethnic identification of the millions of individuals who make up this category.

For persons in multiethnic societies, the ethnic group becomes a key source of social-psychological attachment and serves as an important referent of self-identification. Put simply, people feel naturally allied with those who share their ethnicity and identify themselves with their ethnic group. Their behavior is thus influenced by ethnicity in various areas of social life.

However, just as ethnicity will differ in scope and consequence at the group level, so too for individuals it will play a varying role. For some, it is a major determinant of behavior, and most social relations will occur among those who are ethnically similar. For others, ethnicity may be insignificant, and they may remain essentially devoid of ethnic consciousness. For most people in multiethnic societies, however, the ethnic tie is important in shaping primary relations—those that occur within small, intimate social settings such as the family and the peer group. These relations include one's choice of close friends, marital partner, residence, and so on.

There are certainly other groups in modern societies to which people feel a sense of attachment and that provide a source of identification. These include one's social class, gender, age, and occupation. Like ethnic groups, they also become bases of solidarity and societal cleavage. But in multiethnic societies, ethnicity is a primary base of loyalty and consciousness for most people and thus serves as a strong catalyst for competition and conflict. Moreover, ethnicity is usually interrelated and overlaps with these other sources of group identification and attachment.

Most important, ethnicity is a basis of ranking, in which one is treated according to the status of his or her ethnic group. In no society do people receive an equal share of the society's rewards, and in multiethnic societies, ethnicity serves as an extremely critical determinant of who gets "what there is to get" and in what amounts. In this sense, ethnicity is a dominant force in people's lives whether or not they are strongly conscious of their ethnic identity and regardless of the degree to which ethnicity shapes their interrelations with others.

RACES

Without question, *race* is one of the most misunderstood, misused, and often dangerous concepts of the modern world. It is not applied dispassionately by laypeople or even, to a great extent, by social scientists. Rather, it arouses emotions such as hate, fear, anger, loyalty, pride, and prejudice. It has also been used to justify some of the most appalling injustices and mistreatments of humans by other humans.

The idea of race has a long history, extending as far back as ancient civilizations. It is in the modern world, however—specifically, the last two centuries—that the notion has taken on real significance and fundamentally affected human relations. Unfortunately, the term has never been applied consistently and has meant different things to different people. In popular usage, it has been used to describe a wide variety of human categories, including people of a particular skin color (the Caucasian "race"), religion (the Jewish "race"), nationality (the British "race"), and even the entire human species (the human "race"). As we will see, none of these applications is accurate and meaningful from a social scientific standpoint. Much of the confusion surrounding the idea of race stems from the fact that it has both biological and social meanings. Let us look briefly at the difference between the two.

Race as a Biological Notion

The essential biological meaning of *race* is a population of humans classified on the basis of certain hereditary characteristics that differentiate them from other human groups. Races are, in a sense, pigeonholes for categorizing human physical types. Efforts at classification, however, have created a virtually hopeless disagreement among social and biological scientists. The biological understanding of race has led to an enormous variation in thought and almost no accord among biologists, geneticists, physical anthropologists, and physiologists concerning either the term's meaning or its significance. Although it is impossible to do justice to the controversies surrounding the notion of race in a few pages, we will outline several of the more apparent problems attached to this most elusive of ideas.

Genetic Interchangeability To begin with, the difficulty in trying to place people into racial categories on the basis of physical or genetic qualities stems from the fact that all members of the human species, *Homo sapiens,* operate within a genetically open system. This means that humans, regardless of physical type, can interbreed. If genes of different human groups were not interchangeable, the idea of race as a biological concept might have some useful meaning. But because this is not the case, we see an unbounded variety of physical types among the peoples of the world.

The Bases of Racial Classification That a person from one genetic population can interbreed with a person from any other population creates a second difficulty in dealing with the notion of race: answering the question, "What are the characteristics that differentiate racial types?" Again, there has been little agreement among social scientists. Physical anthropologists distinguish three major categories of human traits: anatomical features such as skin color, hair texture, and body and facial shape (*phenotypes*); internal physiological traits such as metabolic rate, genetic diseases, and hormonal activity (*genotypes*); and blood composition. Races have traditionally been classified chiefly on the basis of easily observable anatomical traits; internal and blood traits have been deemphasized or disregarded. Today, physical anthropologists tend to emphasize blood characteristics, but no one of these sets of traits alone can be used consistently to distinguish human groups. Moreover, attempts to clearly categorize humans have proved futile because differences among individuals of the same group (or "racial type") are greater than those found between groups (Marks, 1995).

Anthropologist Ruth Benedict fittingly remarked that "in all modern science there is no field where authorities differ more than in the classifications of human races" (1959:22). If researchers are in agreement about anything concerning race, it is that racial classification systems are by and large arbitrary and depend on the specific objectives of the classifier. All agree that "pure" races do not exist today, and some even question whether they have ever existed (Dunn, 1956; Fried, 1965; Pettigrew, 1964).

Physical differences among people obviously exist, and these differences are statistically clear among groups. It is true that, through a high degree of inbreeding over many generations and as adaptations to different physical environments, groups with distinctive gene frequencies and phenotypic traits (that is, observable physical features) are produced. There are evident differences, for example, between a "typical" black person and a "typical" white person in the United States. People may be said, therefore, to fall into statistical categories by physical type.

But these statistical categories should not be mistaken for actual human groupings founded on unmistakable hereditary traits. Racial categories form a continuum of gradual change, not a set of sharply demarcated types. Physical differences between groups are not clear-cut but instead tend to overlap and blend into one another at various points. Petersen aptly notes that humans are not unique in this regard: "It follows from the theory of evolution itself that all biological divisions, from phylum through subspecies, are always in the process of change, so there is almost never a sharp and permanent boundary setting one off from the next" (1980:236).

The popular division of the human population into three major racial groupings—Caucasoid, Mongoloid, and Negroid—is thus imprecise and largely arbitrary. Moreover, this scheme excludes large populations that do not easily fit into a simple tripartite arrangement. Where, for example, shall we place East Indians, a people with Caucasian features but with dark skin? Or where do groups thoroughly mixed in ancestry, like most Indonesians, fit? Because all human types are capable of interbreeding, there are simply too many marginal cases like these that do not easily conform to any particular racial scheme, regardless of its complexity. At best, the human species is divisible into subpopulations (Gould, 1984; Marks, 1995).

Race and Social Traits Perhaps the most troublesome aspect of the idea of race concerns the relation of racial inheritance to social and personality traits. As with other features of the race controversy, there is no unanimity of thought on the issue of whether and to what degree one's racial inheritance affects intelligence, temperament, and other individual characteristics.

Particular controversy has arisen over the issue of race and intelligence. This argument reflects the more general debate concerning the effects of heredity and environment on human behavior, or what has been referred to as the "nature versus nurture" question. Although it is generally understood that heredity and environment are interrelated and affect individual behavior and aptitude in complementary ways, the dispute revolves around the degree to which each is consequential. As to intelligence, whereas hereditarians argue that differences among racial groups are attributable most basically to genetic factors, environmentalists contend that social variables such as class, family, language, and the development of cognitive skills are most significant. The debate is made relatively meaningless, however, by the fact that there is little consensus regarding the very meaning of intelligence and how it can be adequately measured. Do we mean by intelligence innate potential, educability, or the capacity to think abstractly? And how can any of these be measured correctly by "intelligence" tests? None of these questions has been answered satisfactorily.

In any case, the issue of the effects of heredity and environment on intelligence must remain unresolvable if for no other reason than that environmental factors are never exactly the same for any two individuals. People of any society are products of different social classes and subcultures and thus are afforded different life chances, education in particular. Some are born into well-to-do families and thereby receive the maximum opportunities to develop their talents and skills, and others are born into less fortunate circumstances, resulting in fewer chances to fully realize their capacities.

If all people were to start at an equal point in the social hierarchy and were afforded truly equal opportunities for social advancement, the differences they subsequently displayed might be attributed to nonenvironmental causes. However, the opportunity structure of any society is hardly equitable, and people do not begin their quest for social success with the same resources. If runners begin a race at different points on the track, the winner is not necessarily the fastest athlete. The same is true in trying to determine intelligence. That some prove themselves more intelligent than others may indicate only that they have begun with more favorable opportunities. Thus, until the opportunity structure can be made equitable—a remote possibility in any society—the relative effects of heredity and environment on intelligence will not be scientifically measurable with much accuracy.

The issue of race and intelligence was given new impetus when educational psychologist Arthur Jensen (1969) published a report suggesting that heredity was the major determinant in accounting for the collective differences in IQ between blacks and whites. Jensen's theory, methods, and interpretation of findings were all quickly and thoroughly challenged (Deutsch, 1969; Montagu, 1975; Rose and Rose, 1978).

More recently, the publication in 1994 of Richard J. Herrnstein and Charles Murray's book, *The Bell Curve,* refueled the debate regarding the impact of race on mental ability and set off much controversy. Herrnstein and Murray present a kind of Social Darwinistic thesis: intelligence, as measured by IQ, is in large part genetic. They see strong relationships between IQ and various social pathologies. Thus, those with lower IQs have a greater proclivity toward poverty, crime, illegitimacy, poor educational performance, and other social ills. Because IQ is mostly genetic, they argue, there is no way to change the condition of those with low intelligence through educational reforms or welfare programs. Because lower-intelligence people are reproducing much faster than higher-intelligence people, the society is faced with the possibility of a growing underclass, increasingly dependent on the more intelligent and productive classes.

Perhaps the most controversial aspects of Herrnstein and Murray's book concern the linkage of IQ and race. The racial dimension is introduced when the authors point out that the average IQ of blacks is fifteen points lower than the average IQ of whites, a difference, they claim, that holds, regardless of social class. Moreover, even when change in average IQ for groups is acknowledged, the differential remains intact. The implication, then, is clear: blacks are inferior to whites and are thus apt to remain in a state of dependency on the nonpoor and continue to engage in antisocial activities. Herrnstein and Murray therefore question the value of welfare payments, remedial educational programs, affirmative action, and other efforts designed to raise the social level of the poor who, in the United States, are disproportionately black.

As with Jensen earlier, Herrnstein and Murray's thesis, findings, and conclusions have been overwhelmingly rejected by mainstream social scientists, who claim that their methods are flawed and their reasoning specious (Fraser, 1995; Jacoby and Glauberman, 1995). Moreover, many have viewed the book as much a statement of the political leanings of the authors as a work of social science. Specifically, Herrnstein and Murray are challenged on a number of points. For one, IQ has been shown to measure only certain kinds of intelligence. Furthermore, IQ is not fixed but is subject to variation within one's lifetime and, for groups, subject to change over generations. Perhaps most important, the authors do not place sufficient weight on the environmental factors that enable people to express their intelligence, regardless of IQ. Also, the authors reify "race," referring to whites and blacks as if these were clear-cut, distinct genetic groups, ignoring the countless variations within designated racial categories.

Theoretical views like Jensen's and Herrnstein and Murray's have appeared periodically, but the understanding that environment, not racial inheritance, is the key influence in shaping social behavior clearly predominates in social science thinking today (Marks, 1995; Montagu, 1972). As we will see, however, this has not always been the case.[3]

It is important to bear in mind that the issue of race and social behavior is significant only to the extent that a society makes it so. As Berry and Tischler note, "It only assumes importance if people have racist views and want to use the arguments as ammunition for their cause" (1978:84). It is of significance, in other words, only where people are dealt with not as individuals but as members of racial categories. This distinction suggests the social meaning of race.

The Social Meaning of Race

Many popular ideas are of very dubious scientific validity; race is certainly among these. But as André Béteille has pointed out, "Sociological analysis is concerned not so much with the scientific accuracy of ideas as with their social and political consequences" (1969:54). Whether the idea of race is meaningful in a biological sense remains a controversial and seemingly unresolvable issue. But whatever its biological validity, the importance of race for the study of intergroup relations clearly lies in its social meaning.

Most simply, people attach significance to the concept of race and consider it a real and important division of humanity. And, as long as people *believe* that differences in selected physical traits are meaningful, they will act on those beliefs, thereby affecting their interrelations with others. Sociologist W. I. Thomas observantly asserted that "if men define situations as real they are real in their

[3]For a succinct discussion of the controversy surrounding race and intelligence, see Berry and Tischler (1978:63–86). The issue of intelligence tests and differences among racial groups is well covered in Samuda (1975) and Gould (1983). A simplified explanation of Jensen's controversial position and a rejoinder by a prominent geneticist are found in Jensen (1973a) and Dobzhansky (1973). The arguments surrounding *The Bell Curve* are detailed in Fraser (1995) and Jacoby and Glauberman (1995).

consequences" (Thomas and Znaniecki, 1918:79). If, for example, those classified as black are deemed inherently less intelligent than those classified as white, people making this assumption will treat blacks accordingly. Employers thinking so will hesitate to place blacks in important occupational positions; school administrators thinking so will discourage blacks from pursuing difficult courses of study; white parents thinking so will hesitate to send their children to schools attended by blacks; and so on.

The creation of such categories and the beliefs attached to them generate what sociologists have called the "self-fulfilling prophecy" (Merton, 1968). This refers to a process in which the false definition of a situation produces behavior that, in turn, makes real the originally falsely defined situation. Consider the aforementioned case. If blacks are considered inherently less intelligent, fewer community resources will be used to support black schools on the assumption that such support would only be wasted. Poorer-quality black schools, then, will inevitably turn out less capable students, who will score lower on intelligence tests. The poorer performance on these tests will "confirm" the original belief about black inferiority. Hence the self-fulfilling prophecy. The notion of black inferiority is reinforced, and continued discriminatory treatment of this group rationalized.

Anthropologist Robert Redfield has noted that "it is on the level of habit, custom, sentiment, and attitude that race, as a matter of practical significance, is to be understood. Race is, so to speak, a human invention" (1958:67). The scientific validity of race, then, is of little consequence; rather, it is the belief system of a society that provides its significance. Each heterogeneous society takes whatever are perceived as important physical differences among people and builds a set of racial categories into which those people are placed. But these categories are fully arbitrary. Different societies will use different criteria with which to assign people racially, thereby creating classification systems that may have little or no correspondence from one society to the next. Omi and Winant use the term *racial formation* to describe "the process by which social, economic, and political forces determine the content and importance of racial categories, and by which they are in turn shaped by racial meanings" (1986:61). The social meaning of race, they explain, is constantly subject to change through political struggle.

The arbitrariness of racial categorizing can be seen easily when we compare different societies, each with numerous physical types. The same individual categorized as "black" in the United States, for example, might be categorized as "white" in Brazil. The racial classification systems in these two societies do not coincide. As we will see in Chapter 13, Brazilians do not see or define races in the same way that Americans do, nor do they necessarily use the same physical characteristics as standards with which to categorize people (Harris, 1964; Pitt-

Rivers, 1987; van den Berghe, 1978). Obviously, different criteria and different categories of race are operative in each society. Indeed, so subject to cultural definition is the idea of race that the selected physical attributes used to classify people need not even be obvious, only the *belief* that they are evident. In Northern Ireland, for example, both Protestants and Catholics sometimes say they are able to identify members of the other group on the basis of physical differences, despite their objective similarity.

What is perhaps most important regarding the social classification of races is that the perceived physical differences among groups are assumed to correspond to social or behavioral differences. Thus, blacks are assumed to behave in certain ways and to achieve at certain levels because they are black; whites are assumed to behave and achieve in other ways because they are white; and so on. As van den Berghe notes, "What makes a society multiracial is not the presence of physical differences between groups, but the attribution of social significance to such physical differences as may exist" (1970:10). Redfield has drawn an apt analogy: "If people took special notice of red automobiles, and believed that the redness of automobiles was connected inseparably with their mechanical effectiveness, then red automobiles would constitute a real and important category" (1958:67).

It is most critical, then, that we look not simply at the racial categories that different societies employ but also at the social beliefs attached to those categories. Such beliefs are the product of racist thinking, which we will consider shortly.

Race and Ethnicity: A Synthesis

As should now be obvious, the term *race* is so charged and misconceived that it is very difficult to employ in a useful analytic manner. Adding to the confusion is the fact that many groups now defined as ethnic groups were in previous historical periods defined as races. Immigrant groups in the United States, for example, representing different nationalities or religions (Slavs, Italians, Jews, and so on) were classified as races during the early part of the twentieth century by laypersons and many social scientists as well.

Another difficulty arising from the use of the term race is that it is often applied too sweepingly, encompassing many specific ethnic groups. For example, to speak of "blacks" in the United States in the aggregate is to assume a homogeneity that does not exist. Overlooked is the ethnic variety among U.S. blacks. Although African Americans, those whose ancestry is traceable to American slavery, predominate, today Haitians, Jamaicans, other Caribbeans, many Latinos, and even a growing population of African immigrants are also part of the American "black" population. Culturally these groups may have little in common.

Moreover, to speak of a global racial commonality is even more misleading. As Glazer and Moynihan have noted,

> it is hardly likely that Moslem, Swahili-speaking blacks of Zanzibar would find much in common with the black institutions and culture that are now being built up in this country. They would not have any predilection for soul music or soul food, would find the styles of dress, hair, walk, and talk that are now popular as defining blackness distinctly foreign (1970:xxxix).[4]

Likewise, the term *white* has no significance beyond a reference to white ethnic groups in the aggregate. To speak of Polish Americans, Jewish Americans, and Irish Americans as part of a common group is to falsely meld groups whose cultural traditions are quite distinct.

As a result of its confusing usage and its questionable scientific validity, many sociologists and anthropologists have dispensed entirely with the term *race* and instead use *ethnic group* to describe those groups commonly defined as racial (Berreman, 1972; Gordon, 1964; Schermerhorn, 1970; Shibutani and Kwan, 1965; Williams, 1979). In the United States, African Americans, American Indians, Chinese Americans, Japanese Americans, and Mexican Americans have all the earmarks of ethnic groups—unique culture, consciousness of kind, ascriptive membership, and in some cases even territoriality—at the same time that most members of these groups are physically distinct from Americans of European origin. Classifying all these groups as ethnic seems most reasonable because, in addition to their physical traits, there are always consistent and significant cultural traits that set them off from other groups. Sociologist Richard Alba has offered a definition of *race* as a variant of ethnicity: "A racial group is...an ethnic group whose members are believed, by others if not also by themselves, to be physiologically distinctive" (1992:576).

For the sake of simplicity and clarity, then, we will use the term *ethnic group* in this book in a broad manner so as to include groups and organizations identified by national origin, cultural distinctiveness, racial characteristics, or religious affiliation. As we will see, ethnic groups in most modern societies comprise combinations of these national, cultural, physical, and religious traits. For those groups that are particularly divergent physically from the dominant group, such as African Americans, Richard Burkey (1978) has suggested the term *racial-ethnic group,* and that term will be used accordingly.

---

[4]From Nathan Glazer and Daniel P. Moynihan, *Beyond the Melting Pot,* 2d ed. Copyright © 1970 by MIT Press, Cambridge, Mass. Reprinted by permission.

Though the impact of hereditary features on social behavior has been shown by social science to be, at the most, questionable and, at the least, minimal, many people continue to be guided by racist thinking. As Benedict put it, "Any scientist can disprove all its facts and still leave the *belief* untouched" (1959:99).

Racist thinking involves principles that lead naturally and inevitably to the differential treatment of members of various ethnic groups. As we will see, in no society are valued resources distributed equally; in all cases, some get more than others. But in multiethnic societies, ethnicity is used as an important basis for determining the nature of that distribution. Ethnic groups are ranked in a hierarchy, and their members are rewarded accordingly, creating a system of ethnic inequality. Groups at the top compound their power and maintain dominance over those lower in the hierarchy. Such systems of ethnic inequality require a belief system, or ideology, to rationalize and legitimate these patterns of dominance and subordination, and racism has usually served this function.

The Ideology of Racism

As a belief system, or ideology,[5] racism is structured around three basic ideas:

1. Humans are divided naturally into different physical types.
2. Such physical traits as they display are intrinsically related to their culture, personality, and intelligence.
3. On the basis of their genetic inheritance, some groups are innately superior to others (Banton, 1970; Benedict, 1959; Montagu, 1972; Shibutani and Kwan, 1965).

In sum, *racism* is the belief that humans are subdivided into distinct hereditary groups that are innately different in their social behavior and mental capacities and that can therefore be ranked as superior or inferior. The presumed superiority of some groups and inferiority of others is subsequently used to legitimate the unequal distribution of the society's resources, specifically, various forms of wealth, prestige, and power.

[5]On varying definitions of *ideology,* see Lane (1962), Mannheim (1936), Plamenatz (1970), and Shils (1968).

Racist thinking presumes that differences among groups are innate and not subject to change. Intelligence, temperament, and other primary attitudes, beliefs, and behavioral traits are thus viewed as not significantly affected by the social environment. The failures of groups at the bottom of the social hierarchy are interpreted as a natural outcome of an inferior genetic inheritance rather than of social disadvantages that have accumulated for the group over many generations. In the same manner, the achievements of groups at the top of the social hierarchy are seen as a product of innate superiority, not of favorable social opportunities.

Racist thought is inherently ethnocentric. Those espousing racist ideas invariably view ethnic out-groups as inferior. Moreover, such thought naturally leads to the idea that ethnic groups must be kept socially and, especially, physically apart. To encourage social integration is to encourage physical integration, which, it follows, contributes to the degeneration of the superior group.

Ideologies do not necessarily reflect reality; indeed, they are largely mythical. They comprise beliefs that, through constant articulation, become accepted as descriptions of the true state of affairs. We have already discussed some of the scientifically erroneous or dubious principles concerning race. As anthropologist Manning Nash (1962) has explained, racist ideologies depend on three logical confusions: (1) the identification of racial differences with cultural and social differences; (2) the assumption that cultural achievement is directly, and chiefly, determined by the racial characteristics of a population; and (3) the belief that physical characteristics of a population limit and define the sorts of culture and society they are able to create or participate in.

Racist thought is most prevalent in societies in which physical differences among groups are pronounced, such as differences between blacks and whites in the United States or South Africa. But racism describes any situation in which people's social behavior is imputed to innate, or hereditary, sources. Racist beliefs are therefore not limited to ideas about groups commonly referred to as "races" but can apply to any ethnic group whether distinguishable by race or culture. Racism can pertain to Jews, Italian Americans, Northern Irish Catholics, or French Canadians as much as to African Americans, North American Indians, or other physically salient groups. Members of ethnic groups identifiable primarily by culture rather than physical traits are often described as displaying behavioral characteristics "natural" to their group. Jews in Germany during the 1930s, for example, were physically indistinct from other Germans, but this did not prevent the creation by the Nazis of an elaborate racial ideology pertaining to the Jews. Claims that Italians are by nature flamboyant or that Poles are innately slow witted are similar manifestations of racist thought.

A better term might be *ethnicism* because racism applies to all types of ethnic groups. However, *racism* has become so well established in the sociological lexicon and in everyday language that its continued usage seems inevitable.[6]

Certain beliefs regarding behavioral and personality differences between men and women are based on the same mode of thinking as racism. Men are assumed to be innately better qualified for certain social roles on the basis of their masculinity, and women are seen as quite naturally occupying other roles suited to their femininity. Although most of the behavioral differences between men and women are attributable more to social learning than to biology, the beliefs regarding the congenital nature of these differences are accepted uncritically by many. The belief in the innate behavioral differences between men and women has been referred to as *sexism,* but it is obvious that the foundation of this ideology is similar to that of racism.

The Functions of Racism

The belief in innate differences among groups is used to justify the unequal distribution of a society's rewards. The place of groups at the top of the social hierarchy and of those at the bottom is explained quite simply as "natural." Racist ideology, then, promotes an ethnic status quo in which one group predominates in the society's economy, polity, and other key institutions and thus receives the greatest share of the society's wealth and power.

Why, for example, do whites in the United States occupy a disproportionate number of top positions in all important institutions, own a disproportionate share of wealth, and enjoy inordinate honor and prestige by comparison with blacks? Racist ideology explains such inequalities as a result of the inherent inferiority of blacks and superiority of whites. It is asserted that something in the character of black people themselves is at the root of their socially subordinate position, just as something in the character of white people leads very naturally to their social dominance. The same explanation might be used by Protestants in Northern Ireland to account for their dominance over Catholics in various areas of social life or by South African whites to rationalize their past dominance of nonwhites.

In the racist mode of thought, the different social and cultural environments of groups are not of major importance in accounting for their differences in

---

[6]In recent years, the term *racism* has often been used in a sweeping and imprecise fashion, describing almost any negative thought or action toward members of a racial minority or any manifestation of racial inequality, not simply as an ideology. On the different conceptualizations and applications of the notion of racism, see Blauner (1992).

social achievement. And, as we noted earlier, the perpetuation of beliefs in group superiority and inferiority gives rise to the substantiation of those beliefs through the self-fulfilling prophecy.

If groups are effectively portrayed as inferior, they can be not only denied equal access to various life chances but in some cases enslaved, expelled, or even annihilated with justification. Slave systems of the eighteenth and nineteenth centuries in the Americas were rationalized by racial belief systems in which blacks were seen as incapable of ever attaining the level of civilization of whites. Similarly, the enactment in the 1920s of strict U.S. immigration quotas, favoring northwestern European groups and discriminating against those from southern and eastern Europe and Asia, was impelled by an intricate set of racist assumptions. Northwestern Europeans were seen as innately more adaptable to the American social system, and other groups were viewed as naturally lacking in favorable social and moral qualities. The latter were believed to represent inferior human breeds, whereas northwestern Europeans embodied the most desirable traits of the species.

The response to the arrival of several thousand black Haitians in the United States in 1980 and 1981 illustrates the continuation of similar racist attitudes. Although many were seeking economic betterment, all came as political refugees requesting asylum from a notoriously repressive dictatorship. At approximately the same time, over 100,000 Cubans, most of them white, also arrived as political refugees in much the same manner, many in crude boats.

The treatment accorded these two groups, however, was noticeably different. Unless they had criminal records, most Cubans were released into the community once sponsors had been found. The Haitians were instead placed in detention camps or in federal prisons and treated as felons. They were not released even when sponsors had been located for most. Many observers believed that the intent of the harsh policies of the Reagan administration toward the Haitians was to encourage them to return to Haiti. Late in 1981, President Reagan issued an executive order that, in effect, directed the Coast Guard to turn back at sea all subsequent Haitian refugees.

The same issue arose again in 1991 and 1992. Following a military coup in Haiti that toppled a democratically elected government, thousands of Haitians, most of them in flimsy boats, sought refuge in the United States. In response, the Bush administration claimed that the migrants did not qualify for political asylum even though most were fleeing a general climate of political repression and violence as well as economic deprivation. He subsequently ordered the migrants returned to Haiti and instructed the Coast Guard to intercept at sea any ships with Haitians destined for Florida. At the same time, hundreds of Cubans, making similar claims as the Haitians, were landing in Florida. They were allowed to remain in the United States as legal immigrants and political refugees, receiving help from relatives and volunteer agencies.

Whether racism was the key factor in explaining the discriminatory treatment accorded Haitians or whether political considerations were paramount in this case can only be conjectured. But most observers agreed that the noticeably different response to the Haitians in comparison with the response to refugees from Cuba, from eastern European states, and earlier from Vietnam was tinged with racial overtones.

The Development of Racism

Although beliefs in the superiority and inferiority of different groups have been historically persistent in human societies (recall the universality of ethnocentrism), the belief that such differences are linked to racial types is a relatively new idea, which did not arise forcefully until the eighteenth century in Europe. At that time a number of political and scientific factors came together that seemed to inspire the ideology of racism.

First, European peoples began to have contacts with peoples of the Americas and Africa, who were not only culturally alien but physically distinct as well. During the Age of Discovery, starting in the fifteenth century, lands were conquered by Spain, Portugal, England, France, and Holland, and white Europeans encountered large numbers of nonwhite peoples for the first time. At first, the justification for subjecting these groups to enslavement or to colonial repression lay not so much in their evident physical differences as in what was seen as their cultural primitiveness, specifically, their non-Christian religions (Benedict, 1959; Gossett, 1963).

It was the later development of "scientific" racism, however, that gave impetus to the view that European peoples were superior to nonwhites because of their racial inheritance. Eighteenth-century scholars of various disciplines, including medicine, archaeology, and anthropology, had begun to debate the origin of the human species, specifically, the question of whether the species was one or many. Up to this time, most had viewed all human types as subdivisions of a single genus (Rose, 1968). Thought now turned to the much earlier but generally discounted theory of *polygenesis,* the notion that human groups might be derived from multiple evolutionary origins. No resolution of this debate came until the publication in 1859 of Darwin's theory of evolution, *The Origin of Species* (Benedict, 1959; Gossett, 1963). Darwin was clear in his explanation that differences among humans were superficial and that their more general similarities nullified any idea of originally distinct species or races. Although he recognized racial differences, it was left to others to vigorously pursue the measurement of these differences and to attach social meaning to them (Gossett, 1963).

Scientists of the nineteenth century investigating the idea of race were heavily influenced by Darwin even though he had said little about race per se. Many who studied his ideas, however, drew inferences to human societies from what he had postulated about lower animal species. Darwin's idea of natural selection was now seen as a mechanism for producing superior human societies, classes, and races. Expounded by sociologists such as Herbert Spencer and William Graham Sumner, these ideas became the basis of Social Darwinism, popularly interpreted as "survival of the fittest." Early racist thought, then, was supported by an element of what was then considered scientific validity.

With the scientifically endorsed belief that social achievement was mostly a matter of heredity, the colonial policies of the European powers were now justified. Native people of color were seen as innately primitive and incapable of reaching the level of civilization attained by Europeans. Economic exploitation was thus neatly rationalized. And, because nonwhites represented a supposedly less developed human evolutionary phase, the notion of a "white man's burden" arose as justification for imposing European cultural ways on these people.

In short, the idea of race appropriately complemented the political and economic designs of the European colonial powers. Sociologist Michael Banton notes that "the idea that the Saxon peoples might be biologically superior to Celts and Slavs, and white races to black, was seized upon, magnified, and publicized, because it was convenient to those who held power in the Europe of that day" (1970:20). Banton adds that the coincidental appearance of these theories with the demise of slavery in the early and mid-nineteenth century provided a new justification to some for subordinating former slaves.

Racism should not be thought of as an opportune invention of European colonialism. As van den Berghe explains, racist thinking "has been independently discovered and rediscovered by various peoples at various times in history" (1978:12).[7] But it was in the colonial era that the idea was most enthusiastically received and firmly established as a social doctrine.

The idea that race was immutably linked to social and psychological traits continued to be a generally accepted theory in Europe and America during the early years of the twentieth century, aided by the unabashedly racist ideas of writers such as Houston Stewart Chamberlain and Count Arthur de Gobineau in Europe and Lothrop Stoddard and Madison Grant in the United States. Grant's book, *The Passing of the Great Race,* published in 1916, was essentially a

---

[7]On the intellectual perils of interpreting past racist thought and action in terms of current understandings of the idea, see Banton (1983).

diatribe aimed at restricting immigration of southern and eastern European groups, mainly Catholics and Jews, who were at that time most numerous among American immigrants. Grant divided the European population into three racial types — Alpines, Mediterraneans, and Nordics — and he attributed to the latter (from northwestern Europe) the most desirable physical and mental qualities. Consequently, he warned of the degeneration of the American population through the influx of southern and eastern Europeans, whom he classified as inferior Mediterranean and Alpine types.

The impact of Grant's and Stoddard's books was increased by the development of intelligence testing during World War I by American psychologists, whose findings seemed to confirm the notion of inherent racial differences. On these tests, Americans of northwestern European origin outscored, on the average, all other ethnic groups. This result was taken by many as empirical evidence of the superiority of the Nordic "race." Intelligence was now seen as primarily hereditary even though little attention was paid to environmental factors in interpreting test results. That northern blacks outscored southern whites, for example, was attributed not to social factors such as better educational opportunities in the North but to the selective migration of more intelligent blacks out of the South. This argument proved false, however, because the same differences were shown between northern and southern whites (Montagu, 1963; North, 1965; Rose, 1968).

During the 1920s, cultural anthropologists began to question the biological theories of race and to maintain that social and cultural factors were far more critical in accounting for differences in mental ability. Chief among these theorists was Franz Boas, who pointed out the lack of evidence for any of the common racist assertions of the day. Gossett explains Boas's critical role in reversing social scientific thinking on the notion of race: "The racists among the historians and social scientists had always prided themselves on their willingness to accept the 'facts' and had dismissed their opponents as shallow humanitarians who glossed over unpleasant truths. Now there arose a man who asked them to produce their proof. Their answer was a flood of indignant rhetoric, but the turning point had been reached and from now on it would be the racists who were increasingly on the defensive" (1963:430).

Racism in Modern Thought With few exceptions (such as Jensen's controversial findings in 1969 and Herrnstein and Murray's in 1994), the social scientific emphasis on environmental factors as determinants of most social and mental characteristics has remained firm. The preeminent view is that genetic differences among human groups are of minimal significance as far as behavior and intelligence are concerned. Those scholars not subscribing to this view have, in van den Berghe's words, "been voices in the wilderness (except in societies like Nazi Germany, South Africa, or the American South) and their views have not been taken seriously" (1978:xxii). Yet as William Newman (1973) points out, ironically, science itself created the myth of race that it is today still attempting to dispel. Moreover, for many decades few sociologists questioned the prevailing ideas about racial inequality. "Both popular and educated beliefs," notes James McKee, "provided an unqualified confidence in the biological and cultural superiority of white people over all others not white" (1993:27).

Just as scientific thinking on the issue of race has changed, so, too, has lay thinking. Most people now view the discrepancies in social achievement among ethnic groups more as cultural differences, not as biogenetic ones (Banton, 1970; Schuman, 1982). Because of an inability to adapt to the dominant culture, it is argued, economic and social handicaps persist among certain ethnic groups. These inabilities are traced to a people's way of life — its culture — which hinders conformity to the norms and values of the dominant group.[8]

Even though popular thinking has drifted away from the old biological racist ideas, much inconsistency and confusion remain. As a result, the prevailing social science explanation stressing environment is not necessarily accepted. Instead, many attribute low achievement among certain ethnic groups to lack of individual motivation while still denying both biological *and* environmental arguments (Schuman, 1982). In sum, racism is a belief system that has proved tenacious, though modifiable in form and content, in multiethnic societies. It is a social phenomenon whose consequences continue to be felt by both dominant and subordinate ethnic groups.

SUMMARY

The major aspects of the study of race and ethnic relations are (1) the nature of relations among ethnic groups of multiethnic societies; (2) the structure of inequality among ethnic groups; (3) the manner in which dominance and subordination among ethnic groups are maintained; and (4) the long-range outcomes of interethnic relations, that is, either greater integration or greater separation. In looking at these issues, we will compare the United States with other multiethnic societies.

---

[8]Although the ideology of cultural deprivation is seemingly more benign than "old-fashioned" racism, focusing on biological superiority and inferiority, critics have pointed out that the difference between the two is not so sharp (Ryan, 1975). They argue that cultural deprivation, like racist notions, emphasizes individual and group shortcomings rather than a social system that, through subtle discrimination, prevents the bulk of minority group members from attaining economic and social parity with the dominant group.

An *ethnic group* is a group within a larger society that displays a common set of cultural traits, a sense of community among its members based on a presumed common heritage, a feeling of ethnocentrism among group members, ascribed group membership, and, in some cases, a distinct territory. Each of these characteristics is a variable, differing from group to group and among members of the same group.

Race is an often misused notion having biological and social meanings. Biologically, a race is a human population displaying certain hereditary features distinguishing it from other populations. The idea is essentially devoid of significance, however, because there is virtually no scientific agreement about how many races exist, what the distinguishing features of races are, and what bearing race has on human behavioral traits. The sociological importance of race lies in the fact that people have imputed significance to the idea despite its questionable validity. Hence, races are always socially defined groupings and are meaningful only to the extent that people make them so.

Racism is an ideology, or belief system, designed to justify and rationalize racial and ethnic inequality. The members of socially defined racial categories are believed to differ innately not only in physical traits but also in social behavior, personality, and intelligence. Some "races," therefore, are viewed as superior to others.

The ideology of racism emerged most forcefully in the age of colonialism, when white Europeans confronted nonwhites in situations of conquest and exploitation. The idea that race was linked to social and psychological traits was given pseudoscientific validity in the late nineteenth and early twentieth centuries by American and European observers, a view that was not basically reversed until the 1930s. Today, scientific thought has given little credence to racial explanations of human behavior, stressing instead environmental causes.

Suggested Readings

Banton, Michael, and Jonathan Harwood. 1975. *The Race Concept.* New York: Praeger. Traces the concept of "race" historically and explores the controversies surrounding it.

Benedict, Ruth. 1943. *Race: Science and Politics.* New York: Viking. A brief work, by one of the most eminent American anthropologists, that has had a profound impact on social scientific thinking about race.

Glazer, Nathan, and Daniel P. Moynihan (eds.). 1975. *Ethnicity: Theory and Experience.* Cambridge, Mass.: Harvard University Press. A collection of essays by prominent social scientists that explores the concept of ethnicity.

Isaacs, Harold R. 1989. *Idols of the Tribe: Group Identity and Political Change.* Cambridge, Mass.: Harvard University Press. A finely written account of the strength of ethnic identity and its political consequences in contemporary societies.

Marks, Jonathan. 1995. *Human Biodiversity: Genes, Race, and History.* New York: Aldine de Gruyter. Examines the biological basis of human differences, demonstrating the speciousness of the division of the world's people into racial categories.

McKee, James B. 1993. *Sociology and the Race Problem: The Failure of a Perspective.* Urbana: University of Illinois Press. Traces the development of the sociology of race relations, particularly in the earlier part of the twentieth century.

Miles, Robert. 1989. *Racism.* London: Routledge. Explains the concept of racism and argues for its continued usage in sociological analysis.

Montagu, Ashley. 1974. *Man's Most Dangerous Myth: The Fallacy of Race.* 5th ed. New York: Oxford University Press. A good explanation of the mythical nature of race. Demonstrates that many alleged differences among so-called racial groups do not, in fact, exist or are socially insignificant.

Omi, Michael, and Howard Winant. 1986. *Racial Formation in the United States: From the 1960s to the 1980s.* New York: Routledge & Kegan Paul. Illustrates the idea of race as a social construct, emphasizing how race has been differently conceptualized in the United States in response to changing political circumstances.

van den Berghe, Pierre. 1981. *The Ethnic Phenomenon.* New York: Elsevier. One of many significant works on race and ethnic relations by van den Berghe. Includes his controversial sociobiological interpretation of ethnicity.

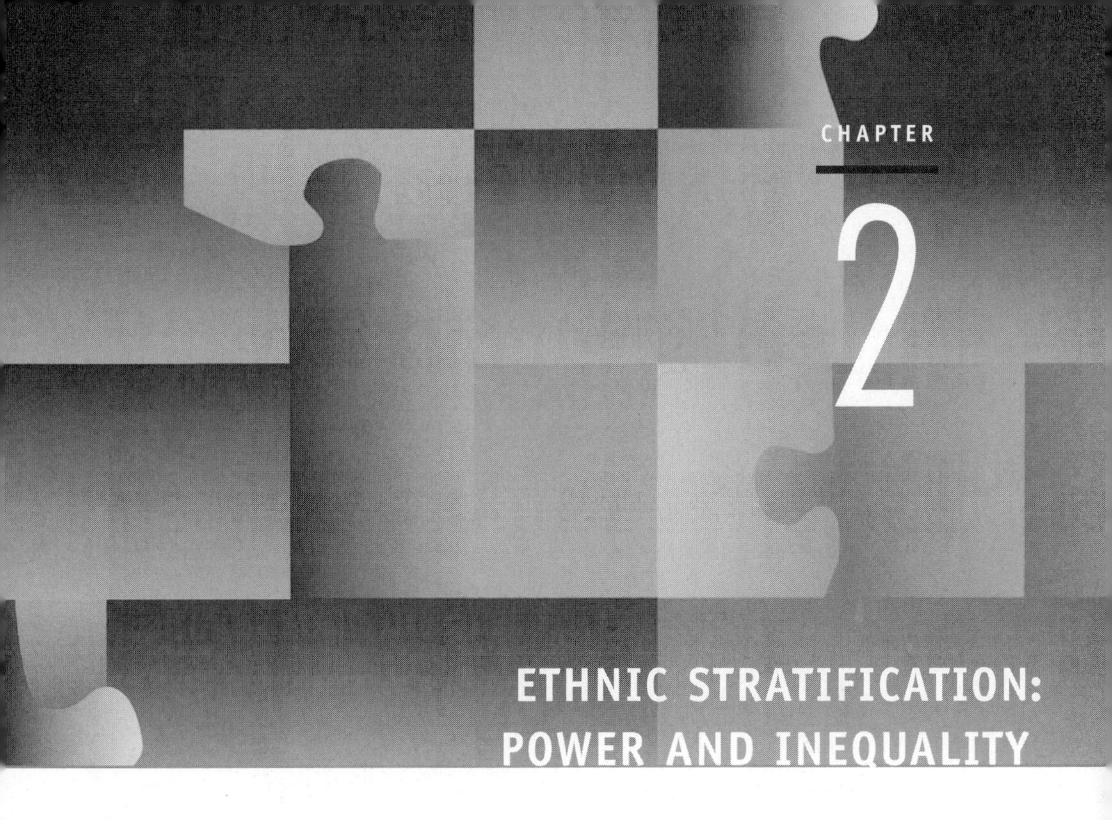

CHAPTER 2

ETHNIC STRATIFICATION: POWER AND INEQUALITY

A PROMINENT SOCIOLOGIST has suggested that the first questions asked by sociology were these: "Why is there inequality among men? Where do its causes lie? Can it be reduced, or even abolished altogether? Or do we have to accept it as a necessary element in the structure of human society?" (Dahrendorf, 1968:152). Such questions remain fundamental to sociological inquiry and are particularly critical in the study of race and ethnic relations.

Humans are unequal, of course, in many ways. They differ in physical features and in mental capacities, talent, strength, musical aptitude, and so on. All these inequalities are a product of both social learning and genetic inheritance though the significance of each of these factors is, as we have now seen, not always clear. Perhaps more important, however, people are also unequal in their access to social rewards, that is, various forms of wealth, power, and prestige. These differences, all primarily of social origin, are of greatest consequence in accounting for who we are and who we ultimately may be as members of our society. It is these inequalities with which sociologists are most concerned.

STRATIFICATION SYSTEMS

In all societies, people receive different shares of what is valued and scarce. This unequal distribution of the society's resources creates a system of *stratification.*

People are grouped on the basis of how much of the society's rewards they receive, and these groups, or strata, are arranged in a rank order, or hierarchy. Those at the top receive the most of what there is to get, and those at the bottom the least. Societies may comprise any number of strata, but in all cases, this system of inequality is structured. This means that stratification is not random, with groups and individuals occupying different positions by chance; rather, social institutions such as government, the economy, education, and religion operate to assure the position of various groups in the hierarchy. Moreover, the system of stratification in all societies is legitimized by an ideology that justifies the resultant inequality. The pattern of stratification in a society therefore remains stable for many generations.

Modern societies are stratified along several dimensions, the most prominent of which is *class* stratification, in which groups are ranked on the basis of income, wealth, and occupation. Multiethnic societies are also stratified on the basis of ethnicity, and it is on this dimension that we will primarily focus. So that we can better understand the specific nature of ethnic stratification, however, we will look first at some of the more general characteristics of stratification systems.

Dimensions of Stratification

Beginning with Karl Marx, many have stressed the economic dimension of social stratification. A social class, according to Marx, comprises those who stand in a common position with regard to the society's productive system. Although he recognized many subclasses, he saw the eventual emergence of two encompassing classes: those who own and control the means of production and those who do not. Those who have access to the society's productive resources—in industrial societies, factories, banks, transportation, communications, and so on—constitute a ruling class. Those who own no productive property, the working class, can offer only their human labor in exchange for material needs. The different economic interests of these two classes become the basis of political struggle and societal change.

Weber's Multidimensional Model Marx's view of stratification is essentially unidimensional: Classes are founded on economic criteria. One of the first scholars to suggest a more elaborate, multidimensional model of stratification was Max Weber. In addition to a hierarchy based on economic factors, Weber denoted hierarchies based on status and on political power, the latter referred to as "party" (Gerth and Mills, 1946).

Weber's notion of class is similar to Marx's: A class comprises those who stand in a similar position with regard to their opportunities to acquire the society's economic rewards. It is in the status and power dimensions, however, that Weber's model is innovative.

The status dimension, according to Weber, consists of groups that display a particular lifestyle and that are aware of differences between themselves and other status groups. Such status communities manifest themselves in common consumption patterns, club memberships, residential areas, and schools and, ultimately, in intermarriage. Weber was careful to explain that economic class and status are generally overlapping dimensions because a particular lifestyle necessitates commensurate wealth. Most sociologists today use the term *socioeconomic status,* a combination of Weber's class and status dimensions, in place of *social class.*[1]

Weber's third dimension, party, denotes political rank, that is, one's standing in a collectivity or organization whose "action is oriented toward the acquisition of social 'power,' that is to say, toward influencing a communal action" (Gerth and Mills, 1946:194). Such organizations may represent either class or status group interests or perhaps both. In addition to formal political organizations like parties, examples in modern industrial societies are labor unions, consumer groups, business and professional associations, and ethnic organizations—in short, any group designed for or capable of involvement in political action. This element of Weber's scheme is often misinterpreted as describing power as a distinct dimension of stratification. Weber notes, however, that all three dimensions—class, status, and party—are "phenomena of the distribution of power within a community" (Gerth and Mills, 1946:191).

Both the Marxian and the Weberian models of stratification are valuable, and they are necessary to fully understand structured inequality in societies. The basic difference between them, however, is that with Weber's multidimensional model, stratification can be seen as emanating from several sources, not simply one's economic position. Though Weber, like Marx, recognized the fundamental nature of people's place in the productive system in determining their overall position in the stratification system, Weber's model provides for cases in which other factors enter into the picture. This is particularly important to ethnic stratification, wherein people are ranked on the basis of their ethnic group membership in addition to their economic position. Although, as we will see, there is generally a close correlation between ethnicity and economic class, the two do not always coincide.

Lenski's Multidimensional Model Gerhard Lenski (1966) has offered another multidimensional model of stratification, which disentangles much of the

---

[1]For an explanation and discussion of the varied usages of stratification terminology in American sociology, see Gordon (1963).

confusion of stratification analysis and is particularly suited to our purposes. Lenski recognizes several class hierarchies, which he calls "class systems," each based on some key social criterion—wealth, occupation, education, political authority, ethnicity, and so on. Each class system comprises a hierarchy of classes that are made up of individuals who enjoy an approximately equal measure of authority, wealth, ethnicity, or whatever criterion of rank is applied. These various hierarchies, or class systems, together make up what Lenski calls the "distributive system," which can be interpreted essentially as the total socioeconomic structure of the society. If we visualize a social system as comprising individuals, classes, class systems, and the distributive system, each represents a different level of organization. They can be seen as telescoping in such a way that individuals are at the basic level and constitute the units within classes, which in turn are the units within class systems. Together, these class systems make up the distributive system. This model is shown in Figure 2.1.

Lenski's model of class systems is advantageous to our purposes for several reasons. First, it shows us the relationship between ethnicity and other dimensions of stratification. How much consistency across the various class systems do individuals and groups display? Do those who rank high in the ethnic dimension rank equally high in others, and if there are inconsistencies, what accounts for them? In most multiethnic societies, there is a clear-cut relationship between one's place in the ethnic hierarchy and in other hierarchies such as occupation and political authority. Those ranking high in the ethnic hierarchy ordinarily rank high in the others. Exceptions do arise, however, among both ethnic groups and their individual members.

A second advantage to Lenski's model is that it may help us determine in different societies the relative importance of each class system or hierarchy, particularly the ethnic one. In some societies, like Brazil and Mexico, ethnicity plays a less consequential role in determining people's life chances than in societies like the United States or South Africa, where it is enormously significant. In the latter, people's ethnic rank may loom large in determining which occupations they will perform, how much income and wealth they can be expected to acquire, and how much political power they will be able to wield.

Power and Stratification In a basic sense, power underlies all forms of stratification. Whether social classes are seen as outgrowths of a society's productive system (à la Marx) or as derivative of other criteria as well (à la Weber or Lenski), they are essentially founded on and maintained by differentials of power (Parkin, 1971). Just as differences in wealth, education, occupation, and prestige are mirrors of a society's power arrangement, so, too, are differences in rank among a society's ethnic groups. Social stratification, then, is a system of unequal distribution of a society's rewards, determined above all by power dif-

Figure 2.1 The Distributive System of American Society. *A, B,* and *C* Represent Three Individuals

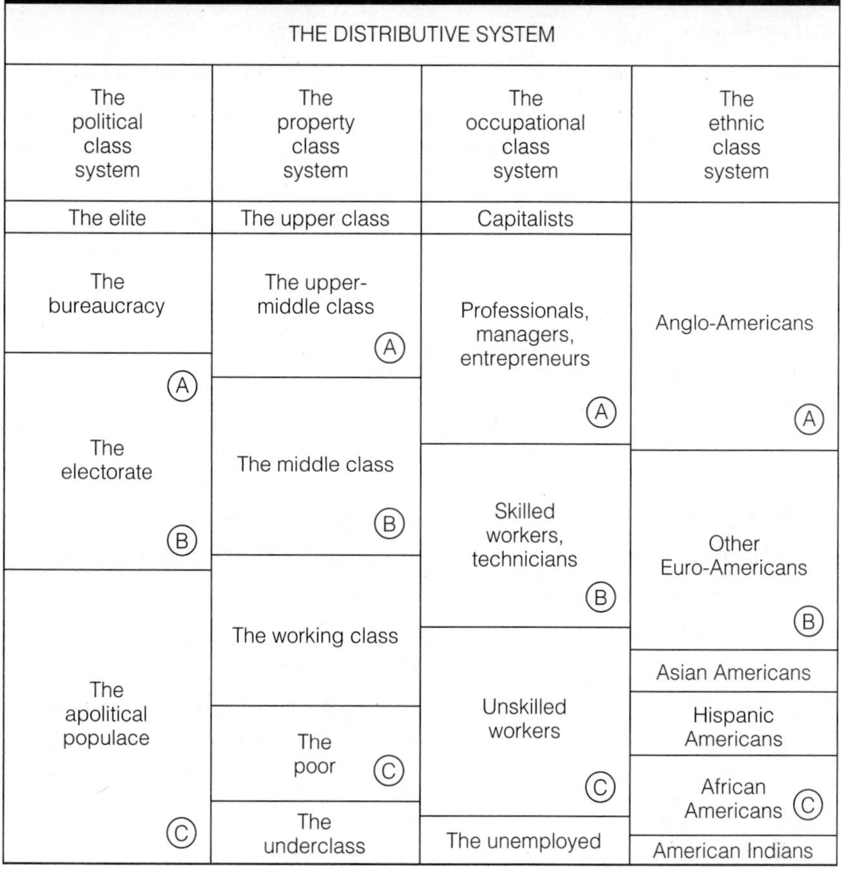

THE DISTRIBUTIVE SYSTEM			
The political class system	The property class system	The occupational class system	The ethnic class system
The elite	The upper class	Capitalists	Anglo-Americans (A)
The bureaucracy	The upper-middle class (A)	Professionals, managers, entrepreneurs (A)	Other Euro-Americans (B)
(A) The electorate (B)	The middle class (B)	Skilled workers, technicians (B)	Asian Americans
The apolitical populace (C)	The working class	Unskilled workers (C)	Hispanic Americans
	The poor (C)	The unemployed	African Americans (C)
	The underclass		American Indians

SOURCE: Adapted from Lenski, 1966.

ferentials. In simple terms, those at the top get more of what is valued because they are more powerful; they possess greater power resources in the form of wealth, property, political office, arms, control of communications, and knowledge. The position of others in the stratification system is determined accordingly on the basis of their ability to amass and apply power resources.

Stratification and Ideology

The power of a dominant class is not simply the power of force but also the power to propound and sustain an ideology that legitimizes the system of inequality. Although coercion is always at the root of obedience to authority, and all dominant groups use force when the need arises, coercive techniques are commonly used only in societies where the prevailing system is not accepted by a significant part of the populace. In South Africa under apartheid or in the antebellum U.S. South, for example, the powerful traditionally enforced their will through blatant forms of repression.

The use of raw force alone, however, cannot be effective in prompting compliance with a system of inequality over long periods. The stability of systems that rely primarily on coercion is always precarious. For government and other supportive institutions of the dominant group to establish and sustain a ruling system that is popularly supported over many generations requires that power be legitimized in less repressive and less direct ways. People must come to see the inequalities in power and wealth as just and even socially beneficial. Only then do systems of social inequality attain stability. When this is accomplished, ruling groups need no longer resort to force as the principal means of assuring their power and privilege. Such long-range stability and legitimacy require the development of an effective ideology and its communication through socialization.

Despite their acceptance—usually reflexively—by both ruling groups and masses, the fundamental ideological values tend to accommodate mostly the interests of the society's ruling groups. In the United States, for example, the dominant explanation for social inequality centers on the belief that the society's opportunity structure is open, providing equal chances for all to achieve material success or political power, regardless of their social origins. This presumably being the case, each person controls his or her placement in the social hierarchy. Social success, then, is explained as the result of one's willingness to work hard; failure is the product of lack of ambition or desire to improve oneself. Differences in wealth and power are not denied, but they are seen as a result of individual factors rather than the workings of a class system that ordinarily engenders success for the wellborn and failure for the poor. In reality, however, the opportunity structure is hardly equal, and the dominant values of individualism, competition, and achievement favor those who are wealthy and can easily avail themselves of the opportunities for success. The social system, then, tends to favor those who are already wealthy and powerful.

Just as there are ideologies that explain and rationalize inequalities in societal wealth, occupation, prestige, and political authority, so, too, there are ideologies that explain the differential treatment of groups on the basis of ethnicity. These ideologies generally comprise some form of racism (or *ethnicism*), the basic components of which were discussed in Chapter 1. And just as most of those who are not wealthy and successful still accept much of the dominant class ideology, so, too, may minority ethnic groups sometimes accept the dominant ethnic group's explanations of superiority and inferiority. They may internalize attitudes of inferiority and even manifest feelings of group self-hatred. Antisemitism has at times been observed among Jews, for example, in which individuals attempt to disguise their Jewish identity and identify instead with the dominant ethnic group by changing their name, altering their behavior, and even changing their physical appearance. The acceptance of dominant ethnic group values is also illustrated in the once-common adoption by American blacks of white standards of beauty. For many generations, blacks used various devices such as hair straighteners and skin bleaches in conforming to white ideals. Women, too, in the past generally accepted prevailing explanations of their inferiority vis-à-vis men, and only in recent decades did they begin to question the inevitability of male dominance in most areas of social life. Such acceptance by minority groups of the prevailing standards, values, and beliefs further solidifies the power of the dominant group and contributes to the maintenance of the status quo.

To summarize, social stratification is a system of structured inequality in which people receive different amounts of the society's valued resources. This inequality is relatively stable and gives rise to social classes, that is, groups of people of approximately equal income and wealth. In multiethnic societies, ethnicity becomes an additional basis of stratification. Differential power underlies all forms of inequality, and the system is underwritten by an ideology—propounded by the dominant group but generally accepted by other classes—that justifies differences in social rewards.

ETHNIC STRATIFICATION SYSTEMS

Ethnic stratification, like other forms of stratification, is a system of structured social inequality. In almost all multiethnic societies, a hierarchical arrangement of ethnic groups emerges in which one establishes itself as the dominant group, with maximum power to shape the nature of ethnic relations. Other, subordinate, ethnic groups exert less power, corresponding to their place in the hierarchy, extending down to the lowest-ranking groups, which may wield little or no power.

Group rank is determined mainly on the basis of distance from the dominant group in culture and physical appearance. Those most like the dominant group are more highly ranked, and those exceedingly different are ranked correspondingly low. A system of ethnic stratification, then, is a rank order of groups, each made up of people with presumed common cultural or physical characteristics interacting in patterns of dominance and subordination. Sociologists ordinarily refer to ethnic stratification systems as *majority–minority,* or *dominant–subordinate,* systems. Let us look more closely at the nature of both minority and majority groups.

Minority Groups

Differential Treatment Minority groups are those groups in a multiethnic society that, on the basis of their physical or cultural traits, receive fewer of the society's rewards. In a classic definition, Louis Wirth defined a minority group as "a group of people who, because of their physical or cultural characteristics, are singled out from the others in the society in which they live for differential and unequal treatment, and who therefore regard themselves as objects of collective discrimination" (1945:347). Members of minority groups thus occupy poorer jobs, earn less income, live in less desirable areas, receive an inferior education, exercise less political power, and are subjected to various social indignities. These inequalities are the result of their social mark, that is, the physical or cultural features that distinguish them. Moreover, as Wirth pointed out, minority group members are conscious of the fact that they are differentially treated.

In addition to the fact that they receive fewer of the society's rewards, several additional features are fundamental to minority groups.

Social Definition First, it must be understood that the physical or cultural traits on which minority status is based are socially defined. Thus, any characteristic may serve as the basis of minority status as long as it is perceived as significant. Let us suppose that hair color was considered a meaningful distinguishing feature. If blond hair was deemed more desirable than black or brown, those with black or brown hair might be singled out accordingly and treated differentially for no other reason than that their hair was not blond.

Differential Power Second, minority groups are afforded unequal treatment because they lack the power to negate or counteract that treatment. Blond-haired persons could continue to treat nonblonds as less than desirable types and withhold various forms of social rewards only if they maintained sufficient power to do so. As Yetman notes, "Majority-minority relations do not appear until one group is able to impose its will on another" (1991:1). Minority status, then, is above all a reflection of differential power. James Geschwender's definition of minority includes this important power factor. He defines a *minority* as "any group that is socially defined as different from the dominant group in society, is at a power disadvantage, receives less than its proportionate share of scarce resources due to its power disadvantage, and finds its differential treatment justified in terms of socially defined differences" (1978:17).

Categorical Nature Third, *minority* denotes a group, not an individual, status. All those who are classified as part of the group will experience differential treatment, regardless of their personal characteristics or achievements. Thus, people cannot voluntarily remove themselves from their minority position.

Sociological and Numerical Meaning Finally, it is important to understand that the sociological meaning of *minority* is not the same as the mathematical definition. Numbers have no necessary relation to a group's minority status. For example, nonwhites in South Africa, as we will see in Chapter 12, make up over 85 percent of the population, yet traditionally have constituted a sociological minority. Until recently, they had almost no access to political and economic power, were assigned the lower occupational positions, and were afforded grossly inferior opportunities in all areas of social life by comparison with South African whites, who make up less than 15 percent of the population. Rather than relative size, it is a group's marginal location in the social order that defines it as a minority.

Types of Minorities

In its sociological meaning, the term *minority* can be applied to a variety of social groups. In addition to ethnic traits, many other physical or behavioral characteristics are sufficient to set off groups of people from the society's mainstream, resulting in differential treatment. To equate minority groups only with highly visible ethnic groups fails to account for the many other types of minorities found in complex societies in various social contexts.

Sex, for example, is a clearly distinguishable physical characteristic that in most societies serves to single out one group—ordinarily, women—for unequal treatment. It is in this sense that women constitute a minority group. Traditionally, women have rarely occupied positions of great political or economic power, have been barred from entrance into many occupations, and have been excluded from numerous areas of social life. Only in recent years have these deeply rooted patterns of sex discrimination begun to change.

Physically handicapped people—the blind, the deaf, those confined to wheelchairs, and so on—are another evident minority. On the basis of their physical distinctions, they are singled out and given differential treatment in many social areas. Until recently, efforts had rarely been made to accommodate such people in public buildings, for example, and they had rarely been afforded equal educational and occupational opportunities.

Age constitutes another physical feature that serves to set groups apart for differential and unequal treatment. In all societies, of course, certain limitations are placed on people according to their age. The very young are not expected to fulfill adult roles, and at the opposite end of the life cycle, the aged are not expected to perform as younger persons do. But in some cases, this necessary age differentiation exceeds any rational explanation, and people of a particular age are singled out for differential treatment merely on the basis of their age. Often workers are forced into retirement at 65, even if they remain capable of carrying out their occupational duties and wish to do so. Or certain occupations are denied those who exceed an arbitrarily selected age requirement. Cases such as these constitute age discrimination.

Certain groups are also singled out for differential treatment on the basis of their past or current behavior. Homosexuals are a very obvious example, as are those labeled mentally ill and ex-convicts. Those acknowledging homosexuality, past treatment for mental illness, or a prison record risk social exclusion and discrimination by employers, landlords, and law enforcement agencies.[2]

For each of these groups, a belief system, or ideology, explains and justifies their differential treatment. Women are alleged to be physically or mentally inferior to men or to be unable to perform at the level of men; the elderly are perceived as slow, incompetent, or senile; homosexuals are "sick" or "depraved"; and so on. Although there may be little or no truth in these assertions, they serve as devices to sustain unequal treatment.

Ethnic Minorities

Our chief concern is with *ethnic* minorities, those groups singled out and treated unequally on the basis of their cultural or physical differences from the dominant group. Ordinarily, our attention is directed to the most conspicuous ethnic groups, those with especially marked differences in skin color or those that maintain unusually divergent cultural beliefs and behavior. In the United States today, these are blacks, American Indians, Asians, and Hispanics. These are the groups that experience the most blatant and consequential forms of discriminatory treatment and about whom mythical belief systems have been created to justify such treatment.

But we should not think of ethnic minorities as always indelibly marked or as only those that experience overt and severe prejudice and discrimination. There are also ethnic minorities whose identification is not so clear because they do not experience such flagrant, institutionalized forms of discrimination and because many individuals in these groups have gained access to important occupational and political positions. Ruth Glass refers to such groups as "hidden minorities" who "do not wear a striking label of inferiority." But as she explains, "If one looks closely, one can see that there is still a blank patch on the collar where such a label can be pinned on; and at any time of general social stress, it may indeed be stuck on again" (1964:150). The case of Japanese Americans illustrates this idea. Although they had long been targets of particularly vicious prejudice and discrimination in West Coast communities, as a collectivity they had made a strik-

---

[2]On these nonethnic minorities, see Sagarin (1971).

ingly successful adjustment to American society. But in 1942, thousands of Japanese Americans were forcibly removed from their residences in California, Oregon, and Washington and interned in prison camps, where they were kept almost until the end of World War II. Similarly, Jews often experience the manifestations of such underlying hostility. German Jews in the 1930s, for example, were considered the most thoroughly assimilated among all European Jews, but this did not prevent the Nazi holocaust, which resulted in the almost total destruction of the German Jewish population.

In any case, the treatment of ethnic minorities will vary from group to group and from time to time. Some groups may be consistently singled out and deprived of social rewards, and others may experience only minimal discriminatory treatment. Even for any single group, minority status is not a simple matter to explain. Prejudice and discrimination may be quite strong at certain times and diminished at others depending on various economic, political, and social conditions. Moreover, members of a particular group may experience rejection and denial in some areas of social life but not in others.

One feature of ethnic minorities that distinguishes them from other types of minorities is *endogamy,* which refers to marriage within one's social group and is regulated on the basis of social class, educational level, religion, or other social characteristics. Middle-class people are encouraged to marry other middle-class people, Catholics are encouraged to marry other Catholics, and so on. But especially strong enforcement of endogamy in multiethnic societies concerns ethnicity, particularly among physically distinct groups.

Such endogamous patterns may be enforced by the dominant group or by the minority group itself. Jews, on the one hand, have traditionally been more endogamous than other ethnic groups, and most of the compulsion to marry within the group has been dictated by the group itself. Prohibition of interracial marriage in the United States, on the other hand, has been rigidly enforced primarily by the dominant white group. In fact, not until 1967 were statutes proscribing marriage between blacks and whites in many Southern states struck down by the Supreme Court.

Dominant Groups

In the study of ethnic relations, it has been most common for sociologists to focus primarily on minority groups. Ethnic relations, however, involve not only the problems of those at the bottom of the ethnic hierarchy but also the manner in which those at the top maintain their dominance. Obviously, the existence of minority groups implies a majority group. In this regard, Hughes and Hughes have noted: "It takes more than one ethnic group to make ethnic relations. The relations can be no more understood by studying one or the other of the groups than can a chemical combination by study of one element only, or a boxing bout by observation of only one of the fighters. Yet it is common to study ethnic relations as if one had to know only one party to them" (1952:158).

In the United States, for example, for many years (and even today) black–white relations were portrayed as "the black problem in America," and most studies dealt with the social and psychological problems faced by blacks as a result of their minority status. Only in the 1960s did sociologists begin to show that intergroup relations between blacks and whites were mostly a *white* problem because it was the dominant white group that controlled the character and course of those relations more than did blacks themselves. Our major concern is not the unique character of particular ethnic groups per se as much as it is the social situation within which different groups interact. It is thus necessary to look carefully at the nature and functions of dominant as well as minority ethnic groups.

Most sociologists refer to a society's majority group as the *dominant* group, and that usage is adopted here. This avoids the tendency to think in numerical terms in cases where the dominant group may not be a numerical majority. If consistent, we would also refer to minority groups as *subordinate* groups, implying the existence of a power relationship rather than a numerical one. However, because the term *minority* is so firmly fixed in the sociological literature as well as in popular usage, we will continue to use it.

Political, Economic, and Cultural Dominance Schermerhorn defines a dominant group as "that collectivity within a society which has preeminent authority to function both as guardians and sustainers of the controlling value system, and as prime allocators of rewards in the society" (1970:12–13). The term *dominant,* or *majority,* then, has both cultural and political connotations, in each case denoting the power of one group over others. Let us look more closely at each element of the dominant group's power.

Most simply, the dominant, or majority, ethnic group is the group at the top of the ethnic hierarchy, with maximal access to the society's power resources, particularly political authority and control of the means of economic production. This power advantage in the political and economic realms enables it to acquire a disproportionate share of the society's valued resources and thereby to further sustain its dominance. This does not mean, of course, that all those classified as part of the dominant ethnic group enjoy equally great power advantages. Rather, it means only that members of the dominant ethnic group occupy disproportionately—in some cases totally—such positions. In the United States, for example, white Anglo-Saxon Protestants (WASPs), or Anglo-

Americans, historically have been, and to a great extent remain, the dominant ethnic group.[3] But not all members of this category are part of the decision-making orders of the polity or the economy. They are, however, disproportionately holders of the most important positions in these institutions and are able to render decisions that favor the interests and values of Anglo-Americans generally. Thus, compared with other ethnic groups, Anglo-Americans own a greater share of the society's wealth, earn more income, acquire more and better education, work at higher-ranking and more prestigious occupations, and generally attain more of the society's valued resources.

In addition to the dominant ethnic group's greater economic and political power, its norms and values prevail in the society as a whole. In short, the cultural characteristics of the dominant group become the society's standards. As Schermerhorn explains, "When we speak of a 'dominant group' we mean that group whose historical language, traditions, customs, and ideology are normative for the society; their preeminence is enforced by the folkways or by law, and in time these elements attain the position of cultural presuppositions" (1949:6).

The cultural supremacy of the dominant ethnic group in a multiethnic society applies to *major* norms and values rather than to every element of the society's culture. In any heterogeneous society, certain cultural traits of minority ethnic groups are bound to seep into the mainstream. In the United States, for example, egg rolls, bagels, tacos, and pizza are standard "American" fare. But items such as foods represent only minor cultural components. There is no similar acceptance of variety regarding the more basic aspects of culture such as language, religious values, political practices, and economic ideology. On these counts, ethnic groups are expected to acculturate to the dominant group's customs and ideals. In the United States, for example, it is assumed that all groups will speak English, will maintain Judeo-Christian ethics, will abide by democratic principles, and will accept capitalist values. Those who do not remain outside the societal mainstream and may be subjected to severe disdain and oppression. Although changes even in these areas of culture are usually evident over the long run, such changes are slow, deliberate, and accompanied by controversy and open conflict. Bilingual education in the United States, for example, has met with strong resistance.

Control of Immigration The dominant ethnic group, given its political, economic, and social power, is able to regulate the flow and composition of further immigration to the society and to determine the social treatment of other

[3]In Chapter 5, the specific makeup of the dominant ethnic group in the United States will be discussed in greater detail.

groups after they have entered the society. Once its preeminence has been established, the dominant group becomes synonymous with the "host" or "receiving" society, or what Porter calls the "charter group":

> In any society which has to seek members from outside there will be varying judgements about the extensive reservoirs of recruits that exist in the world. In this process of evaluation the first ethnic group to come into previously unpopulated territory, as the effective possessor, has the most say. This group becomes the charter group of the society, and among the many privileges and prerogatives which it retains are decisions about what other groups are to be let in and what they will be permitted to do (1965:60; see also Gordon, 1964).

No matter how liberal the immigration policies of a multiethnic society may be, certain restrictions or quotas are always imposed on groups that are highly unlike the dominant group in culture or physical appearance. And in the same manner, those immigrants whose origins are culturally and physically close to the dominant group's enjoy not only easier entrance but also more rapid and less impeded upward social mobility.[4]

It must be emphasized that dominant group power is never absolute. Although the dominant group is at the top of the ethnic hierarchy, its power and influence are relative, not complete. Members of that group are *disproportionately* in positions of power and influence. Likewise, cultural dominance should not be interpreted as complete. The dominant group has a disproportionate influence on shaping the society's cultural mold, but, as already noted, minority groups obviously make contributions and compel changes. The United States today is hardly a WASP model in literature, music, the arts, politics, education, and most other aspects of culture. On the really fundamental aspects of culture, however, such as language, law, and religion, the dominant group's influence remains much stronger and less vulnerable to minority influences.

It is equally important to understand that the ethnic hierarchy is not fixed. Rather it is always under challenge, to some degree, and therefore groups' positions may change and their power either enhanced or diminished. To speak, for example, of the Anglo group as dominant in American society today is not to imply that its position is comparable to what it was in an earlier time when its

---

[4]Petersen (1980) notes that the dominant group may not always be large and powerful enough to act as a "host" to immigrants. He points to modern Israel as a case in which immigrants have been acculturated not to a host population but to an ideology of Zionism. Similarly, Argentina's development in the past 100 years has been marked by the economic and political ascendance of immigrants.

power was far more thorough. As minority groups over the years have acquired more economic, political, and social resources, WASP power has been diluted.

Middleman Minorities

Certain ethnic groups in multiethnic societies sometimes occupy a middle status between the dominant group at the top of the ethnic hierarchy and subordinate groups in lower positions. These have been referred to as *middleman minorities* (Blalock, 1967; Bonacich, 1973; Bonacich and Modell, 1980; Turner and Bonacich, 1980; Zenner, 1991).

Middleman minorities often act as mediators between dominant and subordinate ethnic groups. They ordinarily occupy an intermediate niche in the economic system, being neither capitalists (mainly members of the dominant group) at the top nor working masses (mainly those of the subordinate groups) at the bottom. They play such occupational roles as traders, shopkeepers, moneylenders, and independent professionals. Middleman minorities therefore serve a function for both dominant and subordinate groups. They perform economic duties that those at the top find distasteful or lacking in prestige, and they frequently supply business and professional services to members of ethnic minorities who lack such skills and resources.

Given their intermediate economic position, such groups find themselves particularly vulnerable to out-group hostility, emanating from both dominant and subordinate groups. In times of stress, they are, as Blalock (1967) describes them, natural scapegoats. They are numerically and politically lacking in power and therefore must appeal to the dominant group for protection, which will be provided as long as it is felt that their economic role is necessary. But the role may still be seen as tainted, thus prompting feelings of revulsion and discriminatory actions. Subordinate groups will also view middleman minorities with disdain because they often encounter them as providers of necessary business and professional services. Such entrepreneurs therefore come to be seen as exploiters. In the United States, for example, Jews often operated businesses in black ghetto areas of large cities, a role increasingly assumed today by new immigrants, especially Koreans. Conflict between these businesspeople and their neighborhood customers has been common and, at times, severe.

Because they stand in a kind of social no-man's-land, middleman minorities tend to develop an unusually strong in-group solidarity and are often seen by other groups as clannish (Bonacich and Modell, 1980). Such in-group solidarity as well as their business success creates resentment and antipathy, which in turn prompt continued ethnic solidarity.

A number of groups that seem to fit the characteristics of middleman minorities have been evident in almost all parts of the world. Jews in Europe have historically been a classic illustration. As moneylenders in medieval times, they were a generally despised group. They assumed this economic role, however, because for Christians, moneylending was regarded as sinful. Such moneylending activity was nonetheless necessary, and as non-Christians, Jews naturally came to fill this occupational place.

The Chinese in various Southeast Asian societies, sometimes called the "overseas Chinese," have played a similar role (Freedman, 1955; Zenner, 1991). In the Philippines, for example, the Chinese have traditionally been highly successful in business but have also been the target of much prejudice and discrimination by Filipinos (Hunt and Walker, 1974). Originally, they acted as commercial intermediaries between the native Filipinos and, first, the Spanish and, then, the American colonialists. Their business activities were protected by the dominant group because they were valuable to the development of the colonial economy. Following Philippine independence, the Chinese held a significant place in the society's commerce and, as a culturally despised and politically powerless minority, they became a readily available target of both official and informal discrimination. Like other such middleman groups, the Chinese in the Philippines remain relatively segregated in housing, education, and religion and are viewed by Filipinos as a cohesive and clannish group (Hunt and Walker, 1974).

Obviously, many individuals within groups that have been labeled middleman minorities do not conform to the preeminent characteristics of this type, and even the group as a whole may, at different times, display few of them. But the idea of middleman minorities forces us to consider the many variations among minority groups of a multiethnic society. Moreover, as Bonacich and Modell suggest, it "makes us look at the relationship between an ethnic group and its societal context" (1980:24).

The Relativity of Dominant and Minority Status

Whether a person is part of the dominant group or a minority group depends, in all cases, on the social situation. In some instances, people will be part of one, and in other instances, they may find themselves part of another. For example, Jews in American society are ordinarily designated a minority group. Yet from the standpoint of blacks, who generally rank lower in economic class and prestige, Jews are part of the dominant white group. Similarly, Appalachian whites who have migrated to northern cities are often discriminated against and derisively referred to as "hillbillies"; they are, then, a minority in that context even though they are usually white, Protestant, and Anglo-Saxon in origin, that is, indisputably part of the society's dominant ethnic group (Killian, 1985; Philliber and McCoy, 1981).

Group membership in a modern, complex society is rarely a simple matter of "either-or" but instead takes the form of combinations that often yield confusing and ambiguous statuses. Different ethnic designations alone can produce equivocal statuses. For example, are American Irish Catholics responded to more generally as whites (that is, part of the dominant group) or as Catholics (that is, part of a minority group)? In different social contexts, either response may be expected. Among blacks, Irish Catholics will be seen and interacted with primarily as white, and their religion or national origin will be of no significance. Among white Protestants, however, their religious identity will be stressed, and their racial identity unimportant.

THE ORIGINS OF ETHNIC STRATIFICATION

How does ethnic stratification arise? And why does it seem so inescapable in multiethnic societies? In other words, what are the forces that seem to lead almost always to patterns of dominance and subordination when ethnic groups come into contact with one another? Sociologists have explored these questions and have suggested several factors critical to the emergence of ethnic inequality.

Forms of Contact

To begin with, all systems of ethnic stratification are products of the contact of previously separated groups (Shibutani and Kwan, 1965). Put simply, the composition of multiethnic societies depends on diverse groups coming together in some manner. This entails the movement of people from one area to another but, more fundamentally, from one political unit to another. People may move great distances and remain within the confines of the same political unit, as when a family in the United States moves from New York to California. Ethnic stratification systems, however, are created by the movement of people across national boundaries, usually bringing with them different languages and cultural systems, or by the establishment of new political boundaries. Multiethnic societies are formed through one or a combination of several contact patterns.

Conquest Conquest is a form of contact in which people of one society subdue all or part of another society and take on the role of the dominant group. European colonialism of the eighteenth and nineteenth centuries best exemplifies this pattern. Through greater military technology, the British, French, Spanish, Dutch, Belgians, and Portuguese conquered peoples in a variety of world areas, including Asia, Africa, Australia, and North and South America. Indigenous groups were brought under colonial rule and became economic appendages of the mother country.

Annexation A political occurrence in which a part or possibly all of one society is incorporated into another is annexation. As Burkey explains, "If the incorporating society has a dominant group, then the ethnic groups within the incorporated, or annexed, society become subordinate at the point that sovereignty is transferred" (1978:72). Such annexation may occur in a peaceful or a violent manner. The United States' acquisition of the Louisiana Purchase from France in 1803 illustrates a peaceful annexation. More commonly, however, such acquisitions of territory are made through successful military ventures. For example, the United States acquired most of what is today its Southwest through an expansionist war with Mexico in 1846. After the transfer of territory at the conclusion of the war, Mexicans living in these annexed areas became a minority, subject to the dominance of American cultural and political institutions.

Voluntary Immigration The most common patterns by which ethnic groups come into contact involve immigration. The immigration of peoples from one society to another may be either voluntary or involuntary. Voluntary immigration has been the chief source of ethnic heterogeneity in the United States, Canada, Australia, and New Zealand. All white ethnic groups in these societies have been the product of voluntary immigration from European societies during the nineteenth and twentieth centuries. In the United States and Canada, immigrants from Asia, Latin America, and the Caribbean have, in recent decades, contributed even greater diversity to these societies' ethnically varied populations. In addition, many previously ethnically homogeneous societies, especially those of Western Europe, have been infused with new racial and ethnic groups as a result of voluntary immigration since the end of World War II.

The chief objective of people who emigrate from their home society is ordinarily economic betterment though sometimes political or religious considerations play an important role. Demographers who study migration patterns refer to factors of "push and pull" that motivate people to leave their original society and migrate to one that promises improved conditions of life. During times of economic hardship, people will be encouraged to emigrate if they perceive more favorable economic opportunities in another society. This is the "pull." Depressed economic conditions, involving minimal job opportunities and low wages, along with a low expectation of betterment of such conditions, constitute the "push" (Bouvier, 1979; Lee, 1966).

Most European migration to North and South America is explained as a coming together of push and pull factors, specifically, hard economic times in

Europe coupled with the labor needs of growing, industrializing societies of the New World. The clarity of push–pull factors can be seen in the case of Irish immigration to North America during the 1840s. At that time, a disastrous potato famine left thousands of people starving and created an enormous impulsion to leave Ireland. Additional push factors were the increase in evictions by landlords and the unlikelihood of any major political changes that would have improved the economic situation. On the pull side, the most appealing societies were those in need of unskilled labor, like the United States and Canada, which were then in the primary stages of industrialization. It is estimated that, between 1845 and 1854, over 1.8 million people emigrated from Ireland, most going to North America (Kennedy, 1973).

Similar economic factors, though perhaps not as harsh as in the case of the Irish, accounted for most of the immigration of European groups to the United States and other immigrant-receiving societies during the latter nineteenth and early twentieth centuries. Such factors continue to play the most significant role in international migration.

Involuntary Immigration Involuntary immigration involves the forced transfer of peoples from one society to another. Such forced movements are best exemplified by the slave trade of the eighteenth and nineteenth centuries, which brought millions of blacks from Africa to work the cotton and sugar plantations of the United States, Brazil, and the West Indies. This will be discussed further in Chapters 8 and 13.

Consequences of Contact

Lieberson's Model The nature by which diverse ethnic groups initially meet has been shown to be a critical factor in explaining the emergence of ethnic inequality and the specific patterns it subsequently takes. Sociologist Stanley Lieberson (1961) distinguishes two major types of contact situations: those involving subordination of an indigenous population by a migrant group and those involving subordination of a migrant population by an indigenous racial or ethnic group. The first type, *migrant superordination,* is illustrated by various colonial conquests in which a technologically and organizationally more powerful migrant group subdues the native population. The second, *indigenous superordination,* is characteristic of most voluntary and involuntary immigrations such as those to North America; in these cases, the arriving groups are initially made subordinate to a resident dominant group.

Lieberson maintains that long-term conflict is more likely in societies where the indigenous population at initial contact is subordinate. Native groups less powerful than the arriving colonials are left with few options other than resistance to the new social order imposed on them. This hostility is further strengthened when the conquering group, over time, becomes itself an indigenous group.

In each of these situations, it is the relative power of the migrant and indigenous groups that determines the eventual nature of ethnic stratification. Where an invading group is successful in subduing the native population, the political and economic systems of the new group are imposed, and warfare and general conflict are likely to result quickly. Van den Berghe (1976) also points out that in contact situations between a conquering group and a weaker native group, there is usually a territorial factor, with the defeated group retaining an indigenous area. Indian reservations in North America are an example. Such a territorial base may provide the foundation for a separatist movement, an option not available to immigrant groups who enter as subordinates and are typically dispersed.

Situations in which the native group wields greater power and immigrant groups enter as subordinates produce less overt conflict initially. The indigenous group retains control over the size and character of immigration and may encourage quick assimilation, as in the case of most European immigrants to the United States. Moreover, conflict is diminished by the fact that if the immigration is voluntary, dissatisfied immigrants may return to their society of origin.

Noel's Model Although the nature of initial group contact may be important in giving rise to and shaping the eventual system of ethnic stratification, Donald Noel (1968) has pointed out three additional factors: ethnocentrism, competition for scarce societal resources, and an unequal distribution of power. Let us look more closely at Noel's theory.

On initial contact, divergent groups will judge each other in terms of their own culture, that is, ethnocentrically. Given the nature of ethnocentrism, such evaluations will usually be negative. The extent of these negative judgments will, however, depend on the degree of difference between the groups: The more dissimilar they are, the more negative the judgment. Studies measuring the degree of acceptance of members of different ethnic groups in the United States, for example, have consistently shown those groups closest in culture and physical appearance to the dominant Anglo group (such as northwestern Europeans) to be ranked more favorably than southern and eastern Europeans such as Italians and Poles and considerably more favorably than African Americans, Mexican Americans, and American Indians (Bogardus, 1959). When culturally or physically dissimilar groups meet, then, *ethnocentrism* can be expected to typify intergroup attitudes.

However, ethnocentrism alone, explains Noel, is not sufficient to produce ethnic stratification. Groups may view one another negatively without the necessary emergence of dominant–subordinate relations among them. An additional prerequisite is *competition,* structured along ethnic lines. When groups strive for the same scarce resources, their interrelations take on the characteristics of competition and conflict. Noel posits that the more intense such competition, the greater the likelihood of the emergence of ethnic stratification. Within the competitive arena, those groups with the greatest capacity to adapt to the social and physical environment will end up higher in the ethnic hierarchy. Thus, in the American case, groups emigrating from Europe arrived with different skills, which initially determined their occupational place and subsequently enabled them to climb upward at different rates. In no case, however, did any European group place below blacks because the latter were kept in a basically noncompetitive situation.

Differential power among the various groups is, according to Noel, the final prerequisite for the development of ethnic stratification. Unless one can overpower another, there is no basis for a stable rank order of ethnic groups, even if there is competition and ethnocentrism among them. When there is a particularly wide power gap between competing and ethnocentric groups, the emergent stratification system is likely to be quite durable. Power breeds more power, and once established, the dominant group uses its power to obstruct the competition of other groups and to solidify its dominance. In the end, then, differential power among the various groups is the most critical of the requirements for the emergence of ethnic stratification (see also Wilson, 1973).

In sum, Noel's theory postulates that competition for scarce resources provides the motivation for stratification, ethnocentrism channels this competition along ethnic lines, and differential power determines whether one group will be able to subordinate others (Barth and Noel, 1972).

Ethnic Strata: Clarity and Mobility

Ethnic stratification, like socioeconomic forms of stratification, is founded on the power of one group over others. The ensuing relations among groups are, therefore, ordinarily relations of conflict (Olsen, 1970). But if the theme of differential power is common to ethnic and socioeconomic forms of stratification, there are also several important differences between the two, namely, the clarity of and mobility between strata.

Mobility Between Strata In modern industrial societies, socioeconomic classes are relatively open, with boundaries between them indiscrete. Possibilities thus exist for individuals to move upward from a lower to a higher class or vice versa. Sociologists refer to such movement as *social mobility.* In reality, social mobility from one generation to the next or within one's own lifetime is ordinarily limited. People inherit their class position, and most do not move extensively upward or downward. Nonetheless, equalitarian ideologies proclaim and encourage class mobility. Those born into lower economic classes are, at least theoretically, able to advance.

Ethnic stratification, however, is a system in which the boundaries between strata (ethnic groups) are far more distinct. For most people, ethnicity is, as we have seen, an ascribed status; ethnic group membership is assigned at birth and is ordinarily not subject to fundamental change. As a result, ethnic consciousness is more strongly developed among members, and the competition and conflict among ethnic groups is more sharply focused.

In most multiethnic societies, the degree of mobility between ethnic strata is minimal. Particularly where physical differences, like skin color, are pronounced, extremely wide schisms develop between groups, and the lines of ethnic division remain rigid and relatively impermeable for many generations. This is so even when cultural differences are slight. African Americans are well assimilated into the dominant culture, yet they remain more rigidly segregated than others. In general, the more visible the differences between ethnic groups, the more clear-cut and inflexible will be the divisions between them. Moreover, physical differences ordinarily relate closely to the degree of inequality between the groups. Wilson notes that "there are no known cases of racial groups in advanced nation states having established equalitarian relationships" (1973:18).

Though it is nowhere common, ethnic mobility differs in degree in various multiethnic societies. In a few, the movement from one ethnic group to another is not unusual. In Mexico and Peru, for example, people who may be Indians can move into the mestizo group, the culturally and politically dominant group, merely by dropping the use of their Indian language and adopting Spanish in its place, by wearing shoes rather than the peasant *huaraches,* or sandals, and generally by practicing non-Indian ways (van den Berghe, 1978, 1979b). Physical visibility is not a critical factor that prevents such movement. To a lesser degree, the same is true of Brazil. Even in those societies with more rigid boundaries between ethnic groups, mobility may be evident among those close to the dominant group in culture and physical appearance, and such groups may be absorbed into the dominant group over a few generations. But in general, where groups remain culturally or physically distinct, mobility between ethnic strata is limited.

Caste In some cases, ethnic stratification takes on the characteristics of a *caste system,* the most rigidly static type of stratification, in which movement from one group to another is highly restricted by custom and law. The caste system of India is perhaps the most noted, but it is marked not so much by physical distinctions between groups as by people's social descent (Béteille, 1969). As it is used in non-Indian settings, the idea of caste more generally refers to "a major dichotomous division in a society between pariahs and the rest of the members of the society" (Berreman, 1966:292; see also van den Berghe, 1981). Pariahs are those who are stigmatized on the basis of some physical or cultural feature and, in most multiethnic societies, that stigma is skin color.

Where ethnic groups are racially defined, relations among them tend toward caste. Endogamy within castes is strongly encouraged, and interaction between them in intimate social settings such as peer groups, clubs, and neighborhoods is minimized. Subordinate castes are usually exploited occupationally by the dominant group and experience little or no change in their collective social position.

The United States and South Africa provide the best examples of castelike systems. In both societies, the lines of division between black and white groups are sharply drawn on the basis of perceived physical differences, and endogamy is the norm. Relations between blacks and whites remain chiefly of a secondary nature, that is, in settings that do not call for close personal and intimate contacts.

One interesting case of caste exists in Japan, where a pariah group, the Burakumin, are indistinguishable physically from the rest of the population. Nonetheless, the extreme discrimination imposed on this group has been justified in terms of race. As De Vos and Wagatsuma explain, there is a commonly shared social myth that the Burakumin "are descendants of a less human 'race' than the stock that fathered the Japanese nation as a whole" (1966:xx). In the past, Burakumin were required to wear unique identifying garb, were strictly segregated residentially, and were limited to low-status occupations (Aoyagi and Dore, 1964). Although they are today not as stigmatized as in the past, their social and economic standing remains low (Kristof, 1995). Here, then, is an example of how presumed group differences—though objectively slight or even nonexistent—can form the basis of interethnic attitudes.

Despite cases such as the Burakumin, placement and subsequent mobility within an ethnic stratification system are ordinarily affected greatly by visibility. Generally, those who are very distinct culturally and, most important, physically from the dominant group tend to remain at the bottom of the ethnic hierarchy. Those who can be stigmatized on the basis of easily perceived differences are more easily discriminated against than those whose differences are relatively indistinct. It is for this reason that the Nazis forced German Jews, physically indistinguishable from the rest of the German population, to wear yellow Stars of David on their clothing.

Not all minority groups in a multiethnic society react similarly to their subordinate status. Some may begrudgingly accept their place and wait for a more just world in the future, and others may struggle relentlessly to reverse their position. Wirth (1945) suggested several types of minority response, each of which we will briefly consider.

Pluralistic Minorities

Pluralistic minorities seek to maintain their cultural ways at the same time as they participate in the society's major political and economic institutions. Groups that enter multiethnic societies as voluntary immigrants usually adopt this position, at least for some time after their arrival, and appeal to the dominant group to tolerate their differences.

Some groups may carry the pluralistic idea further, opting out almost completely from the larger cultural, economic, and political systems. Certain religious groups in the United States and Canada such as Hutterites, Amish, and Hasidic Jews have chosen to segregate themselves even though they have not necessarily been rejected by the dominant group. These groups see contact with the larger society as a kind of contamination and a threat to their cultural integrity. Thus, they may not use the public schools or participate in mainstream political processes. Even economic matters may be largely self-contained within these groups (Bennett, 1967; Hostetler, 1993; Poll, 1969; Rubin, 1972). Such groups have achieved an accommodation with the dominant group in their society, permitting them to practice cultural patterns that are clearly aberrant in terms of the mainstream culture. Consider, for example, this description of the Amish:

> At best, the Amish seek accommodation, or a state of equilibrium in which working arrangements can be developed whereby they may maintain their unique group life without conflict; in short, they seek a kind of antagonistic cooperation. The Amish group's aim, then, is merely tolerance for their differences, and they are not at all interested in assimilation, "Americanization," or anything that would tend to merge them with the American culture and society (Smith, 1958:226).

Assimilationist Minorities

By contrast with pluralistic minorities, *assimilationist minorities* seek integration into the dominant society. Wirth explained that such groups crave "the fullest opportunity for participation in the life of the larger society with a view to uncoerced incorporation in that society" (1945:357–58). Thus, whereas the pluralistic minority will usually insist on endogamy and will enforce adherence to the norms and values of the in-group, assimilationist minorities aim for the eventual absorption of the group into the larger society.

Most European ethnic groups in American society have maintained assimilation as their long-range objective though some have been more desirous than others of retaining their ethnic heritage for several generations. Those groups closely similar to the dominant group in culture and physical appearance will ordinarily assimilate more easily and thoroughly than others. Some may even reach the point of complete absorption into the dominant group. In the United States, for example, northwestern European ethnic groups (Dutch, Scandinavian, and German) have been so completely assimilated over several generations that today they are virtually indistinguishable as ethnic units. However, such cases are exceptional. Rarely are minority ethnic groups completely absorbed into the dominant group, totally relinquishing their culture or physical identity. Moreover, even where assimilation is the group's objective, the prejudice and discrimination the group encounters will often retard such efforts. In Chapter 4, we will discuss at greater length the process of assimilation and the factors that promote it.

Secessionist Minorities

Secessionist minorities, according to Wirth, desire neither assimilation nor cultural autonomy. Their aim is a more complete political independence from the dominant society. Such groups are usually what were earlier referred to as nations, not only aspiring to some degree of political autonomy but also maintaining territorial integrity. The separatist movement in the Canadian province of Quebec represents a contemporary secessionist minority. In this case, a substantial element of the Quebec populace seeks not only to retain the French language and French-Canadian culture but also to establish an independent Quebec state. Most nationalist movements are the outgrowths of dissatisfaction on the part of ethnic groups stemming from their minority status and treatment.

In addition to seeking separation from the larger society, Wirth pointed out, a secessionist minority may desire integration with another group or society to which it feels a closer cultural and political similarity. For example, Northern Ireland's minority Catholics seek to unite the six counties of Ulster (Northern Ireland) with the Catholic-dominated Irish Republic.

Militant Minorities

Militant minorities seek as their ultimate goal not withdrawal, as do secessionist minorities, but dominance over other groups in the society. In a sense, their objective is to establish themselves as the society's dominant group. In Israel, for example, European Jews, having escaped the segregation, discrimination, and, ultimately, genocide of pre–World War II Europe, have assumed the role of a dominant ethnic group, often clashing on matters of culture and politics with later-arriving Jews from Middle Eastern and North African societies and with indigenous Arabs. If militant minorities are successful in their efforts, of course, they are no longer minorities.

Wirth maintained that these four types of response represent general stages in the life cycle of a minority group. At first, a group will seek toleration for its cultural differences and, if successful, may then, over several generations, seek incorporation into the dominant group. Blockage of such assimilation, however, will produce secessionist tendencies, which in turn may lead to the objective of domination, abetted by militant tactics.

It is important to consider that these variable responses of minority groups are not totally or, in some cases, even largely voluntary. Rather, they very much depend on the power of the dominant group to accept or reject minority group aims (Schermerhorn, 1970). It is not only a question of what a minority group initially desires, but what it desires in conjunction with what the dominant group desires for it. A minority group may seek assimilation, for example, but be repelled by the dominant group's aim to keep it isolated. Minority objectives may also change depending on the dominant group's responses. Continual frustration of assimilationist goals, for example, is likely to eventually create a pluralistic or secessionist response.

One shortcoming of Wirth's typology is its concentration on the cultural elements of ethnicity and its neglect of other critical factors, particularly race, as they shape dominant–minority relations. The nature of ethnic relations is not dependent simply on whether and in what degree the minority ethnic group becomes culturally "like" the dominant group. In Chapter 4, we will show that assimilation and pluralism are multifaceted processes. They involve not only the group's adoption or rejection of mainstream cultural traits but also its power relations and other interactions with members of the dominant group. As Schermerhorn (1970) points out, in multiethnic societies where groups are distinguished on the basis of physical as well as cultural traits, the dominant group is often concerned not with whether the minority group will adopt its cultural ways but only with maintaining sufficient social distance between itself and the out-group. Ethnic groups may assimilate culturally yet remain excluded from significant participation in the dominant institutional structure. This is the case for African Americans. Although they have adopted the major elements of the dominant culture, this has not affected their place at the bottom of the ethnic hierarchy, with its consequent prejudice and discrimination.

THE RELATIONSHIP BETWEEN CLASS AND ETHNICITY

Although ethnicity and class are distinct dimensions of stratification, they are closely interrelated. In almost all multiethnic societies, people's ethnic classification becomes an important factor in the distribution of social rewards and, hence, in their economic and political class positions. Put simply, people receive different amounts of what is valued—jobs, education, wealth, and so on—in some part as a result of their ethnicity.

That ethnic groups are hierarchically ordered would mean nothing if that rank order were not tied to the distribution of the society's wealth, power, and prestige. As Lenski and Lenski explain, as long as ethnic groups have no effect on how the society's benefits are distributed, they are not a part of the stratification system. "But when membership in such a group has an appreciable influence on an individual's access to those benefits the group becomes a part of the system" (1982:330). The fact that members of one or a few ethnic groups maintain most of the important positions of political and economic power, own an inordinate share of the society's wealth, and enjoy the most social prestige is due neither to chance nor to their greater motivation or innate capabilities. Rather, it is a consequence of the integral link between ethnic stratification and other forms of stratification in the general distributive system. In short, the ethnic and class systems are in large measure parallel and interwoven. Where people begin their quest for the society's rewards and what they ultimately achieve very much depend on their ethnicity.

If a person is a member of the dominant ethnic group, his or her way will not necessarily be unimpeded; success is by no means assured by a favored ethnic background. People's class position at birth, even for those of the dominant ethnic group, is an overarching factor in determining their eventual wealth, power, and prestige. Thus, not all Anglo-Americans, for example, are doctors, lawyers, corporation executives, and high-ranking politicians. But for dominant group members, the ethnic factor is removed as an impediment to upward mobility; other factors, both individual and structural, will affect their fortunes, but ethnicity will not.

In the same way, of course, we should not think that minority status means that a person is automatically relegated to the bottom rungs of the wealth, occupational, educational, political, and other class hierarchies. But for minorities, the chances of winding up at the bottom are much greater. As we proceed down the ethnic hierarchy, we find increasing political powerlessness, lack of economic opportunity, and social discrimination and exclusion. The closer to the bottom of the hierarchy, the more difficult the path to social success in whatever form, regardless of other nonethnic social traits.

In sum, the effect of ethnicity is that minority ethnic group members encounter barriers to the attainment of the various rewards of their society that dominant ethnic group members do not face.

This does not mean, however, that all minorities are affected to the same extent. For those who are members of low-ranking groups with high ethnic visibility — African Americans, for example — minority status may be the overriding determinant of one's economic and political class position. For other minorities more highly ranked and less visible — for example, Polish Americans or Irish Americans — ethnicity will be much less consequential. It is important to remember that there are *degrees* of minority status. The ethnic hierarchy in any society is rarely a simple two-part structure with a dominant group at the top and subordinate groups at the bottom. Rather, minority groups occupy places on a continuum, and the impact of ethnicity varies among them. The question for minority group members is, then, To what extent does ethnicity become a factor in the allocation of jobs, education, wealth, political power, and all other vital life chances?

Minority status does not necessarily indicate where in the class hierarchy a group's members enter the society but rather where they remain long after they have entered. Indeed, most groups, even those who immigrate voluntarily, enter collectively at or near the bottom. Some may rise very quickly, whereas others remain trapped at the bottom, generation after generation.

Let us look more closely at how ethnic groups as collectivities relate to other forms of stratification.

Ethnicity and Economic Class

Members of various ethnic minorities are found at different points in a society's occupational hierarchy. In the United States, for example, we find African-American doctors, Jewish-American construction workers, Mexican-American engineers, and unemployed Anglo-Americans. But these combinations of ethnicity and occupation are unusual. More commonly, we find Jewish-American doctors, Mexican-American construction workers, Anglo-American engineers, and unemployed African Americans. The relationship between economic class and ethnicity is usually very evident in the society's occupational patterns. Ethnic groups tend to concentrate in particular areas of work, especially as they enter the society. In the United States in the 1830s and 1840s, Irish workers built the canal system. In the second half of the nineteenth century and the early twentieth century, Italians were heavily represented in building construction, Jews in the garment trades, and Slavs in the steel industry. Racial-ethnic groups like blacks remained overwhelmingly in the least advanced sectors of the economy as agricultural and service workers. Though ethnic groups today are more occupationally dispersed, the remnants of these earlier patterns are still evident. Newer ethnic groups, moreover, display much occupational concentration.

In addition to occupation, other aspects of economic class—education, income, and wealth—are closely linked to ethnicity. Those most highly educated, those earning the highest incomes, and those possessing the greatest wealth are statistically more likely to be members of the dominant ethnic group or members of ethnic groups closest to the dominant group in culture and physical appearance.

Members of minority ethnic groups are never randomly scattered throughout the economic class structure but are disproportionately clustered at certain points. If ethnicity had no effect on the allocation of wealth and income, we would find a broadly proportionate distribution of members of various ethnic groups in all economic classes. If, for instance, a group constitutes 10 percent of the general population, we would ideally expect to find 10 percent of each economic class (lower, working, upper-middle, upper, and so on) made up of members of that group. Of course, other social variables such as age, sex, and region also affect people's class position, but ethnicity is a particularly critical factor. In Part II, we will see this class–ethnicity relationship more clearly as it pertains to specific American ethnic groups.

A similar class–ethnicity relationship characterizes most other multiethnic societies. The differentials among groups may be more or less extreme than what is seen in the United States, but in any case, the pattern is usually evident.

Ethnicity and Societal Power

Because all aspects of stratification are founded on power, whatever changes occur in the distribution of wealth, income, education, and other life chances necessarily depend on the makeup of the society's power elites, those who formulate policies, guide the activities, and decide the significant issues of the government, the corporation, education, and other major societal institutions. In analyzing the relationship between decision-making power and ethnicity, the key question to be asked is, How open are the power elites to members of different ethnic groups? To the extent that they are closed, the status of these groups in other dimensions of stratification will remain consistently low.

In most multiethnic societies, there is an apparent relationship between ethnicity and access to important power positions. Some groups are favored over others, and much ethnic conflict centers on the process of filling these elite posts. In the United States, the ideology of equal opportunity does not always conform to the reality of ethnic discrimination in elite recruitment. With slight changes from time to time and from one elite sector to another, historically, the dominance of white Protestants has been the general rule at the highest levels of government and the economy.

Similar patterns can be discerned in other multiethnic societies. Israel, for example, is a multiethnic society in which one group is clearly preeminent in societal power. Oriental Jews—those from North Africa and the Middle East—are a numerical majority, but political and economic power remain within the control of Jews of European background, who are also generally higher ranked in the society's income, occupational, and educational hierarchies (Cohen, 1968; Simon, 1978; Smooha, 1978, 1988). Israeli Arabs, about 13 percent of the Israeli population, are also severely limited in access to positions of influence and prestige.

Other cases demonstrating the relationship between societal power and ethnicity will be examined in Part III.

Class and Ethnicity Reconsidered

At this point, observations of the social standing of some members of dominant and minority groups might give rise to wonderment at what ethnic dominance and subordination really mean. Do we not see in the United States, for example, Jews in important power positions or blacks in high-ranking occupations? And, conversely, are not some Anglo-Americans continually at the bottom of the economic class hierarchy? If members of certain minority ethnic groups achieve upper-class standing or significant power, and similarly, if members of the dominant ethnic group remain at the bottom of the economic and political hierarchies, does this not negate the idea of ethnic dominance and subordination? How, in short, can we account for these inconsistent cases?

First, it must be emphasized that the link between class and ethnicity in any society is not perfect; obviously, some do achieve significant upward mobility despite the handicap of low ethnic rank. But we cannot generalize about groups as collectivities on the basis of the achievement of an outstanding few. Popular interpretations of the relative success or failure of ethnic groups, however, often rely on such faulty generalizations.

For example, a widespread explanation for the inability of certain ethnic groups to rise collectively in economic and political standing is "self-motivation" (Schuman, 1982). It is assumed that individuals themselves, not social forces, are mainly responsible for their social placement. The poor are poor, in this view, because they lack the motivation to improve themselves, and the wealthy are successful because they have greater incentive. That a few from low-ranking minority groups do actually achieve great social success, and that many from the dominant group do not, serve to "confirm" this oversimplistic explanation.

Individual talent, motivation, and ability are, of course, important factors in social success but only after the competitive field has been severely thinned out by ascribed characteristics such as class origin, gender, and ethnicity. Those with advantages by birth are assured relatively high achievement regardless of their individual capacities. Many in low-ranking groups may have abilities that are necessary to high social achievement, but unless they can gain access to the proper staging of those abilities through quality education, a good first job, and the nurturing of key social connections, they will go unnoticed. One of the basic features of minority groups, it must be remembered, is the social discrimination their members face in important life chances such as these.

Second, no matter how many exceptional cases sneak through the ethnic barriers, minority groups remain underrepresented in the society's top wealth, power, and prestige classes. This underrepresentation will vary in degree from group to group, but the lack of equal access to the society's rewards is precisely what defines all such groups as minorities. Glass (1964) suggests that *minority* designates a group not totally part of the society's mainstream. The extent to which a group will be "outside" will vary, of course. Italian Americans, for example, are obviously less removed from the mainstream of American society than are African Americans. Thus, we would expect to, and do in fact, find more Italian Americans penetrating the higher ranks of wealth, prestige, and power than African Americans.

Moreover, even when members of a minority group are increasingly scattered throughout the socioeconomic and political class systems, the status of individual members at the higher levels is associated with their ethnic group membership and is therefore tenuous. As Glass explains, the economic and political status of a minority ethnic group may be neither uniformly high nor uniformly low, but in any case, it is "invariably a vulnerable one, associated with notions of inferiority" (1964:142). In the United States, a city mayor who happens to be black, for example, is not seen or responded to simply as a mayor but always as a *black* mayor.

Third, in accessing the society's rewards, members of the dominant ethnic group—even those at the bottom of the economic and political hierarchies—often retain privileges not enjoyed by members of minority groups, even those of upper-class standing. For example, before the 1960s in most cities of the American South, black entertainers receiving huge salaries still could not stay at the very hotels in which they were performing. A white person, regardless of income or occupation, would not have experienced such exclusion on the basis of skin color. Today, African Americans with high status occupations—doctors, lawyers, managers—commonly describe the suspicion their presence arouses in suburban, primarily white, neighborhoods or prestigious shopping areas.

Finally, the empirical link between class and ethnicity is complex and subject to frequent change. The class position of particular groups or large segments of those groups may be altered significantly from one generation to another. For example, most members of white ethnic groups in the United States today are part of the middle class or working class though most of their forebears entered the society at an appreciably lower level. A large proportion of African Americans have also displayed a substantial change in class standing in the past two decades.

Given such changes and the intricacies of the relationship between class and ethnicity, we should not expect to find all members of a particular ethnic group with the same economic and political rank or the same economic and political concerns and attitudes. There is a range of classes *within* each ethnic group even though, for the most part, group members remain concentrated at particular levels of the society's general class system.

SUMMARY

Social stratification is a system of structured social inequality in which groups receive different amounts of the society's wealth, power, and prestige and are hierarchically arranged accordingly. In multiethnic societies, ethnicity is an additional basis of stratification. There is, therefore, a hierarchy of ethnic groups in which one, the dominant group, maintains maximum power to determine the nature of interethnic relations. Minority, or subordinate, groups rank in different places below the dominant group, depending on their cultural and physical distance from it.

Dominant–minority relations are relations of power. Although there are other bases for minority status such as gender and age, our focus is on ethnicity. On the basis of their physical and cultural characteristics, ethnic minorities are distinguished from others in the society and given differential treatment. The power of the dominant group is based on its control of the society's political and economic resources and its ability to shape the society's major norms and values.

Ethnic stratification is the product of contact between previously separate groups. Initial contact may be in the form of conquest, annexation, voluntary immigration, or involuntary immigration. The manner in which ethnic groups meet is a critical factor in explaining the shape of the system of ethnic inequality that ordinarily ensues. Following contact, groups engage in competition, view one another ethnocentrically, and, ultimately, one imposes its superior power over the others, emerging as the dominant group.

There are four types of ethnic minorities: pluralistic minorities, seeking to maintain some degree of separation from the larger society; assimilationist minorities, aiming for full integration into the dominant society; secessionist minorities, seeking political autonomy from the dominant society; and militant minorities, trying to establish dominance themselves.

Ethnic rank commonly parallels closely rank in the society's class system. Members of minority ethnic groups are never randomly scattered throughout the economic and political class hierarchies but tend to cluster at specific points.

Suggested Readings

Béteille, André. 1977. *Inequality Among Men.* Oxford: Basil Blackwell. Examines various forms of stratification, including racial and ethnic, using examples from different societies.

Kessner, Thomas. 1977. *The Golden Door: Italian and Jewish Immigrant Mobility in New York City 1880–1915.* New York: Oxford University Press. A sociohistorical comparison of the upward mobility of two ethnic groups, Italians and Jews, following their entrance into American society.

Lenski, Gerhard E. 1966. *Power and Privilege: A Theory of Social Stratification.* New York: McGraw-Hill. A classic work that examines the basis of structural inequality in its various forms.

Shibutani, Tamotsu, and Kian M. Kwan. 1965. *Ethnic Stratification: A Comparative Approach.* New York: Macmillan. Develops a theory of ethnic stratification, drawing on case studies from around the world.

Steinberg, Stephen. 1989. *The Ethnic Myth: Race, Ethnicity and Class in America.* Updated and expanded ed. New York: Atheneum. Deals with the issue of ethnic versus class factors in group mobility, leaning strongly toward the latter.

Wilson, William J. 1973. *Power, Racism, and Privilege: Race Relations in Theoretical and Sociohistorical Perspectives.* New York: Free Press. Using a power-conflict approach, explains the development of racial stratification with a comparison of the United States and South Africa to illustrate.

Zenner, Walter P. 1991. *Minorities in the Middle: A Cross-Cultural Analysis.* Albany, N.Y.: State University of New York Press. Examines middleman minorities in a number of historical and cultural settings.

CHAPTER

3

TECHNIQUES OF DOMINANCE: PREJUDICE AND DISCRIMINATION

WE HAVE NOW seen that in most multiethnic societies, ethnic groups are arranged in a hierarchy in which the dominant group receives a disproportionate share of the society's rewards because of its greater power, specifically, political authority and control of the means of economic production. It is not enough, however, to simply proclaim that the dominant group is more powerful though this is, of course, the crux of the matter. We must also look at the specific techniques by which dominance and subordination are maintained and stabilized in systems of ethnic stratification.

To enforce its power and sustain its privileges, the dominant ethnic group employs certain tools, which can be subsumed under the categories of prejudice and discrimination. Widely held beliefs and values regarding the character and capacities of particular groups are necessary to assure the long-range durability of ethnic inequality. These beliefs and values take the form of *prejudices,* that is, negative ideas regarding subordinate ethnic groups and ideas expressing the superiority of the dominant group. These beliefs often come together in a cohesive ideology of racism or another deterministic notion, but at other times, they are applied to groups in a somewhat disparate, unsystematic fashion.

In addition to prejudices, the dominant group applies various actions against minority ethnic groups, including avoidance, denial, threat, or physical attack. At different times, all these forms of coercion may be used, depending on how threatening the minority group is perceived to be. These actions are collectively termed *discrimination*.

In this chapter, we look at some theories and research findings regarding particular forms of prejudice and discrimination in multiethnic societies. This is an area of ethnic relations that, unlike others, has been the focus of much research, yielding a great amount of empirical data.

In looking at prejudice and discrimination, we will necessarily investigate some of the social psychology of ethnic relations—that is, ethnic relations at the individual, or interpersonal, level. To this point, we have been concerned primarily with ethnic relations at the group, or structural, level, and our approach to prejudice and discrimination will continue to emphasize the group dynamics and consequences of these social phenomena. But here, more than in other aspects of ethnic studies, we must pay closer attention to individual attitudes and motives, recognizing that, as C. Wright Mills so aptly put it, "Neither the life of an individual nor the history of a society can be understood without understanding both" (1956:3).

PREJUDICE

Most simply, *prejudice* can be defined as an arbitrary belief or feeling toward an ethnic group or its individual members. Specifically, prejudice involves a judgment, "based on a fixed mental image of some group or class of people and applied to all individuals of that class without being tested against reality" (Mason, 1970:52). It is, in other words, a generalized belief, usually unfavorable and rigid, applied to all members of a particular group. Although often defined as a prejudgment or preconcept founded on inadequate evidence (Klineberg, 1968), prejudice is, as Berry and Tischler (1978:235) have pointed out, "more emotion, feeling, and bias than it is judgment."

Ethnic prejudices are characterized by several specific features. First, they are *categorical,* or generalized, thoughts. Individuals are judged on the basis of their group membership, not their personal attributes. A prejudicial attitude may be directed at a particular person, but it is the person's group and its alleged traits that evoke this attitude rather than his or her individual actions and qualities. Once the person's group is known, his or her behavioral traits are inferred. Thus, prejudice violates "the norms of rationality" (Pettigrew, 1980:821).

Second, ethnic prejudices are *inflexible.* As Gordon Allport explains, "Prejudgments become prejudices only if they are not reversible when exposed to new knowledge" (1958:9). A prejudice is not simply an error in thought, but one not subject to correction. Individuals develop emotional attachments to certain beliefs and will not discard them in the light of contrary evidence. People may proclaim that "some of their best friends" are members of a particular ethnic group that they generally view adversely. The implication is that such persons do not exhibit the negative qualities ordinarily attributed to members of their ethnic group. But instead of refuting the belief, which logic would dictate, they serve only as "exceptions that prove the rule." Such contrary evidence is recognized but excluded from the generalization; it thus has no correcting effect (Allport, 1958).

Third, ethnic prejudices are usually *negative* in content. That is, the specific traits ascribed to targeted groups are inferior and socially undesirable. Of course, prejudices may be positive as well as negative. Ethnic group members maintain an overly favorable image of their own group, just as they maintain overly unfavorable images of certain out-groups. Indeed, all ethnic groups express ethnocentric notions regarding their unique character. In the area of ethnic relations, however, sociologists and psychologists have been concerned almost exclusively with negative prejudices.

Finally, prejudices are based on erroneous or inadequate group images called *stereotypes.* Because they compose the chief content of ethnic prejudices, we might look more closely at these generalized group images.

Stereotypes

Stereotypes were first suggested in 1922 by Walter Lippmann, who described them as "pictures in our heads" that we do not acquire through personal experience. In the case of ethnic stereotypes, distinctive behavioral traits of an ethnic group are selected by out-group members who exaggerate them to construct what Shibutani and Kwan call "a shorthand depiction" of the group (1965:86). Ehrlich refers to stereotypes as "a special language," which functions to "reinforce the beliefs and disbeliefs of its users, and to furnish the basis for the development and maintenance of solidarity for the prejudiced" (1973:21). Mackie defines stereotypes as "those folk beliefs about the attributes characterizing a social category on which there is substantial agreement" (1973:435).

Rarely will people claim to dislike someone of another ethnic group merely because he or she is a member of that group. Instead, the adverse view will be couched in "rational" terms. One dislikes Jews because they are "shrewd," blacks because they are "lazy," or Italians because they are "loud and uncouth." These mental images of groups thus serve as supports for the negative beliefs that constitute prejudice. Once we learn the stereotypes attached to particular groups, we tend to subsequently perceive individual members according to those generalized images.

Clearly, however, generalizing about groups or objects is a most common pattern of thought. Indeed, it is the very mental technique that facilitates social interaction, particularly in modern, complex societies where we cannot possibly know all the personal characteristics of those whom we encounter daily. Thus, on the basis of some identifying marks, such as ethnicity, sex, age, or occupation, we generalize and make judgments about people. In a sense, generalizing on the basis of group membership is a kind of predictive mechanism we use in various social situations.

Consider an example familiar to college students. On the first day of classes, students meet Professor X. Professor X is known only as part of the group called professors, about which there are certain general ideas. Professors ordinarily lecture, give examinations, grade students, and so on. Armed with this understanding of the category *professors,* students naturally expect Professor X to behave accordingly. The chances are very great that Professor X will conform to their expectations. In the same way, of course, Professor X may meet his or her students for the first time knowing nothing more about them but that they are students. With this bit of knowledge, however, Professor X can "predict" the actions of the students, using the same technique of categorizing as they have used. In this situation, people are expected to perform the roles of professor and student according to the social "script" attached to each.

In most cases our expectations or predictions about how people will act in different social situations will be correct. As a result, we are enabled to interact with people about whom we know nothing more than their social identifications, that is, their group memberships. Individuals occasionally fail to perform in the manner expected, and our predictions are then incorrect. But these are not the usual cases. For the most part, we understand correctly what is expected of people in various social roles—professor, student, doctor, father, Catholic, and so on—and pattern our behavior toward them accordingly. This is the nature of rational thought and behavior. How, then, do ethnic stereotypes differ from these common forms of generalized or predictive thought?

Stereotypes differ from rational generalizations in that they are oversimplistic and overexaggerated beliefs about a group, generally acquired secondhand and resistant to change (Harding, 1968; Pettigrew, 1980). Thus, stereotypes are sustained despite numerous individual cases that clearly refute their validity.

The characteristics attributed to various ethnic groups are established in popular beliefs and are reinforced through selective perception. This means that people take note of those cases that confirm their stereotypical pictures and overlook or ignore those that refute them. Those who believe blacks to be lazy or Jews to be deceitful will take special notice of those blacks or Jews who do in fact exhibit such traits but will fail to notice the many more blacks and Jews who do not. Moreover, these stereotypical traits will be inferred, even if they are not evident, so that the observers of an ethnic group will interpret the actions of group members in terms of their preconceived image.

In addition, stereotypes fail to show how members of the dominant group may share the same negative traits imputed to minority groups or how the dominant group, through the self-fulfilling prophecy, may contribute to the very creation of these unfavorable traits (Simpson and Yinger, 1972).

Studies of Stereotyping The pioneer study of ethnic stereotyping in the United States was conducted in 1933 by Katz and Braly, who questioned 100 Princeton undergraduates on the prevailing stereotypes of various ethnic groups. Their findings indicated a very high degree of uniformity, in some cases as high as 75 percent. Jews, for example, were consistently described as "shrewd," "mercenary," and "ambitious"; blacks as "superstitious," "happy-go-lucky," and "lazy"; and Germans as "scientifically minded," "industrious," and "stolid." Katz and Braly noted that the students had had little or no contact with members of most of the groups they described, an indication that they had acquired the stereotypes by absorbing the dominant culture.

Gilbert replicated this study in 1951 and found a prevalence of many of the same stereotypes. However, he also detected a marked change in attitude. Many students now expressed reluctance to categorize people whom they did not know. A third generation of Princeton students was questioned in 1967, and the researchers in that case also encountered irritation and resentment among those asked to generalize about ethnic groups. Further, they found that the actual content of the stereotypes had now changed considerably, with some groups such as blacks and Jews being assigned more favorable traits than had been the case in 1933. But the uniformity of the application of stereotypes had not changed. In other words, some traditional stereotypes had declined in frequency, but they had been replaced by others, some more favorable than previously. The researchers concluded that although the students in 1967 appeared to be more tolerant and less receptive to stereotyping, the application of ethnic stereotypes was still evident.

One noteworthy study of an international nature was conducted by Buchanan and Cantril in the early 1950s. Respondents from eight countries were asked to describe people of other countries by choosing from a list of descriptive adjectives. (This was similar to the technique used by Katz and Braly and others.) The general findings of this study were that people in all eight countries displayed a tendency to use stereotypes in describing other national groups, that their own compatriots were always described in flattering terms, and that the choice of either complimentary or derogatory adjectives depended largely on the current state of relations between the nations.

Table 3.1 Stereotypes of Russians Held by Americans

TRAIT	1942	1948
Hardworking	61%	49%
Intelligent	16	12
Practical	18	13
Conceited	3	28
Cruel	9	50
Brave	48	28
Progressive	24	15

SOURCE: Buchanan and Cantril, 1953. Reprinted by permission.

On the basis of their evidence, Buchanan and Cantril also concluded that stereotypes, rather than preceding people's reaction to a certain group, ordinarily do not exist until objective events demand their creation. Thus, they stressed that stereotypes should not be thought of as causative but as symptomatic. As they put it, "Perhaps their important function is the wartime one of providing a rationale within which men are able to kill, deceive, and perform other acts not sanctioned by the usual moral code" (1953:57).

A clear illustration of the creation of stereotypes as a means of rationalizing events is the dramatic change after World War II in the stereotypes of Japanese and Germans held by Americans. During the war, negative images of these two groups—as evil, hostile, and cruel—prevailed, but by the 1960s the groups were seen as clean, efficient, and industrious. The image of Russians was also altered, but in the opposite direction. As World War II allies, the Russians were portrayed in generally positive terms, but this image was changed in the late 1940s with the onset of the Cold War and the emergence of the Soviet Union as the United States' chief ideological foe. Note in Table 3.1 the striking increases in the adjectives *conceited* and *cruel* and the decrease in *brave.*

In the same way that wartime stereotypes are used to rationalize hostility toward enemy nations, stereotypes are used by dominant groups in multiethnic societies to sustain their competitive advantage over challenging or threatening ethnic groups. The negative images of blacks held by whites in the United States or South Africa have their counterparts in Northern Ireland, where Protestants hold adverse images of Catholics, or in Israel, where Jews of European origin often express negative beliefs about North African or Middle Eastern Jews. In each of these cases, the persistence of negative group images can be explained as a rationale on the part of the dominant group for keeping the minority ethnic group or groups in a subordinate position. As long as groups are perceived as undeserving, their social disadvantages can be justified.

The competitive use of stereotypes may often explain their very content. Simpson and Yinger note, for example, that those groups that have successfully competed with the dominant group cannot be labeled lazy or unintelligent, "so they are pictured as too ambitious, and with a crafty kind of self-interested intelligence" (1972:155). This can be seen clearly in the case of Jews in the United States and other societies where they have exhibited an uncommon ability to achieve economic success. Allport (1958) compares the admirable traits of Abraham Lincoln with the disliked traits of Jews and finds them quite the same. Both are generally described as thrifty, hardworking, eager for knowledge, ambitious, devoted to the rights of the average man, and eminently successful in climbing the ladder of opportunity. The key difference, explains Allport, is that the terms used to describe the Jews are often disparaging. Thus, thrifty becomes "tightfisted," hardworking becomes "overambitious," ambitious becomes "pushy," and concerned about human rights becomes "radical" (1958:184). Much the same semantic reversals of positive traits have been used in describing Chinese and East Indians in various Asian and African societies where they constitute economically successful minority groups (Hunt and Walker, 1974).

Although stereotypes are commonly applied in any multiethnic society, their content may change periodically and sometimes radically, depending on different economic, political, and social circumstances. Whatever their content, however, stereotypes are conveyed in subtle but effective ways through various socialization agencies, including the family, the school, and the mass media. Thus, stereotypes cannot be seen as irrational pictures of ethnic groups held by a numerically insignificant part of the general population. Rather, they are part of the society's heritage, and as Ehrlich (1973) explains, no person can grow up in a society without having learned them.

The Mass Media and Stereotypes In modern societies, the mass media—television, radio, newspapers, magazines, and motion pictures—are extremely critical agents of socialization. They are one of the chief means by which people interpret the structures and events of their society, and they provide role models to which individuals might not otherwise be exposed. Because their techniques are so pervasive, the media as socializers may supersede in some ways even the family and the school. Although there is still debate over the nature of the media's exact effect on socialization, no one denies that they play an important role for most people (Marger, 1993; Wright, 1986).

With their enormous impact, the media—television and motion pictures, in particular—are key conveyors of ethnic stereotypes. Until recently, blacks in American films, for example, traditionally played subservient characters and rarely were given starring roles except in all-black productions (Brown, 1981). Other racial and ethnic minorities have fared little better. American Indians have been portrayed customarily as savages and Hispanics as untrustworthy villains (Engelhardt, 1975; Moore and Pachon, 1985). In American television, the 1950s were marked by an almost total exclusion of minorities from the screen except for stereotyped roles in programs like "Amos 'n' Andy." The absence of blacks was due mainly to the fear of networks and program sponsors of offending white viewers, especially in the South (U.S. Commission on Civil Rights, 1977). This situation was modified somewhat in the 1960s with the inclusion of a few blacks and other minorities as actors and as news reporters.

The 1970s marked a significant change in television's presentation of blacks, who were increasingly incorporated into situation comedies and even became the chief protagonists in several. Many of these roles presented a new racial consciousness to television, priding themselves "not on their similarity to the mainstream, but on their differences from it" (Lichter et al., 1987). Critics pointed out, however, that blacks in these programs remained stereotyped and were almost always portrayed as comical figures, not featured in serious parts (MacDonald, 1983; U.S. Commission on Civil Rights, 1979). Moreover, minorities other than blacks, particularly Hispanic Americans, American Indians, and Asian Americans, were rarely seen in any role on American television.

The heightened racial and ethnic diversity of the 1970s gave way in the 1980s to more integrated casts and characters. Communications researcher Robert Lichter and his associates (1987) reported that from 1975 to 1986, nearly one in ten characters on television was black. Moreover, blacks played in more starring roles and portrayed more positive characters than in the past. Bill Cosby, for example, was among the most prominent television personalities of the decade. Programming with blacks in major roles, especially sitcoms, continued to proliferate in the 1990s. This was in some measure a response on the part of advertisers and producers to the realization that blacks were more frequent television watchers than other ethnic groups. Although Hispanics were given far less exposure in prime-time roles, the rapidly growing Hispanic-American population augured a similar expanded presence for them in television programming (Wilson and Gutiérrez, 1995).

One of the more apparent negative ethnic images portrayed by the media in recent years concerns Arabs. One researcher documented more than 100 different pop-entertainment programs, cartoons, and major documentaries relative to Arabs and concluded that negative stereotypes were virtually the only TV images of this group. The TV portrayal of Arabs, he found, perpetuated four basic myths: "they are all fabulously wealthy; they are barbaric and uncultured; they are sex maniacs with a penchant for white slavery; and they revel in acts of terrorism" (Shaheen, 1984:4).

Most attention has been focused on how the mass media sustain negative ethnic stereotypes, but they may also convey exaggerated positive images. De Roche and de Roche (1991) studied episodes of five police series popular in the 1980s, comparing characterizations of black and white officers. They found that black men in these dramas had been "counterstereotyped." That is, not only were these men not described with negative ghetto stereotypes, but they were portrayed in overly favorable terms in comparison with white officers. "In the collection of characters on which we focused, black men are modeled for us as more bourgeois than their white counterparts, more self-directed and affectively well-managed, and more tasteful.... They clearly confirm the traditional ideal of a stable, financially responsible husband/father, and this in an era when female-headed, single-parent families are becoming increasingly common in all social sectors" (de Roche and de Roche, 1991:86).[1]

Although most observers agree that the electronic media are important transmitters of ethnic stereotypes and attitudes, both positive and negative, studies have indicated that their effect may be only to reinforce ideas already acquired before exposure. Vidmar and Rokeach's study (1974) of viewers of a popular television sitcom of the 1970s, "All in the Family," demonstrates this point. This program set a precedent in its candid use of ethnic humor and its general treatment of ethnic-related issues. The chief protagonist, Archie Bunker, is a humorous bigot who voices all the well-established ethnic slurs and stereotypes and expresses blatant racist ideas. The producers of the program argued that by openly dealing with bigotry and making the major character a ridiculous figure, it would serve to reduce societal prejudice. Poking fun at bigotry, it was felt, would force viewers to recognize their own prejudices and thus reduce them. Others, however, argued that the program encouraged prejudice by introducing people, particularly children, to ideas they might not otherwise be exposed to.

Vidmar and Rokeach hypothesized that a process of selective perception would cause viewers to react differently to Archie Bunker's bigotry, depending on their prior attitudes. Viewers who were not prejudiced to begin with, they felt, would see Bunker as a bigot and would understand the show's satirical messages, whereas prejudiced viewers would see Bunker as "telling it like it is." Two groups of respondents, one made up of American adolescents, the other of Canadian adults, were asked their reactions to the program and were, in addition, given a set of attitudinal questions designed to measure their level of prejudice. The study's findings confirmed the selective-perception hypothesis: Reactions to the program varied in relation to prior attitudes. Those less prejudiced

[1]The researchers found, however, that blacks in these police dramas tended to remain support characters, not as qualified as whites to head up their organizations.

recognized Archie Bunker as bigoted, rigid, and domineering, and those more highly prejudiced recognized him as down-to-earth, honest, and hardworking.

Vidmar and Rokeach also hypothesized that those high in prejudice would not watch the program as frequently as those less prejudiced. This hypothesis was based on the well-researched notion that people expose themselves to social stimuli that are compatible with their established views and attitudes. For example, political scientists have demonstrated that campaign speeches by political candidates are listened to or watched mainly by those who are already committed to those candidates. In much the same way, it was thought that, because the show depicted satirical bigotry, the more frequent viewers would be those low in prejudice. However, Vidmar and Rokeach suggested that if many viewers did not see the program as satire, it would be just as reasonable to predict the opposite, that is, that the more frequent viewers would be the most highly prejudiced. Their data supported the latter view: The program seemed more appealing to those more highly prejudiced.

This and other studies (see, for example, Sarlin and Tate, 1976) suggest that television may only bolster ethnic stereotypes and attitudes already conveyed in other socialization settings. Though the electronic media are unarguably pervasive, their effectiveness in conveying the same messages to all viewers is by no means certain.

Social Distance

Prejudice involves not simply mental perceptions of ethnic groups (sociologists and psychologists refer to this as the cognitive dimension of prejudice) but also emotions and a preparedness to act in a certain way toward members of those groups (the affective and conative dimensions). If people believe blacks to be lazy and shiftless, for example, they will also probably feel resentment over welfare payments to blacks, who, it is felt, obviously do not deserve such assistance. Similar payments to poor whites, however, may go unnoticed or may be seen as merited.

The affective and conative dimensions of prejudice are reflections of what sociologists call *social distance.* Park (1924) first defined social distance as a degree of intimacy that people are prepared to establish in their relations with others. Feelings of social distance are, according to Williams, "feelings of unwillingness among members of a group to accept or approve a given degree of intimacy in interaction with a member of an outgroup" (1964:29). It is, in a sense, an indication of how acceptable or objectionable are various ethnic groups.

In 1925, sociologist Emory Bogardus constructed a technique to measure social distance between specific American ethnic groups. This social distance scale has subsequently been used commonly by sociologists as a general measure of ethnic prejudice. Respondents are asked to indicate whether they would accept a member of an ethnic out-group in varying social contexts, extending from very close encounters to very remote ones. Bogardus asked his respondents to indicate their willingness to interact with members of particular groups in the following situations:

- as close kin by marriage
- as fellow club members
- as neighbors
- as workers in my occupation
- as citizens of my country
- as visitors to my country
- as persons to be excluded from my country

Each situation represents a lower degree of social intimacy, and by assigning an increasing numerical value to each, a score can be computed for each ethnic group. Table 3.2 indicates the American social distance scales compiled by Bogardus and others over several generations. What is striking is the relative consistency of the group ranking from year to year. WASPs are at the top of the scale, followed by other northwestern European groups and, in descending order, southern and eastern Europeans, Jews, and various racial-ethnic groups.

Studies of social distance in non-American societies indicate the similar construction of a hierarchy (Bogardus, 1968; Lever, 1968; Pettigrew, 1960), but the basis of social distance may vary from one society to another. Whereas ethnic differences, particularly those with a physical basis, are the most significant criteria of social distance for Americans, in other societies, religion, class, or political ideology may be more important factors in separating people (Banton, 1967).

Feelings of social distance and ethnic prejudicial attitudes in general are difficult to accurately determine, for individuals will not always think or act in real situations as their verbal expressions seem to indicate. In short, attitudes are elusive and not always subject to clear-cut measurement, no matter how sophisticated the technique. What people tell pollsters, for example, may reflect only what they think is the socially acceptable response, not necessarily their real beliefs and feelings. That is, the pressure to conform to community and other group standards may force individuals to express attitudes that are not genuine. Moreover, the way in which poll questions are worded on racial and ethnic issues may strongly influence the results (Langer, 1989).

A study by LaPiere (1934) demonstrated the unreliability of people's statements regarding ethnic prejudice. Between 1930 and 1932, LaPiere traveled throughout the United States with a Chinese couple and was refused hotel and dining service only one time. Shortly after his travels, he sent a questionnaire to hotels and restaurants asking whether they would accept Chinese as guests. To his surprise, LaPiere received mainly negative replies from those who responded. A similar finding was obtained in the early 1950s using a black couple as the test case (Kutner, Wilkens, and Yarrow, 1952).

In sum, prejudice is multidimensional, and the negative attitudes people maintain toward various ethnic groups vary not only in intensity but also in consistency. As we will see, changing situational contexts may force people to change their attitudes accordingly.

DISCRIMINATION

Whereas prejudice is the attitudinal element in enforcing ethnic stratification, discrimination is the active, or behavioral, element. Discrimination involves behavior aimed at denying members of particular ethnic groups equal access to societal rewards. Thus, it goes well beyond merely thinking unfavorably about members of certain groups. Feagin and Feagin define discrimination as "actions or practices carried out by members of dominant groups, or their representatives, which have a differential and negative impact on members of subordinate groups" (1978:20–21). They further note that discrimination entails both effects and mechanisms. That is, there are impacts of discriminatory behavior, and there are modes of operation or techniques by which discrimination is carried out.

Although there are links between the two, prejudice and discrimination must be dealt with as distinct phenomena. There is ordinarily a tendency for prejudicial attitudes to accompany discriminatory behavior, but, as we will see, one may be evident without the other. Moreover, although they may be causally related in some instances, there need be no cause–effect relationship between the two in others. In any case, as Simpson and Yinger (1972) note, prejudice and discrimination are most frequently mutually reinforcing.

It is important to stress that, like prejudice, discrimination is applied on the basis of group membership, not individual attributes. Antonovsky explains that discrimination is a situation in which "individuals are denied desired and expected rewards or opportunities for reasons related not to their capacities, merits, or behavior, but solely because of membership in an identifiable out-group" (1960:81). Or, as Pettigrew defines it, discrimination is "an institutional process of exclusion against an out-group, racial or cultural, based simply on who they are rather than on their knowledge or abilities" (1980:821).

Table 3.2 Changes in Social Distance Rankings in the United States, 1926–1977

1926		1946		1956	
1. English	1.06	1. Americans	1.04	1. Americans	1.08
2. Americans	1.10	2. Canadians	1.11	2. Canadians	1.16
3. Canadians	1.13	3. English	1.13	3. English	1.23
4. Scots	1.13	4. Irish	1.24	4. French	1.47
5. Irish	1.30	5. Scots	1.26	5. Irish	1.56
6. French	1.32	6. French	1.31	6. Swedish	1.57
7. Germans	1.46	7. Norwegians	1.35	7. Scots	1.60
8. Swedish	1.54	8. Hollanders	1.37	8. Germans	1.61
9. Hollanders	1.56	9. Swedish	1.40	9. Hollanders	1.63
10. Norwegians	1.59	10. Germans	1.59	10. Norwegians	1.66
11. Spanish	1.72	11. Finns	1.63	11. Finns	1.80
12. Finns	1.83	12. Czechs	1.76	12. Italians	1.89
13. Russians	1.88	13. Russians	1.83	13. Poles	2.07
14. Italians	1.94	14. Poles	1.84	14. Spanish	2.08
15. Poles	2.01	15. Spanish	1.94	15. Greeks	2.09
16. Armenians	2.06	16. Italians	2.28	16. Jews	2.15
17. Czechs	2.08	17. Armenians	2.29	17. Czechs	2.22
18. Indians (American)	2.38	18. Greeks	2.29	18. Armenians	2.33
19. Jews	2.39	19. Jews	2.32	19. Japanese Americans	2.34
20. Greeks	2.47	20. Indians (American)	2.45	20. Indians (American)	2.35
21. Mexicans	2.69	21. Chinese	2.50	21. Filipinos	2.46
22. Mexican Americans	—	22. Mexican Americans	2.52	22. Mexican Americans	2.51
23. Japanese	2.80	23. Filipinos	2.76	23. Turks	2.52
24. Japanese Americans	—	24. Mexicans	2.89	24. Russians	2.56
25. Filipinos	3.00	25. Turks	2.89	25. Chinese	2.68
26. Negroes	3.28	26. Japanese Americans	2.90	26. Japanese	2.70
27. Turks	3.30	27. Koreans	3.05	27. Negroes	2.74
28. Chinese	3.36	28. Indians (from India)	3.43	28. Mexicans	2.79
29. Koreans	3.60	29. Negroes	3.60	29. Indians (from India)	2.80
30. Indians (from India)	3.91	30. Japanese	3.61	30. Koreans	2.83
Arithmetic mean of 48,300 reactions	2.14	Arithmetic mean of 58,500 reactions	2.12	Arithmetic mean of 61,590 reactions	2.08
Spread in distance	2.85	Spread in distance	2.57	Spread in distance	1.75

1966		1977	
1. Americans	1.07	1. Americans	1.25
2. English	1.14	2. English	1.39
3. Canadians	1.15	3. Canadians	1.42
4. French	1.36	4. French	1.58
5. Irish	1.40	5. Italians	1.65
6. Swedish	1.42	6. Swedish	1.68
7. Norwegians	1.50	7. Irish	1.69
8. Italians	1.51	8. Hollanders	1.83
9. Scots	1.53	9. Scots	1.83
10. Germans	1.54	10. American Indians	1.84
11. Hollanders	1.54	11. Germans	1.87
12. Finns	1.67	12. Norwegians	1.93
13. Greeks	1.82	13. Spanish	1.98
14. Spanish	1.93	14. Finns	2.00
15. Jews	1.97	15. Jews	2.01
16. Poles	1.98	16. Greeks	2.02
17. Czechs	2.02	17. Negroes	2.03
18. Indians (American)	2.12	18. Poles	2.11
19. Japanese Americans	2.14	19. Mexican Americans	2.17
20. Armenians	2.18	20. Japanese Americans	2.18
21. Filipinos	2.31	21. Armenians	2.20
22. Chinese	2.34	22. Czechs	2.23
23. Mexican Americans	2.37	23. Chinese	2.29
24. Russians	2.38	24. Filipinos	2.31
25. Japanese	2.41	25. Japanese	2.38
26. Turks	2.48	26. Mexicans	2.40
27. Koreans	2.51	27. Turks	2.55
28. Mexicans	2.56	28. Indians (from India)	2.55
29. Negroes	2.56	29. Russians	2.57
30. Indians (from India)	2.62	30. Koreans	2.63
Arithmetic mean of 78,150 reactions	1.92	Arithmetic mean of 44,460 reactions	1.93
Spread in distance	1.56	Spread in distance	1.38

SOURCES: Bogardus, 1968; Owen, Eisner, and McFaul, 1981. Reprinted by permission.

The discriminatory actions that create disadvantages for minority group members may vary widely both in form and degree. The use of derogatory labels in referring to members of ethnic groups (*kike, wop, nigger*) or phrases with pejorative ethnic references (to *Jew down,* to *gyp, Indian giver*) are relatively minor forms of ethnic discrimination and in many cases may not even be understood by the user as disparaging. Nonetheless, such terms and phrases contribute to the perpetuation of ethnic stereotypes and render psychological damage of some nature to those who are their subjects.

More serious forms of discrimination with much greater injury to minorities involve the denial of access to various life chances such as jobs, housing, health, education, justice, and political participation. Minority ethnic groups are placed in disadvantageous positions with regard to these societal rewards and end up receiving less than they would if no ethnic barriers were established. Indeed, the very notion of *minority* implies differential treatment in vital areas of social life.

The most severe forms of discrimination involve acts of aggression against minority ethnic groups, ranging from isolated incidents of violence to the deliberate destruction of an entire group. Examples of the full range abound in the modern world. In the United States, attacks on ethnic minorities have a long tradition, occurring as both selective actions undertaken by individuals or communities (such as lynchings, beatings, and bombings) and calculated public policies (such as Indian removal in the nineteenth century and Japanese-American internment in the 1940s). But the United States holds no monopoly on ethnic aggression. Comparable actions typify the history, current and past, of most multiethnic societies. Indeed, the nineteenth and twentieth centuries have witnessed extremes of ethnic violence and destruction on a massive scale in a variety of world areas, including, among many others, the annihilation of native peoples in Australia, South Africa, and North America, the slaughter of over a million Armenians by Turks, the systematic murder by the Nazis of six million Jews,[2] and, more recently, massacres of entire populations in Rwanda and Bosnia.

Discrimination is not always overt, nor does it always entail intentional actions of denial or aggression. Furthermore, there is a vast difference between isolated actions of individuals and the rational policies of institutions in creating and sustaining patterns of discrimination. The behavior of one café owner in refusing to serve members of a particular ethnic group is hardly the equivalent of a state policy requiring separate eating facilities for members of that group. The different social contexts in which it may occur and the often concealed and unintentional forms it may take require that we more precisely outline discrimination.

---

[2]The literature on the Nazi holocaust is enormous, but two of the more comprehensive works are Dawidowicz (1975) and Hilberg (1979). On policies of genocide, see Kuper (1981).

To simplify matters, we can classify two general types of discrimination: individual and institutional. Actions carried out by individuals or small groups constitute one type, and actions carried out as a result of the norms and structures of organizations and institutions constitute the other. The latter may be further broken down into direct and indirect types.

Individual Discrimination

Actions taken by individuals or groups of limited size to injure or deny something to members of minority ethnic groups are perhaps the most easily understood form of discrimination. The employment manager who refuses to hire Asians, the judge who metes out unusually harsh sentences to blacks, and the homeowners' group that agrees not to sell houses in the neighborhood to Jews are examples of discriminators at this level. In these cases, actions are taken by one or a few with the intent to harm or restrict in some way members of minority groups. Notice that the actors are not part of a large-scale organization or institution but operate within a relatively isolated context.

In cases of individual discrimination, the actions taken against minority group members are intentional. Moreover, they appear to be the implementation of prejudicial attitudes. At first glance, we might assume that the employment manager thinks unfavorably of Asians, the judge of blacks, and the homeowners of Jews. This may in fact be the motivating force behind the discrimination in all these cases, but we cannot be certain until we understand more fully the context in which these actions occur. The employment manager, for example, may have no ill feeling toward Asians but may feel compelled to carry out what he perceives to be the unwritten but generally understood company policy of not hiring Asians. The judge may feel that sentencing blacks more harshly will gain her votes among her predominantly white constituency in the next election. And the members of the homeowners' group may simply be responding to what they feel are neighborhood pressures to conform. Thus, prejudice need not be at the root of even such blatant instances of intentional discrimination. In any of these instances, of course, whether or not the actors' beliefs and attitudes are consistent with their actual behavior does not negate the detrimental effect on those who are the victims of the discriminatory actions: The Asian is still not hired, the black still serves a longer sentence, and the Jew is still denied a home.

Institutional Discrimination

The preceding cases pertain to the actions of individuals and small groups usually acting in violation of the society's norms. Discrimination, however, may be legal or customary, in which case it is not socially unexpected or disapproved but is legitimized. Such *direct institutional discrimination* is not limited to specific cases of negative actions taken against members of particular groups but is firmly incorporated in the society's normative system. Social conventions exist in which members of particular groups are legally or customarily denied equal access to various life chances.

In the United States before the 1960s, a well-institutionalized system of discrimination served to effectively block the access of blacks to the same economic, political, and social opportunities afforded whites. In the South, an elaborate system of custom and law legitimized segregated and unequal schools, housing, transportation, and public facilities; kept blacks in low-paying and less desirable jobs; and essentially prevented them from participating in the political process either as voters or officials. Most of this system was formally established and maintained by specific discriminatory laws, but much was also based on the development of customary practices.

South Africa is another case in which a formal and entrenched system of racial segregation and discrimination traditionally served to assure the power, wealth, and prestige of one group—whites—at the expense of another—nonwhites. As we will see in Chapter 12, this system, called *apartheid,* has been officially renounced but many of its essential features are still supported by customs that validate and enforce discriminatory policies and practices. Clearly, cases such as South Africa under apartheid are exceptional in the modern world. In most multiethnic societies, discrimination is more subtle, less obvious, and more indirect in application.

The form of discrimination most difficult to detect in modern multiethnic societies is *indirect institutional discrimination,* or structural discrimination. That it is so covert also makes it particularly difficult to eradicate. Unlike other forms of discrimination, it is unintended. It cannot be attributed to the prejudicial beliefs or conforming pressures of individuals or to the deliberate establishment of a set of rules seeking to withhold privileges or injure members of particular ethnic groups. Rather, it exists as a product of the normal functioning of the society's institutions. Because of past discrimination of an overt, intentional nature, or because of the spillover effect of direct discrimination in one institutional area into another, certain groups find themselves perpetually at a disadvantage in the society's opportunity and reward structures.

The structural form of discrimination is difficult to perceive easily because it does not use ethnicity as the subordinating mechanism, but uses other devices only indirectly related to ethnicity. This can be illustrated with a few current examples from the United States.

In recent decades, most new industrial jobs have been created not in central cities, where they had been concentrated in the past, but in outlying and suburban areas, where factories can be built on large expanses of land, where transportation lines, particularly highways, are more accessible, and where taxes are cheaper. The outlying location of these jobs, however, handicaps nonwhites who might qualify for them but who reside mainly in central cities. Qualified nonwhites are therefore less likely to secure these jobs, but for reasons that do not necessarily involve direct discriminatory practices by industrial employers. Executives may choose to locate their factories in the suburbs not because they wish to avoid hiring nonwhites but simply because it is economically more advantageous. When combined with the more overt discriminatory practices in housing that create the concentration of nonwhites in central cities, the outlying location of industries has the effect of discriminating against nonwhites. In other words, the more overt discrimination in one institutional area—housing—has created an indirect discriminatory effect in another—employment.

Feagin and Feagin refer to this kind of situation as "side effect discrimination." As they explain, "Intentional discrimination by persons in one sphere can result in unintentional discrimination by those in another sphere, because most societal spheres (the economy, the polity, etc.) are intimately linked to one another. Discrimination has an inter-institutional character" (1978:32).

This interinstitutional nature of discrimination can also be seen clearly in the area of education. Given the poorer quality of inner-city schools, attended primarily by nonwhites, those who attend them are automatically placed at a disadvantage in qualifying for well-paying and promising jobs. Similarly, entrance into top-ranking colleges and universities will be more difficult for them because they will not be adequately prepared to meet the rigid academic requirements. In both cases, there is no necessary intention to discriminate against nonwhites. Decisions by employment managers or college admissions officers may be quite rational, made not on the basis of the applicants' skin color but on objective employment and academic standards. What is overlooked here is that nonwhites are, as a group, less qualified than whites because inner-city public schools have inadequately prepared them for better jobs and colleges.

Such unintentional, yet effective, ethnic discrimination is repeated in a variety of areas. Bankers who hesitate to lend to black entrepreneurs because they fear they are poor credit risks are not necessarily antiblack. They are simply engaging in practical business tactics. The grocery chain that charges higher prices in its inner-city stores than in its suburban stores will explain such practices as necessary to offset the higher operating costs in the downtown stores, due to more expensive rents or insurance costs. Similar patterns emerge in health services, administration of justice, and other key areas of social life.

Perhaps most significant is that such discriminatory practices are not only unintentional but also largely unconscious. As Baron notes, "The individual only has to conform to the operating norms of the organization and the institution will do the discriminating for him" (1969:143). Discrimination, in other words, does not depend in these cases on the actions of specific individuals or even organizations. Instead, it is simply a function of the normal operating procedures of societal institutions.

Because of its unintentional and largely unwitting nature, structural discrimination remains difficult to detect, and even when it does become apparent it is not easy to determine who is ultimately responsible. Paradoxically, it is ordinarily carried out by individuals and groups who do not consider themselves discriminators. Thus, many who have purged their own behavior of discriminatory actions may come to feel that it is minority ethnic groups themselves who are mostly responsible for their depressed social condition. Poverty, crime, and other social problems among them are interpreted as the deficiencies of people who have failed to adjust to a basically neutral, not discriminatory, social system. "If we don't discriminate against them, it must be *them*." If ethnic minorities are poor, it is because of their lack of motivation, not because of structural discrimination impeding their mobility; if ethnic minorities unduly commit crimes, it is because of their criminal nature, not because of a discriminatory system's failure to provide them with the conventional means of achieving "success." The burden of responsibility for social problems, then, is placed on the individual or the group, not on the patterns of discrimination built into the institutional structure. Some have referred to this as "blaming the victim" (Ryan, 1975).

THEORIES OF PREJUDICE AND DISCRIMINATION

How do prejudice and discrimination arise, and why are they seemingly so inevitable in societies where diverse ethnic groups live side by side? Common-sense explanations, and even some scientific thought of earlier times, have accounted for out-group antipathy as a natural, or innate, pattern of human thought and action. The pervasiveness of conflict between different cultural and racial groups and the universality of ethnocentrism among peoples of the world seem to validate the notion that antagonism toward strangers or members of out-groups is simply part of the human psyche. To dislike or fear those who are different from us is, according to this view, both natural and unavoidable.

Today, however, there is general agreement among sociologists and psychologists that prejudice and discrimination are not innate human characteristics (Westie, 1964). A closer look at historical evidence indicates that people of different cultures and physical attributes do not always relate to each other antagonistically and that, when they do, the nature and intensity of their animus are highly variable. Given the variety of form, content, and extent of prejudice and discrimination, efforts to explain them as universal phenomena fall short. Further evidence that ethnic antagonism is not innate lies in social-psychological studies that have traced the various phases of socialization from early childhood through later stages of the life cycle.[3] All indicate that prejudice and discrimination are learned patterns of thought and action.

But if they agree that prejudice and discrimination are not inborn human traits, sociologists and psychologists differ in their explanations of how they arise and are sustained in multiethnic societies. The psychological tradition emphasizes prejudice as the key component in ethnic interrelations; prejudicial thought, therefore, is the focus of most psychological theories. Sociologists have stressed situational factors and power structures as the bases of both negative thought and action toward ethnic groups; prejudice is thus seen not necessarily as the source of hostility but as its outgrowth. Whereas psychologists are likely to see prejudice and, by implication, discrimination as characteristic of certain personalities, sociologists tend to see them as part of the society's normative order, to which individuals are socialized. In the remainder of this chapter, we look at several different theories, both psychological and sociological. Because prejudice and discrimination are multifaceted, each theory may contribute to our understanding of these phenomena.

Psychological Theories

Psychological theories of prejudice have focused on the manner in which antipathy toward out-groups either satisfies certain psychic needs or complements the general personality structure of certain people. In each case, the source of prejudice is traced primarily to individuals rather than to the social forces weighing on them or the groups within which they interact.

Frustration-Aggression One of the earliest psychological theories explains prejudice as a means by which people express hostility arising from frustration. This has also been referred to as scapegoating. The essential idea is that a person who is *frustrated* in his or her efforts to achieve a highly desired goal tends to respond with a pattern of *aggression.* Because the real source of frustration is either unknown or too powerful to confront directly, a substitute is found on whom the aggression can be released. The substitute target is a scapegoat, a person or group close at hand and incapable of offering resistance. The aggressive behavior is thus displaced (Dollard et al., 1939). Minority groups in multiethnic societies have served as convenient and safe targets of such displaced aggression. Allport (1958) explains that racial, religious, or ethnic groups can be blamed unfairly for a variety of evils because they are permanent and stable and can be

---

[3]For an explanation and discussion of these processes, see Allport (1958:271–321).

easily stereotyped. Therefore, they serve as more of an "all-purpose" scapegoat than those groups or individuals who are blamed for specific frustrations.

The frustration-aggression, or scapegoating, theory of prejudice seems at first glance convincing. It is, as Allport notes, easily understood because of "the commonness of the experience" (1958:330). All of us at times suffer frustration of our needs and desires, whether simple ones, like seeing our favorite football team lose a key game, or significant ones, like failing to gain a job promotion. When these events occur, we are sometimes inclined to strike out at substitute targets such as our spouses or children. If the frustration is continuous, we may begin to blame more remote groups or institutions like "the government," "bureaucrats," "blacks," "Jews," or "gays."

Despite the common experience of frustration-aggression, however, this theory leaves many questions unanswered. First, under what conditions will frustration *not* lead to aggression? Psychologists have shown that the frustration-aggression sequence is not necessarily inevitable (Allport, 1958; Ehrlich, 1973). Nor is aggression always displaced; sometimes it is directed inward, and other times it may be thrust on the real source of frustration. This theory does not tell us how and why these responses may be produced.

Another obvious shortcoming of the frustration-aggression theory is its failure to explain the choice of scapegoats or targets of displaced aggression. Why are some groups chosen rather than others? The notion that scapegoats are always "safe goats," that is, defenseless and easily used, does not hold up when we consider that minority group members may themselves harbor great antipathy for the dominant group or for other minorities. A national survey in 1995, for example, documented the strong prejudices of American minority groups toward each other. It indicated, in fact, that minorities are more likely than whites to agree to negative stereotypes about other minority groups (National Conference, 1995).

Finally, we must consider the fact that the displacement of aggression on substitute targets can bring at best a very short-lived relief of an individual's anxiety. Unless the actual source is attacked, the feeling of frustration will quickly recur or even intensify. As Allport has put it, concerning frustration, "Nature never created a less adaptive mechanism than displacement" (1958:332).

The Authoritarian Personality The question of why prejudice exerts a strong influence on certain individuals but plays a relatively minor role for others is addressed by the theory of the authoritarian personality. The essence of this theory is that there is a personality type prone to prejudicial thought. In the same way, according to this view, there are basically tolerant personalities.

The authoritarian-personality theory was developed after World War II by a group of social scientists who were determined to trace the psychological foundations of such destructive and regressive movements as Nazism. It subsequently became one of the most widely tested and debated ideas in the social sciences (see Kirscht and Dillehay, 1967). In their studies, T. W. Adorno and his associates (1950) found evidence to support the notion that prejudice and political extremism are more generally characteristic of a definite personality type. Such people, they maintained, are highly conformist, disciplinarian, cynical, intolerant, and preoccupied with power. They are particularly authority-oriented and are thus attracted to sociopolitical movements that require submission to a powerful leader. Such personality traits extend well beyond people's political beliefs and are reflected in all aspects of their social life. In the family, for instance, authoritarians will subject their children to strong disciplinary action, and in their religious beliefs, they will emphasize submission and obedience. In sum, such people strongly support conservative values and resist social change. They are thus more likely to display prejudicial thought and to discriminate when given the opportunity.

Other scholars, although not necessarily subscribing to the notion of an authoritarian personality, suggest that prejudice is a general way of thinking for some people. Hartley (1946), for example, found that when purely fictitious groups were presented in a social distance test, people prejudiced toward other groups tended to express prejudice toward the fictitious groups as well. Allport also maintains that "the cognitive processes of prejudiced people are in general different from the cognitive processes of tolerant people" (1958:170).

In any case, theories suggesting that certain personality types are generally prone to prejudice and discrimination suffer several critical shortcomings. Like the frustration-aggression theory, they fail to tell us how ethnic prejudice and discrimination arise in the first place. We see the prejudiced person in action but not the social conditions that create *ethnic* hostility rather than other forms.

Moreover, the emphasis of this genre of studies has been on patently and intensely prejudiced people, such as members of the Ku Klux Klan. Most prejudice and discrimination, however, are more subtle and less intense and are characteristic of people who cannot be categorized as extremists. For example, although the Nazi movement certainly drew many adherents from among the social misfits and authoritarian personalities of German society, studies have shown that the movement had appeal across a wide social spectrum (Gerth, 1940; Peukert, 1987). Undoubtedly, prejudice and discrimination play a vital part in the thought and actions of many people, but it is necessary to carefully delineate the varying degrees and forms of ethnic antipathy displayed by different individuals.

The chief criticism of psychological theories in general is that the situations within which people think and act are not given sufficient attention as variables that fundamentally affect the nature of that thought and action. Schermerhorn cautions us to consider that "if research has confirmed anything in this area, it is that prejudice is a product of *situations,* historical situations, economic situations, political situations; it is not a little demon that emerges in people simply because they are depraved" (1970:6).

However, psychological theories such as frustration-aggression and the authoritarian personality have been popular not only because they are more easily understood than sociological theories emphasizing structural conditions but precisely because they focus the blame for ethnic antagonisms on disturbed individuals, those who are pathological or overtly irrational in behavior and thought. This focus deflects attention from the society's normally functioning institutions, which may compel people to think and act negatively toward members of particular ethnic groups. Indeed, as will be seen shortly, prejudice and discrimination are in most cases *conforming* thoughts and actions, not those of a few maladjusted persons. As long as ethnic antipathy is thought to be characteristic only of the sick few rather than a proper response to the expectations of the community or society as a whole, it can be seen as eradicable simply by treating those few, not by painfully reexamining established societal institutions.

Whatever their effect on prejudice and discrimination, psychological components must be seen in conjunction with political and economic structures out of which intergroup relations develop and are sustained. William Wilson asserts that psychological explanations of prejudice and discrimination "prove to have little predictive value when social factors are taken into account" (1973:39). These social factors are the basis of normative theories, to which we now turn.

Normative Theories

Why do we often compliment a friend on his or her new hairdo or clothes when in fact we think they are quite unattractive? Similarly, why do we many times feel obliged to contribute to a class discussion or a business meeting when we really have nothing meaningful to say? Sociologists explain such apparently inconsistent thought and action as a product of situational norms by which we feel compelled to abide. We understand that such actions are expected of us, and in most cases, we conform to these expectations even if we have a real desire to ignore or disobey them. *Norms* are group standards that define how people are expected to act in particular social situations. There are positive sanctions for conforming to and negative sanctions for deviating from them. Because there are norms pertaining to all social situations in which we find ourselves, these social "rules" enable us to predict others' behavior, and in doing so they facilitate interaction. In a real sense, they provide the society with order.

Prejudice and discrimination can be explained within the framework of social norms. Rather than the thoughts and actions of a deviant few, they are conforming responses to social situations in which people find themselves. When negative thoughts about particular ethnic groups and discriminatory behavior toward them are expected, individuals will feel compelled to think and act accordingly. Thus, it is to individuals' social environment—the groups to which they belong, the cultural and political norms operative in their society and community, and the processes of socialization—that prejudice and discrimination can be traced. Obviously, such explanations are considerably different from psychological theories, which focus not on the group contexts of individual thought and action but on the individuals themselves.[4]

In this view, bigots emerge out of the social experiences to which they are exposed. Frank Westie has succinctly explained the essence of normative theory: "Individuals are prejudiced because they are raised in societies which have prejudice as a facet of the normative system of their culture. Prejudice is built into the culture in the form of normative precepts—that is, notions of 'ought to be'—which define the ways in which members of the group ought to behave in relation to the members of selected outgroups" (1964:583–84). Normative theories thus concentrate primarily on the transmission of ethnic prejudices through the socialization process and on the social situations that compel discriminatory behavior.

Socialization Prejudice and discrimination can be seen as part of a society's social code, which is passed down from generation to generation. Fear of, dislike for, and antipathy toward one group or another are learned in much the same way that people learn to eat with a knife and fork rather than with their bare hands or to respect others' privacy in personal matters. These standards of behavior are the product of learning processes of which we ordinarily have little cognizance. Socialization is subtle and works in a largely unconscious manner. Thus, prejudice and discrimination need not be taught directly and intentionally. If these are the norms and values of the society or community within which the individual interacts, the chances are very great that they will be adopted with little specific instruction. As Westie puts it, "The child soon learns which people are 'good' and which are 'not so good' in the eyes of his society" (1964:583).

Parents are sometimes puzzled by certain expressions and attitudes of their children because they are sure they did not impart these. In their puzzlement, they discount the informal and often undetected ways in which the society's culture is transmitted by the various agents of socialization outside the family. Most

---

[4]Many studies showing that people's behavior toward different ethnic groups is dictated by social circumstances rather than their personal or psychodynamic makeup are cited in Dean and Rosen (1955) and Williams (1964).

learning is accomplished not through direct teaching methods but through observation and imitation. Children—and adults—take cues from their peers and other important reference groups as well as the mass media. Social psychologists have shown that among American children 4 years of age, ethnic values and attitudes are already beginning to crystallize (Aboud, 1987; Goodman, 1964; Porter, 1971; Ramsey, 1987). By this age, children have been exposed to the society's ethnic hierarchy, particularly the white-over-black element.

Prejudice and discrimination, then, are no more indicators of a defective personality than one's taste in food or fashion. They are simply products of socialization. If prejudice and discrimination are pervasive in the society or community, the more logical question may not be "Why do some people display prejudice and discrimination?" but rather "Why do some people *fail* to display prejudice and discrimination?"

Consider South Africa under apartheid or the pre-1960s American South. Using the normative approach, white prejudice and discrimination against blacks in those settings can be explained not as the product of deviant individuals but as the natural outgrowth of a whole system of racist norms, learned early and thoroughly, that guided people's actions and attitudes. In that system, blacks were not to be thought of or treated in the same way as whites. For whites to avoid almost any contact with blacks beyond the most purely functional (supervising them in a work situation, for example) was correct, expected, and positively sanctioned by societal norms. For whites to deviate from such behavior would have been unusual and responded to with negative sanctions. John Stone (1973), in a study of British immigrants to South Africa in the 1970s, discovered that they frequently changed their attitude toward the segregationist policy of apartheid. Before leaving Britain, the majority was either opposed to or had no opinion about it; after living in South Africa for a time, however, an even larger majority stated that it favored the policy. Stone concluded that the change reflected not the manifestation of latent racist personalities but the need of immigrants to adapt to the ways of their new society: "We are not witnessing the mass attraction of bigoted racialists to a segregationalist's dream, rather we are observing how ordinary people, confronted by a particular social structure, will tend to conform to the attitudes, values, and norms implicit in it" (1973:253).

Reference Groups Even where societal norms dictate fairness toward different groups, prejudice and discrimination may typify the behavior of some people who have been exposed to reference groups that strongly prescribe such behavior. *Reference groups* are those that provide individuals with standards by which they shape their own patterns of action and from which they adopt important beliefs and values. In a sense, they serve as models of thought and action. We ordinarily think of the family as a reference group, but there are many others to which we may look for behavioral guidance, even some of which we are not members. In the latter case, we may aspire to membership and thus take on the ways and attitudes of the group. This is most apparent in the early stages of socialization, when children begin to identify with particular occupational groups—firefighters, doctors, nurses, and the like—or with sports and entertainment groups they admire. As might be expected, studies have shown that individuals tend to adopt beliefs and values congruent with those of the groups with which they identify.

Applying the concept of the reference group to prejudice and discrimination, we see such thoughts and actions as normal responses of individuals when called for by their reference groups. No one is immune to the pressures applied by family, friendship cliques, or other significant groups to conform; the fear of group rejection is constantly present and serves as an effective disciplinary mechanism. Again, this process is subtle, and the individual may not see such conformity as a response to external coercion. As Vander Zanden explains, "Through conformity individuals achieve pride, self-identity, a sense of security—products of belonging to a group. It is *their* group, *their* norms" (1983:104).

If the person's reference groups change, attitudes and actions can be expected to change accordingly. College students, for example, often face a challenge to their well-formed values when they encounter new ideas from their instructors and classmates. These new ideas often involve social issues like ethnic prejudice. Pearlin's (1954) study at a southern women's college demonstrated the effect of a change in reference groups on students' racial attitudes. Pearlin started with the assumption that white students would find in the college environment sentiments more favorable to blacks than they had found in their precollege experiences. But merely being exposed to new and positive ideas about blacks, he believed, would not in itself reduce prejudicial attitudes. Rather, modification of attitudes was believed to depend more on changes in social relationships. Thus, attitudinal changes were likely to come about only if the students began to identify with new groups holding those favorable attitudes toward blacks. Pearlin's hypotheses were confirmed: Those most prejudiced toward blacks remained most strongly affiliated with their precollege membership groups, and those least prejudiced experienced a weakening of such ties and an increasing identification with their new college groups. In short, Pearlin's findings showed that people tend to take the attitudes of those groups with which they most strongly identify.

Merton's Paradigm Given the social, rather than the personality, origins of ethnic prejudice and discrimination, changing social situations can produce fluctuations in individual thought and behavior. Ethnic prejudice and discrimination are thus not constant and unchanging but variable, depending on a number of situational factors: the person's definition of the situation, the compulsion to conform to societal and reference group norms, and the rewards—economic, prestige, political—to be gained by acting and believing in such a manner. That attitudes and actions toward members of particular ethnic groups may fluctuate within different social contexts is demonstrated by sociologist Robert Merton (1949) in a well-known paradigm.

By combining the prejudicial attitudes or lack of such attitudes with the propensity either to engage in discriminatory actions or to refrain from them, Merton suggested four ideal types. First, he denoted unprejudiced nondiscriminators, whom he called "all-weather liberals." These are people who accept the idea of social equality and refrain from discriminating against ethnic minorities. Their behavior and attitudes are thus consistent. A second type, also consistent in behavior and attitude, is prejudiced discriminators, whom Merton labeled "active bigots." Such people do not hesitate to turn their prejudicial beliefs into discriminatory behavior when the opportunity arises. Members of organizations such as the Ku Klux Klan or neo-Nazi parties in the United States or the National Front in Britain exemplify such people. Both of these types, consistent as they are in belief and behavior, might indicate by themselves support for the psychological perspective; that is, there are prejudiced people or tolerant people, who may be expected to act accordingly.

Merton's third and fourth types, however, demonstrate the situational context and the effect it may have on people's behavior. In these cases, behavior and attitude are not consistent. Prejudiced nondiscriminators, or "timid bigots," as Merton called them, maintain negative beliefs and stereotypes toward ethnic minorities but are precluded from acting out those beliefs by situational norms. If a situation requires fair treatment toward ethnic groups who are viewed negatively by such people, fair treatment will mark their behavior. For example, whites traveling from the American South to northern states before the 1960s would find that the laws and customs of the North required that they interact with blacks in a manner unheard of in their home states. Lewis Killian described how white working-class southerners who had migrated to Chicago responded to blacks in their new environment: "The 'hillbillies' constantly praised the southern pattern of racial segregation and deplored the fact that Negroes were 'taking over Chicago.' In most of their behavior, however, they made a peaceful, if reluctant, accommodation to northern urban patterns" (1953:68). The hotel and restaurant keepers encountered by LaPiere in his previously cited study would also fall into this category.

Unprejudiced discriminators, whom Merton called "fair-weather liberals," also adjust their behavior to meet the demands of particular circumstances. When discrimination is normative in the group or community, such people abide by those patterns of behavior even though they may harbor no prejudicial feelings toward members of the targeted group. To do otherwise would jeopardize their social standing and might even constitute violations of the law. Such a situation was faced by many whites in the pre-1960s American South who did not share the racial animosity of their neighbors. In a more recent example, the owner of a pharmacy in Tifton, Georgia, dismissed a black student pharmacist who had been placed in the store as part of her training at the University of Georgia's College of Pharmacy. She was dismissed, the owner explained, because he feared negative customer reaction (*New York Times,* 1987).

It must be remembered that each of Merton's four cases is an ideal type and thus does not reflect perfectly any individual's behavior and attitudes. More realistically, people can be expected to display higher or lower degrees of each. We should also remember that prejudice and discrimination directed at one ethnic group do not necessarily imply the same attitudes and behavior toward others. Those who are antiblack are not necessarily anti-Catholic, and so on. One may be a fair-weather liberal in one instance and a timid bigot in another.

Situational explanations of prejudice and discrimination like Merton's demonstrate that there is no necessary causal relationship between the two. Common thought has generally assumed that prejudice leads to or causes discrimination. Abundant sociological evidence, however, has shown not only that this sequence need not occur but also that the very opposite is more usual (Pettigrew, 1979; Raab and Lipset, 1971). "What we call prejudices," writes anthropologist Marvin Harris, "are merely the rationalizations which we acquire in order to prove to ourselves that the human beings whom we harm are not worthy of better treatment" (1964:68). Prejudice, then, is used to rationalize discriminatory behavior *after* the fact.

This is an important observation, for it seriously challenges the idea that eliminating discrimination requires a change in attitude, that is, the elimination of prejudice. This was the generally shared opinion of both scholars and policymakers in the United States before the 1960s. Reeducating people was therefore the most frequently proposed remedy for alleviating ethnic hostility. It was thought that if people's faulty ideas about race and ethnicity could be corrected, they would, as a result, be induced to change their behavior. In the past few decades, however, it has become obvious that prejudicial attitudes may have little or no bearing on whether people discriminate against particular groups. Instead, people appear to be motivated to change their behavior toward ethnic groups by laws and other social mechanisms that seriously alter their social situation vis-à-vis those groups. After such situational changes occur, individuals seem to adjust their ideas to fit these new modes of behavior (Ehrlich, 1973; Pettigrew, 1980). Hence, changes in ethnic relations must be impelled not by trying to change attitudes but by changing the structure of those relations.

Power-Conflict Theories

Though the normative theories of prejudice and discrimination appear to go well beyond the earlier psychological theories, both approaches are deficient in one important regard: They basically explain only how these social phenomena are transmitted and sustained, not how or why they arise in the first place. To begin to understand the origins of ethnic antagonism, we must turn to power-conflict theories.

Most simply, these theories view prejudice and discrimination as emerging from historical instances of intergroup conflict (Bernard, 1951; Newman, 1973). In this view, discrimination serves as a means of injuring or neutralizing out-groups that the dominant group perceives as threatening to its position of power and privilege. Negative beliefs and stereotypes, in turn, become basic components of the dominant group's ideology, which justifies differential treatment of minority ethnic groups. When prejudice and discrimination are combined, they function to protect and enhance dominant group interests. And once established, prejudice and discrimination are used as power resources that can be tapped as new conflict situations demand.

In short, prejudice and discrimination, in this view, are products of group interests and are used to protect and enhance those interests. To understand negative ethnic beliefs and behavior thus requires a focus not on individual personalities or even on the constraints and demands of different social situations but on the economic, political, and social competition among groups in a multiethnic society.

Economic Gain Chief among power-conflict theories are those that emphasize the economic benefits that derive from prejudice and discrimination. Simply put, prejudice and discrimination, in this view, yield profits for those who engage in them. Different groups may be targeted as they present or are perceived as presenting a threat to the economic position of the dominant group. In addition to the economic advantages that emanate from such thought and behavior, benefits may also take the form of enhanced prestige or political gain.

Colonial and slave systems, buttressed by elaborate racist ideologies, are obvious cases in which economic benefits accrue to a dominant group from the exploitation, both physical and mental, of minority groups. We need not look at such historically distant examples, however, to understand the relation between ethnic antagonism and economic gain. In his study of black–white relations in the American South of the 1930s, John Dollard showed that in every sphere of social life—work, health, justice, education—the white middle class realized substantial gains from the subordination of blacks. In exploiting blacks, explained Dollard, southern whites were simply "acting as they have to act in the position within the social labor structure which they hold, that is, competing as hard as they can for maximum returns" (1937:115). Later studies (Glenn, 1963, 1966; Thurow, 1969) concluded that prejudice and discrimination against blacks in the United States continued to benefit at least some segments of the white population.

The strong, often violent, resistance to school busing, residential desegregation, and affirmative action in the past three decades can be interpreted as the negative reaction of whites who see their economic gains threatened by blacks and other minorities seeking upward social mobility. Similar reactions can be expected whenever economic advantages appear to be challenged by lower-ranking or more recently arrived ethnic groups. In 1981, for example, white fishermen in the shrimping grounds in Galveston Bay, Texas, encountered competition from immigrant Vietnamese fishermen. About 100 Vietnamese shrimpers had come to the area during the previous two years, challenging the economic dominance of the whites. Although the situation was eventually resolved, for several months white fishermen, with support from the Ku Klux Klan, engaged in acts of intimidation against the Vietnamese, including physical attacks and arson.

Marxian Theory Class theorists, in the tradition of Marx, have conventionally held that in capitalist societies, ethnic antagonism serves the interests of the capitalist class, those who own and control the means of economic production, by keeping the working class sufficiently fragmented and thus easier to control. The basic idea is "divide and rule." One ethnic element of the working class is pitted against another, and as long as this internal discord can be maintained, the chances of the working class's uniting in opposition to the interests of the capitalists are reduced. Capitalists are able to foster ethnic division and ethnic consciousness among the workers, thereby curtailing the development of worker solidarity and class consciousness.

In the United States, for example, the conflict between black and white workers has been construed by Marxists as deflecting attention from the common anticapitalist interests of both groups (Allen, 1970; Cox, 1948; Reich, 1978; Szymanski, 1976). Ethnic prejudice, therefore, is viewed as a means of sustaining a system of economic exploitation, the benefits of which accrue to the capitalist class. Oliver C. Cox described prejudice as "the social-attitudinal concomitant of the racial-exploitative practice of a ruling class in a capitalistic society" (1948:475). Though capitalists may not consciously conspire to create and maintain racist institutions, they nonetheless reap the benefits of racist practices and therefore do not seek to completely dismantle them.

The Split Labor Market Theory Whereas conventional Marxist thought holds that the profits of ethnic hostilities redound primarily to the owners of capital, others maintain that it is workers of the dominant ethnic group who are the chief beneficiaries of prejudice and discrimination. If ethnic minorities are kept out of desired occupations, the favored workers, rather than the capitalists, are viewed as gaining the most from discriminatory institutions. This is the crux of sociologist Edna Bonacich's split labor market theory (1972, 1976).

According to Bonacich, there are three key groups in a capitalist market: businesspeople (employers), higher-paid labor, and cheap labor. One group of workers controls certain jobs exclusively and gets paid at one scale, and the other group is confined to jobs paid at a lower rate. Given the imperatives of a capitalist system, employers seek to hire workers at the cheapest possible wage and therefore turn to the lower-paid sector when possible as a means of maximizing profits. Recent immigrants or ethnic groups migrating from rural areas in search of industrial jobs ordinarily make up this source of cheap labor. These groups can be used by employers as strikebreakers and as an abundant labor supply to keep wages artificially low. Because these groups represent a collective threat to their jobs and wages, workers of the dominant ethnic group become the force behind hostile and exclusionary movements aimed at curtailing the source of cheap labor. Thus, wage differentials do not arise through the efforts of capitalists to prevent working-class unity by favoring one group over another but through the efforts of higher-paid laborers to prevent lower-paid workers—mainly lower-ranking ethnic groups—from undercutting their wages and jobs. This goal is effected by various forms of prejudice and discrimination.

The split labor market theory is supported by historical evidence in American society. Successive waves of European immigrants during the nineteenth and early twentieth centuries traditionally served as a source of cheap labor and became the targets of nativist movements, usually supported heavily by labor unions. Following the cessation of European immigration, northward migrating blacks from the rural South assumed a similar role, touching off periodic racial violence in many cities. Depending on how threatening they were perceived by native workers, various groups at different times were the objects of worker-inspired hostility. For example, efforts in the nineteenth century to restrict the Chinese to particular occupations and to limit their immigration was spurred largely by white workers fearing a deluge of cheaper labor. Lyman notes that after 1850, "It was the leadership of the labor movement that provided the most outrageous rhetoric, vicious accusations, and pejorative demagoguery for the American Sinophobic movement" (1974:70).

Makabe (1981) studied Japanese immigrants in Brazil and Canada and found support for the split labor market theory. Japanese immigrants entering Canada, specifically British Columbia, in the pre–World War II years experienced an extremely harsh reception from native Canadians. Makabe explained that among white workers the rejection of the Japanese was unusually cruel. This is accounted for by the fact that the Japanese entered the Canadian economy at the bottom, enabling employers to pay them lower wages and thereby undercutting the more highly paid native workers. Striving for upward mobility, Japanese found themselves in direct competition with those immediately above them in economic position—white workers. The result was discrimination against the Japanese in the work force and pressure to halt all Japanese immigration. In Brazil, however, the Japanese experienced a significantly different situation. Rather than entering the labor force in competition with higher-paid workers, they found themselves with skills and financial resources superior to those of most native workers, who themselves were mostly severely disadvantaged former slaves. Little competition and, hence, little conflict arose between them because they did not seek similar occupational positions.

Illustrations of the notion that prejudice and discrimination arise out of economic competition can also be seen at the higher levels of the class hierarchy. Williams (1964) found the greatest frequency of prejudice against Jews among people of higher occupational rank. He found this compatible with the supposition that prejudice reflects a sense of threat. Upper-class Gentiles are likely to see Jews as more threatening to their economic position than other less affluent and less educated minorities. Williams also found that upper-class people were less likely to be prejudiced against blacks. He interpreted this as logical because blacks do not represent an economic threat to them. McWilliams (1948), too, concluded that the development of strong antisemitism in the late nineteenth century in the United States was largely a creation of the wealthy—specifically, the industrial magnates of the day, who used prejudice and discrimination against Jews as a "mask for privilege." We will discuss antisemitism in some detail in Chapter 7.

Status Gains In addition to economic benefits, status privileges may derive from ethnic antagonism. People may enjoy more prestige simply from being a member of the dominant group, regardless of their social class. In the American South, working- and lower-class whites could traditionally take comfort in knowing that they were part of the dominant ethnic group even though economically they were in much the same position as blacks. As Dollard described it, white subordination of blacks consisted of "the fact that a member of the white caste has an automatic right to demand forms of behavior from Negroes which serve to increase his own self esteem" (1937:174). Van den Berghe described the same well-understood racial etiquette in South Africa: "Non-Europeans are expected to show subservience and self-deprecation, and to extend to the whites the titles of 'Sir,' 'Madam,' or '*baas*.' The Europeans, as a rule, refuse to extend the use of titles and other forms of elementary courtesy to non-whites, and call the latter by first names (real or fictitious), or by the terms 'boy' and 'girl'" (1967:142).

Political Gain Prejudice and discrimination directed at certain ethnic groups may also be the result of political circumstances, regardless of the economic or status relationship between dominant and minority groups. In the United States, ethnic antagonism has often been tied closely to domestic and foreign political issues. Anti-German feelings were extremely high before and during World War I, for example, with charges of disloyalty leveled at German-American immigrants and even those of the second generation. This case demonstrates how prejudicial beliefs and discriminatory behavior can change both rapidly and radically. Before 1915, when the anti-German campaign began, Germans had been considered the most assimilable and reputable of all immigrant groups and were repeatedly praised for their law-abiding and patriotic behavior (Higham, 1963).

The political basis of negative feelings and actions toward members of particular ethnic groups is well illustrated by the anti-Arab sentiments expressed by Americans in recent years. Starting in the 1970s, a series of key political events—namely, the rise of the Organization of Petroleum Exporting Countries (OPEC) cartel, the activities of Palestine Liberation Organization (PLO) terrorists, and the holding of American hostages in Iran—created a generally anti-Arab public mood and, in the case of Iranians, hastily conceived government policies to expel those living or studying in the United States.[5] The development of negative Arab stereotypes was now evident, with the popular media portraying Arabs as people given to ostentatious displays of wealth or to political fanaticism (see Said, 1981; Shaheen, 1984). Arab Americans, though directly involved in none of these events, were nonetheless affected by the adverse feelings they aroused. The Persian Gulf War and the bombing of the World Trade Center in New York in 1993 reinforced anti-Arab sentiment and negative stereotypes. When the federal building in Oklahoma City was bombed in 1995, the perpetrators were assumed almost immediately by the media and public officials to be Arab terrorists. Although this assumption quickly proved false, it exposed how firmly this stereotype is applied to Arabs.

Political leaders have long recognized the value of exploiting ethnic divisions for attaining and enhancing their power. For example, until blacks became an

---

[5]Though technically not part of the Arab world, Iran was generally linked with the Arab states because of its Muslim heritage, its location in the Middle East, and its political involvement with the Arab nations.

electoral factor of some significance in the 1970s, racist politicians in the American South effectively manipulated white fears of blacks to their own ends. After losing the Alabama gubernatorial election in 1958 to a candidate even more avowedly racist than himself, George Wallace declared that he would not be "out-nigguhed again" (Frady, 1968). These fears continue to be exploited by politicians. In the 1988 presidential campaign, George Bush used the case of Willie Horton, a convicted murderer who raped a woman while on furlough from a Massachusetts prison, to portray his Democratic opponent, Michael Dukakis, as soft on crime. As a black man, Horton's image was intended to elicit white fears of black crime. Similarly, in recent years, politicians in France, Germany, and other western European countries have stirred anti-immigrant feelings, particularly against Muslims from North Africa and the Middle East, in appealing to voters.

Power-Conflict Theories in Sum In all these cases, prejudicial beliefs and discriminatory actions are used by the dominant ethnic group, sometimes directly and other times indirectly, to secure its power and privileges. Herbert Blumer (1958) has pointed out that prejudice is always a protective device used by the dominant group in a multiethnic society in assuring its majority position. When that group position is challenged, prejudice is aroused and directed at the group perceived to be threatening.

Wilson (1973) also notes that when the system of ethnic stratification is challenged, that is, when minority groups no longer accept their group position, strong prejudices founded on a racist ideology emerge. Through this ideology, members of the dominant group can, as Wilson explains, "claim that they are in a superior position because they are naturally superior, that subordinate members do not possess qualities enabling them to compete on equal terms" (1973:43). The dominant ideology, incorporating key negative stereotypes of minority ethnic groups, thus reinforces the sense of group position, aids in maintaining patterns of subordination, and serves as a philosophical justification for exploitation.

This does not mean that in all multiethnic societies the nature and intensity of prejudice will be the same. It may differ, depending on such factors as the size of the targeted group, its cultural and physical characteristics, how much of a threat it is perceived to be, the ideology that has grown up around it, and the historical circumstances out of which its position in the ethnic hierarchy has been determined.

Discrimination, too, as we have seen, can vary in degree and form. In some cases, it may consist of little more than name-calling or the telling of ethnic jokes, but in others, the negative actions may be more serious, including the denial of various life chances to members of certain groups and even the physical destruction of human life and property. But whatever its form or intensity, it is beneficial in some way to the dominant group and is sustained on that basis.

Here we might note that, paradoxically, prejudice and discrimination may serve certain functions for minority groups themselves. Sociologists have recognized that conflict between groups has a unifying effect on the members of each. External threats tend to strengthen group ties and create a sense of solidarity that might not otherwise exist (Coser, 1956). The ability of Jews to survive in various societies in which they were persecuted, for example, has often been attributed to the continued antisemitic hostility itself. As constant targets of antagonism, Jews have strengthened their resolve to maintain a group identity and cohesiveness.

Continued prejudice and discrimination directed at a minority group may also contribute to a sense of psychological security for its members. Even though their place is at the bottom, they may take consolation in the certainty and predictability of their social relationships with the dominant group (Levin and Levin, 1982). Moreover, the minority individual's self-esteem may be protected by attributing personal failures to abstract notions like "the system" or "racism," rather than to individual shortcomings.

Prejudice may also serve as a release of frustration for minorities, just as it may for those of the dominant group. Indeed, we should not think that prejudice is characteristic only of dominant groups. Although sociologists have been reluctant to deal with it, prejudice is commonly displayed by minority groups as well, not only toward the dominant group—which seems entirely logical—but also toward other minority groups. Recall the strong antipathy toward each other expressed by American minority ethnic groups, mentioned earlier.[6] If prejudice is normative in the society, minority group members socialized to those norms will be affected in much the same way as members of the dominant group.

The benefits to minority ethnic groups that derive from prejudice and discrimination, however, should not be overdrawn. Clearly, the primary beneficiaries of ethnic antagonism are members of the dominant group.[7]

---

[6]Westie (1964) asserts that the prejudice of minority group members has been largely ignored by social scientists. He suggests that this may be a result of the sympathy social scientists usually display for social underdogs. Moreover, interethnic conflict is usually perpetrated by members of the dominant group, and the prejudices of minority group members are seen mainly as responses to these actions. Westie maintains that, however well intended this view may be, it has produced social science literature "which gives the impression that the minority person can 'do no wrong'" (1964:605).

[7]There are also certain negative effects of prejudice and discrimination on the dominant group. See, for example, Bowser and Hunt (1981).

Theories of Prejudice and Discrimination: An Assessment

As we have now seen, the explanation for prejudice and discrimination in multiethnic societies is complex and by no means agreed on by theorists and researchers. It may very well be that a full investigation of these phenomena requires a multidimensional approach using different aspects of psychological, normative, and power-conflict theories. All may have some validity, depending on which aspect or level of ethnic antagonism is focused on.

In this book, power-conflict theories of prejudice and discrimination are favored because the structural rather than the psychological or small-group dynamics of race and ethnic relations are emphasized. At this level of intergroup analysis, psychological and normative theories are less appropriate. In Parts II and III, therefore, the analysis of prejudice and discrimination in the United States and other multiethnic societies will view these ethnic attitudes and actions as tools of dominance, developed and used by one group over others in competition for the society's resources.

Although power-conflict theories are stressed, it is important to bear in mind that prejudice and discrimination are multifaceted, and therefore other theories cannot be disregarded. Power-conflict theories will not entirely explain, on the one hand, why some people will not discriminate even when it is profitable to do so or, on the other hand, why some will continue to discriminate when it is no longer beneficial. For such cases, psychological or normative theories may offer additional insight. As Simpson and Yinger point out, not all prejudice and discrimination can be explained by structural variables alone; individuals' responses to group influences are conditioned by their personality, and vice versa. Therefore, "The task is to discover how much of the variance in prejudice and discrimination can be explained by attention to personality variables, how much by social structural variables, and how much by their interaction" (1972:29).

SUMMARY

Prejudice and discrimination are techniques of ethnic dominance. *Prejudice* is the attitudinal dimension of ethnic antagonism. Prejudices are categorical, inflexible, and negative attitudes toward ethnic groups, based on simplistic and exaggerated group images called stereotypes. *Discrimination* is the behavioral dimension and involves actions designed to sustain ethnic inequality. Discrimination takes various forms, ranging from derogation to physical attack and even extermination. Two types of discrimination are *individual* and *institutional,* the former carried out by single persons or small groups, usually in a deliberate manner, the latter rendered as a result of the norms and structures of organizations and institutions, often in an unwitting and unintentional manner.

There are three major theoretical traditions in explaining the origins and patterns of prejudice and discrimination. *Psychological* theories focus on the ways in which group hostility satisfies certain personality needs; prejudice and discrimination, in this view, are traced to individual factors. *Normative* theories explain that ethnic hostilities are conforming responses to social situations in which people find themselves. *Power-conflict* theories see prejudice and discrimination as products of group interests and as tools used to protect and enhance those interests. Focus is placed not on the individual or even the immediate group but on the dynamics of political, economic, and social competition among a society's ethnic groups. The latter view has been adopted in this book.

Suggested Readings

Allport, Gordon W. 1958. *The Nature of Prejudice.* Garden City, N.Y.: Doubleday. Presents perhaps the most complete examination of the origins and forms of prejudice, stressing psychological factors. A classic work.

Gioseffi, Daniela (ed.). 1993. *On Prejudice: A Global Perspective.* New York: Anchor. A collection of essays and literary pieces by social scientists, writers, and poets that focus on ethnic and racial prejudice and its consequences.

Knowles, Louis L., and Kenneth Prewitt (eds.). 1969. *Institutional Racism in America.* Englewood Cliffs, N.J.: Prentice-Hall. Describes historical and modern-day patterns of institutional discrimination in the United States.

Levin, Jack, and William Levin. 1982. *The Functions of Discrimination and Prejudice.* 2d ed. New York: Harper & Row. Examines prejudice and discrimination by focusing on the consequences of these thoughts and actions for both dominant and minority groups.

Pettigrew, Thomas F. et al. 1980. *Prejudice.* Cambridge, Mass.: Harvard University Press. Clearly explains the psychological and social dimensions of prejudice.

Schuman, Howard, Charlotte Steeh, and Lawrence Bobo. 1985. *Racial Attitudes in America: Trends and Interpretations.* Cambridge, Mass.: Harvard University Press. Through a review of relevant survey data, traces the changing racial attitudes of Americans during the past five decades.

Simpson, George Eaton, and J. Milton Yinger. 1972. *Racial and Cultural Minorities: An Analysis of Prejudice and Discrimination.* 4th ed. New York: Harper & Row. Explains the psychological and sociological origins and effects of prejudice and discrimination, with a comprehensive review of significant studies regarding these phenomena.

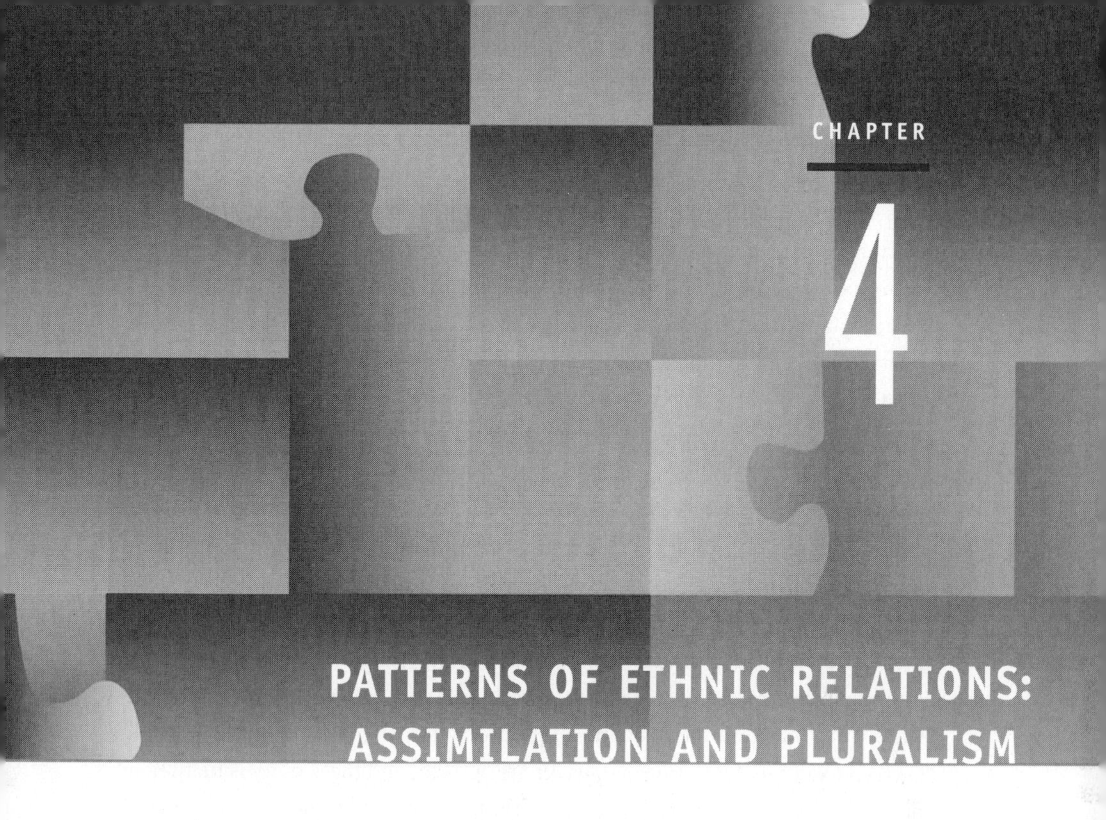

CHAPTER 4

PATTERNS OF ETHNIC RELATIONS: ASSIMILATION AND PLURALISM

THE CHIEF QUESTION we pursue in this chapter is "What is the nature of relations among diverse groups in a multiethnic society?" Specifically, when various ethnic groups meet, what is the outcome of that contact? We have already seen that the emergence of a system of ethnic stratification, with its attendant prejudice and discrimination, is most common. But there are other dimensions and processes of intergroup relations that create considerably different patterns between dominant and minority groups.

CONFLICT AND ORDER

Judging from news accounts of violence and general hostilities in multiethnic societies, it would seem that ethnic relations are as a rule conflictual. Indeed, almost all theorists have viewed conflict as a fundamental, if not permanent, feature of relations between ethnic groups. As Donald Young has declared, "Group antagonisms seem to be inevitable when two peoples in contact with each other may be distinguished by differentiating characteristics, either inborn or cultural, and are actual or potential competitors" (1932:586).

Though it may not always be overt and extreme, conflict is the primary underlying facet of systems of ethnic inequality. In this regard, Berry and Tischler note that "if we recognize the fact that conflict involves also those subtle, restrained forms of interaction wherein one seeks to reduce the status of one's opponents, and not to eliminate them entirely from the conflict, then perhaps it is true that conflict invariably occurs when unlike peoples meet" (1978:123). No matter how apparently stable and peaceful dominant–minority relations may seem, potential conflict always remains beneath the surface and reveals itself in periodic and unanticipated eruptions. Ethnic relations, then, are by definition relations of conflict. The dominant group may mobilize its power through force, ideology, or both to assure its dominance, and minority groups will respond with counterforce, accommodation, or submission.

Although conflict does characterize interethnic relations as a rule, certain qualifications to this maxim must be kept in mind. First, ethnic conflict is not maintained at a constant rate, does not take the same form, and is not based on the same factors from one society to another or even within the same society. It is obviously more intense and sustained in some cases than in others. Ethnic conflict can be reduced, and some societies, though rarely, may even achieve a system whereby groups live side by side for long periods in a generally harmonious state (see, for example, Gillen, 1948; Redfield, 1939). Switzerland provides such an example. Here four languages are spoken, and the society is further subdivided along two main religious lines. Yet linguistic and religious tolerance have characterized Swiss ethnic relations, and serious conflict has been avoided for many generations. In Part III, we will investigate several societies that represent a broad range of ethnic relations, extending from relative tranquillity to deep-seated and extreme hostility among groups.

We should also understand that conflict is not characteristic only of multiethnic societies, nor is *ethnic* conflict the only or even chief form in all societies, even those that are ethnically heterogeneous. Societal conflict is based on power relations that stem from differences in class, age, gender, and numerous other social factors in addition to ethnicity.

Finally, no matter how discordant or peaceful they may be, intergroup relations in multiethnic societies do not remain fixed. Internal migrations, immigration from other societies, political events, and varying economic conditions continually force revisions in public policies and create the social conditions that lead to change. At particular times, societies may move in the direction of either more harmonious relations among groups or greater separation and more intense conflict.

Sociologists have often proposed theoretical models that describe patterns of intergroup relations in multiethnic societies (Barth and Noel, 1972; Burkey, 1978; Marden and Meyer, 1978; Schermerhorn, 1970) or cycles of relations through which such societies presumably pass (Bogardus, 1930; Gordon, 1964;

Park, 1924). These theories have usually suggested that ethnic groups follow one of two paths: They either increasingly blend together or remain segregated. Most simply, groups may become more alike culturally and interact with one another more freely — this is *assimilation;* or they may remain culturally distinct and socially segregated — this is *pluralism.* The latter is of two types, depending on the distribution of political and economic power among groups. In the first, culturally and structurally distinct groups are relatively balanced and proportionate in political and economic resources; in the second, they are unequal politically and economically.

As models of interethnic relations, assimilation and pluralism are related to the two broad theoretical paradigms in sociology, *order* and *conflict.* These paradigms pertain not simply to specific parts or aspects of societies (like ethnic relations) but to *all* social structures and relations. They are intended to explain generally how societies are sustained and how they change.

Order theorists, whose tradition is most heavily influenced by the late-nineteenth-century French sociologist Émile Durkheim, see society as a relatively balanced system made up of differently functioning but interrelated parts. In this view, society is held together and social order maintained through a consensus of values among its groups and through the imperatives of functional interdependence. In contrast, conflict theorists, beginning with Karl Marx, have seen societies as held together not by broad agreements among groups but by the power of dominant classes and ruling elites to impose their will on others. Stability and order are maintained through coercion, not consensus. Whereas order theorists stress the manner in which societies maintain cohesion and balance, conflict theorists emphasize the distintegrative aspects of societies and the manner in which they change. As many have pointed out, societies are, of course, neither wholly ordered nor wholly in conflict, and we must therefore account for both (Dahrendorf, 1959). Nonetheless, sociologists have favored one or the other of these broad theoretical perspectives in their analyses.

Assimilation and pluralism, as models of interethnic relations, correspond quite closely with these two general sociological paradigms. Order theorists have stressed the assimilation side, emphasizing the ways in which different groups progressively become more unified and indistinct. Conflict theorists, in contrast, emphasize the inequality among ethnic groups and the patterns of dominance and subordination that develop among them. They have thus preferred a pluralistic model of ethnic relations that underscores the persistence of group differences in culture and structure.

For many years, American sociologists traditionally favored the assimilationist model (Metzger, 1971). The prevailing assumption was that multiethnic societies like the United States tend to gradually but inevitably move toward a fusion of diverse groups. American and world patterns of ethnic conflict in recent decades, however, seriously undermined this assimilationist bias. Sociologists today increasingly recognize the complexity of interethnic relations and seem to have accepted the inevitability of some degree of pluralism and conflict.

The investigations of American ethnic relations in Part II and of several other multiethnic societies in Part III generally employ a conflict approach; thus, the pluralistic aspects of these societies are highlighted. Theoretical frameworks in ethnic relations vary in utility, however, depending on the specific problems to be explained. An assimilationist model may therefore be applicable to some societies and groups and a pluralistic model to others. In general, assimilation, stressing progressive cohesion, applies to groups in multiethnic societies that have entered as voluntary immigrants, whereas some form of pluralism is relevant to those groups that have entered through involuntary immigration, conquest, or expansion. In the remainder of this chapter, we look more closely at these two general processes and the more specific levels and forms of each.

ASSIMILATION

Assimilation, as sociologist J. Milton Yinger defines it, is "a process of boundary reduction that can occur when members of two or more societies or of smaller cultural groups meet" (1981:249). Similarly, Harold Abramson defines it as "the processes that lead to greater homogeneity in society" (1980:150). Both of these definitions stress process. Rather than a fixed condition or state of relations, assimilation is best seen as a path, or trajectory, on which ethnic groups may move. The end point of this homogenizing process is stated by Barth and Noel as "the biological, cultural, social, and psychological fusion of distinct groups to create a new ethnically undifferentiated society" (1972:336).

Ideally, at the point of complete assimilation, there are no longer distinct ethnic groups. Rather, there is a homogeneous society in which ethnicity is not a basis of social differentiation and plays no role in the distribution of wealth, power, and prestige. This does not mean, of course, that other forms of social differentiation and stratification such as age, gender, and class do not exist; it means only that the ethnic forms are no longer operative. In essence, a society in which all groups have perfectly assimilated is no longer a multiethnic society.

This complete form of assimilation, however, is rarely achieved either for the society as a whole or for specific groups and individuals. Instead, assimilation takes different forms and is evident in different degrees. In other words, it is a variable that "can range from the smallest beginnings of interaction and cultural exchange to the thorough fusion of the groups" (Yinger, 1981:249). Therefore, in examining the assimilation of ethnic groups, the question that must concern us most is not simply "Are groups becoming more alike?" but "To what extent and in what ways are they becoming more alike?"

Dimensions of Assimilation

Assimilation can be seen at four distinct, though related, levels or dimensions: the cultural, the structural, the biological, and the psychological. Our concern is mostly with the first two levels, but let us briefly examine each of the four.

Cultural Assimilation The cultural dimension of assimilation involves the adoption by one ethnic group of another's cultural traits—language, religion, diet, and so on. Some have referred to this process as *acculturation* (Gordon, 1964; Yinger, 1981). Almost always, weaker (that is, minority) groups take on the cultural traits of the dominant group, though there is ordinarily at least some exchange in the opposite direction as well. This results because of the superior power of the dominant group but also because of the social advantages for the subordinates in adapting to the dominant group's ways. As van den Berghe explains, "It often pays to learn the ways of the rich, the powerful and the numerous; in the process one becomes more like them and, by that token, often becomes more acceptable to them" (1981:215). The end point of the process of cultural assimilation implies a situation in which the previously distinct cultural groups are no longer distinguishable on the basis of their behavior and values.

Yinger (1981) has noted that when groups are not highly antagonistic or culturally very disparate, acculturation can be additive rather than substitutive. That is, one group may augment its native culture with selective elements of the other's rather than substituting entirely. If the most basic cultural components (like language and religion) are not exchanged, however, the assimilation process can proceed only to a minimal point.

Structural Assimilation Whereas cultural assimilation refers to a blending of behaviors, values, and beliefs, structural assimilation refers to an increasing degree of social interaction among different ethnic groups. Specifically, with structural assimilation, members of minority ethnic groups are dispersed throughout the society's various institutions and increasingly enter into social contacts with members of the dominant group.

Structural assimilation may occur at two distinct levels of social interaction: the *primary* (or informal) and the *secondary* (or formal). Primary relations are those that occur within relatively small and intimate groups, in particular, the family and friendship cliques. Relations among members of these groups are affective, and the group's purposes extend well beyond merely instrumental goals. There is, most simply, an emotional bond among group members, and relations are therefore close and long lasting. Secondary relations, by contrast, are chiefly within large, impersonal groups like the school, the workplace, or the polity. These groups are purposeful, designed to fulfill some practical and specific social need; relations among members are thus formal and nonaffective.

Structural assimilation at the primary level implies interaction among members of different ethnic groups within personal networks—entrance into clubs, neighborhoods, friendship circles, and ultimately, marriage. In short, people interact in close, personal relations without regard for one another's ethnic identity. Other social traits, such as class, and individual characteristics become more critical than ethnicity. To measure the degree of informal, or primary, structural assimilation of a particular group, we would look at such indicators as rate of intermarriage, club memberships, and residential patterns. The higher the level of interaction with members of the dominant group in these areas of social life, the greater the extent of structural assimilation.

At the secondary level, structural assimilation entails equality of access to power and privilege within the society's major institutions—the economy, the polity, education, and so on. That is, jobs, housing, schooling, and other key life chances are not distributed with regard to people's ethnic affiliation. We might say, then, that structural assimilation at the secondary level involves, in its ultimate stage, the elimination of minority status. To measure the degree of secondary structural assimilation, we would look at the extent to which a minority ethnic group is approaching parity with the dominant group in the allocation of income and wealth, political power, and education (Hirschman, 1975).

Secondary structural assimilation has often been referred to as *integration* (Burkey, 1978; Davis, 1978; Hunt and Walker, 1974; Simpson, 1968; Vander Zanden, 1983), and we can consider these terms to be synonymous. The essential idea is that people of diverse ethnic groups come to interact in the institutions of the larger society free of the constraints of ethnicity. Hunt and Walker refer to integration as a situation in which "all citizens of the nation, or possibly even all members of the society regardless of citizenship, participate freely in all forms of social interaction without concern for ethnic affiliation" (1974:8).

This level of assimilation involves essentially a legal termination of group discrimination based on ethnicity. Inequality exists, of course, but is founded on bases other than ethnic group membership. With integration, ethnic groups may remain distinct, and members may continue to identify with them. Similarly, out-groups may continue to recognize and respond to them as ethnic groups. What has been achieved, however, is a measure of political and economic equality.[1]

The distinction between primary and secondary levels of structural assimilation is important because it is clear that the entrance of ethnic minorities into formal relations with the dominant group must precede relations within intimate social settings. Groups may achieve a significant degree of secondary structural assimilation without moving beyond this level into the primary type. Most white ethnic groups in American society, for example, are approaching a point at which they enjoy relatively equal access to jobs, political authority, and other important life chances. They have, in other words, begun to enter into full participation in all institutional areas, even though for some, primary relations may remain limited largely to the ethnic group. For blacks and other racial-ethnic groups, the distinction between secondary and primary structural assimilation is even more obvious. Although in recent years they have begun to achieve substantial integration in the economy, polity, and education, they do not significantly interact with whites in clubs, cliques, and other primary groups.

Increasing degrees of structural assimilation produce concomitant biological and psychological assimilation.

Biological Assimilation Biological assimilation, or *amalgamation,* represents the ultimate stage in the assimilation process. At this point, intermarriage has occurred to such an extent that there is a biological merging of formerly distinct groups. They are indistinguishable not only culturally and structurally but physically as well. Some degree of amalgamation is a common byproduct of group contact and interaction. But the complete biological fusion of diverse groups is an unusual outcome of interethnic relations. Mexico, for one, is a society that seems to have moved far along toward this long-range objective (van den Berghe, 1978), and Brazil, as we will see in Chapter 13, has also progressed in this direction.

---

[1]Schermerhorn (1970) has used the term *integration* in a somewhat different but related way. As he defines it, integration is "a process whereby units or elements of a society are brought into an active and coordinated compliance with the ongoing activities and objectives of the dominant group" (1970:66). The key to integration in this sense is a mutual acceptance of the scope and nature of activities and group objectives by both the dominant group and minority groups. Thus, societies may be integrated around a system in which extreme segregation and inequality are accepted by all parties. Schermerhorn postulates that "when the ethos of the subordinates has values common to those in the ethos of the superordinates, integration (coordination of objectives) will be facilitated; when the values are contrasting or contradictory, integration will be obstructed" (1970:172). See also Kuper (1968).

Psychological Assimilation Our focus is primarily on how various groups interrelate, and our interest, therefore, is chiefly in the cultural and structural dimensions of assimilation. There is, however, an individual dimension of assimilation in which attention falls on particular members of an ethnic group rather than on the group as a whole. This is part of the social psychology of ethnic relations and concerns the extent to which individuals have been absorbed into the larger society and identify with it.

With psychological assimilation, members of an ethnic group undergo a change in self-identity. To the extent that individuals feel themselves part of the larger society rather than an ethnic group, they are psychologically assimilated. As psychological assimilation proceeds, people tend to identify themselves decreasingly in ethnic terms. Whereas first-generation immigrants and their children will define themselves as Italian American or Irish American, by the third generation, ethnicity for most is no longer a major component in the response to the question "Who am I?" This level of assimilation consists not simply of becoming culturally like members of the mainstream society but also of "accepting that society as the home base, the prime focus of allegiance and the place where personal ambitions are formed, achieved, and enjoyed" (Wilkie, 1977:88).

Psychological assimilation is a process of change not only in self-identification but also in identification by others. Although individual members of an ethnic group may see themselves as simply part of the larger society rather than as ethnics, outsiders may continue to identify them as members of their group, thereby impeding psychological—and structural—assimilation. As Yinger notes, "Prejudice on the part of a dominant group may prevent the granting of full membership in a society to members of minority groups, even though the latter think of themselves only in terms of the larger society" (1981:253). Visibility is, of course, critical here. Those with salient marks of ethnic identity—skin color, in particular—are unable to fully achieve out-group recognition as "nonethnics."

Some individuals may find themselves unable to feel fully part of the larger society or the ethnic group. Such people have been referred to as "marginal men" (Park, 1928; Stonequist, 1937). First-generation immigrants often find themselves in a situation where they are pulled in the direction of the culture of the new society but remain culturally and psychologically tied to the old. Those who are not clearly part of one racial category or another also may be marginal people. In American society in recent years, such individuals have begun to be recognized by themselves and others as of a unique "mixed-race" or "multiracial" category (Wright, 1994).

Although sociologist Robert E. Park, in a classic work (1928), described marginal people as experiencing psychological turmoil, not all of them find difficulty in social adaptation. Some successfully adapt to two cultures, shuttling between them, and others may use their marginality to advantage by serving as middlemen between dominant and minority groups. Moreover, where they constitute a large population, marginal people may form communities of their own, occupying an in-between status in the society's ethnic hierarchy (Shibutani and Kwan, 1965). Recall our discussion of middleman minorities in Chapter 2.

The social psychology of ethnic identity is a complex matter, the details of which go well beyond our present purposes.[2] Suffice it to say that within any multiethnic society, individuals may vary widely in the extent to which they identify with an ethnic group or choose to disregard ethnicity as part of their self-identity. Much depends not only on visibility but also on the political and social conditions that may affect the individual costs and values of an ethnic identity. Most people are permitted much latitude in their control of personal identity. Thus, for any single person, ethnic identity "expands and contracts, it changes over time, it looms large or small depending on the particular context and possibly on the advantage to be gained from manipulating the ethnic label" (Elkin and Handel, 1978:101–2).

Two Theories of Assimilation

Most analysts of assimilation have dealt with societies like the United States or Australia, in which sizable groups of voluntary immigrants, mainly from Europe, have made up the largest element of the ethnic mosaic (Taft, 1963). As a result, their findings and the hypotheses they suggest are applicable only to a narrow range of societies and specific ethnic groups. We now look at two assimilation theories that, though American in origin, have somewhat broader applications and have been especially influential. Park's is a classic theory, and Gordon's is more recent.

Park's Race Relations Cycle American sociologists in the 1920s, led by Robert Park, were concerned with the ethnic polyglot that made up cities of the Northeast and Midwest, and the focus of their studies became the forces that brought these groups together or sustained their differences. Park was one of the first to suggest a cycle of race or ethnic relations through which groups would pass in a sequence of stages, leading ultimately to full assimilation.

Park explained that groups first come into *contact* through migration and subsequently engage in *competition,* often characterized by conflict. Out of such competition eventually emerges some form of *accommodation* among the groups,

[2]The issues of ethnic identity at the individual level are detailed in De Vos and Romanucci-Ross (1975), Horowitz (1975), and Shibutani and Kwan (1965).

leading finally to *assimilation.* Park maintained that this four-stage cycle pertained to race relations everywhere, not simply the United States. Moreover, he saw the sequence as "apparently progressive and irreversible" (1950:150).

A view of ethnic assimilation not unlike Park's race relations cycle has been generally popular in American society. The prevailing thought is that, over several generations, group boundaries break down, and the society becomes more homogeneous. However, Park's model has been subject to much sociological criticism over the years. Some have noted the cycle's lack of applicability to many groups (Lyman, 1968a). Although it has seemed to describe fairly accurately the experiences of many European immigrant groups in the United States (as well as Australia, Canada, Argentina, and other immigrant societies), it does not conform to the patterns displayed by more salient, that is, physically distinct, groups and those that have entered involuntarily. Critics of Park's and other cyclical theories of ethnic relations have also pointed out that such cycles are rarely complete; that is, there are too many truncated instances (Shibutani and Kwan, 1965). Barth and Noel (1972) point out that interethnic contact can lead to stable outcomes such as exclusion, pluralism, or continued ethnic stratification that do not lead inevitably to assimilation. Finally, some have criticized the model's claim of irreversibility (Barth and Noel, 1972; Berry and Tischler, 1978). They note that the cycle may be terminated at any point, and groups may even revert to earlier stages.[3]

Gordon's Stages of Assimilation Park's model, despite its shortcomings, stands as a precursor to subsequent, more sophisticated theories of the assimilation process. Perhaps the most precise and compelling of these is Milton Gordon's (1964).

Like Park, Gordon explains assimilation as a series of stages, or steps, through which various groups pass. But rather than following a straight line leading from contact to eventual absorption, groups may remain indefinitely at one or another of these stages. The seven stages that Gordon outlines extend from cultural assimilation, the least intense, to civic assimilation, the most complete (Table 4.1). The first two, cultural and structural assimilation, are, however, the most important.

As we have already seen, cultural assimilation, or acculturation, denotes the adoption by a minority ethnic group of the dominant group's (or, as Gordon prefers, the host society's) cultural patterns—language, political beliefs, and so

---

[3]Geschwender (1978) disputes the interpretation of Park's race relations cycle as absolute, maintaining instead that it is "situationally specific." He sees Park's cycle as an ideal type, not to be taken as inevitable.

Table 4.1 Gordon's Stages of Assimilation

STAGE	CHARACTERISTICS
Cultural or behavioral assimilation (acculturation)	Change of cultural patterns to those of host society
Structural assimilation	Large-scale entrance into cliques, clubs, and institutions of host society on primary group level
Marital assimilation (amalgamation)	Large-scale intermarriage
Identificational assimilation	Development of sense of peoplehood based exclusively on host society
Attitude receptional assimilation	Absence of prejudice
Behavior receptional assimilation	Absence of discrimination
Civic assimilation	Absence of value and power conflict

SOURCE: Milton M. Gordon, *Assimilation in American Life: The Role of Race, Religion, and National Origins,* New York: Oxford University Press, 1964, p. 71. Copyright © 1964 by Oxford University Press, Inc. Reprinted by permission.

on. But acculturation, though a prerequisite, does not assure movement to the next phase. Gordon explains that groups may become very much like the dominant group in behavior and values but still remain structurally segregated.

Structural assimilation, the second stage, is seen by Gordon as essentially what we have previously referred to as primary, or informal, structural assimilation. This stage is the most critical, explains Gordon, for it is the key to all subsequent stages: "Once structural assimilation has occurred, either simultaneously with or subsequent to acculturation, all of the other types of assimilation will naturally follow" (1964:81). Presumably, as people of minority and dominant groups interact within close, intimate social settings, the other stages of assimilation necessarily occur, much like falling dominoes: Minority group members increasingly intermarry with those of the dominant group, relinquish their ethnic identity, no longer encounter prejudice and discrimination, and fully agree with the dominant group on issues involving values and power conflicts. Indeed, the remaining stages, three through seven, in Gordon's scheme can be subsumed under structural assimilation. The dominant group's acceptance of members of a minority ethnic group into primary relations implies, for example, the absence of prejudice and discrimination and the likelihood of increased intermarriage.

It is with structural assimilation, then, that full assimilation involving all other stages becomes inevitable (though Gordon suggests no timetable or further conditions that affect the rate at which they are achieved). Cultural assimilation, the first stage, does not necessarily lead to structural assimilation, however, but may take place independently of any of the others and may continue to be the extent of assimilation for many generations (Gordon, 1964:77; see also van den Berghe, 1981:216; Wagley and Harris, 1958:288). Minorities may take on all or most of the cultural ways of the dominant group but still be refused entry into primary relations with its members. African Americans are an evident illustration of this. Though they have adopted in large measure the major elements of the dominant culture, they remain unassimilated at the structural level (specifically, as we previously noted, the *primary* structural level). As we will see in Chapter 12, the same condition is evident among Coloreds in South Africa, who, culturally, are indistinguishable from the dominant white group but remain segregated in most institutional spheres. In short, groups may become culturally alike yet remain in relatively segregated subsocieties.

One serious shortcoming of Gordon's assimilation model lies in its understanding of structural assimilation, stage two, as entailing interaction with the dominant group only at the primary level. Intergroup relations, however, occur at the secondary level as well and indeed are antecedent to primary relations in any significant degree. To what extent do members of a minority ethnic group enter into positions of power in the society's economic, political, and other key institutions? To what extent are they afforded equal opportunities in employment and education? These are significant measures of structural assimilation—at the secondary level—that Gordon does not consider (Marger, 1979). African Americans, for example, appear to have realized substantial integration in recent years in the areas of work and government. They have, in other words, experienced increasing secondary structural assimilation. That they have not accomplished an equivalent level of interaction with whites at the primary level would, in Gordon's view, imply that little structural assimilation had occurred.

Moreover, Gordon seems to suggest that if minorities do not enter into primary relations with the dominant group, it is because the dominant group has held them out. But such social segregation may be largely voluntary.

Finally, Gordon's model has, like Park's, been criticized by some as implying a unidirectional movement whereby groups, though perhaps delayed indefinitely at one stage, do not return to earlier stages. The assimilation process is, in this view, invariably progressive in direction.

Despite these omissions, Gordon's model, with modification, is valuable to the analysis of interethnic relations primarily because it spells out the intricacies of the assimilation process and the various forms and degrees it manifests. Assimilation, as Gordon demonstrates, is clearly not a simple, straightforward movement, as earlier theories had seemed to imply.

Factors Affecting Assimilation

Why do some groups in multiethnic societies display a rapid and almost complete assimilation into the larger society, whereas others remain segregated and are the constant targets of prejudice and discrimination? Where assimilation is the prevailing model of intergroup relations, several factors are important in shaping the experience of different groups. These are how and when a group enters a society, its size and dispersion, its cultural similarity to the dominant group, and its visibility.

Manner of Entrance As we have already seen in Chapter 2, the way in which a group enters the society is critical in determining its place in the ethnic hierarchy; it is also important in accounting for the nature of the group's long-range societal adjustment. Except for those groups that maintain unfaltering pluralistic aims, those that enter voluntarily always make a less conflict-ridden adjustment than those that enter involuntarily or those that are conquered by more powerful invaders. Involuntary immigrants or conquered groups remain in a condition of segregation to one degree or another.

The United States presents a clear illustration of this tendency. European ethnic groups, having entered by choice, were able to assimilate culturally at a pace determined in large part by themselves, no matter how strong the pressures to conform. Furthermore, the option of returning to their society of origin was always present and was, in fact, taken by many. African Americans, American Indians, and many Mexican Americans were absorbed into the society involuntarily and as a result could not follow similar paths no matter what their intentions or long-range goals (Blauner, 1972). As we noted in Chapter 2, indigenous groups that are made subordinate at the entrance of an invading group (for example, North American Indians) or who enter involuntarily (for example, African Americans) are left with few options other than resistance to the new social order that is imposed on them. Long-term conflict is thus the usual outcome. Subordinate voluntary immigrants (like Europeans to America), in contrast, are more rapidly assimilated (Lieberson, 1961; Schermerhorn, 1970; van den Berghe, 1976; Wilkie, 1977).

Time of Entrance In general, the more recent a group's entry into the society, the more resistance there is to its assimilation (Mack, 1963). Other things being equal, the simple factor of time will tend to ease the fear and suspicion that accompany the entrance of strangers. Groups with alien ways are seen differently after they have lived in the society for several generations. Examples abound in the United States, beginning with the influx of large numbers of Irish immigrants in the early nineteenth century. Given the very substantial assimilation of

Irish Americans today (such that they are approaching ethnic extinction, save for their Catholicism), the virulent and rabidly hostile reception encountered by the first generation is easily forgotten.

Demographic Factors The degree and rate of assimilation for minority ethnic groups is also affected by their size and the concentration of their population (Blau, 1977; Frisbie and Neidert, 1977). The entrance and assimilation of groups relatively small in number will be resisted less forcefully than groups that represent a competitive threat. Van den Berghe (1981) suggests that smaller groups are assimilated more easily because they have fewer resources and therefore depend on the larger society and because they necessarily interact more frequently with out-group members.

The concentration or dispersion of ethnic groups also bears on their assimilation. Concentration in particular neighborhoods and geographical areas or in certain occupations tends to retard assimilation because the group is better able to retain its cultural ways and resist intrusions of the dominant group (van den Berghe, 1981). Dispersal, on the other hand, leads to unavoidable contact and interaction with the dominant group and thus speeds up the assimilation process.

Cultural Similarity No matter how or when the group enters the host society or what its demographic patterns, assimilation is favored if the group is culturally similar to the dominant group (Berry and Tischler, 1978; van den Berghe, 1981). Those groups in the United States that have followed the assimilation route furthest have, predictably, been culturally closest to Anglo Protestants. In general, the more compatible the culture of the minority group with the dominant group's, the greater will be the force and speed of assimilation.

Visibility In almost all multiethnic societies, the most critical factor in determining the degree and rate of assimilation of ethnic groups is visibility. Where physical differences are obvious, manner of entrance, temporal factors, demographic patterns, and even cultural similarity are of relatively slight consequence. For racial-ethnic groups, structural separation remains far more persistent than for groups who are only culturally distinct (Wirth, 1945; Yinger, 1981). Observers of American ethnic relations have long interpreted the retarded structural assimilation of blacks, for example, as a product chiefly of visibility (Park, 1950; Warner and Srole, 1945). It is, of course, not visibility per se that deters assimilation of racial-ethnic groups but rather the beliefs connected with their physical distinctness.

In short, physical differences delay the process of assimilation more than other factors. Harold Isaacs has poignantly described this dilemma: "An individual can change his name, acquire a new language, ignore or conceal his origins, disregard or rewrite his history, abandon his ancestral religion or convert to another one, adopt a different nationality, embrace new mores, ethics, philosophies, styles of life. But there is not much he can do to change his body" (1989:46). Thus, the more visible the group or individual, the longer and more difficult is the process of structural assimilation.

Assimilation as Public Policy

Assimilation (and, as we will see, pluralism as well) can be seen as a goal, or ideal, for which multiethnic societies aim. As such, it is sometimes the basis of public policies designed to reduce the cultural and structural divisions between groups. In the United States, for example, measures intended to lessen segregation in various public spheres (housing, schools, work) and to equalize access to power and privilege (affirmative action programs, voting rights) can be understood as outgrowths of a societal commitment to the eventual achievement of complete assimilation for all ethnic groups. But the form and ultimate objective of assimilation may vary in the minds of policymakers as well as members of both dominant and minority groups.

Abramson (1980) points out that there are three possible forms of complete assimilation for which societies may strive, each involving a different path and a somewhat different objective. First, minority ethnic groups may assimilate into the dominant ethnic group. This, as we shall see in Chapter 5, has been the chief form and process of assimilation in the United States. Second, groups may assimilate into an entirely new ethnicity. This is the popular notion of the "melting pot," in which all groups surrender their ethnic heritage but in the process create a hybrid society with no dominant group as such (see Gordon, 1964). Israel is a modern multiethnic society on such a course. Third, minority ethnic groups may assimilate into another nondominant ethnic background or group. For example, some have suggested a "triple melting pot" in the United States, in which numerous ethnic groups representing diverse national origins have seemed to converge into three encompassing religious bodies, Protestantism, Catholicism, and Judaism (Herberg, 1960; Kennedy, 1944, 1952).

It is the first form of complete assimilation, however, that remains the major objective in most multiethnic societies where groups are encouraged to pursue an assimilationist course. The answer to the question "Assimilation to whose values and traits and to whose institutions?" is almost always "The dominant group's culture and social structure." This is nothing more than a manifestation of power. As long as one group disproportionately controls power resources in the economic and political realms, that group is enabled to dictate the shape and direction of minority group adjustment. It is the dominant group that, in the main, mandates or approves policies of acculturation and integration. In the United States, for example, there has never been any question of whose language, whose religious and political principles, and whose institutions will prevail. All have been those of the dominant Anglo core group and continue to be so. As William Newman has succinctly put it, "Regardless of the number of groups involved, the theory of assimilation is a theory of majority conformity" (1973:53). Ideologically, societies may advocate some kind of ethnic melting pot wherein all groups contribute in proportionate amounts to form a new social system, but such a cultural and, particularly, structural fusion is a chimera.

As previously noted, however, we should not think that assimilation is a one-way process whereby minority ethnic groups seek out and become like the dominant group, with no change occurring in the opposite direction. Obviously, the assimilation process will be to some extent mutual, with many aspects of minority cultures becoming part of the dominant culture (Greeley, 1974; Wagley and Harris, 1958); even social interaction is at times initiated by members of the dominant group. But the exchange is far short of equal. On the major elements of culture—language, religion, political beliefs, economic practices, and so on—there is little evidence of mutuality, and at the structural level the exchange is even more lopsided. These aspects of culture and social structure remain firmly controlled by the dominant group, and minority ethnic groups must adjust to them. Indeed, so powerful is the dominant group's model that it is often internalized by minority group members, who, in the process of assimilation, "tend to conceive of themselves as *inferior* objects and look with envy upon members of the dominant group" (Shibutani and Kwan, 1965:507).

PLURALISM

Like assimilation, pluralism entails several dimensions and forms.[4] In all cases, however, the retention or even strengthening of differences among ethnic groups is presumed. Thus, in a general sense, pluralism is the opposite of assimilation. Abramson defines pluralism as "conditions that produce sustained ethnic differentiation and continued heterogeneity" (1980:150). In short, pluralism is a set of social processes and conditions that encourages group diversity and the maintenance of group boundaries.

---

[4]Pluralism as applied to *political* systems refers to the relative dispersion of power among various interest groups in a society. This is different from its usage in describing ethnic relations. See Marger (1987).

Just as assimilation occurs in different degrees and at different stages, so too pluralism must be understood as a variable for groups and societies. Ethnic pluralism never entails an absolute separation of groups. Here we might recall the definition of *ethnic group* as a distinguishable group *within a larger society*. Thus, in a pluralistic society there is always some common political or economic system that binds various ethnic groups together. If this were not so, there would be not a multiethnic society but several distinct societies in themselves. Beyond the recognition of a common political or economic system, however, groups may differ widely.

As with assimilation, we can delineate several dimensions of pluralism, the most important being the cultural and the structural. *Cultural* pluralism implies the maintenance of many varied cultural systems within the framework of a common economic and political system (Gordon, 1964). *Structural* pluralism connotes not simply differences in culture but also the existence in some degree of segregated ethnic communities within which much of social life occurs for group members. These ethnic subsocieties, or communities, comprise institutions — schools, businesses, churches, and the like — that duplicate to some extent those of the dominant group. Notice how cultural and structural pluralism are basically opposing counterparts of cultural and structural assimilation.

Two forms of ethnic pluralism are evident in modern multiethnic societies. In the first, groups maintain cultural and structural autonomy but remain relatively equal in political and economic power; moreover, their separation is mainly voluntary. This form we will refer to as *equalitarian pluralism*. In the second, groups maintain structural segregation and perhaps cultural distinctness as well but are unequal in political and economic power; further, group separation in these cases is ordinarily involuntary. This form we will refer to as *inequalitarian pluralism*.

Equalitarian Pluralism

Where equalitarian pluralism characterizes interethnic relations, groups retain their cultural and, for the most part, structural integrity while participating freely and equally within common political and economic institutions (Barth and Noel, 1972; Shibutani and Kwan, 1965). Some have referred to this condition as *accommodation*, wherein the minority group "desires equality with, but separation from, the dominant group and the dominant group agrees to this arrangement" (Kurokawa, 1970:131). Technically, of course, if equality with the dominant group is reached, there are no longer dominant–minority relations; there are relations among ethnic groups, but they are not hierarchical and invidious.

Equalitarian pluralism corresponds, like assimilation, to the order model of society, in which balance and cohesion are emphasized. Differences among groups are recognized and even encouraged, but within the framework of a larger set of agreed-on principles. All groups give allegiance to a common political system, participate in a common economic system, and understand a common set of broad ethical values (Williams, 1977). In a sense, ethnic groups become political interest groups that compete for the society's rewards (Glazer and Moynihan, 1970). But these competitive differences do not lead necessarily to serious cleavages and conflict; rather, they are dealt with by a reasonable give-and-take within the context of the consensual rules of the society.

There may be vast cultural differences among ethnic groups, but their essence is not threatened because tolerance of such differences is integral to the social order. Ronald Taft explains that "pluralism implies that beyond the acceptance of supraordinate national values essential to the nation's existence there need be no agreement between immigrants and native citizens excepting that their cultural differences be mutually tolerated and preserved" (1963:279). Relations between ethnic groups are thus confined mainly to functional areas such as government and the marketplace, not affective ones like the family or friendship circles.

The equalitarian form of pluralism has been manifested somewhat differently in the United States than in other multiethnic societies.

Cultural Pluralism: The American Case Order theorists of pluralism have concentrated chiefly on the United States and its system of ethnic heterogeneity, in which groups continue to express elements of their ethnic cultures despite the forces of assimilation. Those who have analyzed American society with such a model have noted the persistence of ethnic groups over many generations after their immigration and the tendency for individuals to continue to conduct their primary group relations within an ethnic context. The choice of close friends or marital partners, it is pointed out, is still strongly dictated by ethnicity.

Ethnicity in such a setting is more a matter of individual choice than a collective imperative. Gordon (1975, 1981) has referred to this condition as "liberal pluralism," in which there is no formal recognition of ethnicity in the allocation of government offices or economic rewards and in which individuals are free to express their ethnic identity to whatever extent they choose.

Many have also advocated cultural pluralism, rather than assimilation, as a desirable end product of American interethnic relations. Gordon explains:

> The presumed goal of the cultural pluralists is to maintain enough subsocietal separation to guarantee the continuance of the ethnic cultural tradition and the existence of the group, without at the same time interfering with the carrying out of standard responsibilities to the general American

civic life. In effect, this demands keeping primary group relations across ethnic lines sufficiently minimal to prevent a significant amount of intermarriage, while cooperating with other groups and individuals in the secondary relations areas of political action, economic life, and civic responsibility. Within this context the sense of ethnic peoplehood will remain as one important layer of group identity while, hopefully, prejudice and discrimination will disappear or become so slight in scope as to be barely noticeable (1964:158).[5]

Although this ideal has fairly accurately described the process by which some degree of cultural and structural pluralism has persisted in American society for white ethnic groups, it does not account for the maintenance of gross inequalities in the pluralistic system, particularly as it works for racial-ethnic groups. Moreover, though this form of American pluralism is most commonly referred to as *cultural* pluralism, it is more accurately *structural* pluralism because most groups have adopted the key elements of the mainstream culture after the second immigrant generation.

In the United States, structural pluralism exists to a greater or lesser extent depending on either the visibility of the ethnic group or the desire of the group to remain apart from the larger society. Particularly at the level of primary relations, ethnic groups continue to display at least minimal internal cohesion by encouraging endogamy, a sense of community, and at least in symbolic form, ethnic identification. For the more salient groups such as African Americans, Puerto Ricans, and American Indians, separation extends well into the level of secondary relations, where segregated neighborhoods and schools persist and job opportunities are circumscribed. There is, then, a severely unequal outcome among ethnic groups despite the society's adherence to an ideology of political equality and equality of economic opportunity. The United States has, as we will see in Part II, retained many characteristics of an ethnically pluralistic society but in ways closer to the inequalitarian form than to the equalitarian form stressed by order theorists.

Corporate Pluralism: Non-American Cases American sociologists have paid far less attention to other pluralistic societies, in which a much greater degree of structural and cultural separation is sustained by political authorization. In societies such as Switzerland, Belgium, Malaysia, and to some extent, Canada,

---

[5]From Milton M. Gordon, *Assimilation in American Life: The Role of Race, Religion, and National Origins.* Copyright © 1964 by Oxford University Press, Inc.; renewed 1992 by Milton M. Gordon. Reprinted by permission.

the structural and cultural differences among ethnic groups are protected by the state, and institutional provisions are made to encourage an ethnically proportionate distribution of societal rewards. Gordon (1975, 1981) refers to such cases as *corporate* pluralism.

In corporate pluralistic systems, ethnic units are formally recognized by the government, and political and economic power is allocated on the basis of an ethnic formula. Thus, in the political arena legislative seats and other government offices may be apportioned on the basis of ethnicity. Not only is proportionality in the distribution of political benefits assumed, but most important, there is cooperation among leaders of all the significant segments of the plural society (Lijphart, 1977). Moreover, on local matters each group maintains a great deal of political autonomy. In the economic realm as well, the objective is an equal distribution of income and jobs among the various groups, proportional to their makeup in the national population. Such systems are sometimes referred to as *consociational* (Lijphart, 1977; McRae, 1974; van den Berghe, 1981).

In this type of pluralism, cultural and structural separation are emphasized, not discouraged, and multilingualism is officially sanctioned. As Gordon explains, the position of corporate pluralism is "that the various racial and ethnic groups have the right and, indeed, should be encouraged to retain their ancestral languages, that there is no reason why there must be only one official language, and that all members of the national polity should be encouraged, perhaps even compelled, to become bilingual or multilingual" (1981:187). Switzerland with its four official languages — German, French, Italian, and Romansh (the latter spoken only by a very small number) — is perhaps the most obvious and successful multilingual system (McRae, 1983). Canada, with its French- and English-speaking groups, is another familiar multilingual (in this case, bilingual) society, though, as we shall see in Chapter 14, multilingualism has not been accepted enthusiastically and without serious problems.

In societies characterized by corporate pluralism, ethnic groups consist mainly of homogeneous, territorially concentrated peoples who have long historic roots in their native area. They have become part of a larger national society either through conquest or by voluntarily relinquishing sovereignty to a central state in order to secure economic and political benefits. Such societies are not at all similar to multiethnic societies like the United States, in which ethnic groups have been formed primarily by voluntary immigrants who came from distant societies, severed most of their native roots, and scattered geographically throughout the society.

The cultural divisions among ethnic groups in corporate pluralistic systems are much sharper than those among groups in the United States, where the English language tends to become a great commonizing factor. Structural separation is also stronger, given the territorial concentration of groups. In Switzerland, for example, each ethnic group remains geographically distinct, and in Canada over 80 percent of French-speaking people reside in one province, Quebec. By contrast, ethnic groups in the United States are more dispersed though regional concentrations of particular groups are noticeable (such as Mexican Americans in the Southwest or Swedish Americans in the upper Midwest). Ethnic concentrations in the United States more often take the form of urban pockets or ghettos.

With their retention of language and their territorial base, ethnic groups in such corporate pluralistic systems are essentially "subnations" (Petersen, 1980). They are integrated only in their mutual allegiance to a larger national government and the need to participate in a national economic system. Ideally, no single group is dominant, and each is afforded an approximately proportionate share of the society's rewards. As Wagley and Harris note, "A satisfactory pluralistic adjustment implies not only that the dominant group will accept a minority's distinctive traits without prejudice, but also that the minority will have a more equal share in the material and nonmaterial wealth of the nation" (1958:288). Corporate pluralistic societies, of course, do not meet this ideal though some have achieved a greater degree of group equality than others.

Belgium and the Netherlands are societies that are organized along such corporate pluralistic, or consociational, lines. Belgium is divided mainly on the basis of language: Walloons speak French, Flemings speak Dutch. Each group is also regionally concentrated. Traditionally, French was the dominant language, and Flemings, if they expected to rise in social standing, had to adopt that language as well as other aspects of Francophone culture. This changed in the post–World War II era as the economic position of the Flemish region improved markedly. The language division became more acute, however, and ethnic separation more pronounced. In response, Belgian social and political institutions, already with consociational features, were further adjusted to accommodate this division (McRae, 1986). Language and educational rights were afforded each group and, most important, a political power-sharing formula was created. The national cabinet, for example, must comprise equal numbers of Dutch- and French-speaking members (Heisler, 1991). Comparable power-sharing institutions prevail in the Netherlands, where the major division is religious, between Protestants and Catholics (Kruijt, 1974).[6]

---

[6]Lijphart (1977) notes, however, that consociationalism in the Netherlands has declined since the 1960s, with parties expressing attitudes more in line with majority rather than proportional rule.

Switzerland, with its four linguistic groups, is another striking example of ethnic autonomy and balance. Some have explained the Swiss success in maintaining concordant interethnic relations as due to the lack of contacts between groups and the decentralization of decision making in matters of ethnic relevance, such as the use of language (Connor, 1976).

Equalitarian pluralism is more a product of group choice than the largely involuntary structural pluralism of racial-ethnic minorities in the United States or of the inequalitarian form of pluralism, which is discussed below. But if we can judge from contemporary societies that have attempted this path to ethnic harmony, the results are not always more benign. The breakup of Yugoslavia is a tragic illustration. Here, several ethnic nationalities, living for the most part in distinct territories, made up a state that provided roughly proportional political power for each. But the system collapsed in 1991, giving rise to one of the most brutal episodes of ethnic warfare that has been witnessed in modern Europe. We will discuss this case in more detail in Chapter 16. Another such case is Canada, where, as we will see in Chapter 14, French and English groups have sustained conflict (though nonviolent) for two centuries. Cases like Belgium and Switzerland, where interethnic relations are less discordant, appear to be exceptional, not common, and even in those societies periodic group conflict is evident. The assertion of Wagley and Harris (1958) that pluralistic aims perpetuate some degree of conflict and the subordination of one group by another is well taken. Indeed, pluralism, no matter how seemingly benign, seems almost always productive of conflict.

Inequalitarian Pluralism

Both assimilation and equalitarian pluralism are most characteristic of societies that have blunted to some extent the acuteness of ethnic stratification and have made the reduction of ethnic inequality a well-established commitment of the state, legitimated by an equalitarian ideology. As public policies, assimilation and equalitarian pluralism are aimed at reducing the level of ethnic conflict. Inequalitarian pluralism, however, describes outcomes and processes that are clearly inequitable for the society's various ethnic groups. Although most characteristic of classic colonial societies and modern racist regimes like South Africa during the period of apartheid, inequalitarian pluralism can also be seen to some degree in other multiethnic societies where assimilation or equalitarian pluralism is the prevailing type of interethnic relations.[7]

---

[7]This model of ethnic relations is favored by power-conflict theorists, who stress the divisiveness of multiethnic societies and the competition among groups. The classic statements of the inequalitarian school of pluralism are Furnivall (1948) and Smith (1965).

Under inequalitarian pluralism, ethnic groups are not only separated structurally and perhaps culturally but also exist in a state of highly unequal access to power and privilege. Indeed, the authority and power of the dominant group are the key coordinating mechanisms of such systems (Kuper and Smith, 1969; van den Berghe, 1978). In an equalitarian pluralistic system, the various ethnic groups are held together through the consensual allegiance to a common state. In inequalitarian pluralistic societies, however, the state holds the different groups together not through a mutually recognized legitimacy but through coercion. Basically, the state acts to protect the interests of the dominant group. Anthropologist M. G. Smith (1969:33) explains that in a plural society of this type, the state is the agent of the ruling group only. Others have no rights or protection. The majority of people are "subjects, not citizens." Whereas equalitarian pluralism assumes a progressive equalization of political and economic power among groups, the assumption here is quite the opposite. Sustained or increased inequality among groups is a built-in feature of the system, with the dominant group retaining all political authority and the bulk of material wealth.

Social relations between dominant and minority groups are typified by extreme polarization, supported by high levels of prejudice and discrimination. Basically, it is only within the impersonal confines of the economic and political systems that dominant and minority group members interrelate, and these relations are limited to purely functional contacts such as work and government administration. As Kuper and Smith (1969:11) describe it, "Economic symbiosis and mutual avoidance, cultural diversity and social cleavage" characterize the social basis of this type of plural society.

Although the dominant group holds sway, ethnic groups are in a state of economic interdependence. The dominant group needs subordinate ethnic groups to perform physical and menial laboring tasks necessary to economic production, and given their relative powerlessness, subordinate groups must meet these demands. Occupational roles are assigned strictly on the basis of ethnicity, with the subordinate group or groups delegated those most onerous and least prestigious (Rex, 1970). The dominant group, as van den Berghe explains, "rationalizes its role in an ideology of benevolent despotism and regards members of the subordinate group as childish, immature, irresponsible, exuberant, improvident, fun-loving, good humored, and happy-go-lucky; in short, as inferior but lovable as long as they stay in their place" (1978:27).

Institutional separatism and duplication characterize inequalitarian pluralistic systems (Smith, 1965; van den Berghe, 1978). This means that each group maintains its own schools, churches, businesses, and so on. In each case, however, there is a great disparity in the quality of dominant and minority group institutions.

In its most extreme form, inequalitarian pluralism resembles a caste system in which strict segregation is enforced in all areas of social life. J. S. Furnivall describes it as follows:

> Each group holds by its own religion, its own culture and language, its own ideas and ways. As individuals they meet, but only in the marketplace, in buying and selling. There is a plural society, with different sections of the community living side by side but separately, within the same political unit. Even in the economic sphere there is a division of labour along racial lines.... There is, as it were, a caste system, but without the religious basis that incorporates caste in social life in India (1948:304–5).

Where the extreme of caste is reached, the social distance between groups is maximized in all social situations. What emerges is a system of what van den Berghe (1978) calls "paternalistic" race relations. All people understand their social place and, as long as the subordinates do not deviate from their ascribed role, stability is assured. In such systems there is at least some acquiescence of subordinate groups to their inferior positions. Actual physical distance need not be enforced in all situations so long as the *social* distance between groups is understood and adhered to. Slavery in the American South, for example, was characterized by a good deal of physical proximity, particularly among slaves who performed household duties (Stampp, 1956).

Competitive Race Relations Inequalitarian pluralistic relations are realized in full only in slave or classic colonial systems (Rex, 1970), neither of which is evident in the contemporary world (though South Africa, until recently, closely approximated them). In those industrialized societies where ethnic divisions are based primarily on race rather than culture, ethnic stratification bears some resemblance to a caste or colonial system but is less extreme in the segregation of social institutions and relations. Van den Berghe (1978) has referred to such cases as "competitive" race relations. With the society's economic base changed from agrarianism to industrialism, such competitive relations replace paternalistic relations. Industrialization requires that social roles be assigned more through competition than ascription. As a result, there are no longer "master–servant" relations as in a caste or paternalistic system; however, competition between the subordinate group or groups and the working-class element of the dominant group emerges (van den Berghe, 1978).

The gap between ethnic groups in such a competitive system is narrowed, and each group itself becomes more class diversified. In a slave or colonial situation, ethnicity and class are essentially synonymous: One group has almost all the society's power, wealth, and prestige, and the others have relatively none. In a competitive system, however, class hierarchies emerge within each ethnic group. Of course, as we saw in Chapter 2, members of particular groups continue to cluster at well-defined points in the overall class system of the society.

Because the norms of social distance, so well institutionalized in the paternalistic condition, now break down, physical segregation becomes more rigid. Ethnic groups tend to reside and carry out much of their daily activities in homogeneous ghettos. The dominant group's image of the subordinate group or groups also changes, explains van den Berghe, "from one of backward but ingratiating grownup children to one of aggressive, insolent, 'uppity,' clannish, dishonest, underhanded competitors for scarce resources and challengers of the status quo" (1978:30). Accordingly, conflict and bigotry become more virulent. As we will see in Chapter 8, the plight of blacks in the United States following slavery generally conformed to this pattern.

Internal Colonialism Internal colonialism is a type of inequalitarian pluralism characteristic of societies like the United States where ethnic relations otherwise follow the assimilation model. Here, racial-ethnic groups are treated in a colonial fashion. In the traditional form of colonialism practiced by the European powers in the seventeenth, eighteenth, and nineteenth centuries, a conquering group—a numerical minority—established political and economic domination over a native people. Internal colonialism, by contrast, is a condition in which both the dominant group and subordinate groups are indigenous; moreover, the dominant group is a numerical majority. Classic colonialism, explains Robert Blauner, "involved the control and exploitation of the majority of a nation by a minority of outsiders. Whereas in America the people who are oppressed were themselves originally outsiders and are a numerical minority" (1969:395).

Although the situations from which they emerge are different, classic and internal colonialism nonetheless have common features. Blauner (1969:396) states four basic components of the "colonization complex": (1) colonization begins with a forced, involuntary entry; (2) the colonizing power alters basically or destroys the indigenous culture; (3) members of the colonized group tend to be governed by representatives of the dominant group; and (4) the system of dominant-subordinate relations is buttressed by a racist ideology.

Using this model, Blauner and others (Blackwell, 1976; Carmichael and Hamilton, 1967; Clark, 1965) have interpreted the African-American experience in the United States as basically a colonial one and ghetto uprisings as anticolonial movements. This internal colonialism model has also been applied to other racial-ethnic groups in the United States, namely, Mexican Americans and American Indians (Blauner, 1972; Moore, 1970; Snipp, 1986; Thomas, 1966), and with some variation, to groups in other societies as well (Zureik, 1979). We will discuss internal colonialism as it pertains to African Americans and Mexican Americans in Part II.

Particularly in societies like the United States, where in addition to voluntary immigration, conquest or involuntary immigration has contributed to the society's multiethnic character, the internal colonial model represents an alternative for explaining the inability of racially distinct groups — those that have usually entered involuntarily or who have been conquered — to follow a course of gradual assimilation like that followed by white European immigrant groups. As we shall see in Part II, however, not all sociologists have accepted the notion that blacks and other racial-ethnic groups in American society are essentially different from white ethnic groups and that their social experiences require different models of explanation.[8]

Annihilation or Expulsion Inequalitarian pluralism may reach an extreme form in the expulsion or even the annihilation of minority ethnic groups. Neither of these outcomes is without precedent in recent Western history. The deportation of Chinese immigrants from the United States in the nineteenth century and of Germans from Canada during World War I, and the internment of Japanese in both the United States and Canada during World War II, are notable examples of expulsion. As for annihilation, destruction of native groups by white settlers in North America, Australia, and South Africa in the nineteenth century as well as in Latin America earlier are all relevant examples. A more deliberate and methodical case is the genocidal policy of the Nazis, which resulted in the destruction of the European Jews. In the 1930s, German Jews were subjected systematically to an almost complete expulsion from every phase of the society's life. Later, Jews were impelled to leave Germany through a terroristic campaign that included physical attacks and the appropriation of their homes, businesses, and wealth. Finally, Nazi policies culminated in the establishment of death camps to which the German and subsequently other European Jewish populations were sent to be slaughtered. The campaign of "ethnic cleansing" by Serbs against Muslims in Bosnia is a more recent case of a combination of expulsion and annihilation.

THE VARIABILITY OF ETHNIC RELATIONS

The three major patterns of ethnic relations we have examined are outlined in Figure 4.1. Several points should be kept in mind in looking at each.

[8]A somewhat different application of the idea of internal colonialism is made by Michael Hechter (1975), who describes it as a condition in which a dominant ethnic group in the industrialized core of a nation subjugates an industrially backward ethnic group in the nation's periphery.

Figure 4.1 Relations Among Three Ethnic Groups, Beginning with (1) Initial Contact, Followed by (2) Emergence of Competition and Stratification, and (3) Movement in One of Three General Directions

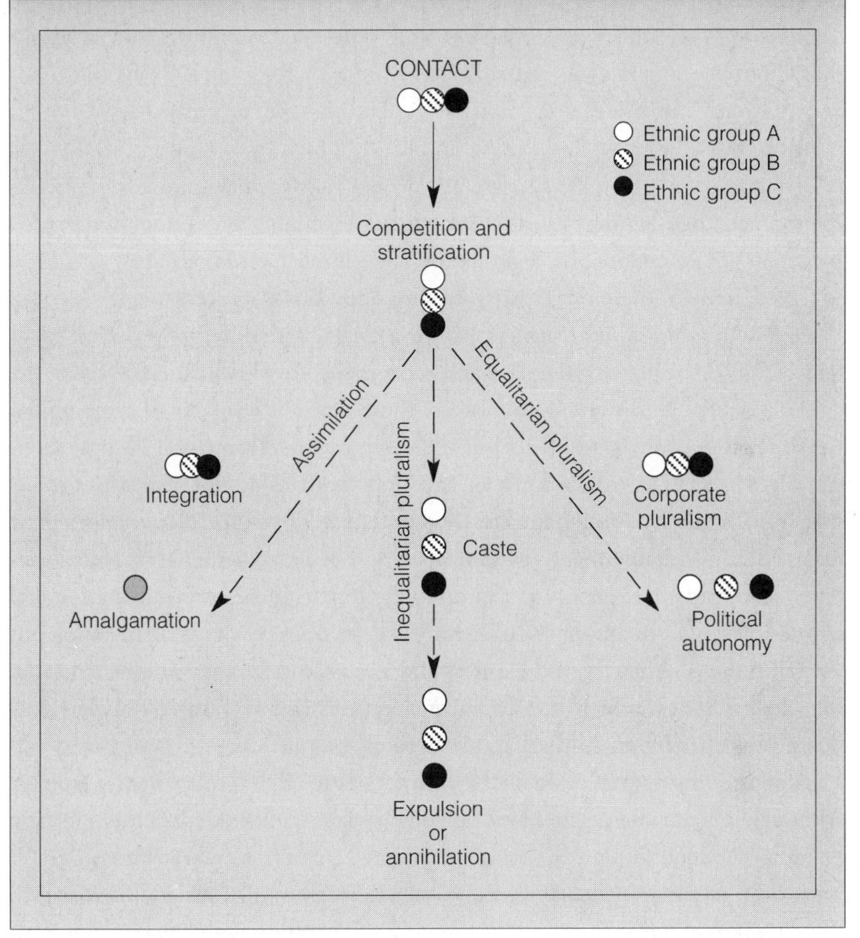

First, in no case do any of these models by themselves characterize perfectly what occurs in any multiethnic society. Realistically, most societies will exhibit features of two or possibly even all three simultaneously. The United States illustrates this quite well. As we will see in Part II, cultural assimilation has progressed steadily for most white ethnic groups so that only symbolic traces of the ethnic culture remain in evidence past the third immigrant generation. Structural assimilation has also proceeded steadily for most of these groups though more slowly and far less thoroughly. For blacks and other racial-ethnic groups, however, inequalitarian pluralism, specifically, the internal colonial model, seems more appropriate. These groups remain highly segregated socially despite substantial cultural assimilation, particularly in the case of blacks. Thus, in societies like the United States, Canada, and South Africa where there are physically distinct minorities alongside those that differ only in culture, there may be two or more sets of ethnic relations in place, requiring different theoretical models for analysis.

Second, these different processes and outcomes of interethnic relations should not be thought of as mutually exclusive, even for a single group in the society. Groups can move along two or more paths concurrently. As we noted, blacks in the United States have accomplished a high level of acculturation yet continue to display features of caste. Or groups may vacillate, moving in one direction and then another at different times, depending on various factors. As Newman (1973) points out, assimilation and pluralism are not linear.

Third, none of these three processes or conditions is irreversible. If we recall Wirth's four types of minorities (Chapter 2), each with different goals, all "contain within themselves tendencies and movements in which we can discern the characteristic features of one or more of these types" (1945:364). Group objectives may thus change at different historical moments. Traditional cyclical theories like Park's or even Gordon's assimilation variables do not deal with groups that initially resist assimilation but then accept it or groups that reassert their distance from the dominant group following substantial assimilation. Such theories cannot easily explain, for example, the shifting goals of acculturation and cultural nationalism among African Americans (Wilson, 1973) or the apparent revitalization of ethnicity in the 1960s and 1970s among white ethnic groups in the United States, which for several generations had seemingly traveled well down the path of both cultural and structural assimilation.

Fourth, some degree of conflict is characteristic of all three general patterns. Although it is obvious in the case of inequalitarian pluralism, conflict is a feature in more subdued fashion of assimilation and equalitarian pluralism as well. It must be remembered that these types of interethnic relations are *movements* in the direction of either homogeneity or ethnic separation with equality, not realized conditions. In either case, some degree of ethnic stratification will continue. Inequality among groups is present, and conflict, whether potential or active, is therefore present as well. In societies where assimilation or equalitarian pluralism is more prevalent, the ethnic hierarchy will, of course, be less inflexible and the divisions between groups less acute than in societies where inequalitarian pluralism is the predominant form.

Fifth, any of these processes and outcomes, whether a form of assimilation or pluralism, ultimately depends on the objectives of both the dominant group and minority groups. As noted in Chapter 2, minority groups have different goals regarding their place in the society and the nature of their relations with the dominant group, but this is only one side of the issue, perhaps not the most important. We must also know what goals the dominant group wishes for the minorities (Schermerhorn, 1970). When the objectives of dominant and minority groups are congruent, conflict is reduced; when they are in opposition, conflict is unavoidable.

Schermerhorn (1970) has suggested that multiethnic societies may experience centripetal and centrifugal forces in regard to dominant–minority group relations. Centrifugal forces are those that drive groups apart, whereas centripetal forces thrust them toward closer relations. In both cases, there may or may not be agreement among dominant and minority groups. An assimilationist goal of a minority group, for example, may be at odds with the segregationist goal of the dominant group. Such opposing goals have been at the crux of black–white relations for most of the past century in the United States, resulting in continual conflict of one form or another. The reverse situation may also be evident. That is, the dominant group may insist on assimilation, and the minority group may resist relinquishing its ethnic culture and social structure. The relationship between American Indians and whites has been typified by this set of incongruent goals. Indians did not seek assimilation but desired to remain on their traditional lands as independent nations. But whites responded to Indian objectives either by destroying the social and economic base on which their communities were built or by demanding assimilation into white values and lifestyles to which Indians did not aspire.

Moreover, the dominant group may use manipulation of ethnic relations for political advantage and thus dictate sharp reversals of policy. As Leo Kuper notes, "Politicians readily find in racial and ethnic cleavages a resource for political exploitation" (1969:485). Societies that advocate and display harmony and progressive assimilation among ethnic groups may revert to pluralistic policies when political conditions demand. The Nazis, for instance, determinedly reversed the quite thorough assimilation of Jews into the mainstream German society. Such policy reversals may occur in the opposite direction as well. For example, following the civil rights movement in the American South, many politicians, ordinarily staunchly segregationist, advocated moderately integrationist positions when it became obvious that the black vote might be a decisive electoral factor.

It must be understood, of course, that ethnic groups do not speak and act as one, and thus the implementation of goals and policies may not always reflect the wishes of all group members. In all cases, then, we can expect some individuals of both the dominant group and minority groups to deviate from general group processes and objectives.

In sum, all these processes and outcomes of ethnic contact are relations of power—power of minority groups and, of course, power of the dominant group. In any situation of ethnic contact, one must first ask, "Who has power and to what ends can they apply it?" On that question hinges the eventual nature of relations among groups.

A TYPOLOGY OF MULTIETHNIC SOCIETIES

In Chapter 2, we explored the manner in which multiethnic societies are stratified, noting that the degree of ethnic stratification can vary from society to society. These different levels of ethnic inequality can now be combined with the processes of assimilation and pluralism to shape a typology of multiethnic societies (Table 4.2). These descriptions are not meant to serve as pigeonholes into which particular societies can be conveniently placed. Any society will probably exhibit some features of all; in broad terms, however, it will be more characteristic of one or the other. The typology will, then, point out major features of multiethnic societies and will serve as a reference that we can consult in our analyses of specific groups and societies in Parts II and III.

Colonial Societies

Inequalitarian pluralism is the chief feature of societies that can be called colonial. A dominant group exerts maximum political and economic power and is thereby enabled to shape the nature of interethnic relations in such a way as to sustain its interests.

In these societies, the dominant group has ordinarily entered as a conqueror of physically distinct indigenous groups or has brought the minority groups to the society as slaves. Such societies are ordinarily agrarian or preindustrial, with labor-intensive economies calling for a large supply of unskilled workers; minority groups assume that role. Undergirding this exploitative system is a racist ideology in which the subordinate place of minorities is deemed natural.

Social segregation between dominant and minority groups is maximized, and the rules of interaction are explicitly defined and enforced by both tradition and law. The dominant group and minority groups develop parallel and duplicative institutions (education, religion, recreation), and only within the polity and economy do they interact with frequency. In these contacts, however, the dominant group is clearly in command.

The socioeconomic and ethnic hierarchies closely overlap in these societies so that, for the most part, defining one's ethnicity is tantamount to defining one's class. Indeed, stratification is more castelike, with members of groups automatically afforded privileges (in the case of the dominant group) or inferior status and low access to the society's rewards (in the case of minorities). There is little or no mobility for individuals or groups within this system.

Table 4.2 Three Types of Multiethnic Society

FEATURE	TYPE OF SOCIETY *Colonial*	*Corporate Pluralistic*	*Assimilationist*
Initial contact between dominant and minority groups	Conquest of indigenous groups by dominant group or involuntary migration of minorities	Annexation or voluntary immigration	Mainly voluntary immigration but involuntary immigration and conquest for salient minorities
Relations between dominant and minority groups	Paternalistic or competitive	Ideally equalitarian, but often competitive	Competitive
Nature of stratification	Caste or castelike; caste and ethnicity overlap closely	Class hierarchy within each ethnic group	Class; class and ethnicity generally overlap, but minority group members are dispersed throughout general class system
Segregation between groups	Very rigid; explicitly defined and enforced by tradition and law	Voluntarily rigid; groups ordinarily concentrated in distinct territories	Mild and largely voluntary for groups culturally and physically similar to dominant group; rigid and involuntary for salient minorities
Institutional separation among ethnic groups	High except in economy	High except in economy and central government	Low in polity and economy; variable in other areas
Physical and cultural differences between dominant and minority groups	Sharp physical differences; sharp cultural differences, at least initially	Usually slight or no physical differences; key cultural difference usually language or religion	Broad range of physical types; sharp cultural differences initially
Main objectives of ethnic policy	Inequalitarian pluralism	Equalitarian pluralism	Assimilation; some degree of unofficial structural pluralism for salient minorities
Degree of conflict among ethnic groups	High eventually though usually subdued for long periods	Relatively low except on matters pertaining to cultural and political rights	Variable but generally high between racially distinct groups
Examples	Antebellum U.S. South, colonial India, South Africa under apartheid	Switzerland, Malaysia, Canada (partially)	United States, Brazil, Israel

Given the extremely wide power differentials and the well-entrenched racist ideology, conflict usually remains submerged for long periods, and attempts to upset the system are rare and short-lived. Eventually, however, the forces of both coercion and ideology break down, and the degree of internal conflict becomes great.

The colonial societies and slave societies of the seventeenth, eighteenth, and nineteenth centuries are, of course, the prime examples of this type. In recent times, among the world's nations only South Africa has reflected some of the characteristics of a colonial society.

Corporate Pluralistic Societies

Corporate pluralistic societies are also characterized by cultural and physical separation of ethnic groups, but the key difference between these and colonial societies is the extent to which one group is dominant. In corporate pluralistic societies, the groups are relatively balanced in political and economic power, and no single one is able to exert its will on all vital societal issues. Moreover, group segregation is mainly voluntary. As Hunt and Walker describe it, "Group relationships are so ordered that the extremes of inequality found in segregation are avoided and each group feels that it is gaining some benefit from association in the common society" (1974:15).

A basic assumption on which such societies are founded is that ethnicity is not to be discouraged; on the contrary, ethnic groups are expected to retain their identity, uniting only within a common political and economic system. Indeed, the pluralistic basis of the society is usually built into the political framework. Territorial concentration of groups contributes to the retention of ethnic differences and interests.

Given the territorial clustering of groups and the relatively balanced power situation among them, conflict in these societies is ordinarily held to a minimum. Where issues of cultural or territorial integrity arise, however, competition and conflict may become quite intense. In Canada, for example, the dominant English-speaking group is not entirely comfortable with the cultural and political rights of Quebecers. This, as we shall see in Chapter 14, has created constant tension between the two groups, at times subdued and at other times heightened.

Assimilationist Societies

The assimilationist society differs from either of the first two types in that there is no recognized obligation or objective in protecting the retention of ethnicity. If ethnic groups survive, they do so because of voluntary actions by the groups themselves or because of informal patterns of prejudice and discrimination, not through the designs of political institutions, as in a corporate pluralistic society, or the segregationist dictates of a dominant group, as in a colonial society.

In these societies, there is a dominant group and a large number of minority ethnic groups in various stages of cultural and structural assimilation. The dominant group encourages cultural assimilation — to its ways — and most groups retain only expressive aspects of their ethnic cultures beyond two or three generations. Structural assimilation, however, is a different matter. Much social interaction, particularly at the primary level, continues to take place within ethnic subsocieties. Highly visible ethnic groups are most segregated in social relations and thus least structurally assimilated. Those closer to the dominant group in culture and physical appearance are in more advanced stages of structural assimilation, including intermarriage.

Ethnic relations are competitive. Groups compete in the political and economic arenas for power and jobs and in other areas of social life as well.[9] The nature of socioeconomic stratification is class, not caste, and members of various ethnic groups can thus rise (or fall) on their individual merits, at least theoretically. In reality, ethnic groups tend to cluster at certain points in the socioeconomic and political hierarchies so that class and ethnicity overlap to a greater extent than the society's ideology may proclaim. Indeed, for some groups, namely, racial-ethnic groups, castelike features of stratification are often apparent.

In assimilationist societies, a racist ideology explaining the superiority of the dominant group is more restrained in expression and is not accepted by minority group members. Any claims to its legitimacy therefore create conflict.

As a result of the more open and competitive stratification system and the lack of an officially sanctioned racist ideology, conflict among ethnic groups may be more prevalent in assimilationist societies than in either of the other types. In the colonial society, people "know their place" and stick to it; in the corporate pluralistic society, conflict is held in check by a relative balance of political and economic power. But in the assimilationist society, minority ethnic groups compete with the dominant group for social positions and power. In this competition, of course, the dominant group retains substantial power resources and thus great advantages.

SUMMARY

What is the nature of relations among different groups in a multiethnic society? Although intergroup conflict seems intrinsic, groups follow one of two general

[9]Many of the features of assimilationist societies are described by van den Berghe (1978) as a type of competitive race relations.

paths: increasing integration or increasing separation. The former is called *assimilation,* and the latter, *pluralism.* Each is a process through which groups pass as well as an outcome of group relations.

Assimilation can be viewed at four distinct but related levels: cultural, structural, biological, and psychological. For our purposes, emphasis is placed on the first two. *Cultural* assimilation is the adoption by one ethnic group of another's (usually the dominant group's) cultural traits. *Structural* assimilation is the increasing social interaction between different ethnic groups at both the primary and secondary levels.

Pluralism occurs in two different forms in modern multiethnic societies, *equalitarian* and *inequalitarian.* In the former, groups retain their cultural and much of their structural distinctness but participate on an equal basis in a common political and economic system. In the latter, ethnic groups are also structurally separated, but they do not participate in the society's political and economic institutions in an equitable manner.

Multiethnic societies can be placed in a typology on the basis of several characteristics: the form of contact that brought the different groups together, the character of relations among them, the main objectives of the society's ethnic policies, and the degree of conflict among groups. *Colonial* societies are those in which groups meet through conquest or involuntary migration of minorities. Paternalistic relations within a castelike stratification system are in effect, and segregation between groups is rigid. The aim of the dominant group is inequalitarian pluralism, and conflict between groups may become intense. In *corporate pluralistic* societies, annexation of territory or voluntary migration brings previously separate groups together. Segregation between them is high, usually in distinct areas, though it is mainly voluntary. Equalitarian pluralism is the basis of the society's ethnic policies, and intergroup conflict is minimized. *Assimilationist* societies emerge mainly from voluntary immigration of groups, though other contact situations may be apparent as well. The degree of segregation between groups varies on the basis of cultural and physical visibility, as does the level of institutional separation and intergroup conflict. Assimilation is the long-range societal objective.

Suggested Readings

Banton, Michael. 1983. *Racial and Ethnic Competition.* Cambridge: Cambridge University Press. Explores the competitive relations among racial and ethnic groups using different societies as illustrative cases. Proposes an application of "rational choice theory," commonly used by economists, to race and ethnic relations.

Blalock, Hubert M., Jr. 1967. *Toward a Theory of Minority-Group Relations.* New York: Wiley. Constructs theoretical propositions that explain competition and conflict between dominant and minority groups.

Gordon, Milton M. 1964. *Assimilation in American Life: The Role of Race, Religion, and National Origins.* New York: Oxford University Press. The most complete explanation of ethnic assimilation in its various forms. Presents a theoretical model that remains influential in the study of race and ethnic relations.

Horowitz, Donald. 1985. *Ethnic Groups in Conflict.* Berkeley: University of California Press. Examines the sources of ethnic conflict in multiethnic societies, and the possibilities for reducing it, with special emphasis on the developing nation-states of Asia and Africa.

Montville, Joseph V. (ed.). 1991. *Conflict and Peacemaking in Multiethnic Societies.* New York: Lexington Books. A collection of essays by social scientists that explores the varieties of conflict in multiethnic societies in different parts of the world today and the prospects for controlling ethnic strife.

Schermerhorn, R. A. 1970. *Comparative Ethnic Relations: A Framework for Theory and Research.* New York: Random House. Not an easy book to read, but a valuable theoretical work that tries to analyze race and ethnic relations with a set of universally relevant principles.

van den Berghe, Pierre. 1978. *Race and Racism: A Comparative Perspective.* 2d ed. New York: Wiley. A brief, but significant, work that presents a theoretical framework and a useful typology for analyzing race relations. Four multiracial societies are compared: the United States, Mexico, Brazil, and South Africa.

PART

II

ETHNICITY IN THE UNITED STATES

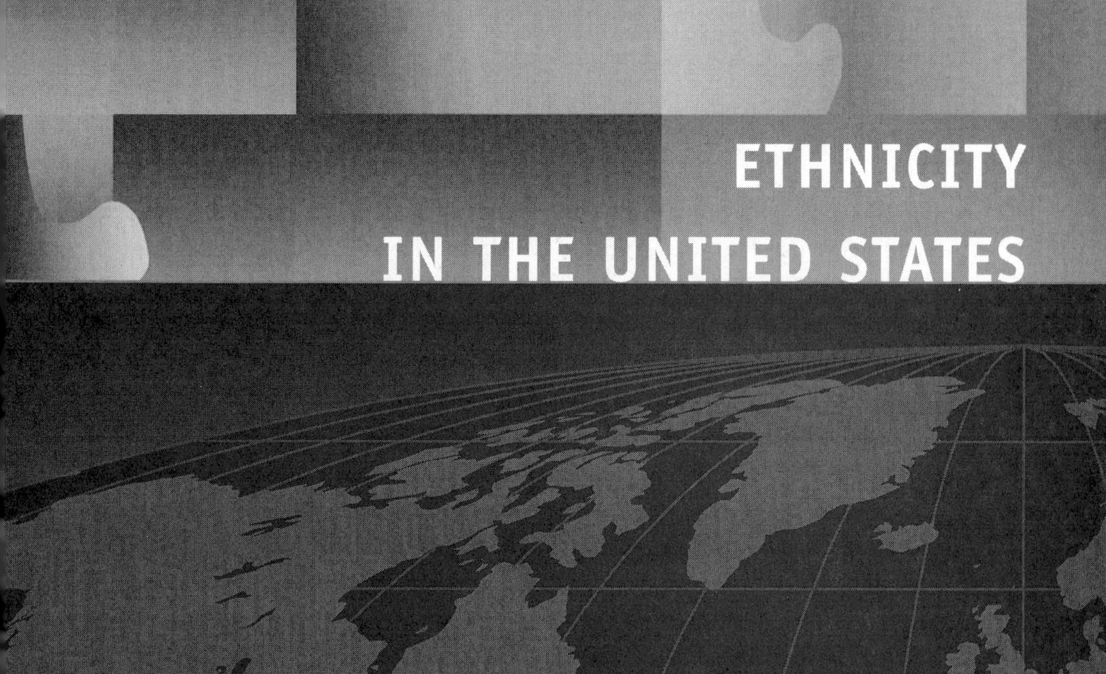

Now that the basic concepts and theories of ethnic relations have been outlined, we will investigate several contemporary multiethnic societies. In Part II, our focus is on the United States. So ethnically diverse is American society that many volumes would be required even to begin to describe its various groups and the relations among them. For this reason, several groups, each of which is broadly representative of a particular American ethnic experience, are dealt with in depth. In Chapter 5, the two initial groups of the American ethnic hierarchy, white Anglo-Saxon Protestants and American Indians, are discussed. Separate chapters are then devoted to Italians, Jews, African Americans, Latinos, and Asians. These five (the latter two are really broad categories that each include a number of distinct ethnic groups) represent a range of American ethnic groups in their immigration and settlement patterns, their past and present place in the society's stratification system, the extent to which they have been targets of prejudice and discrimination, and the degree to which they have assimilated into the mainstream society.

As we have discussed, the manner in which an ethnic group enters the society has enormous implications for its subsequent status and the character of relations between it and the dominant group. Those who enter involuntarily, who are conquered, or whose territory is annexed remain in a subordinate position for many generations and maintain a high level of conflict with the dominant group. Relations between the dominant group and subordinate groups who enter voluntarily, however, are less conflictual and result in more rapid and complete assimilation. The American cases examined in the following chapters demonstrate this principle quite clearly. Blacks entered mostly as slaves, Indians were conquered, Mexicans in the nineteenth century were absorbed through the annexation of their land, and Europeans and Asians came as voluntary immigrants. Except for the Asians, whose physical mark proved a more enduring handicap, the voluntary immigrants progressed steadily toward assimilation, though at noticeably different rates. As we will see, conflict with the dominant group for them was often profound, but their place in the society was not questioned after a generation or two. Dominant–minority relations have been far more tortuous and conflictual for the groups entering involuntarily.

A second key factor accounting for differences in group placement in the stratification system and the severity of conflict with the dominant group is, as we have seen, physical and cultural visibility. The more distant the group is in appearance and behavior from the dominant group, the lower its social standing will be and the more intense the hostilities it must confront. American ethnic groups display a vast range of cultural and physical traits, and the historical patterns of conflict and accommodation among them are very clear. The highest levels of prejudice and discrimination and the greatest resistance to the attainment of political, social, and economic equality with the dominant group have been experienced by racial-ethnic groups—blacks, Indians, Asians, and to some extent, Latinos.

Even some European groups were racially perceived when they entered the society—namely, those from southern and eastern Europe—and their treatment for a time, especially for the immigrant generation, was similar in form to that afforded nonwhites. But how they were perceived changed as their social position improved. As sociologist Herbert Gans has so aptly observed, "Although Italians, Jews, and other participants in the 'new immigration' were treated as dark races when they first came here, their swarthy complexions turned, or were perceived to turn, whiter as their incomes rose" (1982:291). Similar perceptual changes have been far slower for nonwhite groups.

The analysis of each group in the following chapters will be organized around four key issues: the group's sociohistorical development as a minority, its status in terms of several measures of stratification, current and past patterns of prejudice and discrimination directed against it, and the extent to which pluralist and assimilationist models are applicable to it.

CHAPTER

5

FOUNDATIONS OF THE AMERICAN ETHNIC HIERARCHY: ANGLO-AMERICANS AND NATIVE AMERICANS

Historian Oscar Handlin professed that when he began to write a history of immigrant groups in the United States, he discovered that the immigrants *were* American history. With the exception of American Indians and some Mexican Americans, all ethnic groups in the United States trace their origins to other societies. But the manner in which they entered the United States, the nature of the dominant group's response to them, and their eventual placement in the ethnic hierarchy varied enormously.

In this chapter, we trace the formation of the United States as a multiethnic society, concentrating in some depth on its two initial ethnic groups, American Indians and white Anglo-Saxon Protestants.

THE ESTABLISHMENT OF ANGLO DOMINANCE

Before the 1830s, Americans were a relatively homogeneous people in terms of national origin, religion, and physical type. By the third decade of the nineteenth century, the indigenous peoples had been reduced to subordination, and the only other significant non-European group was black slaves who had been brought to the society beginning in the late seventeenth century. Because of their cultural, but especially their physical, distinctness from the European settlers, blacks and American Indians (as well as Mexicans in the Southwest) were, from the first, relegated to the bottom of the emerging ethnic hierarchy; their position would not basically change afterward.

Except for the Native and slave populations, then, the vast majority of Americans in the early nineteenth century were of English and Protestant background. Other, smaller, groups were mainly northwestern Europeans, including Scots, Welsh, Scotch-Irish, Dutch, Scandinavians, and German Protestants, all of whom were culturally and racially close to this charter, or core, group and who, as a result, were relatively quickly absorbed into it. Thus, although the English were clearly its majority component, the core group represented a blend of relatively similar cultural and racial elements. It was this group, the defining ethnic features of which were its northwestern European and Protestant origins, that became the host group and that subsequently set the tone of the society and established its major economic, political, and social institutions.

All subsequent groups would be required to adapt to this Anglo-Protestant social and cultural framework. It became the standard, notes historian Arthur Schlesinger, Jr., "to which other immigrant nationalities were expected to conform, the matrix into which they would be assimilated" (1992:28). The establishment of a dominant ethnic group in American society with the power to select those who came after it is vividly described by Glazer and Moynihan:

> The original Americans became "old" Americans or "old stock," or "white Anglo-Saxon Protestants," or some other identification which indicated they were not immigrants or descendants of recent immigrants. These original Americans already had a frame in their minds, which became a frame in reality, that placed and ordered those who came after them. It was important to be white, of British origin, and Protestant. If one was all three, then even if one was an immigrant, one was really not an immigrant, or not for long (1970:15).[1]

Although the Anglo core group has been diluted by successive immigrant waves over many generations and its cultural imprint likewise colored by other groups, its social dominance has remained unwavering throughout American history. "The white Anglo-Saxon Protestant," notes sociologist Lewis Killian, "remains the typical American, the model to which other Americans are expected and encouraged to conform" (1975:16). The makeup of the dominant group itself, however, has changed somewhat as other white Protestant groups have melded with it. Today, the dominant ethnic group in American society may be said to broadly comprise white Protestants because those Protestants from

---

[1]Nathan Glazer and Daniel P. Moynihan, *Beyond the Melting Pot,* 2d ed.

other northwestern European societies, namely, the Scandinavian nations, Germany, and the Netherlands, have blended almost imperceptibly with those of British origins. White Protestants, then, have varied national roots, but their common Protestantism and racial character have neutralized any meaningful national differences among them. References to the WASP, Anglo, or Anglo-American group, then, should be understood to mean "white Protestants of various national origins." These terms are used interchangeably to denote the dominant American ethnic group and its core culture.

Some might question whether white Protestants constitute an ethnic group in the sense that other more culturally or physically distinct groups do. On this point, Charles Anderson has noted that "as Catholics, Jews, and blacks press in around them, and as internal differences decline, white Protestants increasingly perceive that they, too, constitute a definable group with distinctive social and familial networks, and psychological and cultural moorings" (1970:xiii-xiv).

Let us look at how ideologies and public policies concerning American ethnic relations have historically reflected the cultural preferences and the economic and political interests of the WASP, or Anglo, core group.

Dominant Cultural Values

The preeminence of Anglo cultural values has consistently underlain public policies in education, language, law, welfare, and even religion. Thus, the ascendancy of the English language, the English legal system, and, with few exceptions, the Christian faith, was never seriously challenged. From the beginning, the expectation held sway that entering groups—immigrant, conquered, or enslaved—would conform to this core culture.

At certain times, especially early in this century when ethnic diversity seemed to reach a high point, the idea of the *melting pot* called into question the inevitability as well as the desirability of Anglo cultural dominance. Some believed that the cultural differences among the many immigrant groups, so conspicuous at that time, might somehow disappear through a gradual blending process. Different ethnic cultures, it was felt, would eventually fuse into a single "American" culture as a kind of hybrid creation. Although the notion of a melting pot remained symbolically popular for many decades, it never found manifestation in public policy or gained widespread allegiance. In any case, the idea was incomplete because the place of nonwhites in this ideal social brew was never fully dealt with. The melting pot was belied as well by the fact that the boundaries among the European groups did not disappear, nor were their cultures ever entirely diluted by WASP norms and values.

Another competing ethnic ideology that has gained prominence at times is *cultural pluralism,* the notion that various ethnic cultures and communities should be tolerated and protected within a system of political equality. As explained in Chapter 4, its objective, rather than a fusion of diverse ethnic groups, is the preservation of each. This ideology, too, has never found more than token acceptance at the level of public policy.

A more prevalent ethnic ideology, one expressed most forcefully in governmental policies in the United States, is assimilation into the dominant group, or what Gordon (1964) has called *Anglo-conformity.* Groups are expected to shed their ethnic uniqueness as quickly and as completely as possible and take on the ways of the core culture. This expectation always guided the prevailing social thought and policy regarding new ethnic groups and continues to do so even today despite the pluralistic rhetoric of recent years.[2]

Dominant Economic and Political Interests

The prevalent social thought and policy regarding ethnic groups has most commonly been an expression of the dominant group's economic and political interests. Thus, pressures on ethnic groups to assimilate culturally have not been complemented by corresponding pressures to assimilate structurally, that is, to enter into full and equal relations with the dominant group in all institutional areas. Although Anglo-conformity has generally been expected of ethnic groups in their behavior and appearance, a pluralistic outcome has more often than not typified the dominant group's expectations of interethnic relations. The more distinct a group has been, the more its members have been encouraged to remain "among their own" in the intimate areas of social life—marriage, residence, social clubs, and the like (see Table 5.1).

Most important, the efforts of minority ethnic groups to attain economic and political equality have commonly been met with resistance. Through immigration quotas and exclusionary measures, Indian-removal acts, slave laws, institutionalized segregation, antilabor regulations, voting restrictions, and an array of other measures, dominant interests have traditionally been protected. Throughout American history, of course, these policies have all generated great controversy and conflict, and minority challenges have often met with success. But concessions wrested from the dominant group have always been slow, costly, and incremental.

In sum, for most of American history, public policies and publicly proclaimed ideologies have generally corresponded to the dual aims of, on the one hand, cultural assimilation (specifically, Anglo-conformity) and, on the other, structural pluralism, the latter in a form assuring Anglo dominance. These objectives

---

[2]A complete discussion of the three belief systems regarding interethnic relations in the United States—melting pot, cultural pluralism, and Anglo-conformity—is found in Gordon (1964).

Table 5.1 Degree of Assimilation of American Ethnic Groups			
GROUP	CULTURAL ASSIMILATION	SECONDARY STRUCTURAL ASSIMILATION[a]	PRIMARY STRUCTURAL ASSIMILATION[b]
Anglo-Protestants (core group)	High	High	High
Northwestern Europeans	High	High	High
Germans	High	High	High
Irish Catholics	High	High	Moderate
Southern and Eastern European Catholics	Moderate	Moderate	Moderate
Jews	Moderate	Moderate	Moderate
Asians	Moderate	Moderate	Low
Hispanics	Moderate	Low	Low
American Indians	Moderate	Low	Low
Blacks	High	Low	Low

[a]As indicated by entrance into economic, political, and other social institutions at various levels.
[b]As indicated by entrance into primary relations with the dominant group in areas such as residence, club memberships, and intermarriage.

have reflected the American system of ethnic stratification. Those groups least visibly different from the Anglo core group were received with less hostility and were presented with an opportunity structure more open and less limited than were those unmistakably nonwhite and non-Protestant.

The American Ethnic Hierarchy Today

The analysis that follows in this and the next six chapters suggests that the American ethnic hierarchy today can be viewed broadly as divided into three comprehensive ranges:

1. The top range comprises white Protestants of various national origins, for whom ethnicity has no real significance except to distinguish them from the remainder of the ethnic hierarchy.
2. The intermediate range comprises white Catholics of various national origins, Jews, and many Asians, for whom ethnicity continues to play a role in the distribution of the society's rewards and continues to influence social life, but in both instances, decreasingly so.

3. The bottom range comprises racial-ethnic groups — blacks, Hispanics, American Indians, and some Asians — for whom ethnicity today has the greatest consequences and for whom it continues to shape the basic aspects of social life.

This ethnic hierarchy has remained amazingly resilient for the past century and a half. Although the distance between many of the groups has been reduced, their rank order has, with few exceptions, not been basically altered.

Arranging ethnic groups in such a rank order, of course, masks the more specific class and cultural differences among them as well as the internal differences within each. But in a very general sense, each of the country's ethnic groups can be placed within one of these ranges on the basis of a combination of the factors at which we will look: the collective place of the group in the society's economic and political hierarchies; the extent to which prejudice and discrimination remain significant for the group; and the extent to which group members have entered into full social participation with those of the dominant group.[3]

Perhaps the most important aspect of this three-part American ethnic hierarchy is that the gap between the bottom range of groups and the other two is much greater than the gap between the top and intermediate ranges. Thus, except for Asians and some Hispanics, the American ethnic system today seems to be increasingly dichotomized — white and nonwhite. The overriding issues of American ethnic relations, therefore, remain focused on the economic, political, and social disparities between whites and nonwhites and the policies intended to reduce them. The crux of the matter is relatively simple: Will the gap between whites and nonwhites continue in its present form, will it be reduced substantially, or will it perhaps widen? The issues of American race and ethnic relations are largely subsumed by this pivotal question.

NATIVE AMERICANS: THE FIRST ETHNIC MINORITY

Conquest and Settlement

When Europeans crossed the Atlantic beginning in the late fifteenth century, they found two continents inhabited by perhaps 75 million people (Thornton,

[3]Using a combination of these factors means that some will offset others. For example, although Jews and Asians rank higher in their aggregate economic class position than northwestern Europeans, they still do not fully interact at the primary level with those groups, and the extent of prejudice and discrimination directed against them is much greater. Hence, when all these factors are considered, Jews and Asians are part of the intermediate range of groups rather than the top.

1995).[4] These "Indians," as the Europeans called them, were by no means homogeneous but were practicing a vast array of cultures, some quite advanced in social organization and technology. Different they were from the Europeans, however, and these physical and cultural differences were interpreted ethnocentrically by the white invaders as proof of inferiority and lack of civilization.

Evidence of the Indian presence in North America extends back at least to 35,000 B.C. Estimates of the Indian population north of the Rio Grande at the time of Columbus's arrival range from one million to eighteen million, organized into hundreds of tribes or nations (Thornton, 1995; Wax, 1971). Like the more numerous natives of what is today Latin America, these peoples spoke many different languages, conformed to disparate cultural systems, and were organized in a variety of economic and social forms. There were, then, no "American Indians" per se but rather a great number of diverse ethnic groups, considering themselves to be as different from one another as they were from the European invaders (Lurie, 1991).

The establishment of European supremacy over the Indians conforms closely to the theories of the emergence of dominance and subordination explained in Chapter 2. Indians are the one group in the American ethnic picture whose subordination came about wholly through conquest. As we noted, an indigenous group invaded by a more powerful settler group can only capitulate or, more commonly, resist. This ordinarily makes for a high level of conflict and rejection of assimilation into the dominant group.

The four major conditions for the emergence of ethnic stratification—contact, ethnocentrism, competition, and differential power—are clearly seen in the case of Indian–white relations at the outset of European settlement and afterward.

The Native Americans' physical and cultural distance from European standards inevitably produced an ethnocentric white view. Of particular importance to the North American settlers was that the Indians were "heathen." Moreover, they were preliterate and at a technological level far below that of the Europeans. A combination of these traits contributed to the image of Indians as savage and inferior and, consequently, to the rationale for their exploitation and annihilation.

Contact and even ethnocentrism among alien ethnic groups will not in themselves create dominant–subordinate relations, however. Competition over valued resources is a third factor, and in the case of Indian–white relations, land provided the necessary ingredient. The objective of the settlers was to secure land for agricultural production. From the outset, divergent cultural perceptions of "property" led to misunderstandings and hostilities between Indian and

---

[4]The number of indigenous people inhabiting North and South America at the time of European discovery is disputed among anthropologists and historians. Estimates range from less than ten million to more than 100 million (Utter, 1993).

settler. Indian lands were held communally, not as private property. Whites consequently perceived them as basically unoccupied. Furthermore, that most Indians did not engage in European-style farming led to the view that these lands were underutilized and thus justifiably occupied by those who would cultivate them in a "proper" manner.

In the end, however, it was the differential power between Indians and settlers that proved the crucial element in the establishment of dominance and subordination. Given their superior arms, it was inevitable that whites would prevail in the struggle for resources. Indian lands were ceded irreversibly by treaties, usually entered into by Indians under duress or in ignorance of their meaning, through fraudulent schemes perpetrated by whites, or by sheer force. Where there was desirable land, whites eventually took it. Treaties themselves provided no protection of those lands left to the Native Americans because renegotiations and violations became standard practice whenever desired land was in question.

Displacement of Indian tribes was underwritten by government edict, and by the late nineteenth century, most of the remaining Native population had been forcibly resettled in reservations west of the Mississippi. It was not coincidental that reservations were located on lands deemed by whites virtually worthless for farming or grazing.

Indian societies were reduced physically through a combination of exposure to European diseases, armed conflict, starvation, and the breakup of cultural systems that had traditionally provided for social and material needs. Though Indians did make many cultural adaptations to the white presence, this did not prevent the eventual destruction of their social forms and the succession of white dominance (Wax, 1971). As for diseases, perhaps nothing more severely impacted the physical condition of Indians than the epidemics brought by Europeans, which literally decimated many tribes from the very outset of their contact with settlers and explorers. These continued, repeatedly, through the nineteenth century (Thornton, 1995; Washburn, 1975). Native populations lacked immunity to diseases like smallpox, measles, cholera, typhoid, malaria, and others, and their effects as a result were devastating.

Stephen Cornell (1988) has suggested that the "Indian problem" for Euro-Americans from the time of settlement to the present has had three aspects: economic—how best to secure Indian resources, especially land;[5] cultural—how best to assimilate Indians into the dominant, non-Indian, culture; and political—how best to control Indians so as to bring about solutions to the first two problems. The counterpart to the Indian problem was a "Euro-American

---

[5]Cornell (1990) explains that at the outset of contact between Indians and Europeans, it was not land but other resources, namely furs, that Europeans sought. From the late eighteenth century on, however, land was the major objective.

problem" for Indians. "In its essence," notes Cornell, "this problem seems to have been tribal survival: the maintenance of particular sets of social relations, more or less distinct cultural orders, and some measure of political autonomy in the face of invasion, conquest, and loss of power" (1988:7). It is around these two conflicting problems that Indian–white relations have revolved for the past three centuries.

Government Policies Toward American Indians

From the beginning of European settlement to the end of the eighteenth century, Indian–white relations centered primarily on the fur trade. Through the trade, Indians were integrated into the emergent North American economy, and as a result, their forms of social organization and cultural institutions were radically changed.

At the start of the nineteenth century, Indian–white relations evolved into a confrontation in which the insatiable quest of white settlers for land led to ever-greater efforts to dispossess Indians of lands they were occupying. Two paths to this objective became apparent: Indians would be separated from their lands either through assimilation, that is, by encouraging them to abandon their communal patterns of landholding in favor of private property, or through removal, that is, relocating Indians through negotiated purchase of land or through conquest (Cornell, 1988). These two paths were not mutually exclusive; rather, they were complementary, not only in a practical sense but also as a way of satisfying the objectives of settlers—who wanted land—and reformers—who sought to "civilize" the Natives with American culture. This dual objective, notes historian Robert Berkhofer, was a constant pattern throughout most of American history: "From the founding of the nation until recent times, and some would include today as well, United States policy makers placed two considerations above all others in the nation's relation with Native Americans as Indians: the extinction of native title in favor of White exploitation of native lands and resources and the transformation of native lifestyles into copies of approved White models" (1979:135).

Indian tribes were dealt with as sovereign nations, and the U.S. government entered into treaties with them accordingly. The Indian Removal Act of 1830 was perhaps the most devastating single action taken by the federal government in destroying Indian cultures and societies. The act called for the relocation of all tribes living in the eastern United States to lands set aside for them west of the Mississippi River. This policy was pursued vigorously by President Andrew Jackson, who represented the sentiments of white settlers who saw Indian occupation of desired lands as an impediment to the development of white dominance. Removal was rationalized by the view that Indians were not using land as "God had intended," that is, to farm in the way of the white man. As Governor William Henry Harrison asked, "Is one of the fairest portions of the globe to remain in a state of nature, the haunt of a few wretched savages, when it seems destined by the Creator to give support to a large population and to be the seat of civilization?" (quoted in Hagan, 1961:69). This ten-year movement marked a time of great hardship and suffering for Indians and resulted in the virtual devastation of many tribes, particularly those that resisted removal. Perhaps the most infamous episode involved the forced migration of the Cherokees in 1838, an affair referred to as the "Trail of Tears." In the summer of that year, the U.S. Army rounded up all 16,000 Cherokees and held them for months in disease-infested camps. The march west to Oklahoma extended over the autumn and winter, and by its end, 4,000 had died (Wright, 1992). As one historian has described it, the removal of Indians, at best, "resembled closely the pioneering experience of thousands of their white contemporaries" and, at worst, "approached the horrors created by the Nazi handling of subject peoples" (Hagan, 1961:77).

By the latter part of the nineteenth century, armed resistance to federal policy had ended, and government officials and reformers now advocated a policy that they felt would once and for all solve the Indian problem: Reservation lands would be broken up and allotted to individual tribal members. Unallotted lands would then be sold to non-Indians. This policy was formalized in 1887 with congressional passage of the Dawes Act (also called the General Allotment Act). As a result of the policy of allotment, those lands remaining in Indian possession were further diminished, leaving Indians with even fewer productive resources. At the start of the twentieth century, Indians were impoverished and virtually at the mercy of the federal government, whose paternalistic policies continued to reflect white ethnocentrism. By the end of the century, the Indian population had been reduced to fewer than 250,000. Indians were now officially wards of the federal government, having been awarded full citizenship under the Dawes Act, and were thus no longer seen or dealt with as separate nations.

In the 1930s, an effort was made to reverse the decades of neglect and exploitation that had characterized government policies vis-à-vis American Indians. It was also an attempt to create a more pluralistic system of Indian–white relations. John Collier, enlightened and sympathetic toward Native causes, was appointed to head the Bureau of Indian Affairs (BIA) during President Franklin D. Roosevelt's first term. Collier proposed major changes in Indian policy that included tribal self-rule and efforts to encourage Native cultural preservation. Most important was his proposal to end the allotment system created by the Dawes Act, which had resulted in the loss of much Indian land. Congress passed the Indian Reorganization Act (IRA) in 1934, which included a number of Collier's recommendations. It was not met with enthusiasm by all Indians, however. Some were committed to assimilation and others were not prepared to accept at face value any measure offered by the federal government. Still others saw the act as simply one more form of paternalism. Most tribes, however, approved the Act. The IRA, though failing to fundamentally alter the status and condition of Native Americans, did result in halting some of the more egregious aspects of past Indian policy. It prohibited further allotments of tribal lands and encouraged Indians to establish self-governing systems within their tribes. Moreover, Collier reduced the heavy-handed control of Indian life by the BIA and reduced efforts aimed at forced cultural assimilation.

Once again, in the 1950s, a reversal in the government's approach to Indian affairs emerged with a new policy known as "termination." In essence, the objective now was, once again, assimilation. Moreover, the tribal organization of Indian life was seen as a relic of the past and it was felt that Indians now should be dealt with on an individualistic, not a communal, basis. The idea was that Indians should be treated no differently from other citizens. To accomplish this, it was proposed that the reservation system be dismantled and the federal government's unique role toward Indians be severed. No longer would treaty obligations be recognized, and the sovereign status of tribes would be eliminated. Most Indians strongly resisted termination, realizing that the ultimate result would be the loss of tribal lands as well as Native cultures. The failure of termination led to a return to a policy more in line with the IRA during the next three decades. Reversing its position again, federal policy toward American Indians in the 1970s moved in the direction of greater autonomy for tribal governments. In 1974, the Indian Self-Determination and Education Assistance Act was passed by Congress, permitting tribes to take control of numerous federal programs on reservations.

In the 1980s, the Reagan administration made severe cuts in federal assistance to American Indians. As with all other federal agencies, the Reagan approach was to remove the federal government from activities and turn these over to the private sector. In 1983 alone, aid was cut by more than one-third, affecting programs on all reservations (Worsnop, 1992).

Recent politics of American Indians have focused primarily on political self-determination and on the protection of remaining Indian lands. Indians have, in the past two decades, generally experienced greater empowerment than at any time since the nineteenth century. Much of this power has been the result of the leverage that tribes have been able to exert in their control of valuable minerals and other resources that are now part of Indian-controlled lands. Increasingly they have made effective legal challenges to what they have interpreted as incursions into their lands. Moreover, many tribes, as we will see, have become adept at business affairs and have developed numerous enterprises on reservations.

A certain theme seems to run through the history of government policy toward Native Americans. At all times, whatever the issues or whichever administration formulated policy, Indians themselves lacked the political power to defend their interests. They were always reactors to, never initiators of, policies affecting their welfare. At certain times, the object of government measures was to assimilate Indians to the cultural ways of whites, whereas at other times, Indians were seen as unassimilable and thus to be left alone. In either case, with few exceptions, gross neglect typified the attitudes of policymakers, and the consequences of that neglect have been felt by each succeeding generation of Native Americans.

American Indian Demographics Today

It is difficult today to measure with accuracy the Indian population because definitions of who is "Indian" are not clear and unequivocal. As with other ethnic groups, the U.S. census counts anyone an Indian who declares himself or herself so. Within the Native-American population itself, Indian identity is left to each tribe to define. Thus a variety of different criteria may be used in establishing a person's tribal membership. Often blood quantum will be a qualification, but in some cases, one need only trace his or her lineage to earlier tribal members (Thornton, 1987).

Despite the problems in accurately determining the Indian population, there is little doubt that Indians are a relatively young and growing ethnic category. The "official" Indian population today (including Alaska natives) is 2 million, which is less than 1 percent of the total U.S. population. Nonetheless, this represents a 38 percent increase over the 1980 recorded Indian population of 1.4 million and a quadrupling from 550,000 in 1960.

Much of this increase is accounted for by a high birthrate among American Indians but also in some part by a growing tendency for people to declare their Indian affiliation. By the late 1960s, "Indianness" had risen in popularity. As a result of the Red Power movement, concomitant with the black civil rights movement, Indian activism inspired new pride among Native Americans, as did the more favorable portrayal of Indians by the mass media (Nagel, 1995). In addition to the new positive public perception of this group, many now claimed Indian ancestry in order to be included in lucrative judgments being awarded to tribes in land claims cases. "Every announcement of a large judgment," notes historian William Hagan, "seemed to trigger the memories of some Americans that their family trees included an Indian, usually a chief's daughter, a princess" (1992:284). Declaring oneself "Indian," many discovered, also resulted in eligibility for various federal assistance programs, including health and education. Indian identity, then, remains problematic, with no systematic way in which the validity of genuine Indianness can be verified. This is hardly a recent problem, however, extending well back in American history (Hagan, 1992).

Table 5.2 Largest American Indian Tribes

TRIBE	NUMBER	PERCENT DISTRIBUTION
Cherokee	308,132	16.4
Navajo	219,198	11.7
Chippewa	103,826	5.5
Sioux	103,255	5.5
Choctaw	82,299	4.4
Pueblo	52,939	2.8
Apache	50,051	2.7
Iroquois	49,038	2.6
Lumbee	48,444	2.6
Creek	43,550	2.3

SOURCE: U.S. Bureau of the Census, 1995b.

Today there are more than 300 Indian tribes in the United States that are officially recognized by the federal government and over 100 others that are seeking official recognition. Not all American Indians, however, are actually enrolled members of these tribes. Table 5.2 lists the ten largest American Indian tribes.

The American Indian population is split between those living on and those living off the 278 federal Indian reservations (U.S. Department of the Interior, 1991). The largest of these reservations is the Navajo, covering 16 million acres of land in Arizona, New Mexico, and Utah, with a population of 143,000. Most other reservations, however, are much smaller. The second largest is the Pine Ridge reservation in South Dakota and Nebraska, with a population of only 11,000.

Most nonreservation Indians live in urban areas. Indeed, the movement of American Indians to cities has been a continuous process. Federally sponsored urban relocation programs begun during the 1950s impelled much of this movement. While urban Indians are less impoverished than reservation Indians, they remain concentrated near the bottom of the class system, experiencing high rates of unemployment and difficulty in adjusting to urban economies that no longer can absorb workers with low skills and education. The urban migration of American Indians is not likely to decline and thus will continue to render profound effects on social and cultural aspects of Indian life.

Tables 5.3 and 5.4 show those states and metropolitan areas where Indians are most populous. As is evident, Indians are concentrated in a few states, mostly in the West. The largest populations are in California, Oklahoma, Arizona, and New Mexico.

Table 5.3 States with the Largest American Indian, Eskimo, or Aleut Populations

STATE	NUMBER
California	277,000
Oklahoma	276,000
Arizona	251,000
New Mexico	159,000
Alaska	102,000
Washington	101,000
North Carolina	90,000
Texas	67,000
South Dakota	62,000
Michigan	61,000
Minnesota	58,000
New York	57,000
Montana	55,000

SOURCE: U.S. Bureau of the Census, 1995b.

Table 5.4 Metropolitan Areas with the Largest Indian Populations

METROPOLITAN AREA	INDIAN POPULATION	PERCENT OF TOTAL
Los Angeles–Anaheim–Riverside	87,000	0.6
Tulsa	48,000	6.8
New York–Northern New Jersey–Long Island	46,000	0.3
Oklahoma City	46,000	4.8
San Francisco–Oakland–San Jose	41,000	0.7
Phoenix	38,000	1.8
Seattle–Tacoma	32,000	1.3
Minneapolis–St. Paul	24,000	1.0
Tucson	20,000	3.0
San Diego	20,000	0.8

SOURCE: U.S. Bureau of the Census, 1991a.

Table 5.5 The Socioeconomic Status of Native Americans

	NATIVE AMERICANS	TOTAL POPULATION
Education		
% high school graduate	65.6	77.6
% college graduate	9.4	21.3
Median family income	$21,619	$39,066
% families in poverty	27.2	10.1

SOURCE: U.S. Bureau of the Census, 1995b.

American Indians in the Stratification System

Wherever American Indians reside, on reservations or in urban areas, they are below the national average on most socioeconomic measures—income, education, occupation, employment, health care, mortality, and housing. Table 5.5 indicates some of the economic discrepancies between Indians and the general population. As can be seen, median family income is well below average, and the percent of families in poverty is two-and-a-half times greater than average. The social and economic conditions on reservations are especially dismal and remain resistant to significant change. Three of the ten poorest counties in the United States are Sioux reservations in South Dakota.

Of particular consequence is the serious Native-American unemployment rate. A combination of poor education and geographical isolation has created extremely severe employment problems both on reservations and in urban areas. The rate of unemployment on Indian reservations nationally is about 40 percent (Morris, 1988). And in urban areas, too, Native Americans commonly find themselves without the skills and qualifications to secure steady jobs. Thus the impoverished status of many Native Americans continues from one generation to the next.

Regarding education, Native Americans again find themselves at or close to the bottom of the social hierarchy. For example, Native students have the highest high school dropout rate among all ethnic groups. Achievement levels are among the lowest as well (U.S. Department of Education, 1991). More than any other factor, the low level of human capital among Native Americans, primarily in the form of education, accounts for the socioeconomic discrepancy between Indian and white populations (Waters and Eschbach, 1995).

Economic insecurity and low educational achievement translate into other measures of deprivation. Although improved, the health of American Indians is still worse than that of the general population by almost every indicator. About 33 percent of Indians die before age forty-five, compared to 11 percent for the U.S. population as a whole (Garrett, 1994). Much of the lower health standards among Indians is the result of seriously insufficient health-care delivery. On reservations, there are 96 doctors and 251 nurses per 100,000 people, compared with 208 doctors and 672 nurses per 100,000 of the general population (Worsnop, 1992).

Further, suicide and alcoholism rates are many times higher among Native Americans than among the general population. Alcoholism is particularly problematic, reaching epidemic proportions in some areas (McCoy, 1994). Low economic and educational standing, when combined with alcohol abuse, relate to high crime rates among American Indians, who make up about 2.5 percent of federal prison inmates, far out of proportion to their population size (Grobsmith, 1994).

Although the differences between Indians and the remainder of the population are wide, progress has been made in recent years in improving health and educational standards among the Native American population. Life expectancy of Indians has risen significantly, and infant mortality is now about the same as the national average. Nonetheless, in specific cases, conditions remain much more dire for Native Americans. Among the Oglala Sioux on the Pine Ridge Reservation in South Dakota, for example, the infant mortality rate is three times the national average (Kilborn, 1992). In education, the numbers of Native-American students attending college and pursuing graduate and other professional degrees have steadily risen. Between 1984 and 1993, for example, American Indians attending college increased 45 percent from 84,000 to 122,000. Still, a Department of Education task force studying Indian education concluded that "It is evident that the existing educational systems, whether they be public or federal, have not effectively met the educational, cultural, economic, and social needs of Native communities" (U.S. Department of Education, 1991:12).

One of the reasons that changes in the status of Indian tribes over the past two decades have not impacted profoundly on long-term problems of American Indians is that more than half of all Indians today do not live on reservations, but are urban dwellers. New policies and favorable court decisions have clearly improved economic and social conditions for American Indians vis-à-vis the larger American society. However, because most government policies pertain to tribes on reservations, the majority of American Indians may not be affected (Nichols, 1992).

Recent Economic Projects Indian activism of the past three decades has been met with measures designed mostly to quell militance rather than to address the unique socioeconomic problems of Native Americans. Moreover, as vital resources such as oil, natural gas, coal, and uranium have been discovered in recent years on federal Indian reservations, Native Americans have experienced new pressures threatening their control over their remaining lands.[6] These resources, however, have also served as valuable political tools for particular Indian groups in pursuing political and economic goals.

In recent years, many Indian tribes have become more aggressive in pursuing projects designed to foster greater economic development and independence. Because American Indian nations are considered sovereign, the U.S. government does not tax tribes or their wealth. This has provided opportunities to establish businesses of various kinds, some of which have been extremely successful and have resulted in significant economic benefits to Indian reservations. The Lac du Flambeau Chippewa tribe in Wisconsin, for example, distributes Ojibwe brand pizza in seven states, and the St. Regis Mohawk tribe of New York produces widely sold fishing products. In Arizona and New Mexico, the Navajo reservation now grows and sells produce in twenty-one states as well as Mexico and Canada, with annual sales of more than $32 million (Trueheart and McAuliffe, 1995).

The largest and most significant of these enterprises, however, is legalized gambling. The economic revenues accumulated and the jobs created by gambling have far exceeded those of other projects. Gaming has turned what were economically dormant Indian nations into communities with a solid economic base from which other aspects of improvement have been launched. In 1988, the Federal government enacted the Indian Gaming Regulatory Act (IGRA), which provided for the operation of gaming by Indian tribes, with the objective of promoting tribal economic development and self-sufficiency. Today, 97 tribes in 22 states are operating more than 200 casinos and bingo parlors, earning an estimated $2.7 billion in annual profits (Johnson, 1995; Trueheart and McAuliffe, 1995). Ironically, the success of some of these operations has redounded to non-Indians by providing jobs as well as tax revenues to states. The Mashantucket Pequots of Connecticut, for example, are a community of only a few hundred but they operate a casino that employs 10,000 people and is exceeded only by the federal government as a contributor to the state's treasury (Johnson, 1995). Many Indian nations have used their profits from gambling enterprises as a springboard to other forms of tribal economic development. The Pequots, for example, are buying back their land and are investing in nongaming businesses.

The economic benefits that have accrued to many tribes from newly created businesses should not detract from the fact that, as a part of the American ethnic

[6]One-third of the coal in the West, 3 percent of the oil and gas, and 37 percent of the potential uranium are under tribal lands (Fixico, 1994).

hierarchy, Native Americans remain in a very low position. Whether gambling, in particular, proves to be a long-term source of tribal revenue is not at all certain.

Stereotyping

From the start of contact between Europeans and Natives in North America, ethnocentric views prevailed on both sides. For Europeans, the distinctness of specific tribes was blurred by a perception of uniform inferiority. "That almost no account in the sixteenth century portrays systematically or completely the customs and beliefs of any one tribe," notes Berkhofer, "probably results from the newness of the encounter and the feeling that all Indians possessed the same basic qualities" (1979:26). This tendency to see Indians collectively rather than as separate and culturally varied units carried forward to the modern period and has been questioned only in recent times.

American Indians have endured negative stereotypes from the onset of European settlement. These images have endured largely through the portrayal of American Indians in motion pictures. Films have been an especially important source of the creation and maintenance of Indian stereotypes because most Americans, indeed, most people of the world, have acquired their beliefs about North American Indians through them. Though created as entertainment, they nonetheless shaped a perception of Indian culture that has been stubbornly resistant to change. Even American Indians themselves have drawn from these films in constructing their own views of their cultural heritage (Price, 1973).

The savage, hostile Indian, in confrontation with white settlers pushing westward, has been a common movie theme since the introduction of films in the early part of the twentieth century. From the beginning, films have consistently depicted Indians with a set of degrading stereotypes. "The prototype of the Hollywood Indian," notes Rita Keshena, "was treacherous, vicious, cruel, lazy, stupid, dirty, speaking in ughs and grunts, and often quite drunk" (1980:107). In the 1960s, films began to depict a more sympathetic view of American Indians, but one no less founded on inaccurate and simplistic stereotypes.

Not only have American films typically portrayed Indians as brutal and primitive, but they have blurred the cultural distinctions among them. Historically, films have presented Indians not as individuals or even as members of different tribes but as faceless parts of the collectivity labeled "Indians." Moreover, the Hollywood image of Native Americans has been derived mainly from those Indian societies with a historical reputation for violence, particularly the horsemen of the Plains and Southwest. Tribes with more passive and nonviolent reputations have been ignored. As Price (1973) notes, the emphasis on Plains Indians, like Apaches and Cheyennes, is ironic because they did not evolve until after whites introduced the horse and were not similar to most other American Indians. North American Indians did not ride horses before Columbus, and most still did not hunt from horseback in the mid-1800s.[7] Moreover, most were not hunters at all, but agricultural people.

In recent years, the use of Indian names by sports teams has raised additional issues with Indian stereotypes. Team names such as "Braves," "Redskins," "Chiefs," and "Chippewas" have been seen by many as demeaning and contributing to the perpetuation of the image of the Indian as savage and warlike. A so-called Indian war chant, used by fans of the Florida State University athletic teams (called "Seminoles") as well as the Atlanta Braves baseball team, for example, evokes images of belligerence and brutality. Some college teams have voluntarily changed their names and symbols. Stanford University, for example, changed its athletic teams' name from "Indians" to "Cardinal," and the name of St. John's University teams was changed from "Redmen" to "The Red Storm."

One of the unusual aspects of Indian stereotyping concerns the fact that the "savage" characterization has, throughout American history, often been accompanied by admiration, yielding the "noble savage." Thus, on the one hand, Indians could be brave, innocent, and industrious, and on the other, cruel, treacherous, and lazy. But as Berkhofer has explained, "Beneath both the good and bad images used by explorer, settler, missionary, and policy maker alike lay the idea of Indian deficiency that assumed—even demanded—that Whites do something to or for Indians to raise them to European standards, whether for crass or idealistic motives" (1979:119).

However positive some characterizations of Indians were, the dominant white view was always one of inferiority. Thus, in the confrontation between European and Native, the ultimate outcome in the European view was the triumph of "civilization" over "savagery." As Berkhofer notes, from the middle of the eighteenth century to very recent times, for whites, "the only good Indian was indeed a dead Indian—whether through warfare or through assimilation" (1979:30).

Assimilation and American Indians

Cultural Assimilation Throughout the nineteenth and first half of the twentieth centuries, U.S. policy toward American Indians was designed to destroy native cultures and political entities in hopes of creating assimilation among them. A major effort in this regard was the General Allotment Act of 1887. As noted earlier, the thrust of this legislation was to break up Indian tribes into individual farming units, hoping to accelerate assimilation. The driving force of

---

[7] The horse had been introduced by the Spanish, and it diffused northward from settlements in New Spain (today, the American Southwest) in the 1600s.

the policy was the belief that once Indians were fully assimilated, the "Indian problem" would disappear. In addition to authorizing the breakup of reservations into many smaller parcels, the act revitalized the Bureau of Indian Affairs' efforts to encourage assimilation by forcing Indian children to attend boarding schools where they were stripped of their native languages and customs. The BIA commissioner in 1889 bluntly stated his bureau's intent: "The Indians must conform to 'the white man's ways,' peaceably if they will, forcibly if they must" (quoted in Cornell, 1988:56).

From 1887 to 1934, the efforts of the BIA, supplemented by those of Christian churches, which were active on most reservations, were directed energetically to assimilation. Although the annihilation of Indian cultures and of Indian self-governing institutions was not fully realized, the effects on Indian culture and social life were nonetheless very great. During this period, the policy of forced assimilation engendered the deepest distrust of the federal government and the maximum hostility toward whites (Spicer, 1980).

The assimilation objective changed in 1934 with the appointment of John Collier to head the BIA. Collier, as earlier explained, championed the Indian Reorganization Act (IRA), which proved to be a watershed, bringing with it an entirely new approach and direction to Indian–white relations. Collier, remember, encouraged the preservation of Native cultures. As Cornell explains, however, despite the efforts of Collier and what became known as "the Indian New Deal," the basic assumption of Indian policy—eventual assimilation—was not essentially altered. Rather, this new approach was merely "a means of easing the process of change and giving Native Americans some measure of control over it" (Cornell, 1988:93). By the early 1950s, blatant assimilation sentiments again prevailed with enactment of the policy of termination.

During the late 1960s and early 1970s, a Red Power movement took shape, emulating in many ways similar movements among African Americans and Latinos. Red Power brought the grievances and general plight of Native Americans to public awareness through a number of incidents, including the occupation of Alcatraz Island in 1969, the takeover of the BIA headquarters in 1972, and the FBI confrontation with Indian activists in 1973 at Wounded Knee, South Dakota. The latter affair culminated in a standoff between AIM (the American Indian Movement) and federal officials that continued for over two months, with sporadic violence that resulted in several deaths and many injuries. The choice of Wounded Knee was hardly random. This location held great symbolic significance for American Indians because it had been the site at which the U.S. cavalry, led by General George Custer, had massacred hundreds of Sioux men, women, and children in 1890. The militance of Red Power subsided relatively quickly, however, and was replaced with a more legalistic approach to preserving and enhancing Indian cultural and economic interests.

For most of American history, then, forced cultural assimilation, that is, the disintegration of Native cultures, has been the primary objective of dominant political and social institutions. That objective has never been fully accomplished, but the long-term continual assault on Native cultures has resulted generally in a significant dissolution of Indian cultural influences on individuals, especially regarding language. Dozens of tribal languages continue to exist and function, but their use is common mostly on reservations, and the proportion of Indians not also speaking English is low. Quite clearly, English today is the prevailing language among Native Americans. Many of those who are younger and who have migrated to urban areas have been cut off from tribal cultural influences in addition to language, such as religion; hence, in some ways, their ethnic distinctness becomes increasingly difficult to define sharply. The emergence of a strong Pan-Indian movement, however, has emphasized common Indian values and practices, which may serve as a shield against more complete cultural assimilation.

Structural Assimilation In looking at the structural assimilation of Native Americans, we need to keep in mind that this is an ethnic population divided into two demographic parts: those living within and those living outside the reservations. As already noted, Native Americans today remain at or near the bottom of the society's occupational, educational, and income hierarchies, whether on or off reservations. This has stifled rapid movement into mainstream economic institutions. For those who have migrated to urban areas, however, integration into the corporate labor force is, of course, a meaningful option.

As for politics, again, the distinction between reservation and nonreservation Indians is an important consideration. Those on reservations are subject to laws made by the tribal government, and politics there, as a result, are intratribal. Those off reservations do not represent a large enough group in any area to be able to carry much political weight. Thus, American Indian representation in government at all levels has been minimal. One exception in recent years has been Ben Nighthorse Campbell, who in 1992 was elected to the U.S. Senate from Colorado. Previously he had been a U.S. Representative.

Although secondary structural assimilation has not occurred substantially, one aspect of Indian life that seems to augur increasing assimilation in the future is the expansion of education. A growing number of young Native Americans attend college and, as this occurs, more will inevitably take their place in the business and political worlds. In 1978, 37,000 Native-American men and 41,000 women were enrolled in college; by 1992, those numbers had risen to 50,000 and 69,000 (U.S. Bureau of the Census, 1995b).

At the primary level of assimilation, those who have migrated to the cities appear to have begun to integrate into the larger society. Perhaps the most telling indication of this is the high, and increasing, rate of intermarriage among American Indians and whites (Eschbach, 1995). Well over half of married Indians are married to non-Indians, a percentage considerably higher than for any other nonwhite racial-ethnic group. Surprisingly, this extremely high rate of intermarriage characterizes reservation as well as nonreservation Indians (Snipp, 1989). It is particularly strong, however, among those in urban areas. What this high and, apparently, unflagging rate of intermarriage means for the future of American Indians can only be speculated. In combination with the assimilating forces of urban life, it may indicate a decline of Indian ethnicity. Indians can more easily move into the mainstream, relatively undeterred by racial identity, unlike African Americans, Asian Americans, and some Latinos. Intermarriage may also reduce the number of persons who, in the future, will identify themselves ethnically as Indians. However, the existence of tribes and of relatively cohesive Indian communities on reservations, where intermarriage rates remain lower, would seem to preclude the complete disappearance of American Indians as part of the American ethnic system (Eschbach, 1995).

Indianization The idea of an American-Indian ethnic group has, from the outset of European–Native contact, been a construct of whites. The extreme cultural and social diversity of the Native population was dissolved into a collective notion that created a single category. Among Indians, supratribal consciousness, though emergent in the early part of the twentieth century, did not become a well-developed movement until the 1950s and 1960s. Cornell (1988, 1990) attributes this in large measure to Indian migration to the cities. In the urban environment, Indians found that tribal identities and interests no longer held much usefulness as they found themselves in political and economic competition with other ethnic groups. This Indian ethnic consciousness, developed in the urban communities, reached back to the reservations through returning migrants.

In recent years, then, American Indians have begun to coalesce into a coherent ethnic group, in a kind of "cross-tribalism" or "Pan-Indianism." Tribal identities remain, but increasingly Indians have begun to develop an Indian consciousness that transcends tribal divisions. In a sense, what is emerging is an "American Indian" ethnic group (Nagel, 1995). This unity may result in greater political effectiveness and a stronger public awareness of Indian issues and problems.

SUMMARY

Following the WASP group's conquest of the indigenous peoples and enslavement of blacks, it established its social and political dominance and exercised the power of selection over those who came afterward. American ideologies and public policies regarding ethnic groups and issues have consistently reflected WASP cultural preferences and WASP economic and political interests. Based on these thoughts and actions, an ethnic hierarchy has been shaped, and the rank order of groups has not essentially changed. Today the American ethnic hierarchy is made up essentially of three broad ranges: white Protestants of various national origins at the top; white Catholics and Jews of various national origins, along with most Asian Americans, in the middle; and Hispanic Americans, African Americans, and American Indians comprising the bottom range.

The conquest by European settlers of the Native Americans took the form of contact, ethnocentrism, competition over land, and the imposition of white dominance through superior force. Indians were exploited and largely annihilated in the nineteenth century. Attempts were made repeatedly to force their assimilation into the dominant culture though these attempts were never completely successful. As wards of the state, they have been largely neglected in the twentieth century and rank near the bottom of the economic class system. Today, a growing Pan-Indianism has begun to create a more cohesive and politically active ethnic group.

Suggested Readings

Anderson, Charles H. 1970. *White Protestant Americans: From National Origins to Religious Group.* Englewood Cliffs, N.J.: Prentice-Hall. Describes the changing social composition and status of what has been the dominant ethnic group in American society.

Baltzell, E. Digby. 1964. *The Protestant Establishment: Aristocracy and Caste in America.* New York: Vintage. A fascinating social history of the WASP upper class, especially its efforts to maintain its ethnic purity. Argues that the WASP upper class has become a castelike group, whereas societal power has passed to a more ethnically diverse elite.

Cornell, Stephen. 1988. *The Return of the Native: American Indian Political Resurgence.* New York: Oxford University Press. Traces the historical development of Indian–white relations and concludes with a discussion of the emergence of a Pan-Indian ethnic group, one transcending particular tribal identities.

Glazer, Nathan, and Daniel P. Moynihan. 1970. *Beyond the Melting Pot.* 2d ed. Cambridge, Mass.: MIT Press. An influential book at the time of its writing, its major theme is that American ethnic groups do not completely assimilate but evolve into political units that compete for resources in the urban environment. Contains vivid comparisons of the different groups making up the American ethnic hierarchy, using New York as a backdrop.

Higham, John. 1963. *Strangers in the Land: Patterns of American Nativism 1860–1925.* 2d ed. New York: Atheneum. Examines the different expressions of nativism, antisemitism, and anti-Catholicism that characterized American thought and action during the period of massive European immigration to the United States.

Lieberson, Stanley. 1980. *A Piece of the Pie: Blacks and White Immigrants Since 1880.* Berkeley: University of California Press. Addresses the issue of why European ethnic groups and African Americans have experienced different rates of economic and political mobility.

Snipp, C. Matthew. 1989. *American Indians: The First of This Land.* New York: Russell Sage Foundation. A detailed collection of demographic and social statistics on Native Americans.

IN THIS CHAPTER, we look closely at Italian Americans, a group whose origins are chiefly part of the classic, or new, immigration period. Like all American ethnic groups, Italian Americans are a sociological case with many unique attributes. Yet in a number of ways this is the prototypical white ethnic group, and it illustrates well much of the general experience of this part of the American ethnic system.

It is important to look carefully at white (or Euro-American) ethnic groups, specifically those described in the previous chapter as part of the second range of the American ethnic hierarchy, for several reasons. First, they are a substantial element of ethnic America. Together, they represent the ethnic origins of perhaps as much as 20 percent of the total American population.[1] Second, although it is true that today these groups are increasingly less visible as distinct ethnic units and are being assimilated unwaveringly into the mainstream society, it is erroneous to assume that ethnicity for them is no longer of significance. Many observers maintain that ethnicity, though important to these groups in the past, no longer holds great meaning, either culturally or structurally. This view,

[1] This estimate is based on census reports of both single- and multiple-ancestry persons, that is, those who have claimed either one specific ethnic ancestry or a combination.

however, seems premature. None has reached parity with the dominant Anglo group in economic status and political power, none has entered fully into primary relations with the Anglo group, and none has completely lost its cultural forms through assimilation. Hence, although their conflict with the dominant group is muted by comparison with racial-ethnic groups', they are still, to some degree, ethnic minorities in American society.

Finally, it is important to study the white ethnic experience for comparative purposes. If white ethnics today remain minorities, the fact is they are minorities in a far less subordinate position than nonwhite American groups. Recall our model of the American ethnic hierarchy in which groups of the second range—including white ethnics—are increasingly merging socially and economically with those of the first range—the dominant group, creating a dichotomy between primarily white and nonwhite groups. The social and economic disparities between white and nonwhite ethnic groups are a troubling issue of American society and have generated much debate among social scientists, policymakers, and the general public. These disparities in success have been interpreted by some as a product of internal group differences and by others as the result of inequitable and qualitatively different social contexts within which whites and nonwhites have labored. To shed light on this issue, we need to trace both group and societal factors for whites as well as nonwhites. In other words, it is necessary to understand how the characteristics and social experiences of white ethnic groups have been similar to and different from those of the country's racial-ethnic groups. What cultural systems and social and economic skills did each bring with them? What was the nature of the opportunity structure open to each? What obstacles to social integration for each were dictated by the dominant group? To what extent did each seek assimilation or strive to preserve ethnic uniqueness? What was the nature and level of prejudice and discrimination directed at each? As we will see, although there are similarities between white and nonwhite groups on all these questions, there are also enormously consequential differences.

EUROPEAN IMMIGRATION

The relative homogeneity of the United States in its formative years began to change in the early decades of the nineteenth century with the start of European immigration. The variety of groups that would subsequently enter was quite astonishing and would fundamentally reshape the society from that point forward. Indeed, the onset of large-scale immigration marked the beginning of one of the great social experiments in modern history: creating a multiethnic society whose heterogeneity was unprecedented. In the 100 years to follow, more than thirty million immigrants would come to the United States, making it, as the historian Maldwyn Jones called it, "the classic country of immigration" (1960:2).

Before examining the white ethnic experience as illustrated by Italian Americans, let us look briefly at the two major periods of European immigration to the United States. Both contributed great portions of the white population, but the major groups comprising each were markedly different in their national origins and their sociological characteristics. As a result, not all immigrant groups were received by the dominant, or host, group in the same way, and their ensuing absorption into the society also varied.

The "Old Immigration"

The six decades between 1820 and 1880 mark the first significant period of non-British immigration. The two most sizable groups arriving during this era were the Germans and the Irish, though immigration from Britain and other northwestern European societies continued unabated. Between 1847 and 1854, about 1.2 million Irish immigrants came to the United States, and by the end of the Civil War, they constituted 7 percent of the white population. Germans came in two waves; from 1850 to 1857, almost one million arrived, and between 1865 and 1885, two million more. By the late nineteenth century, Germans were the second largest ethnic group in the society, exceeded only by the British.

This period marked the entrance not only of non-British people in large numbers but also for the first time of significant numbers of non-Protestants. Most of the Irish were Catholics, as were about half the Germans. As a result, nativist activities, specifically anti-Catholic actions and rhetoric, became widespread. Catholics were seen by many as a subversive element and Catholicism as basically incompatible with American ways (Dinnerstein, Nichols, and Reimers, 1990). By midcentury, however, anti-Catholic nativism had begun to wane as national attention turned to the issue of slavery and as immigration was temporarily halted by the Civil War. Moreover, immigration generally was welcomed by industrialists and others of the business community who saw immigrants as a source of cheap labor and a growing consumer market.

Despite prejudice and discrimination toward the Catholics among them, especially the Irish, these northwestern European groups were, by comparison with groups that would follow them, advantaged in two ways: they were close enough culturally to the Anglo core group to assimilate within a relatively short historical time, and they were physically indistinct from the Europeans who had preceded them to America, thereby avoiding imputations of racial identity. Thus, as we shall see, their integration into the mainstream society was not severely constrained, and the resistance of the dominant group to them was relatively mild.

The "New Immigration"

Although substantial German and Irish immigration had presented the first real challenge to WASP dominance, the polyglot nature of American society took shape most forcefully through what has been called the "new immigration." During a period extending roughly from the early 1880s until the outbreak of World War I, almost twenty-five million European immigrants came to the United States. This influx represents the archetypal American immigration, and it is this period that we most often think of when we visualize traditional immigrant life. It is from this wave of European immigration that most of today's non-Protestant white ethnics are descended. Indeed, these are the groups ordinarily understood to be, and most commonly referred to as, *white ethnics.* We will use this term accordingly.

These new immigrants not only represented the numerical peak of European immigration to the United States but also changed radically the ethnic character of the society. To begin with, most were from southern and eastern European societies, whose cultures were alien to those of Britain and northwestern Europe. Most important, they were non-Protestant—primarily Catholic and secondarily Jewish. Although many of the earlier Irish were Catholic, in other respects they were not so culturally distant from the Anglo-Americans. Likewise, the Germans may have spoken a different language, but over half were Protestant, and they were not looked on by the dominant British group as cultural inferiors. Finally, the new immigrants were mainly people whose class origins were lower than those of previous groups, except perhaps the Irish. They were chiefly peasants from the poorer states of Europe—southern Italy, Russia, Poland, and the many parts of the Austro-Hungarian Empire.

Their cultural and class differences brought forth a more strongly antagonistic reaction from the dominant group than previous ethnic groups had produced. Americans questioned how the immigrants would ultimately fit into the society and, abetted by notions of Social Darwinism, widely assumed their cultural inferiority. Prompted by the continued influx, nativist feelings and racist theories reached an apex by the early 1920s, resulting in restrictive quotas that curtailed once and for all large-scale European immigration.

Table 6.1 shows the major ancestry groups that currently make up the Euro-American population, which derives primarily from the two main periods, new and old, of European immigration.[2]

---

[2]The diversity of the European component of the American ethnic configuration is greater than is shown here because only the largest ancestry groups are included.

Table 6.1 Largest Euro-American Ancestry Groups

ANCESTRY GROUP	NUMBER[a]	PERCENT U.S. POPULATION
German	57,947,374	23.3
Irish	38,735,539	15.6
English	32,651,788	13.1
Italian	14,664,550	5.9
French	10,320,935	4.1
Polish	9,366,106	3.8
Dutch	6,227,089	2.5
Scotch-Irish	5,617,773	2.3
Scottish	5,393,581	2.2
Swedish	4,680,863	1.9
Norwegian	3,869,395	1.6
Russian	2,952,987	1.2
French Canadian	2,167,127	0.9
Welsh	2,033,893	0.8

SOURCE: U.S. Bureau of the Census, 1995b.
[a]Represents the total number of responses for each group, both single and multiple. Thus, the sum of responses is greater than the total population.

ITALIAN IMMIGRATION AND SETTLEMENT

Small numbers of Italians were part of American society almost from its founding, and a relatively small-scale immigration proceeded throughout the first three-quarters of the nineteenth century. The bulk of Italian immigration, however, occurred as part of the new immigration, from the 1880s until the mid-1920s.

The large-scale movement to the United States after 1880 represented the second great emigration from Italy. Earlier in the nineteenth century, Italians had immigrated mainly to Argentina and Brazil and only secondarily to the United States. Most of these early immigrants, to the United States as well as to South America, were from northern Italy (Foerster, 1919).

Beginning in the 1880s, both the origins and destinations of Italian emigrants changed. With the depletion of coffee plantations in Brazil and political turmoil in Argentina, most Italians now found the United States more attractive. And whereas the northern Italian provinces had previously supplied most of the immigrants, southern Italy now became the major region of emigration. This change in regional origin fundamentally altered the class and cultural character of the Italian population in the United States. By comparison with the more prosperous and literate immigrants from the North, those from the South were poor and uneducated. Further, whereas the northerners had been mostly artisans, shopkeepers, and professionals, those from the South were mainly agricultural peasants.

Southern Italy had historically been an impoverished, backward area, subjected to constant economic and political exploitation by the region's ruling groups, but unusually severe conditions were created in the latter nineteenth century. Depressed wages, little or no opportunity to improve a low agricultural output, little industrialization, poor health conditions, neglect by the central Italian government, and heavy indirect taxation all served to create a strong impetus for emigration. These factors, combined with the call for unskilled labor by a burgeoning American industry, resulted in a veritable exodus of southern Italian peasants. Between 1880 and 1915, four million Italians went to the United States, half of them in the single decade from 1900 to 1910. Between 80 and 90 percent were from the South, that area known to Italians as the *Mezzogiorno,* and about one-quarter of these were from Sicily (Foerster, 1919; Gambino, 1974; Lopreato, 1970; U.S. Bureau of the Census, 1975).

One extremely important characteristic of the Italian emigration of this time was that most who went to the United States did not expect to stay permanently. Most were men unaccompanied by their families. Their objective was to earn enough money to enable them to return to their native villages and purchase land. Many did just that. No other sizable immigrant group displayed such a high rate of return migration. The comparative rates of repatriation of Italian and Jewish immigrants of this period are striking. Between 1908 and 1914, almost one-third of those who arrived from Italy eventually returned or migrated to another society; for Jews, the return rate was only 7 percent. Between 1915 and 1920, the corresponding figures were 56.6 and 4.3 percent for the respective groups, and between 1925 and 1937, 40 and 3.8 percent (Learsi, 1954).

Urban Adjustment

The Italian immigrants of the late nineteenth and early twentieth centuries were in many ways unprepared for life in an industrial, urban society. Most lacked the skills and education that could be converted into rapid occupational mobility and economic success in the United States. As a result, most entered the labor force in unskilled jobs, where they and their children remained. Most settled in the cities of the Northeast, especially New York, where they found work in manual laboring occupations corresponding to their low skill and educational levels. Because cities like New York were at this time establishing their industrial infrastructures—roads, rail lines, subways, factories, housing—opportunities abounded for unskilled labor, and it was here that Italian immigrants found their occupational niche. At the turn of the century, Italians made up three-quarters of the building laborers in New York City, for example, and almost the entire labor force constructing the city's subway system (Glazer and Moynihan, 1970).

Italian immigrants in the cities tended to settle in depressed areas, inhabiting run-down tenements that had been vacated by earlier ethnic groups such as the Irish. The concentration of Italians in these areas was partially the result of simple economic necessity; such housing was available and affordable. The creation of various "Little Italys" in New York and other cities, however, was also the product of the natural attraction of immigrants to areas inhabited by those from their regions in Italy who had preceded them to the United States. Italian immigration to the United States was very much a "chain" phenomenon, in which those from one village gravitated to locations where previous emigrants from the village had settled. There was little identity with the Italian nation per se. Rather, immigrants identified with their province and, even more precisely, their village. As a result, Italian sections of American cities were internally subdivided into "Little Calabrias," "Little Sicilys," and the like, and neighborhoods were often settled by those from a common village or town in the province (Vecoli, 1964).

The Italian-American Population Today

Italian Americans remain one of the largest white ethnic groups. In 1990, almost fifteen million people claimed either single or multiple Italian ancestry. Although the exact number of Italian Americans (or any other white ethnic group) cannot be pinpointed owing to the dilution of the group through ethnic intermarriage, there is no question that this is not only one of the largest but also one of the most heavily urbanized American ethnic groups. Though it is far more dispersed than in earlier times, the Italian-American population continues to reside most heavily in the cities of the Northeast. Italian Americans in New York City, for example, constitute one of the largest elements of that city's extremely diverse ethnic population.

ITALIAN AMERICANS IN THE STRATIFICATION SYSTEM

Italian Americans are well into the third generation, and many, especially those whose ancestors were northern Italian, are even into the fourth generation. Given the length of time since the group's large-scale entry, some evaluation can be made of Italian-American progress on various socioeconomic measures, both in absolute terms and in relation to other American ethnic groups. How far have

Italian Americans come, what means have they employed in seeking upward mobility, and how might we account for their current status in the overall stratification system? We should keep in mind that many of the patterns displayed by Italian Americans are comparable to those of other white ethnic groups.

Social Class

In looking at the place of ethnic groups in the class structure, it is important to remember that these are collective pictures that reflect the general positions of groups, not particular individuals within them. Obviously, Italian Americans, for example, will be found in the entire range of the American class structure, from the very poor to the very affluent. Our concern, however, is with the group as a whole and how it compares generally with others.

By all indications, Italians have "made it" in American society to a greater degree than many ethnic groups and to a far greater degree than blacks, Latinos, and American Indians. Although upward mobility was minimal for the first two generations, Italian Americans of the third and fourth generations are now exhibiting solid evidence of middle-class status in terms of income, occupation, and education. These trends are characteristic of white ethnics in general.

Income On the first measure of social class, income, Italian Americans are above the national average. Especially significant improvement occurred during the 1950s, with only Jews showing a greater increase among white ethnic groups. As is seen in Table 6.2, the median family income of Italian Americans is comparable to that of other Euro-American groups.

Clearly what has occurred in recent years is a convergence in average socioeconomic status among Euro-American ethnic groups. That is, southern and eastern European ancestry groups, like Italians, have reached a point where they are indistinguishable class-wise not only from each other, but, more importantly, from northwest Europeans, that is, the society's dominant ethnic group (Alba, 1995).

Occupation Though above average in income, Italian Americans do not show a commensurate position in the occupational hierarchy. This is because little occupational mobility was evident for this group between the first and second generations (Greeley, 1977; Yans-McLaughlin, 1977). That is, most of the children of immigrant parents remained in working-class jobs. Because children in Italian families were expected to contribute to the family's support, they entered the labor force early, precluding the more extensive education needed for higher-status jobs. Most Italian children thus followed their fathers into manual laboring occupations. Those of the third and fourth generations, however, with

Table 6.2 Socioeconomic Standing of Euro-American Ethnic Groups, 1990

GROUP	MEDIAN FAMILY INCOME	PERCENT COLLEGE GRADUATES
Russian	$58,826	49
Dutch	43,415	18
Greek	43,330	28
Scottish	43,293	34
Hungarian	42,778	27
Italian	42,242	21
Polish	41,700	23
English	40,875	28
Swedish	40,459	27
Slovak	40,072	22
Scotch-Irish	39,816	28
Portuguese	38,370	12
German	38,216	22
Irish	38,101	21
French	36,237	18

SOURCE: U.S. Bureau of the Census, 1993c.

greater education, are now moving in significant numbers into all occupational areas, including the top professions like medicine and law, where they had been severely underrepresented (Alba, 1985; Glazer and Moynihan, 1970).

Although most Italian Americans still remain more solidly working-class than Jews, Irish, and several other white ethnic groups, the extent of their mobility is nonetheless quite extraordinary when one considers how relatively ill equipped they were at the time of their arrival to succeed in the American economic system. Some have argued, however, that third-generation Italian-American mobility is largely a reflection not so much of the group's progress per se as of a change in the society's occupational structure. As more white-collar jobs have been created, more Italian Americans have filled them (Alba, 1985; Gallo, 1974).

Education The relationship between occupation and education is ordinarily very strong, and a group's position on one measure of socioeconomic status should therefore closely reflect the other. This is quite evident in the case of Italian Americans. Those of the first two generations lagged in educational attainment, but the third and fourth generations are now showing educational levels that equal or, in some cases, even exceed those of other white ethnic groups (see Table 6.2).

Clearly, as with other indicators of socioeconomic status, in educational attainment, Euro-American ethnic groups have in recent years converged, with marginal differences among them. Alba (1995) has shown that for men of southern and eastern European ancestry born between 1916 and 1925, there remain some apparent disadvantages when compared to the educational levels of those of northwest European ancestry. But when those born between 1956 and 1965 are looked at (these would be third- or fourth-generation ethnics), southern and eastern European origin men are at or above the average of northwest European origin men.

In sum, the class position of Italian Americans is, on most measures, at or above the average for white ethnic groups. Moreover, for the third and later generations and for those of part-Italian ancestry, parity with the society's core ethnic group, northwestern Europeans, has nearly been reached (Alba, 1985). Overall, the differentials between various European ethnic groups in the stratification system seem to be shrinking (U.S. Bureau of the Census, 1993c).

Political and Economic Power

Having looked at the place of Italian Americans in the economic realm, let us look briefly at the power hierarchy. Although there is commonly a close relationship between economic class and societal power, these are nonetheless two distinct dimensions of stratification. As explained in Chapter 2, the power dimension entails the extent to which people are able to affect others in the society or community through their actions. Ordinarily in American society such power is attached to positions of authority in two key institutions: government and the corporation. The best indicators of societal power, then, are the number and level of top-ranking positions in these institutions held by members of the ethnic group. Where do Italian Americans currently stand, and what has been the extent of their mobility over several generations?

Government Patterns of Italian-American participation at the highest levels of government parallel group patterns in the economic realm: The first two generations made little progress, but in recent years, Italian Americans have begun to achieve important positions in modest numbers. This is a pattern generally duplicated by other groups of the new immigration.

Prior to the 1960s, the Italian-American presence at the highest levels of the federal government was limited. Not until 1950 was an Italian American elected to the Senate, and not until John F. Kennedy's administration in 1962 was an Italian American named to a cabinet post. Since then, Italian-American representation in the Congress and in the higher levels of the executive branch has steadily increased. By 1980, at least twenty-five Italian Americans were serving in the Congress, compared with four in the 1930s and eight in the 1940s (Rolle, 1980). And the choice of a second-generation Italian American as a vice-presidential candidate in 1984 marked a watershed of sorts in national politics for Italian Americans. Moreover, in 1986, for the first time, an Italian American was appointed to the Supreme Court. This marked another ethnic watershed because not only Italians but southern and eastern European Catholics in general had historically been absent from the Court (Lieberson, 1980; Schmidhauser, 1960).[3]

The underrepresentation of *all* Catholic ethnic groups, not simply Italians, in top positions of the federal government has been a historically consistent pattern since the founding of the society. Although a Catholic was elected president for the first time in 1960, Catholics in general and southern and eastern European Catholics in particular have traditionally been underrepresented in the highest offices of the executive, legislative, and judicial branches (Alba and Moore, 1982; Mills, 1956; Stanley, Mann, and Doig, 1967). This historical pattern continues to change, however, with each succeeding generation. Catholics today are represented in Congress approximately in proportion to their share of the total population (Table 6.3). Moreover, the choice of Michael Dukakis, the son of Greek-American immigrants, as the Democratic nominee for president in 1988 seemed to mark a high point in the national political fortunes of southern and eastern European ethnic groups and may indicate their fullest entrance into American politics.

At the local and state levels of government, Italian Americans have attained more substantial power in locales where they constitute a significant proportion of the population, such as New York, New Jersey, Massachusetts, Connecticut, and Rhode Island. In recent decades, Italian-American governors and lower officeholders in those states have been commonplace.

Part of the explanation for the slower development of Italian-American political power lies in the traditionally lower rates of political participation for this group. The tendency to ignore the political world can be seen as partially a product of the distrust of government that so characterized the *contadini* (peasants) of southern Italy. As Lopreato explains, the southern Italian, more than other Europeans, did not participate in politics. "Matters political," he notes, "had been done for him — or more precisely *to* him" (1970:113). Government, therefore, was an institution to be resisted, not embraced. The relationship between successive governments and the southern Italian peasant was characterized most frequently by exploitation or neglect.

---

[3]Five Catholics, including three Irish Catholics, however, had been appointed to the Court before World War II (Baltzell, 1994).

Table 6.3 Religious Makeup of the U.S. Congress

	RELIGION	SENATE[a]	HOUSE[a]	U.S. POPULATION[a]
1961	Protestant	87%	77%	68%
	Catholic[b]	12	21	25
	Jewish	1	3	3
1979	Protestant	76%	66%	64%
	Catholic[b]	15	27	25
	Jewish	7	5	2
1994	Protestant	63%	63%	59%
	Catholic[b]	22	29	25
	Jewish	9	6	2

SOURCES: American Jewish Committee, 1994; *Congressional Quarterly,* 1979, 1994; *Gallup Poll Monthly,* 1994; Stanley, Mann, and Doig, 1967.
[a]Based on those who expressed a religious preference.
[b]Includes Greek Orthodox.

Lower rates of political participation among Italian Americans can also be attributed in some measure to class factors. The relationship between social class and political participation is one of the firmest assumptions of political sociology, backed by a wealth of empirical evidence. Put simply, the higher one's class, the greater the scope, intensity, and level of political participation (Lipset, 1960; Milbrath, 1982; Verba and Nie, 1972). Because they are predominantly working-class, lower participation among Italian Americans generally should be expected, regardless of ethnic factors such as family and regional influences (Greeley, 1974). With progressive upward mobility, however, Italian Americans appear to be behaving politically in much the same manner as is typical of Americans generally (Barone, 1985).

The Corporation Power in American society is by no means limited to the arena of government. In fact, an equally significant locus of societal power lies in the realm of economic activity, specifically, the giant corporation, where decisions are made that shape the society's work force, consumer market, and general economic condition. How extensively have Italian Americans penetrated the highest positions of the corporate world?

Like most white ethnics in American society, Italians have displayed minimal entrance into the top echelons of economic power. Increasing political influence has generally not led to power in other institutional areas, especially the economy. Clearly, the ethnic origins of American business leaders have been and remain primarily northwest European Protestant; Catholics and Jews, especially those of southern and eastern European origin, have been underrepresented.

Although a few individuals of Italian origin have risen to very conspicuous positions in the corporate world in recent years (the chief executive officers of both the General Motors Corporation and the Chrysler Corporation, for example), more commonly, their presence in the executive suites has been below their proportion of the general population. A study of Chicago's largest corporations, for instance, revealed that although Italians made up 4.8 percent of the city's population, they made up only 1.9 percent of these corporations' directors and 2.9 percent of their officers (Barta, 1979). In New Haven, Connecticut, Wolfinger (1966) found that although Italians in 1959 made up 31 percent of the registered voters and held 34 percent of major political offices, they made up only 4 percent of the city's economic elite.

Similar patterns are evident for other white ethnics. Few have entered into the top ranks of economic power. For example, in the previously cited study of Chicago's top corporations, Poles, that city's largest white ethnic group, were shown to be even more severely underrepresented than Italians. Although they made up almost 7 percent of the city's population, they made up only 0.3 percent of these corporations' directors and 0.7 percent of their top officers. The historically consistent domination of the American business elite by those of northwestern European and Protestant origins has been demonstrated repeatedly. Sociologist Suzanne Keller (1953), for example, found that 89 percent of top business leaders in 1900 and 85 percent in 1950 were Protestants, most of British descent; 7 percent in each year were Catholics; and 3 and 5 percent, respectively, were Jews. This pattern has been corroborated by others (Marger, 1974; Mills, 1956; Newcomer, 1955). By the 1970s, little had changed—Catholics made up 14 percent and Jews 6 percent of the corporate elite (Burck, 1976). Alba and Moore in 1982 found a continuation of Anglo dominance of economic power positions.

PREJUDICE AND DISCRIMINATION

Italians have been the objects of long-standing and often severe prejudice and discrimination in American society. The period of large-scale immigration beginning in the 1880s produced negative stereotypes and blatant anti-Italian actions. Though greatly diminished today in form and intensity, these attitudes and acts continue to affect the image and, to some extent, treatment of Italian Americans.

It is important to understand that prejudice and discrimination leveled at white ethnic groups have historically never been as intense, widespread, or institutionalized as that aimed at racial-ethnic groups—African Americans, Latinos,

American Indians, and Asian Americans. Though at times violent in form, discriminatory actions have ordinarily been limited to derogation and denial of access to certain jobs, schools, and neighborhoods. Never have white ethnics been denied legal rights, as were blacks and, to a lesser degree, Asians, nor have they been so completely excluded from the larger society.

The Negative Image of Italian Americans

Anti-Catholic beliefs and actions in America became common beginning in the 1830s, when Irish immigrants began to arrive in large numbers. The Irish were targets of virulent and often militant prejudice and discrimination, which was expressed by the familiar "No Irish Need Apply" notices displayed by landlords and employers. In the 1850s, a major nativist and bitterly anti-Catholic political party arose, the Know-Nothing, or American, party, which gained control of a number of statehouses and elected several congressmen. The anti-Catholic sentiments of Protestant Americans, then, were well established and widely proclaimed several decades before the large-scale influx of southern and eastern European Catholic groups began in the 1880s.

Italians suffered the usual antagonisms directed at Catholics generally, but anti-Italian feelings ran especially high as part of the newly inspired nativism and jingoism that arose in response to the new immigrants of the late nineteenth century. Indeed, with the arrival of hundreds of thousands of southern Italian peasants, Italians now replaced the Irish as the chief target group of the country's well-worn anti-Catholicism.

Much of the prejudice and discrimination was founded on racist ideologies. This was a time in which ideas of Social Darwinism prevailed among scholars and policymakers alike and in which new racial theories were widely espoused and endorsed. Those groups composing the new immigration — overwhelmingly Catholics and Jews — were therefore described in racial terms, that is, as morally and even physiologically inferior to Americans of northwestern European origin. The new immigrants were perceived not simply as different from those who had preceded them but as threatening to the very social and biological fabric of the society (Higham, 1963). Italians were seen as the most degraded of all the entering groups.

The poor and illiterate peasants from southern Italy were received quite differently from the earlier northern Italian immigrants. Whereas the latter had been seen as honest, thrifty, law-abiding, and generally "*molto simpatici* to the American character," the southern Italians were viewed as degenerate, ignorant, lazy, dirty, destitute, violent, superstitious, and criminally oriented (Train, 1912, cited in Moquin and Van Doren, 1974). Northern Italians, too, seemed to share the prevalent American view of their southern counterparts as a generally repugnant people (Gambino, 1974). One journalist of the time put the matter quite bluntly: "What shall we do with the 'dago'?" Note his description of the Italian as almost animalistic in nature:

> This "dago," it seems, not only herds, but fights. The knife with which he cuts his bread he also uses to lop off another "dago's" finger or ear, or to slash another's cheek. He quarrels over his meals; and his game, whatever it is, which he plays with pennies after his meal is over, is carried on knife at hand. More even than this, he sleeps in herds; and if a "dago" in his sleep rolls up against another "dago," the two whip out their knives and settle it there and then; and except a grunt at being disturbed, perhaps, no notice is taken by the twenty or fifty other "dagoes" in the apartment. He is quite as familiar with the sight of human blood as with the sight of the food he eats (Moquin and Van Doren, 1974:261).

The most serious elements in the negative image of Italians in America concerned their "criminal nature" and the existence of a widespread Italian criminal community, known as the Mafia, or the Black Hand. Few accounts of Italian immigrants during the late nineteenth century neglected the presumed criminal organizations among them. By the 1890s, the image of the criminally prone Italians and their association with the Mafia was well established (LaRuffa, 1982).

The strong anti-Catholic sentiments of a large segment of the American populace as well as government leaders aroused fear of Italian immigrants and not infrequently incited hostile actions against them. Anti-Italian antagonism was also partially a response of established white workers, who saw the immigrants as a cheap labor force and thereby a threat to their jobs. Killings and lynchings of Italians, mostly in the South but in other regions as well, were numerous between 1890 and 1915. The most serious of these incidents occurred in New Orleans in 1891. Following the murder of the city's police superintendent, it was assumed that Sicilian immigrants were responsible, and hundreds of Italians were arrested. Nine were subsequently brought to trial, but none was found guilty. Anti-Italian sentiment was so feverish, however, that eleven of the suspects were taken from their jail cells by a mob and murdered; officials made no attempt to intervene. Afterward, local newspapers and business leaders expressed approval of the action (Gambino, 1977; Higham, 1963; Nelli, 1970).

Another especially notorious incident occurred in 1927. This was the famous trial of Sacco and Vanzetti. Two Italian immigrants, Nicola Sacco and Bartolomeo Vanzetti, were implicated in an armed robbery in Braintree, Massachusetts. Numerous witnesses attested to their lack of involvement in the crime, but despite the dubious evidence against them, they were found guilty. After numerous appeals, they were executed in 1927. The case created a storm of protest among civil rights groups throughout the country and remained a cause célèbre for many decades afterward. It emerged not only as an ethnic controversy but as an issue of symbolic class warfare. Those defending the innocence of the two saw their trial as a conflict between the oppressed working classes and the wealthy WASP elite (Baltzell, 1964).

The guilt or innocence of Sacco and Vanzetti is still debated, but historians are generally agreed that their trial was considerably biased and that the judicial process was seriously abused. Indeed, the perversion of justice in the case was finally acknowledged in 1977, when the state of Massachusetts officially proclaimed the trial unfair. In addition to the strong anti-immigrant sentiments of the time, the fact that the two defendants were avowed anarchists undoubtedly contributed heavily to their conviction. This was the era of the Red Scare, when government officials fanatically denounced and harassed union and socialist activities, in which Italians often took part. To many, Sacco and Vanzetti represented an affirmation of the perceptions of Italian Americans as radical, conspiratorial, and criminal that were held by much of the American public. These perceptions seriously colored their arrest, trial, conviction, and finally, execution (LaRuffa, 1982).

This case had an enormous and lasting effect on Italian Americans, and for many years it influenced their view of the state and the judicial process. As Gambino has put it, "The outcome of the affair was simply another confirmation of the ancient belief of the Italian immigrants that justice, a very important part of their value system, had little to do with the laws and institutions of the state" (1974:122).

The Durability of Italian Stereotypes

The negative stereotypes of Italian Americans have been troublingly persistent. Although they are rarely expressed today as virulently and hostilely as they were during the period of immigration, many of the early images remain dynamic elements of American ethnic mythology. Few would be hard put to describe the "typical" Italian American. Gambino observed the consistency of responses of non-Italians to his query about the Italian-American image: "the Mafia; pizza and other food; hard hats; blue collar; emotional, jealous people; dusky, sexy girls; overweight mammas; frightening, rough, tough men; pop singers; law and order; pastel-colored houses; racists; nice, quiet people" (1974:352).

These stereotypes are reinforced by media depictions. Lichter and Lichter's study (1982) showed that negative portrayals of Italian Americans in television outnumbered positive ones by two to one, and a disproportionate number of Italian characters were engaged in criminal activity. Difficulty in speaking English was also a commonly projected Italian-American trait. Comical or foolish characters accounted for a very large proportion of all Italian Americans portrayed on television.

The most enduring negative image with which Italian Americans have had to deal concerns the connection with organized crime. Because the Mafia is synonymous with organized crime and Italian Americans are identified with the Mafia, the logic, to many, becomes simple: "Italians = organized crime" (Iorizzo and Mondello, 1980). Although few deny the considerable involvement of some Italian Americans with organized crime, Italians have held no monopoly on such activities. As many have explained, crime has been a traditional avenue of upward mobility for members of most ethnic groups in American society (Albini, 1971; Bell, 1962; Ianni, 1974). Italian domination of organized crime did not emerge until the late 1920s, when Prohibition made trafficking in liquor profitable. Before that time, much of organized crime was controlled by the Irish. As illicit drugs replaced liquor as the major underworld-supplied item, Italian organized crime turned to more legitimate business areas, leaving the drug traffic to newer urban ethnic groups like blacks, Puerto Ricans, and Cubans (Cook, 1971; Ianni, 1974).

Even though Italians did not introduce organized crime to the United States and, in any case, the percentage of the Italian-American community engaged in such activities now or in the past has been small, the connection of Italian Americans generally with organized crime remains steadfast in the public imagery. Movies and television have been important purveyors of this stereotype (Cortés, 1994). Films about the Italian crime syndicate date back to the 1930s and have continued unabated. Movies such as *The Godfather* and *GoodFellas* and television programs such as "The Untouchables" and "The Gangster Chronicles" are reflective of this genre. The general association of Italian Americans with organized crime is also fostered at times by blatant and insensitive advertising. For example, a pizza restaurant chain called "Godfather's" advertised its product on television by showing a sinister character, wearing the prototypical gangster attire—pinstriped suit, wide-brimmed hat—proclaiming Godfather's as "the pizza you can't refuse." Under pressure from Italian-American groups, the firm agreed to modify its advertising but refused to change its name (Commission for Social Justice, 1982).

Prejudice and Discrimination Today

Surveys have regularly recorded the periodic changes in prejudice toward blacks, Jews, Latinos, and Asians. By contrast, there has been little study of shifting public attitudes toward Italian Americans. As we have seen, certain negative stereotypes remain well entrenched, but it is difficult to determine the nature of other aspects of prejudice as they affect this group. What little evidence exists, however, indicates a significant decline in prejudice against Italian Americans in recent decades.

Discriminatory actions are also difficult to measure. Although overt discrimination against Italian Americans has largely disappeared, subtle rejection of Italians in certain occupations and neighborhoods may become evident from time to time. It was revealed in 1960, for instance, that homeowners in the exclusive Detroit suburb of Grosse Pointe had maintained a rigid point system for screening potential home buyers, based on personal traits such as accent, name, swarthiness, and especially, ethnicity (*Detroit News,* 1990; *Time,* 1960). Out of 100 points, Jews were required to score a minimum of 85, Italians 75, Greeks 65, and Poles 55. Blacks and Asians did not count at all! Antidiscriminatory legislation has made such cases today uncommon, but ethnic discrimination in subtle forms, even for white ethnic groups like Italian Americans, is not entirely something of the distant past.

Although objective measures of anti-Italian attitudes and actions are problematic, we might turn to the perspective of members of the group themselves. How do Italian Americans perceive the levels of prejudice and discrimination against them? In his study of Italian Americans in Albany, New York, Alba found that very few had actually experienced discrimination within the five years preceding his survey, though a much larger percentage had heard about discrimination against Italian Americans and had encountered stereotypes about them (Alba, 1994). Crispino's study provides further illumination of the views of Italian Americans. He found that although they see declining discrimination, a substantial percentage feel that there is still much prejudice against them. More than one-quarter of his third-generation sample reported such feelings. As to discrimination, 86 percent reported never having been refused housing, a job, or a promotion on the basis of ethnicity. The highest rates of discrimination were reported among those in higher-class positions, indicating that the occupational framework itself may affect the level of differential treatment. Crispino suggests that as long as Italian Americans remained within the working class, there was no need for discriminatory measures. Now that they have begun to enter higher occupational positions and to compete with the dominant group for power and privilege, discrimination "has become an important weapon in the fight to exclude them" (1980:118).

It is of note that, in her study of ethnic identity among Euro-Americans, sociologist Mary Waters (1990) found that most of her interviewees, white ethnics of a variety of national origins, had themselves not experienced any type of discrimination or prejudice based on their ethnicity. There was some awareness, however, of past discrimination and prejudice against their particular groups.

Ironically, in recent years, Italian Americans themselves have been accused of racist attitudes and actions against blacks. This has resulted from the fact that the two groups have often found themselves competing for the same housing, jobs, and education in the urban environment. Working-class Italian neighborhoods have been more stable than those of most other ethnic groups, and fewer residents have fled to the suburbs with the advance of blacks into the core cities. Thus, it is Italian Americans, particularly in cities of the Northeast, who have perhaps borne more of the effects of urban change than other ethnic groups. And it is blacks who are seen as largely responsible for this social disruption. Significant cultural differences between Italian Americans and blacks have also contributed to intolerance. Gambino (1974) suggests that although racism is partially the basis of hostility between the two groups, Italian Americans are likely to respond vehemently to any group whose cultural patterns are interpreted as threatening to their family-oriented lifestyle. Blacks in central cities are seen in that light. Despite the apparent urban conflict between working-class Italian Americans and blacks, however, studies have shown that when the racial attitudes of Italians are compared with those of other white ethnic groups, they are not different (Greeley, 1977; Taylor, Sheatsley, and Greeley, 1978).

ASSIMILATION AMONG ITALIAN AMERICANS

As we will see in Chapters 8 and 9, the applicability of either assimilationist or pluralist models of ethnic relations in the cases of African Americans and Hispanic Americans has often been debated. Whereas some see these groups as basically on a trajectory toward assimilation, others see them as excluded from the mainstream of American life; an inequalitarian pluralistic model (specifically, internal colonialism) is, the latter contend, more appropriate. For Italian Americans, however, as well as for other Euro-American groups, there can be little argument that assimilation has been the preeminent pattern of their adaptation to American society and their relations with the dominant group.

Although Italian Americans illustrate quite well the assimilation model of ethnic relations, their collective rate and extent of assimilation have differed in the cultural and structural dimensions. They have substantially adopted the dominant cultural ways but have not entered into equal relations with the dominant group to a corresponding degree at either the primary or secondary level. Nonetheless, movement in this direction seems unabating.

Cultural Assimilation

Italian Americans have, with each succeeding generation, shed Italian cultural ways increasingly in favor of those of the core culture. In this they have exhibited a pattern common to most ethnic groups in American society, regardless of national origin or even physical distinctness. Quite simply, the social history of ethnic groups in the United States has been the adoption of Anglo-American cultural patterns. As we noted in Chapter 4, though some minor reciprocal changes are always evident in the interface of minority ethnic and dominant cultures, for the most part the exchange is one-sided.

Indications of cultural assimilation among Italian Americans are apparent in such areas as family patterns, friendship groups, language usage, and diet. The extent of these changes, however, varies among the generations. Whereas those of the third generation and beyond have substantially disposed of traditional ethnic cultural ways, members of the first and second generations have continued to adhere to them more closely. In his study of Italian Americans in Boston's West End, however, Gans (1982) found that even by the second generation most no longer remained attached to the major features of the southern Italian culture. This generation had retained only traditional cooking and foods as well as some use of the Italian language. At the conclusion of his study in 1962, Gans predicted that with each following generation the traditional Italian culture would be further diluted; subsequent research has confirmed his prediction. In a study of Italian Americans in Bridgeport, Connecticut, Crispino (1980) found that even in such areas as food and cooking, acculturation had taken place to a very great extent in later generations. Whereas 44 percent of first-generation respondents continued the tradition of making one's own pasta (which, Crispino asserts, is "a more discriminating test of Italianness than cooking habits in general"), only 15 percent of the fourth generation did so.

Even intensely strong family ties and loyalties—the sine qua non of the Italian-American cultural system—are weakened by the third generation or have begun to be replaced by more common American patterns (Alba, 1985; Crispino, 1980; Gans, 1982). For example, whereas divorce is either taboo or limited in the first two generations, it becomes more accepted in the third. As Lopreato (1970) notes, those aspects of family behavior peculiar to Italian Americans are present to some degree, but they are overshadowed by those aspects they hold in common with other Americans in corresponding social classes.

Perhaps the most critical factor in the cultural assimilation of European ethnic groups in American society is the loss of the mother tongue. Because language is the means by which group ways are transmitted from one generation to the next, presumably as the language of ethnic origin is discarded and replaced by the host society's language, cultural assimilation will proceed apace. The relationship between loss of mother tongue and degree of cultural assimilation seems very strong among Italian Americans. In Crispino's study (1980), 72 percent of second-generation respondents (both parents born in Italy) spoke Italian, but only 14 percent of the third generation (both parents born in the United States) and 10 percent of the fourth generation (those most thoroughly assimilated culturally) did so. This pattern of language conversion is typical of all

Euro-American groups, including those of southern and eastern European ancestry. While many older members of these groups still speak a European mother tongue in the home, among younger people, usually by the third generation, hardly any do (Alba, 1995).

Observers of the Italian-American ethnic group have noted uniformly the relationship between social class and cultural assimilation (Crispino, 1980; Gambino, 1974; Gans, 1982; Lopreato, 1970; Roche, 1982). As class rises, the level of cultural assimilation rises correspondingly. Today the Italian-American ethnic culture remains essentially a working-class phenomenon. As they continue to leave the working class, moving into middle-class occupations and neighborhoods, Italian Americans tend to become indistinguishable from other middle-class Americans. This is especially so among those of the third generation and beyond. These patterns are generally characteristic of all white ethnic groups in American society.

Structural Assimilation

At the secondary, or formal, level of structural assimilation, we have seen that Italian Americans have begun a large-scale movement into the mainstream of American society. Their occupational, income, and educational levels are at or above the average for the society as a whole. Most important, they have begun to enter positions of power within most institutional realms, though slowly and still in only small numbers at the top echelons. Clearly the process of structural assimilation at the secondary level is proceeding irrevocably.

Within the context of personal social settings (that is, at the primary group level), however, the pace and degree of structural assimilation for Italian Americans are not so high. As with cultural assimilation, much of the difference among individuals in rate and extent of primary structural assimilation is related to social class and generation. For example, Gans (1982) found that second-generation West Enders in Boston maintained social relationships almost wholly within the ethnic community; for later generations, however, particularly those that had experienced upward mobility, close friendships were founded on common class interests and lifestyles, not ethnic or kinship ties (see also Alba, 1994; Crispino, 1980; Roche, 1982).

Residential Patterns A key indicator of primary structural assimilation is residence. The greater the residential integration of members of a minority ethnic group with those of the dominant group, the more completely they are assimilated structurally. As for Italian Americans in this regard, observers have often noted the long-lasting cohesiveness and stability of their neighborhoods. Indeed, some point to the neighborhood as one of the key elements, along with the family, in the chemistry of the Italian-American ethnic group (Glazer and Moynihan, 1970).

This evident stability of residential areas may be mostly a class-related phenomenon, however, rather than something uniquely Italian American. Working-class people are generally more stable residentially primarily because their opportunities for spatial mobility are more limited than those of middle-class people. Therefore, because most first- and second-generation Italian Americans are working class, they have been inclined to remain in their original central-city neighborhoods. When mobility into the middle class occurs, however, there appears to be as much propensity among Italian Americans as among other ethnic groups to leave the ethnic neighborhood for the more heterogeneous suburbs (Crispino, 1980; Gans, 1982). Alba (1995) found that in the greater New York metropolitan region, where they constitute one of the largest ethnic categories, Italian Americans in the 1980s were somewhat less likely than the average white to reside in suburbs rather than the central city, but the continued suburban movement among them was nonetheless evident. By 1990, 70 percent lived in suburbs.

Moreover, the residential solidarity of Italian Americans has often been exaggerated. Studies have shown that in Chicago and Philadelphia, for example, Italians from the outset moved frequently from their original areas of settlement, regardless of their social class. The "Little Italys" of various cities were in any case apparently never as homogeneous as has commonly been assumed (Hershberg et al., 1979; Lopreato, 1970; Nelli, 1970). At an early point of the new immigration, some of the European groups, including Italians, were highly segregated in northern cities, to be sure. But this pattern changed quickly and radically in the early years of the twentieth century (Lieberson, 1980). Today, even where Italian-American ethnic communities are evident, their homogeneity is evaporating. In the greater New York region, Alba found that "The Italians, some of whose ethnic communities are still conspicuous, reside mostly in non-ethnic areas, and their continuing suburbanization is eroding the most ethnic Italian neighborhoods" (1995:12).

Intermarriage Perhaps the most critical indicator of primary structural assimilation is the extent to which members of an ethnic group have intermarried with those of the dominant group or other out-groups. For Euro-American ethnic groups, there appears to be an increasing tendency toward marriage outside the group with each successive generation. Indeed, among whites, most marriages today involve some degree of ethnic intermixing (Alba, 1995). As Waters has noted, "For white ethnics, the longer a group has been in the United States and the greater the percentage of its members in later generations, the lower the in-marriage ratios" (1990:104).

Among Italian Americans specifically, Alba (1981) found that whereas older, mostly first-generation, individuals showed virtually no mixed ancestry, two-fifths of younger Italian Americans, mostly third- and fourth-generation, reported mixed ancestry. In this, Italian Americans were in line with other American Catholics of European descent. Religious endogamy, however, appeared to remain strong. That is, though they married increasingly across ethnic lines, Italian Americans continued to marry primarily other Catholics (Crispino, 1980).[4] By the 1990s, an even stronger tendency toward interethnic marriage was apparent among Italian Americans as well as other southern and eastern European ethnic groups. Almost three-quarters of younger Italian Americans had spouses without Italian ancestry, and for Polish Americans the out-marriage figure was even higher (Alba, 1995).

The Future of Ethnicity Among Italian Americans and Other White Ethnics

It is clear that the strength of ethnicity for Italian Americans and other white ethnic groups like them diminishes in most ways as they move increasingly from the working class to the middle class. With upward mobility, social class rather than ethnicity becomes the most significant determinant of job, school, neighborhood, lifestyle, friendship circle, and even marital partner. For these groups, there is also strong evidence to support what has been called the "straight line theory," which posits that movement toward more complete assimilation proceeds irreversibly with each successive generation (Gans, 1979; Sandberg, 1974).

Does this spell the eventual disappearance of the Italian-American ethnic community and others like it? This development may lie in the distant future, but it is unlikely in the lifetimes of contemporary white ethnics or their children. Whereas certain social forces are propelling assimilation relentlessly, others serve to retard the process.

This does not mean, however, that the form of ethnicity for Italian Americans and other white ethnic groups in the United States will be the same as it was for their parents or grandparents. Although they will probably continue to assert an ethnic identity and retain a few of the major elements of the ethnic culture, their acknowledgment of ethnicity will be largely expressive rather than instrumental. Gans (1979) has suggested the idea of "symbolic ethnicity" to describe the continued and at times even accentuated expression of ethnicity among those of the third and fourth generations. Ethnicity, he explains, becomes a matter of voluntary personal identity, but it is largely devoid of vitality as a determinant of social behavior. It is, as Richard Lambert (1981) has put it, vestigial, something of the past that people can identify with but that has a very limited effect on their daily lives.

---

[4]Alba (1981), however, reports that all Catholic ethnic groups, including Italian Americans, are marrying increasingly across religious lines as well.

Some, however, see the continuation of ethnicity among these groups as more than symbolic. They point to the apparent revitalization of ethnicity among whites during the late 1960s and early 1970s as evidence of the enduring function of the ethnic group in providing a referent of personal identity and a sense of rootedness in an increasingly bureaucratized and depersonalized society (Greeley, 1971; Novak, 1972). Sporting buttons and bumper stickers proclaiming "I'm Polish and proud" or "Kiss me, I'm Irish," third- and fourth-generation whites seemed to develop a newfound ethnic consciousness.

But this "new ethnicity" was short-lived. By the late 1970s, the resurgence of ethnic identity and the prominence of ethnic-related issues among whites had waned considerably. Given the movement toward assimilation of several generations, the ethnic revival was, as Steinberg (1989) contends, "doomed from the outset." Indeed, Alba suggests that ethnicity could be celebrated during the 1970s "precisely because assimilation had proceeded far enough that ethnicity no longer seemed so threatening and divisive" (1981:96).

Its transitory nature also seemed to indicate that the new ethnicity was more a political movement, specifically, a working-class political movement, than a revival of ethnic cultures and communities. Because social class has never served in the United States as an effective political mobilizing force, ethnicity has often functioned in this role. This is in large measure what occurred in the late 1960s and early 1970s in cities where white working-class people—mostly southern and eastern European ethnics—remained in significant numbers (Gans, 1974, 1979). As they perceived the economic and political concessions gained by blacks and other minorities from the federal government and other societal institutions, working-class white ethnics responded with a countermovement, seeking to enhance and protect their interests—jobs, neighborhoods, schools, and so forth. Hence, what were in actuality class issues and conflicts were manifested and interpreted as ethnic ones. In a sense, ethnicity served as a substitute for class-based actions.

This kind of political ethnicity, in which ethnic groups serve as competing interest groups rather than as cultural islands, may be the chief form and function of ethnicity today for those few segments of white ethnic groups that still have substantial core-city communities (Bell, 1975; Glazer and Moynihan, 1970). For those members who with upward mobility continue their residential dispersion, however, ethnicity seems destined to function largely in a symbolic fashion.

Euro-Americans: A New Ethnic Group? As explained in Chapter 5, the sharpest lines of ethnic division in American society today are between whites and nonwhites. As ethnicity for white ethnics continues to decline as a consequential factor in various areas of social life, the boundaries of white ethnic groups become less and less distinct. In fact, Lieberson and Waters (1988) have found that a substantial and growing portion of the white population is apparently unaware of its ethnic origins, does not identify with any of the commonly understood ethnic groups, or simply refuses to identify its ethnic ancestry. They refer to this grouping as "unhyphenated whites."

One of the forces contributing to this blurring of ethnic lines and the tendency for Euro-Americans to deemphasize ethnicity, at least in an instrumental fashion, is the enormous intermixing of ethnic groups that has been occurring continuously for the past 150 years. Today, it is increasingly difficult for people to declare their ethnic ancestry without recording two and often several additional categories. For example, they may declare themselves "Irish," but "Italian" as well. And ethnic intermarriage continues unabated (Lieberson and Waters, 1988). The result is that ethnic identity for Euro-Americans has become largely a matter of choice. And as ethnicity becomes more voluntary, resulting in fading ethnic boundaries, "in a sense," writes Richard Alba, "a new ethnic group is forming—one based on ancestry from *anywhere* on the European continent" (1990:3).

Waters (1990) has studied these fading ethnic divisions and finds that, increasingly, white ethnics, like Italian Americans, make conscious choices about their ethnic identity. Not only can they choose their ethnic affiliation, but they can choose whether or not to emphasize ethnicity in their lives at all. As she explains, "Ethnicity is increasingly a matter of personal preference" (1990:89). Given the strong rate of intermarriage among white ethnics, each generation's ethnic background becomes more complex and difficult to clearly define. Individuals therefore begin to consider the issue of ethnicity and begin to identify themselves in ethnic terms that reflect only a part of what may be their very elaborate ethnic heritage. People tend to stress only one part of their complex ethnic ancestry for any of a number of reasons. They may know more about one parent's ancestry than another, or their surname or physical characteristics may conform to stereotypical images of the chosen group, or they may perceive certain of their ancestral lines as more socially desirable than others (Waters, 1990). In any case, for white ethnics, ethnicity is increasingly voluntary.

Yet, the fact remains that most people *do* choose and continue to affiliate with some ethnic group. Thus, ethnic identity, which may be in large part uncertain or even fictitious, continues to play a symbolic function for many people. However, it has little or no real bearing on their social behavior or economic standing. Thus, they may carefully pick and choose their ethnic affiliation, but, as Waters explains, "It is clear that for most of them ethnicity is not a very big part of their lives" (1990:89). Ethnicity becomes mostly something that provides a feeling of community as well as an identification that may set one off as unique and interesting. It does not impact on vital life decisions, like where to live, whom to associate with, or even whom to marry.

The tendency for individuals to make choices about their ethnic identity demonstrates that ethnicity is fluid and can change in significance within one's lifetime or from one generation to the next. It is not fixed.

In considering these "ethnic options," as Waters refers to them, it is important to understand that the kind of freedom to select the content and strength of one's ethnic identity is not the same for racial-ethnic groups. Whereas for Euro-Americans the question "What are you?" can be answered as the individual chooses, for members of racial-ethnic groups, the answer will have already been given by out-group members. African Americans, Latinos, Asians, and Native Americans do not, therefore, enjoy the same kind of options in the realm of ethnicity. "The reality," notes Waters, "is that white ethnics have a lot more choice and room for maneuver than they themselves think they do. The situation is very different for members of racial minorities, whose lives are strongly influenced by their race or national origin regardless of how much they may choose not to identify themselves in ethnic or racial terms" (1990:157). Failure on the part of Euro-Americans to acknowledge that difference often leads to their belief that ethnicity need play no greater role for African Americans or Hispanic Americans if they so choose. This creates misconceptions about the continuing strength of ethnicity and its impact on the lives of members of racial-ethnic groups.

SUMMARY

Italian Americans are representative generally of groups that derive from the classic (or new) immigration period from the late nineteenth to the early twentieth centuries. Entering the society at the bottom of the class hierarchy, Italian Americans by the third generation had displayed significant upward mobility. Today, they are at or above the national average in income, occupation, and education, and they have begun to enter positions of political and economic power, though slowly and still in small numbers.

Beginning with large-scale immigration in the late nineteenth century, Italians in the United States were the objects of prejudice and discrimination, sometimes of a particularly virulent nature. Perhaps the most persistent negative stereotype with which Italian Americans have had to deal involves their identification with organized crime.

Italian Americans today display a relatively high level of cultural assimilation, having substantially adopted the ways of the core culture. Although it is less thorough and rapid, structural assimilation is also occurring steadily. As this process continues, for Italian Americans as for other white ethnic groups, ethnicity is increasingly reduced to a symbolic function and becomes less a determinant of social behavior.

Suggested Readings

Alba, Richard D. 1990. *Ethnic Identity: The Transformation of White America.* New Haven, Conn.: Yale University Press. Examines the waning strength of ethnic identity among Euro-Americans, suggesting that assimilation has reduced ethnicity for these groups to a largely symbolic form.

Gambino, Richard. 1975. *Blood of My Blood: The Dilemma of the Italian-Americans.* Garden City, N.Y.: Anchor. Describes the culture and social institutions of Italian Americans.

Gans, Herbert J. 1982. *The Urban Villagers: Group and Class in the Life of Italian Americans.* Updated and expanded ed. New York: Free Press. A classic work that is required reading not only for its examination of the Italian-American experience but also for its presentation of issues regarding the retention of ethnic identity and the intersections of ethnicity and social class.

Lieberson, Stanley, and Mary C. Waters. 1988. *From Many Strands: Ethnic and Racial Groups in Contemporary America.* New York: Russell Sage Foundation. Analyzes the current meaning of ethnicity for Euro-American groups, like Italians, drawing on census data.

Lopata, Helena Znaniecki. 1976. *Polish Americans: Status Competition in an Ethnic Community.* Englewood Cliffs, N.J.: Prentice-Hall. A sociological analysis of another Euro-American ethnic group, which like Italian Americans, immigrated mostly in the late nineteenth and early twentieth centuries.

Mangione, Jerre, and Ben Morreale. 1992. *La Storia: Five Centuries of the Italian American Experience.* New York: HarperCollins. An account of the Italian-American experience from colonial times to the present.

Rieder, Jonathan. 1985. *Canarsie: The Jews and Italians of Brooklyn Against Liberalism.* Cambridge, Mass.: Harvard University Press. An ethnographic study of the response of a white ethnic community, largely Italian and Jewish, to the political and economic inroads of blacks.

Waters, Mary. 1993. *Ethnic Options: Choosing Identities in America.* Berkeley: University of California Press. Using census data as well as in-depth interviews, the author reveals how Euro-Americans choose their ethnic identities.

Yans-McLaughlin, Virginia. 1982. *Family and Community: Italian Immigrants in Buffalo, 1880–1930.* Urbana, Ill.: University of Illinois Press. An especially good analysis of the family as the key institution of Italian immigrants in their adaptation to American society.

THE JEWISH EXPERIENCE in American society in some ways resembles the Italian. Both ethnic groups were primarily part of the new immigration, and following large-scale entrance, both found themselves initially at the bottom of the class hierarchy. As non-Protestants who had emigrated from relatively backward societies, both groups were made-to-order targets of the nativist and racist ideas of the turn of the century. Furthermore, because both groups remained concentrated primarily in eastern cities, they did not blend easily into the larger society. The prejudice and discrimination they faced, therefore, proved more pernicious and dogged than those aimed at other white ethnic groups.

But in many ways, as we will see, Jewish Americans are a unique case among American ethnic groups. As people who for centuries had suffered persecution and dislocation in almost all European societies, Jews found the United States a haven that provided for the first time a social atmosphere in which Jewish identity could be retained without fear of official repression and from which there was no need to contemplate flight. Jews in America also found their economic circumstances relatively unconstrained, in sharp contrast to the situation they had faced in Europe. So well nurtured were the opportunities offered them that Jews today represent the prototypical American ethnic success story. In only two generations, Jewish Americans collectively accomplished a truly phenomenal upward mobility that makes theirs a compulsory case for the analysis of American ethnic relations.

Unlike other American ethnic groups, Jews do not identify with a common homeland, having come from many different nations. The broad geographical native region of the majority of Jewish Americans, however, is the same — Eastern Europe — and a common language, Yiddish (a combination of mostly German and Hebrew elements), supersedes most national differences. Jews in the United States are best seen as an ethnic group because they clearly display the key features of such groups: a unique cultural heritage, a consciousness of community, and a group identity. Regardless of their degree of commitment to Jewish religious practices and beliefs (many are nonbelievers), and regardless of their national origins, Jews recognize themselves as a people with common roots.[1] They are seen in that fashion, as well, by the larger society.

Moreover, even the lack of a common homeland does not in itself constitute a significant difference from most white ethnic groups in the United States. In the majority of cases, immigrants from southern and eastern Europe arrived with little national consciousness; regional or village boundaries generally defined their sense of peoplehood. Only in the United States, as a means of social adjustment and in response to out-group recognition, did most adopt a national identity.

IMMIGRATION AND SETTLEMENT

Jewish patterns of immigration and settlement in the United States paralleled in many ways those of the Italians. The bulk of Jewish immigration occurred during precisely those years of heaviest Italian immigration, and like the Italians, Jews initially settled mostly in the large cities of the Northeast. Many of the occupations of the two groups during the major period of immigration also overlapped, as did their residential areas. But important differences between these two groups led, as we will see, to different rates of mobility in the second and third generations and to different positions in the class system.

Sephardic Jews

Jewish immigration to the Americas has a lengthy historical tradition. In fact, a few Jews supposedly accompanied Columbus on his first voyage. The earliest immigrants to the United States were Sephardic Jews, whose ancestors were Spaniards and Portuguese. When the Jews were expelled from Spain and Portugal in the late

[1] A national survey in 1990 found that most Jewish Americans define their Jewish identity in ethnic or cultural, not religious, terms. In fact, the survey found a low level of positive support for the religious group concept. Thus, a majority of Jewish Americans consider themselves Jews primarily as members of an ethnic group (Kosmin et al., 1991).

fifteenth century, many migrated to England, Holland, Brazil, and the Caribbean islands. The first American Jews came from all these locations and even from the Iberian peninsula (Glazer, 1957). The Sephardim are to be distinguished from the Ashkenazic Jews of central and eastern Europe, who constituted almost all Jewish immigration into the United States after the colonial period.

German Jews

A second and more sizable Jewish immigration to the United States was made up of German Jews who came with the great wave of other German immigrants during the 1840s and 1850s. These Jews were socially and economically several notches below the earlier Sephardic Jews, who by that time were relatively prosperous and respected. Although most were merchants, their trade was at a level considerably less significant than that of their Sephardic predecessors. Many, in fact, were peddlers who moved westward with the country's expansion. Indeed, at the middle of the nineteenth century, a majority of America's 20,000 itinerant traders were German Jews (Lipset and Raab, 1995). German Jews dispersed widely to all parts of the country, in both large cities and small towns. Where they settled, they commonly established clothing and dry goods or general stores, the vestiges of which are seen today throughout the United States. A few developed into large national chains. Many familiar department store names, like Macy's, Bloomingdale's, Saks Fifth Avenue, Sears, and Neiman-Marcus, stem from German-Jewish founding families.

Collectively, the German Jews displayed rapid and substantial upward mobility, and by the 1880s, most were middle-class entrepreneurs. Among the most successful of them developed a small, aristocratic upper class, which intermingled socially with the Gentile upper class and even intermarried among them (Baltzell, 1958, 1964). By the turn of the century, German Jews were so thoroughly assimilated that they had lost much sense of ethnic identity. Only their religious affiliation distinguished them from other Americans (Herberg, 1960).

East European Jews

By 1880, Jews were still only a tiny fraction of the American population, numbering no more than 250,000. In the next three-and-a-half decades, however, two million Jews would come to the United States. It is primarily from this wave of immigration that the contemporary Jewish-American population derives.

Jews coming to the United States after 1880 differed markedly from those who had settled earlier. To begin with, most were not from Germany but from eastern Europe, particularly Russia and Russian-occupied Poland. One of the key motivating factors behind the mass emigration of Jews from Russia was the severe oppression they were subjected to under tsarist policies. Following the assassination of Tsar Alexander II in 1881, bloody pogroms were sanctioned, and onerous restrictions were placed on Jews in all areas of social and economic life. Unlike the Italian emigration, which at first comprised mainly single men expecting to return, the Jewish emigration from Russia was a movement of family units that harbored few thoughts of returning to their homeland.

The East European Jews also differed from earlier immigrant Jews in their class origins: They were poorer and less educated. Jewish life in eastern Europe had been confined to all-Jewish villages and towns, and as a result, these immigrants were extremely provincial in contrast to the relatively urbane and worldly German Jews. Striking differences between the two groups also lay in their attitudes toward religion. The East Europeans came with a version of Judaism far more traditional than the modernized and Americanized Reform Judaism of the Germans. To the German Jews, Judaism was simply a faith that was not to interfere with assimilation into the core American culture and society. To the East Europeans, coming from the *shtetl,* or Jewish village, where religion governed all aspects of life, Judaism was an entire social world. As Glazer and Moynihan explain, "In practice, tone, and theology, the Reform Judaism of the German Jews diverged from the Orthodoxy of the immigrants as much as the beliefs and practices of Southern Baptists differ from those of New England Unitarians" (1970:130).

In addition to differences in class and religious practice, East European Jews did not disperse geographically as had the earlier German Jewish immigrants. Instead, they remained in a few large cities, especially New York, where they were concentrated in lower-class ghettos. In this environment, the East European Jews, like the Italian immigrants of this time, experienced strikingly harsh conditions of work and residence. Many labored in garment factories, commonly called sweatshops, where they worked long hours for scant salaries. In 1911, one-third of Jewish heads of household earned less than $400 a year, and in 1914, the average hourly wage of male clothing workers was about 35 cents. Even by 1930, garment workers were still earning an average of only $24.51 a week, less than the pay of workers in stockyards and slaughterhouses (Howe, 1976). As a result, most families became work units in which women and children as well as men labored fifteen or eighteen hours a day, often taking piecework home with them.

Like the Italians who streamed into New York and other large eastern cities at the same time, the East European Jews accepted housing conditions that were frequently brutal and demeaning. They, too, crowded into rundown tenements, abandoned by previous ethnic groups. On the famous Lower East Side, 350,000 people per square mile jammed into squalid rooms and apartments in which sanitation facilities were meager at best. These degraded conditions of life only further confirmed the racist views of those who saw immigrant Jews of this time, along with other southern and eastern Europeans, as innately degenerate.

In sum, the urban environment into which immigrant Jews entered the United States in the late nineteenth and early twentieth centuries required struggle and stamina, which could be sustained only by the expectation that life conditions would improve—possibly for them but surely for their children. "The generation that entered the immigrant ghetto," writes Ben Halpern, "was confronted by one overwhelming task: to get out or enable the next generation to get out" (1958:36). This goal proved attainable for most. Perhaps one important factor in accounting for the determination of immigrant Jews to succeed in American society was their understanding that there was no return to the homeland; for most Jews, the commitment to the United States was complete.

Given their striking differences in class, religion, and culture, it is little wonder that the German and East European Jews were initially antagonistic. The Germans viewed the new immigrants as illiterate, uncouth, and provincial greenhorns, who could only cause embarrassment to the American Jewish community and produce a backlash of antisemitism. By the 1890s, however, the divisions between the two groups were evaporating. Inspired partially by humanitarian motives and partially by the concern that they might be lumped with these poor and religiously traditional Jews in the eyes of their Christian neighbors, German Jews now accepted responsibility for uplifting and assisting their East European cohorts (Wirth, 1956). By the end of large-scale immigration in the early 1920s, the subdivision of the American Jewish community into German and East European groups was no longer apparent (Glazer and Moynihan, 1970; Yaffee, 1968).

Jewish Demography Today

Jewish Americans number about 5.6 million, little more than 2 percent of the total U.S. population (American Jewish Committee, 1994). This represents a declining Jewish population in both absolute and relative terms. The decrease is due to an unusually low birthrate among Jews as well as an increasing rate of intermarriage. Even though they are only a small part of the American population, and a declining one as well, Jews in the United States constitute over 40 percent of the world's Jewish population.[2]

Jews, like Italians, are among the most urbanized of American ethnic groups. In fact, three-quarters of the entire Jewish population in the United States is found in nine large urban areas (Table 7.1).[3] In the early part of this century, over half of all U.S. Jews lived in New York City alone, but today less than 20

[2]Augmenting the Jewish-American population in the last two decades have been immigrant Jews from the former Soviet Union and from Israel. Together, however, these groups constitute only a small portion of Jews in the United States (Gold, 1994).

[3]For a study of Jews in small towns and rural areas, see Rose (1977).

Table 7.1 Metropolitan Areas with Largest Jewish Populations

METROPOLITAN AREA	ESTIMATED JEWISH POPULATION
New York	1,450,000
Los Angeles	574,000
Southeast Florida	533,000
Northeastern New Jersey	359,000
Philadelphia	314,500
Chicago	261,000
Boston	228,000
San Francisco	210,000
Washington, D.C.	165,000
Baltimore	94,500
Detroit	94,000
Rockland County, New York	83,100
Cleveland	71,300
San Diego	70,000
Atlanta	67,500
St. Louis	53,500
Phoenix	50,000
Denver	46,000
Pittsburgh	45,000
Houston	42,000

SOURCE: American Jewish Committee, 1994.

percent do (Ritterband, 1995). Geographically, the American Jewish population is relatively dispersed, though East and West Coast cities predominate. Similarly, at the local level, densely Jewish neighborhoods of the past are largely gone.

JEWISH AMERICANS IN THE STRATIFICATION SYSTEM

That most Jews of the new immigration entered the society in relative poverty made their rapid upward mobility quite remarkable. As a group, Jews rose virtually from the bottom to the top of the stratification system in two generations. On all measures of social class—income, occupation, and education—Jews collectively rank higher than most other American ethnic groups.

Table 7.2 Occupational Distribution of Jewish and U.S. Males

OCCUPATIONS	JEWISH MALES[a]	U.S. MALES[a] (WHITE)
Professionals	39.0%	15.8%
Managers	16.7	14.3
Clerical and sales	24.4	17.5
Workers	19.9	52.5
Total	100.0	100.0

SOURCE: Goldstein, 1992.
[a]Percentages may total more than 100 due to rounding.

The high socioeconomic position of Jewish Americans is quite evident in family income. The median family income of Jewish Americans in the early 1970s was $12,630, more than $2,000 higher than the median for all U.S. families (Massarik and Chenkin, 1973; U.S. Bureau of the Census, 1973). By 1986, 77 percent of Jewish-American households earned more than $20,000, compared to 61 percent of American households in general (Cohen, 1987; U.S. Bureau of the Census, 1988). By 1989, the median Jewish-American household income was $39,000, about $10,000 higher than for the society as a whole (Kosmin et al., 1991; U.S. Bureau of the Census, 1995b).

Commensurate with their relatively high income, Jews display a disproportionate concentration in higher-status occupations, namely, the professions, business ownership, and managerial positions. Although the German Jews had, for the most part, established themselves as part of the middle class by the late nineteenth century, the rise of the East European Jews since the cessation of large-scale immigration in the 1920s has been even more striking (Chiswick, 1984). Only in New York City, containing the largest concentration of Jews in the United States, does there remain a substantial Jewish working class (Glazer and Moynihan, 1970; Goldstein, 1980; Sklare, 1971). Quite simply, as Calvin Goldscheider has stated, "poverty and unskilled occupations are largely uncharacteristic of contemporary American Jews" (1986:4). Table 7.2 demonstrates vividly the concentration of Jews in high-status occupations. Notice the extremely high percentage of Jewish professionals and the comparatively low percentage of workers.

Finally, the extremely high level of educational attainment among American Jews is well recognized, having been noted repeatedly in group studies (Goldscheider, 1986). Indeed, the high income and occupational levels of Jews can be

Table 7.3 Educational Attainment of Jewish and U.S. Adults

EDUCATIONAL LEVEL	JEWISH ADULTS	U.S. ADULTS
Less than college degree	28%	62%
Some college	19	17
College degree	53	21

SOURCE: Goldstein, 1992.

accounted for in large measure by their similarly high educational level. As is seen in Table 7.3, over 50 percent of Jews, compared to 20 percent of adult Americans in general, were college graduates in 1992. If education is the key to upward mobility in American society, Jews are the classic case of an ethnic group having converted educational opportunities into economic betterment in a relatively short time.

Ethnic and Class Factors in Upward Mobility

The saga of Jewish upward mobility is often proclaimed as the great success story of American ethnic history. Indeed, no other ethnic group has seemed to rise so far so quickly. As a collective rags-to-riches case, Jewish Americans are often held up as an example for other groups to emulate. What has provided for this extensive collective mobility in only two generations? Here we might compare the Jews with other white ethnic groups, such as the Italians, that entered the United States on a large scale at precisely the same time yet have not exhibited a similar degree of upward mobility. This comparison may give us strong clues to why different ethnic groups experience varying rates of success in the United States and in other multiethnic societies as well.

Herbert Gans has explained that in looking at the mobility patterns of ethnic groups, it is essential to try to understand whether what happens to them is "ultimately more or less a function of their characteristics and culture than of the economic and political opportunities which are open to them when they arrived and subsequently" (1967b:8). That is, patterns of mobility among Jews, Italians, or any ethnic group can be seen as the result of either the group's cultural traits or the opportunities to which the group is exposed, or perhaps a combination.

Group Culture Some contend that differences in ethnic achievement are attributable mainly to internal characteristics of the groups themselves (Fauman, 1958; Hurvitz, 1958; Lipset and Raab, 1995; Rosen, 1959; Schooler, 1976;

Sowell, 1981; Strodtbeck, 1958). The crux of this argument is that some groups are better enabled to succeed occupationally by virtue of certain cultural values and behaviors. Presumably, those groups that have come to the United States with values similar to or compatible with those of the core group—individual achievement, competition, pragmatism, thrift—are likely to rise more rapidly.

The substance of this view is contained in what Rosen (1959) has termed the "achievement syndrome." Groups differ, he explains, in their orientation toward achievement. Some, like Jews, encourage independence and stress achievement values and high educational aspirations in children, whereas others, like Italians, do not. Italian Americans traditionally stressed accomplishment only as it benefited the family in some concrete fashion. Individual achievement, however, was deemphasized because it destabilized the structure of the family unit and threatened its authority. Glazer and Moynihan write that "that form of individuality and ambition which is identified with Protestant and Anglo-Saxon culture, and for which the criteria of success are abstract and impersonal, is rare among American Italians" (1970:194–95). Among Jews, however, it is preeminent. Thus, instead of plying their father's trade, as did most second-generation Italians, Jews aspired to higher occupational positions. As Philip Gleason explains, "Unlike Catholic peasant immigrants, the Jewish immigrants did not regard intensive education and professional careers as something beyond their experience or capacity; they saw them rather as their natural lines of aspiration" (1964:168).

Perhaps the most frequently cited cultural value in explaining Jewish mobility, particularly as it contrasts with the more limited mobility of other ethnic groups of the new immigration, is the group's attitude toward education. The value of learning has a rich tradition in Jewish culture extending as far back as biblical times, and it permeates all aspects of Jewish life (Fauman, 1958; Gleason, 1964; Strodtbeck, 1958). As a result of this key cultural trait, it is held, Jews have quite naturally achieved proportionately greater success in business, academia, science, and the arts than have other ethnic groups.

The Jewish penchant for education is contrasted sharply with the traditional southern Italian apprehension about this institution. Rather than a means to social and economic advancement, education was viewed by immigrant Italians as having little practical value and in some ways antithetical to family and group interests (Covello, 1967). Hence, they did not avail themselves of the educational opportunities in the United States as quickly and exhaustively as did the Jews. Some have observed that even later generations of Italian-American parents, including those of the middle class, have not overly pressured their children for achievement. "By their norms," concludes one study, "motivation as well as ability should come from the child, not the parent" (Johnson, 1985:189).

Intense family loyalty as part of the southern Italian culture was especially consequential for immigrants and their children. With an allegiance to family above all else, individual Italian Americans found it difficult to pursue occupations that did not concretely improve the family's welfare. The most important work objective was to find a steady, secure job that would contribute to the family's immediate well-being and that would not take the individual outside the realm of the family's influence. Such values restrained upward job mobility (Gambino, 1974; Gans, 1982).

Obviously the choice of fulfilling short-range work goals instead of seeking the long-range advantages of continuing education was not unique to Italian immigrants and their children. But the norm was perhaps more firmly rooted among them. There is little doubt, then, that their skeptical view of formal education contributed heavily to their slower and more limited upward mobility compared to Jewish Americans.

Another culturally based argument holds that, in comparison with other ethnic groups, Jews invest more heavily in their children. Because Jewish families are, on average, smaller than those of most other ethnic groups, Jewish parents, it is argued, are able to devote more time and other resources to child rearing. In turn, Jews are better able to translate their education into prestigious occupations and high labor market earnings. Moreover, more stable family living arrangements (lower rates of divorce, separation, out-of-wedlock births), in this view, additionally improve child quality (Chiswick, 1984).

Group Opportunities Disputing the contention that ethnic groups differ in achievement primarily because of internal group values, some maintain that differences depend mostly on the opportunity structure that groups encounter on entering the society and the skills they bring to that structure (Gans, 1982; Steinberg, 1989). The essence of this view is that class factors in general explain more about mobility than do ethnic cultures.

Although not denying differences in group culture, sociologist Stephen Steinberg (1989) argues that Jewish immigrants came to the United States with an advantageous occupational background that enabled them to cultivate the American economic system better than other ethnic groups could. Russian Jews entering the United States were not peasants but were urban people who had been engaged in manufacturing and commerce. Consequently, although they were certainly poor, they were equipped with skills that would serve them well in the expanding industrial economy. Italians and most other groups of the new immigration were, by comparison, mainly agricultural peasants. Moreover, Jewish occupational skills complemented almost perfectly the needs of the rapidly growing garment industry, where so many Jewish immigrants found employment. "Jews," notes economist Eli Ginsberg, "had the good fortune to be in the right place at the right time" (1978:115). Moreover, as the economy later shifted from agriculture and manufacturing to services, Jews benefited from their background in trade and commerce.

Glazer (1958) also points out that East European Jews who came to the United States had had a long heritage of middle-class occupations, including the professions and buying and selling, by contrast with other immigrant groups. Although tsarist oppression had driven Jews from their traditional trades, their link to them was not broken. They came to the United States with little capital, then, but not with the limited horizons of the working class. In effect, the East European Jews were part of the working class for only one generation. They were neither the sons nor the fathers of workers (Herberg, 1960).

Steinberg also questions the attribution of Jewish occupational success to the high value attached to learning. Although he does not deny that the higher rate of literacy among immigrant Jews compared, for example, to that of southern Italians was instrumental in their more rapid mobility, he maintains that substantial educational advancement *followed* economic mobility rather than the converse. After looking at historical data, Steinberg concludes that for Jews "economic success was a precondition, rather than a consequence, of extensive schooling" (1989:136). Sowell (1981), too, denies that education was the cause of mobility for Jews and, like Steinberg, explains it as the result of economic success. By the time Jews became a substantial part of New York's college population, he notes, mobility was already well under way.

In a similar fashion, the presumably lower educational and occupational aspirations of immigrant Italians can be accounted for by class as much as by ethnic factors. That is, their attitudes toward education and work, particularly among the first and second generations, may have reflected their primarily working-class status, not simply their southern Italian heritage. Gans (1982) explains that many of the behaviors and values he observed among second-generation Italian Americans in Boston's West End were not so much ethnic traits as working-class traits. Thus, the limited mobility aspirations among working-class Italians might be expected as well among, for example, working-class Poles, working-class blacks, or even working-class Jews. As Gans has observed, "I continue to be impressed by how similarly members of different ethnic groups think and act when they are of the same socioeconomic level, are alike in age and other characteristics, and must deal with the same conditions" (1982:277–78).

Looking at Italian-American achievement and attitudes toward education as the products primarily of a unique ethnic trait also fails to take into account the opportunity structure in the old country, which dictated limited aspirations. "To whatever extent Southern Italians exhibited negative attitudes toward education," writes Steinberg, "in the final analysis these attitudes only reflected economic and social realities, including a dearth of educational opportunities" (1981:142).

Historian John Briggs (1978) also discounts the idea of a predisposed negative view of education among southern Italians. He attributes their rejection of schooling in the United States, instead, to their expectation of an imminent return to Italy. It was not a lack of interest in education but the lack of relevance of American schooling that led to its widespread repudiation, he contends. Schooling in the new country could be of little value if one's stay was only temporary.

The Italian pattern also seems to verify the thesis that generally only *after* economic gains have been secured is education ordinarily extended. As we have seen, Italian Americans are only now beginning to attain educational levels in line with those of other ethnic groups even though they earlier attained an above-average level of income (Greeley, 1977). As educational levels rise, of course, the effect is circular, and Italian Americans will no doubt display subsequent rises in occupation and income. Recent data indicate that although Jews have progressed further and faster, Italians and other Catholic ethnic groups are now catching up (Alba, 1981; Greeley, 1977; U.S. Bureau of the Census, 1993).

To put the matter simply, those who interpret different patterns of mobility among Jews and Italians in class rather than ethnic terms stress that, whatever the extent of a group's occupational and educational background or tradition, only when it is combined with a favorable opportunity structure can it enhance mobility. For Jews, there was a fortunate compatibility between the skills and experience they brought with them and the needs and opportunities of the American economic structure. And as Steinberg suggests, "It is this remarkable convergence of factors that resulted in an unusual record of success" (1981:103). Valid comparisons among ethnic groups can be made, then, only when consideration is taken of the society's constantly changing opportunity structure.

Evidence can be marshaled to support both the group culture and opportunity structure arguments regarding ethnic mobility, but as with most issues in sociology, the truth more likely lies somewhere between the two. On the one hand, had Jews not possessed at least some propitious cultural traits, including among them a reverence for learning, it is not likely that such rapid mobility would have been possible even within the context of a favorable opportunity structure. Educational opportunities in the United States existed for other European ethnic groups, but it seems evident that immigrant Jews availed themselves of these opportunities more thoroughly and avidly than did others. Moreover, Jews in other immigrant societies such as Canada, Argentina, and South Africa have displayed, as a group, economic and educational success comparable to that attained in the United States and have pursued similar occupations. On the other hand, there is little question that immigrant Jews brought with them a wider range of occupational skills that automatically placed them a notch ahead of other ethnic groups in the opportunity structure. As we will see in Chapter 9, the rapid economic success of Cuban immigrants in the United States in the past two decades, compared to other contemporary Hispanic groups, poses a quite similar case, as does the upward mobility of many Asian groups.

Political and Economic Power

Jewish participation at the higher levels of government has been relatively greater than that of Italians and most other southern and eastern European groups (Alba and Moore, 1982). Two qualifications to this observation must be made, however. First, Jewish participation in the realm of American political power is a relatively recent phenomenon. Although Jews have historically been in advance of Catholics in their appointment to positions in the federal cabinet and judiciary, this is due to the prominence of many earlier German Jews on the American political scene. Indeed, Jews did not serve in the federal establishment in a really significant way until Franklin D. Roosevelt's administration (Isaacs, 1974). Second, when measured in absolute terms, the Jewish presence in the higher ranks of government, even today, is not oversignificant. Because Jews occupy a disproportionate number of congressional positions that attract much national publicity, the significance of the contemporary Jewish-American political presence can sometimes be misread.

A few Jews were elected to Congress as early as the nineteenth century, but only in recent years have they been elected in meaningful numbers. Before 1945, only fifty-nine Jews had ever served in either house. Today, in light of their minor proportion of the general populace, Jews appear to be somewhat overrepresented in the federal legislature (see Table 6.3). At the national level, they are most strongly represented, however, as political appointees and civil servants (Alba and Moore, 1982).

In a few cases, Jews served as state governors in the late nineteenth century, but these were unusual occurrences (Lieberson, 1980). In recent years, Jews have been elected governor only in a few eastern states where their electoral power is of some consequence. Similarly, at the local level, Jews, like Italians, tend to be prominent mainly in those areas where they constitute a substantial constituency. But this, too, is a recent development. Consider that a Jew was not elected mayor of New York, the city with the largest Jewish-American population, until 1974.

It is important to consider that Jews today experience few impediments to elective office at any level of government. In fact, a 1987 Gallup Poll revealed that 89 percent of Americans would vote for a well-qualified Jewish presidential candidate; only 46 percent felt that way in 1937 (Gallup, 1982, 1988; see also Quinley and Glock, 1979). The acceptance of Jews in higher politics is demonstrated as well by the fact that an increasing number of Jewish candidates are being elected to offices by overwhelmingly non-Jewish constituencies. Jewish

Americans have consistently shown a greater propensity than other ethnic groups for political participation, and this may account in some measure for their increasing prominence in elective offices.

With regard to economic power, despite their collectively high occupational status as well as their significant numbers among the society's affluent, Jews are underrepresented as part of the American corporate elite (Alba and Moore, 1982; Korman, 1988; Zweigenhaft and Domhoff, 1982). The strong Jewish participation in the business world traditionally has been within small or medium-sized firms (most of which are self-owned), not the larger, more powerful corporations that dominate the American economy (Institute for Social Research, 1964; Kiester, 1968; Marger, 1974; Ward, 1965). Furthermore, where they have been successful in the corporate world, Jews more commonly have entered the elite through the growth of their own companies rather than by climbing the organizational ladder of well-established concerns.

The relative underrepresentation of Jews in the corporate elite is ironic in light of the commonly held assumption among most Americans that Jews wield inordinate economic power. Studies beginning in the 1930s have consistently shown this to be a misconception (Burck, 1976; *Fortune,* 1936; Keller, 1953; Newcomer, 1955; Warner and Abegglen, 1963). The notion of excessive Jewish economic power may persist in part because of the extreme visual prominence of those few industries in which Jews do exert a dominant influence, most notably movies and television, retail sales, and real estate. These are businesses with which most people are familiar and with which they have frequent dealings, in contrast to industries like automobiles, steel, insurance, and commercial banking, which remain largely obscure to the general public. The association of Jews with these few highly conspicuous industries may thus contribute to the public illusion of more widespread Jewish influence than is actually the case (Glazer and Moynihan, 1970; McWilliams, 1948; Selznick and Steinberg, 1969).

The historical underrepresentation of Jewish Americans in the top echelons of the corporate world can be explained in part as discrimination and in part as a propensity for Jews to prefer independence in business and the professions. The two are not unrelated. Faced with discriminatory hiring practices, which did not decline significantly until after World War II, Jews naturally turned to self-owned businesses or the independent professions, like law and medicine.

Have traditional patterns of discrimination in the corporate world changed in recent years? Korman (1988) maintains that they have not. Investigating evidence of corporate recruiting, he found that even as late as the 1980s, firms in many of the most important segments of the corporate economy continued to recruit more commonly at schools with few Jewish undergraduates. He concluded that at the managerial/executive level, Jews remain "outsiders" in the corporate world.

Other studies, however, indicate that today the traditional pattern of Jewish absence from the corporate elite appears to be changing, with Jews increasingly entering the top executive positions of major corporations (Klausner, 1988; Silberman, 1985; Zweigenhaft, 1984, 1987). Whatever discrimination in large corporations and law firms continues to exist may be simply the lingering effect of past occupational barriers against Jews that, for the most part, have been formally taken down. Indeed, a study of the late 1980s concluded plainly that "the disadvantage of Jews in the executive suite has all but disappeared" (Klausner, 1988:33).

PREJUDICE AND DISCRIMINATION

For centuries, Jews have been a favorite target of ethnic animosities in Western societies. As non-Christians they were particularly vulnerable to scapegoating during times of social adversity and were generally relegated to the role of economic and political outcast. Jews were made especially ready targets for oppression after the thirteenth century, when they were forced into separate communities in almost every European society in which they resided. The term *ghetto* was first used to describe the area of Venice in which Jews were compelled by custom and law to live. It was subsequently applied to the Jewish quarter of all European cities.[4] Jews were forbidden to own land or join craft guilds and maintained an ever-tenuous civil status. The capricious nature of persecution and the lack of political security forced them to remain constantly vigilant and prepared to flee. As a result of the need to remain mobile, they turned to occupations such as tradesman, moneylender, or professional, which provided for easy movement of capital and skills (Fauman, 1958; Glazer, 1958). These occupational roles often carried social stigmas, which further provoked resentment and hostility from those who were required to deal with them.

The Elements of Antisemitism

A set of distinct and consistent negative stereotypes, some of which can be traced as far back as the Middle Ages in Europe, has been applied to Jews. Among the most common of these are that Jews are monied, dishonest, and unethical; clannish, prideful, and conceited; and power hungry, pushy, and intrusive (Quinley and Glock, 1979). It is the connection of Jews with money, however, that appears to be the sine qua non of antisemitism.[5] As Glock and Stark

[4]On the development of European ghettos, see Wirth (1956).

[5]*Antisemitism* is the term most commonly used to describe hostility toward Jews.

put it, "Perhaps the most constant theme in antisemitism from medieval times down to the present is of the Jew as a cheap, miserly manipulator of money, forever preoccupied with materialism, and consequently possessing virtually unlimited economic power" (1966:109). As late as 1995, over a quarter of non-Jewish whites supported the assertion that "When it comes to choosing between people and money, Jews will choose money"; even larger percentages of African Americans, Latinos, and Asian Americans agreed (National Conference, 1995).

That Jews have more money than most people is, as Selznick and Steinberg (1969) explain, accepted as fact among a majority of non-Jews though this in itself does not necessarily imply antisemitism. Nonetheless, even unprejudiced non-Jews exaggerate the extent of Jewish wealth, and it is, of course, only a short step to interpreting this belief in a negative manner. Although Jews in American society are, as we have seen, collectively higher in socioeconomic status than most other ethnic groups, this does not mean that all Jews are therefore wealthy. Like members of all ethnic groups, Jews will be found at every class level, including the very bottom.[6]

Closely akin to the beliefs regarding the connection of Jews with money are those concerning Jews and business. "Of all the crimes and misdemeanors charged to the Jews," notes Stephen Whitfield, "the worst has been an uncanny or even supernatural capacity to make money, perverted into avarice" (1995:90). The shrewd and unscrupulous businessperson has been a tenacious negative stereotype of Jews for ages. Derivative of this image is the pejorative American phrase, to "Jew down," whose meaning is, roughly, to bargain craftily or to buy something more cheaply than the seller initially offers it. The beliefs regarding Jews and business deal not only with ethics, however, but also with power and manipulation. Jews have been commonly seen as maintaining excessive influence in the business world and as controlling large segments of the economy of the United States and of other societies. The fallaciousness of this image has been confirmed repeatedly, and as earlier noted, Jews have, in the past, been underrepresented in the higher executive positions of the most powerful American and multinational corporations.

The attributions of Jewish power are often blatantly contradictory. One of the staples of crude antisemitism concerns the notion of world conspiracy. Curiously, Jews have been portrayed, on the one hand, as leaders of a world capitalist conspiracy and, on the other, as leaders of a world communist conspiracy. This inconsistency has never deterred antisemites from expressing both beliefs when the situation calls for it. Indeed, as McWilliams so aptly suggested, there

---

[6]For a description of the Jewish poor, a majority of whom are elderly, see Levine and Hochbaum (1974).

is a tone of "demagogic genius" in this contradictory charge, "for it permits an appeal to the dispossessed and a threat to the rich to be voiced in a single sentence" (1948:92).

Antisemitism in the United States

Historically, three general periods of antisemitism in the United States can be defined, each characterized by a different level and content. During the early decades of Jewish settlement, before 1880, anti-Jewish prejudice and discrimination were minimal. With the influx of the East European Jews in the late nineteenth and early twentieth centuries, harsh and overt antagonism developed in response to the much larger number and very different character of these immigrants. The antisemitism that arose during this period peaked in the late 1920s and 1930s and did not abate until after World War II. Since the 1950s, prejudice and discrimination toward Jews have declined considerably though they remain evident in subtle and subdued forms.

The Early Period Although there were anti-Jewish groups in colonial America, they were not excessively vocal or influential (Blau and Baron, 1963). During the era of the old immigration, between 1820 and 1880, German Jews were not essentially distinguished from their Christian counterparts except in religion, and as we noted earlier, they mostly assimilated into the mainstream society. Like the early northern Italians, German Jews (and the earlier Sephardic Jews) were too insignificant numerically to evoke intense hostility. Furthermore, they were geographically dispersed and established themselves in economically secure positions, thereby avoiding the attention of nativist groups, which targeted the much larger, city-concentrated, and destitute Irish Catholics.

This relatively concordant relationship with the Anglo-Protestant majority began to change, however, with the entrance of the East European Jews in the 1880s. A harbinger of the changed attitude toward all Jews occurred in 1877 when Joseph Seligman, a very prominent financier and political adviser, was denied entrance to the Grand Union Hotel in Saratoga, New York, a noted upper-class resort (Baltzell, 1964; McWilliams, 1948). From that point on, Jews were commonly barred from certain resorts, clubs, and college fraternities.

The Development of Popular Antisemitism Like the southern Italians, immigrant East European Jews became prime subjects of the racist ideologies of the late nineteenth and early twentieth centuries. Just as the criminal tendency attributed to Italians was simply part of "their nature," the association of Jews with money was seen as deeply rooted in this group's heritage and perhaps even as genetic. Consider this brief passage from Jacob Riis's influential essay of 1890, *How the Other Half Lives:*

> As scholars, the children of the most ignorant Polish Jew keep fairly abreast of their more favored playmates, until it comes to mental arithmetic, when they leave them behind with a bound. It is surprising to see how strong the instinct of dollars and cents is in them. They can count, and correctly, almost before they can talk ([1890] 1957:84).

This description, it should be noted, was not necessarily written in a derogatory vein. Indeed, the essay was essentially sympathetic to the plight of both immigrant Italians and Jews. Yet its racist assumptions are obvious. The racist element of antisemitism was, of course, carried to its ultimate by Hitler and the Nazis during the 1930s and 1940s, but its genesis can be traced to much earlier ideas and their exponents in the United States as well as Europe.

During the 1920s, American antisemitism was strongly abetted by the activities of the Ku Klux Klan and of Henry Ford, the noted industrialist. The Klan's antisemitic and anti-Catholic propaganda and actions during this time were as widespread and severe as its antiblack campaigns. As for Ford, his antisemitic program lasted from 1920 to 1927, during which time his newspaper, the *Dearborn Independent,* issued virulent attacks on Jews.[7] The "world Jewish conspiracy" idea was a regular component of the *Independent,* which used the notorious *Protocols of the Elders of Zion* as its basis. This was a well-known antisemitic tract circulated originally in the early 1900s by the Russian secret police. Its thesis was a Jewish conspiracy to achieve world domination through control of finance and banking, and it was used by the Russian tsars as a rationale for their oppressive policies and actions against the Jews. Though its contentions were repeatedly disproved, this manifesto was subsequently circulated widely, and its theme became a standard item of the litany of antisemitism. The *Protocols* formed much of the ideological basis of Hitler's murderous anti-Jewish campaign in the Nazi era. Ford later apologized to the American Jewish community for his attacks, saying that he had been misled, but many of the ideas espoused by the *Independent* remained in wide circulation during the 1930s.

Although the antisemitism of this era was extremely vituperative, it rarely generated violence. Most of the discrimination against Jews involved their expulsion or exclusion from various areas of social life, including clubs and resorts, neighborhoods, schools and colleges, and certain occupations (McWilliams,

[7] McWilliams (1948) writes that the rise of the Ku Klux Klan and Ford's antisemitic campaign were not unrelated. The Klan began to attract a mass following in 1920 when Ford launched his attack.

1948). As Jews began to occupy a disproportionate number of seats in elite colleges and professional schools in the early 1920s, restrictive policies were enacted, limiting their entrance. Strict Jewish quotas were established at schools such as Harvard and other Ivy League institutions as well as medical and law schools throughout the country (Baltzell, 1964; Bloomgarden, 1957; Korman, 1988). Jews were also severely discriminated against in the professions, especially medicine, law, and academia. Jewish doctors, for example, were often refused appointments to hospital staffs and were denied clinical privileges. Similar restrictions were imposed by law firms and university faculties.

The rising level of anti-Jewish prejudice and discrimination in the 1920s and 1930s can be seen as a part of the popular anti-immigrant and isolationist sentiments that so characterized that period. But it can also be seen as a response to Jewish upward mobility. Before the early years of the twentieth century, Jews were fairly well removed from the general occupational structure, concentrating mainly in a few industries, like the garment trade, or operating their own small businesses. By the second decade of the century, however, second-generation Jews were beginning to break out of the insular Jewish community and challenge—more times than not, successfully—members of the dominant group, particularly in the more prestigious occupations and educational settings. Artificial barriers, in the form of quotas and other discriminatory policies, were therefore thrown up. The anti-Jewish attitudes among the well-to-do that emerged in the 1870s have been interpreted in much the same manner, as social barriers that arose in response to competition for privilege (Higham, 1963; McWilliams, 1948).

The Post–World War II Period In the 1950s and, particularly, the 1960s, antisemitism in the United States declined precipitously. The anti-Jewish prejudice and discrimination that persisted took on more camouflaged and restrained forms.

The decline of negative Jewish stereotypes among the general American public has been demonstrated by several comprehensive studies and opinion polls (Chanes, 1994; Gallup, 1982; Glock and Stark, 1966; Quinley and Glock, 1979; Selznick and Steinberg, 1969; Stember, 1966; Williams, 1977). The unscrupulous businessperson, for example, is no longer a foremost Jewish image among most non-Jews. Nearly half the respondents in polls taken in 1938 and 1939 believed that Jewish businesspeople were less honest than others. A 1952 poll reported a similar percentage. By 1962, however, this stereotype had diminished considerably; 70 percent felt that Jewish businesspeople were no different from others although the notion that they were shrewd in business remained strong (Stember, 1966). By the early 1980s, those adhering to this stereotype had further declined (Raab, 1989).

A significant change was also registered in the perception of Jewish economic and political power. Public belief in the extensiveness of Jewish influence seemed to peak in the 1930s and 1940s. Forty-one percent of Americans in 1938 thought that Jews had too much power, and 58 percent thought so in 1945 (Quinley and Glock, 1979). In 1952, 35 percent of Protestants and 33 percent of Catholics still answered yes to the question "Do you think the Jews are trying to get too much power in the United States or not?" But in 1979, the percentages of Protestants and Catholics answering yes were only 12 and 13, respectively (Gallup, 1982). Opinion polls throughout the 1980s indicated similarly that only a small percentage viewed Jews as having excessive power and influence (Smith, 1991).[8]

In the realm of personal relations, negative attitudes toward Jews have also declined markedly. In a 1948 survey, 22 percent expressed the view that they would prefer not to have Jewish neighbors. By 1964, only 8 percent answered similarly, a figure that did not essentially change by 1989. Even intermarriage with Jews, perhaps the most sensitive area of interethnic relations, was viewed negatively by only 34 percent of non-Jews in 1981 (*Public Opinion,* 1987; Smith, 1991).

What can account for this noticeable decline in antisemitic thought and action during the past few decades? Though such social trends are never attributable to a single factor, the general rise in educational level among the American populace does stand out as an extremely important agent. Research and public opinion polls indicate consistently that antisemitism is related to educational level. To put the matter simply, the lower the educational attainment, the greater the degree of antisemitism. For example, surveys of attitudes toward Jews have repeatedly found college-educated people more tolerant than those with less formal education (Gallup, 1982; Gilbert, 1988). Thus, as the level of education has increased for Americans generally, anti-Jewish attitudes and actions have waned correspondingly (Glock and Stark, 1966; Lipset, 1987; Quinley and Glock, 1979; Selznick and Steinberg, 1969).

Too much significance should not be given to educational and class levels in explaining the causes of antisemitism, however. More correctly, the forms of antisemitism are different, depending on education and class. Some researchers have shown that those who are economically most secure and most educated are not necessarily more supportive of minority rights than other social classes (Hamilton, 1972; Huber and Form, 1973). Prejudiced members of higher-class groups may be more sophisticated in publicly proclaiming tolerant attitudes while continuing to deny Jews and other minorities full participation in various institutions.

[8] Some recent surveys, however, seem to indicate a reversal of this trend. In 1995, the National Conference reported that 22 percent of non-Jewish whites, 37 percent of Latinos, 35 percent of Asian Americans, and 45 percent of African Americans held the belief that Jews "have too much control over business and the media." Similar percentages of each group felt that Jews "are more loyal to Israel than to America" (National Conference, 1995).

Moreover, despite the clear decline of prejudice and discrimination, Jews continue to be denied entrance into certain fraternal groups, recreational clubs, and similar social organizations. In 1991, for example, the exclusion of Jews by the Kansas City Country Club prompted Tom Watson, the professional golfer, to resign from the club in protest. These restrictive policies constitute an aspect of antisemitism that has been relatively untouched by both antidiscriminatory laws and the general tendency toward more ethnically open social organizations (Braverman and Kaplan, 1967; Selznick and Steinberg, 1969; Slawson and Bloomgarden, 1965). A study of Jewish life in Los Angeles (Sandberg, 1986) showed that Jews themselves perceive more anti-Jewish discrimination in private clubs than in any other area of social life.

Jews and Blacks In recent years, antisemitism has been more noticeable among blacks in American society than among other population elements (Golub, 1990). This is paradoxical in that Jews have traditionally maintained a more sympathetic attitude than other segments of the white population toward black interests and have been consistently supportive of black efforts to secure civil rights (Smith, 1990b). The apparent rise in black antisemitism is accounted for by several factors. First, the emergence of a more militant ideology within the black movement in the 1960s led to a wide-scale purge of whites from black civil rights organizations, within which many Jews had been strongly active. This created some distance between the two groups. Also, militant black rhetoric in the 1960s encouraged blacks to begin to assert control over their neighborhoods and communities. Because Jews in many large cities owned and operated a substantial number of small businesses and apartment houses in black areas, it was very often Jews with whom blacks were forced to deal in economic matters. As a result, Jews served as the closest and most visible symbols of white oppression and the most vulnerable targets of black rage.

A second major factor concerns the implementation of affirmative action programs. The past use of quotas to limit Jews in colleges and professional schools, occupations, and neighborhoods weighs heavily in the resistance by many Jewish leaders and organizations to the establishment of racial quotas today in these same social areas. Even though quotas today are intended to uplift the status of ethnic minorities, many Jews, having themselves experienced their adverse effects, see them as discrimination (albeit a kind of reverse discrimination) against which they have labored for several decades (Glazer, 1975).

Third, Black Muslim leader Louis Farrakhan, along with some black scholars like Leonard Jeffries, have, in recent years, rekindled that antisemitic rhetoric in their speeches and writings. They have portrayed Jews as having been especially instrumental in the slave trade and in the oppression of blacks generally in

America, claims that have been strongly denounced and refuted by mainstream historians and other scholars (Jordan, 1995).

The Tenaciousness of Antisemitism Today, though prejudice and discrimination against Jews remain persistent among certain segments of the American populace, they are mainly random in occurrence, not organized and widespread as in the 1920s and 1930s. Racist anti-Jewish notions are almost nonexistent except among those who are part of the right-wing lunatic fringe of American sociopolitical life. Even black antisemitism does not represent more than what Robin Williams calls "situational antagonisms rather than deep-seated traditionalized hostilities" (1977:6). The antisemitism of most Americans today does not extend much beyond support of mild forms of discrimination and acceptance of certain traditional Jewish stereotypes.

It is important to note that even during those times when it was intense and wide ranging, antisemitism in the United States differed from European forms in that it was never institutionalized, and no official anti-Jewish measures were enforced (Konvitz, 1978). Had this not been so, it is unlikely that Jewish Americans as a group could have attained their current high position in the society's stratification system.

Furthermore, despite the persistence of negative images of Jews, positive images of this group are also common in American society (Martire and Clark, 1982; Quinley and Glock, 1979). Jews are ordinarily seen as intelligent, religious, and family oriented and are often admired for their accomplishments in science and the arts. Indeed, several years ago, C. P. Snow, the famous British scientist and author, suggested that Jews might very well be genetically superior to other ethnic groups. Recognizing this as a thesis of reverse racism, many Jewish leaders were quick to denounce Snow's suggestion. Of course, some of the admiration of Jews expressed by non-Jews may be a manifestation of the tendency to look at the Jewish ethnic group in a negative light while expressing positive ideas about some of its individual members. As was explained in Chapter 3, this is a common pattern of thought among the prejudiced, regardless of the group in question, and is best reflected in the familiar "some of my best friends are...".[9]

---

[9]Although ethnic prejudice is a negative group evaluation, not all the traits alleged to typify a group are necessarily negative. Thus, in the case of the Jews, they may be seen by those prejudiced against them as intelligent, clever, and especially adept at business, but these traits are applied in such a way as to describe Jews as people who nefariously use their intelligence or business acumen to cheat and deceive non-Jews. Glock and Stark (1966:120) point out that a theme in antisemitism "is that the Jew constitutes a grave threat because he is such a worthy opponent."

Although antisemitism in the United States today is probably as mild and subdued as it has been at any time or place in modern Western history, this does not mean that more malignant and intense anti-Jewish thought and action might not be evoked under certain circumstances. As one researcher has explained, "Jews are recognized as an ethnic and religious out-group and are judged and treated in a distinctive manner accordingly" (Smith, 1991:27). Prejudice and discrimination are, it should be remembered, tools applied at varying times and degrees to protect the social, economic, and political advantages of some groups over others. For example, the overrepresentation of Jews in leadership positions of the mass media, particularly television and the movies, has produced frequent charges with antisemitic intimations of Jewish "control" of the media. Even international events may spark latent antisemitism — for instance, an Arab oil embargo or Israeli military actions in the Middle East.

The suddenness with which reinvigorated antisemitism may appear is evidenced by the rise in anti-Jewish incidents starting in the late 1980s. The Anti-Defamation League of B'nai B'rith, a group that assiduously tracks trends in ethnic prejudice and discrimination, announced in 1989 that it had discovered a five-year high in such incidents. The League found a 41 percent increase in reported cases of antisemitic harassment, including threats and assaults against Jews and Jewish institutions, as well as increases in anti-Jewish arson, bombings, and swastika daubings. It also reported a sharp upsurge of anti-Jewish incidents on college campuses. The League attributed the increase in antisemitic actions to the activities of neo-Nazi organizations, especially youth groups calling themselves "skinheads," as well as a response to Israel's actions in quelling the uprising of Palestinians in its occupied territories (*New York Times,* 1989).

These tendencies did not seem to abate in the 1990s. The Anti-Defamation League continued to report a general upward trend both in antisemitic attacks on Jewish-owned property and in physical assaults on people (*New York Times,* 1994; Niebuhr, 1996). The emergence of right-wing extremist groups, rabidly antisemitic, created increasing concern as their propaganda and violent actions seemed to have a public impact that went well beyond their tiny numbers. Moralistic campaigns among leading political figures against the mass media, particularly films, also seemed to take on antisemitic overtones, given the strong influence of Jews in that industry. Even if these increases prove only transitory, they demonstrate how venerable are anti-Jewish stereotypes and how confounding is the phenomenon of antisemitism. Jews themselves perceive the continued potential of serious outbreaks of antisemitism. Surveys indicate that the majority of Jewish Americans feel that antisemitism in the United States either is already or might become a serious problem (Cohen, 1985; Kosmin et al., 1991).

Jewish-American patterns of assimilation, like other features of ethnic life and relations, have to a large degree resembled those of other white ethnic groups. Cultural assimilation has taken place increasingly and irrevocably with each succeeding generation. Structural assimilation, however, has been slower and less profound, especially within the realm of primary relations.

Cultural Assimilation

Each successive generation of Jewish Americans has displayed cultural characteristics less uniquely Jewish than the previous one. And as with other white ethnic groups, much of this assimilation is related to upward class mobility. As they have moved from the working-class occupations and urban ghettos of the first generation to the middle-class occupations and suburbs of the second and third, Jews have steadily adopted lifestyles and values that characterize middle-class Americans generally. Like Italians of the second generation, the children of Jewish immigrants to some extent found themselves wavering between the ghetto and the larger society (Kramer and Leventman, 1961). Those beyond the second generation, however, have clearly chosen the dominant culture.

Perhaps the most significant indicator of cultural assimilation among Jewish Americans is the gradual movement, beginning with the second generation, away from the Orthodox branch of Judaism, encompassing traditional and rigid forms of worship and belief, to the Conservative and Reform branches, each reflecting more Americanized practices and doctrines.[10] It is of note that the division of Jews into three denominational subgroups is, in the main, an American phenomenon. In non-American Jewish communities, people are either Orthodox or nonreligious.[11]

Not only has there been a movement toward less rigorous forms of worship, but religious observance in general has diminished with each generation (Gans, 1958, 1967a; Kleiman, 1983; Sklare, 1969; Wertheimer, 1989, 1993). Although synagogue affiliation (not attendance) remains strong among even those of the third and fourth generations, researchers have cautioned against interpreting this as a resurgence of Jewish religious observance (Gans, 1967a; Sharot, 1973;

[10]Heilman and Cohen's study (1989) suggests, however, that Orthodox Judaism continues to survive and has, in fact, attracted increasing numbers of younger Jews in recent years.

[11]In this regard, Konvitz (1978) notes that although the Reform movement began in Germany, it was most fully developed by German Jews in the United States. Wertheimer (1993) describes a fourth branch of American Judaism, Reconstructionism, subscribed to by only 1 percent of adult Jews.

Sklare, 1969). Rather, the synagogue today serves as the key institution by which Jews can express their ethnic identity, acquaint their children with their Jewish heritage, and interact with others of like ethnicity. Its purely religious aspects are of minor importance. Thus, assimilation with the dominant culture, in a broad sense, continues, while Jews retain their ethnic identity.

Structural Assimilation

Jewish Americans have exhibited steady and in some ways substantial progress in the direction of secondary structural assimilation. We have seen that they have begun to participate fully in almost all areas of American society, exceeding collectively almost all other ethnic groups in levels of income, occupation, and education. Jews have also begun to enter elites in most institutional realms though to a greater extent in some than others.

Structural assimilation at the primary level, however, is a more complicated matter. In general, Jewish Americans have traditionally exhibited strong in-group cohesion in the realm of primary relations (Gans, 1958; Lenski, 1963; Ringer, 1967; Sklare and Greenblum, 1967). This stems from attitudes and actions of the larger society and from the in-group preferences of Jews themselves. Sklare (1978) explains the ambiguity among out-groups regarding Jewish–Gentile social relations. Jews have commonly been seen as "cliquish" when they retreat to their own communities but as "pushy" when they make serious efforts to enter Gentile primary groups. Fearing rejection or at best a marginal acceptance, Jews more often than not were reluctant to leave the safety and psychological comfort of the ethnic group. As a result, concentrated Jewish residential areas, unlike black ghettos, which developed primarily as a result of overt discriminatory actions and policies, evolved through a combination of out-group discrimination and in-group choice. In addition, Jews, like other ethnic groups, initially congregated in relatively homogeneous areas where ethnic institutions were close at hand.

Today, discrimination and the attraction of ethnic institutions no longer play major roles in determining residential patterns, but strongly Jewish neighborhoods are still evident in cities with large Jewish populations. These seem to be a product mostly of housing markets, age, and socioeconomic factors, however, rather than the desire to live in Jewish areas (Goldscheider, 1986). Where they do not constitute a relatively sizable group, of course, Jews tend to be residentially dispersed (Rose, 1977). Moreover, Jewish Americans have generally followed the pattern of other white ethnics in dispersing to suburbs, where ethnic homogeneity is much less apparent. Even there, however, assimilation is not necessarily an unavoidable result. In his study of Boston, Goldscheider found that Jews living in neighborhoods of low Jewish density were not alienated from other Jews and that such individuals were not "on the edge of ethnic survival" (1986:40).

It is in the realm of intermarriage that Jewish Americans today give the greatest indication of accelerating primary structural assimilation. Like Italians and other southern and eastern European groups, Jews are intermarrying increasingly with members of the dominant group as well as other ethnic out-groups.

Until the 1960s, intermarriage among Jewish Americans was very low. Indeed, the rate of out-group marriage among Jews was lower than for other white ethnic groups. Change in this traditional pattern, however, is now evident. Data clearly point to a very substantial increase in Jewish intermarriage in recent decades (Goldstein, 1980; Kleiman, 1983; Massarik and Chenkin, 1973). From 1970 to 1990, the rate of intermarriage more than quintupled, from 9 percent for Jews married before 1965 to 52 percent for those married after 1985 (Steinfels, 1992). Thus, more Jews now marry outside the ethnic group than within.

Increasing Jewish intermarriage is attributed to several factors. Jews in recent years have entered occupational areas and social groups previously closed to them, placing them in social settings in which there is increased interaction with non-Jews. Moreover, as we have seen, cultural assimilation has weakened ethnic traditions among younger Jews. Attitudes toward marriage outside the ethnic community are therefore not so uncompromisingly negative as they were for older generations. Further, as we noted earlier, there is an increasingly tolerant view of the general populace toward intermarriage.

It is important to remain cognizant of the fact that, regarding assimilation, the Jewish-American population, like all other ethnic groups in American society, is internally variegated. Some individuals will remain solidly attached to the ethnic group and retain a strong ethnic identity while others will shed completely their ethnic attachments. Between these two extremes, a wide range of assimilation will be evident. The Jewish-American population today seems to comprise three general components (Cohen, 1995; Liebman, 1973).

The "involved" or "affiliated" are immersed in Jewish life and are religiously observant. They belong to a synagogue, have close Jewish friends, and are self-consciously Jewish despite their adoption of non-Jewish values and behaviors. The "moderately affiliated" or "associated" engage in some Jewish practices and maintain close Jewish friendships but are not deeply involved in organized Jewish life or ritual. These are mainly younger, third- and fourth-generation persons. Finally, the "peripheral" or "nonassociated" are, at best, nominally Jewish. If they maintain Jewish friends, it is only accidental. Such people may be in the arts, professions, or other occupational areas in which a freer, more open social atmosphere prevails.[12] Cohen (1995) estimates the breakdown as "involved," 36 percent, "moderately affiliated," 41 percent, and "peripheral," 23 percent.

---

[12]On this point, see Gordon (1964).

The Future of Ethnicity Among Jewish Americans

Patterns of assimilation and pluralism among Jewish Americans, although in some ways similar to those of other white ethnic groups, have been and will probably continue to be unique in other ways.

The forces of assimilation, barring the unlikely resurgence of overt forms of prejudice and discrimination, seem no less irresistible for Jews than for other white ethnics in American society. In considering the status of Jewish Americans today, Seymour Martin Lipset and Earl Raab have written that, although they are not on the verge of disappearance, nonetheless "current evidence suggests that group identity and cohesiveness are severely eroding for the large majority" (1995:47). With upward mobility, geographic dispersion, and increased intermarriage, what is left for most individuals is ethnicity that is essentially symbolic, having dwindling impact on social life. Indeed, many Jewish leaders have expressed concern in recent years over the threat to the continued existence of the Jewish-American community given the accelerated pace of acceptance of Jews into the mainstream society and the decline of blatant antisemitism. Such fears have been reinforced by declining birthrates and increasing intermarriage among the Jewish-American population. One survey indicated that among Jewish Americans in general, over 60 percent recognized serious dangers to the survival of the ethnic group in light of the current rates of assimilation and intermarriage (Cohen, 1987).

But despite objective indications of continued cultural and structural assimilation, ethnic identity among Jewish Americans has, to some extent, been strengthened in the last several decades. Some, in fact, strongly challenge the view that full assimilation is an eventuality. They see not an erosion of Jewish ethnicity but an adaptation to new conditions, which will assure the future integrity of the group in spite of foreboding demographic trends (Cohen, 1988). Two events of the modern era account in largest measure for the revived sense of Jewish community and identity: the Nazi holocaust and the founding of Israel as a nation-state. Both of these events served to make Jews in the United States more conscious of their heritage and instilled a sense of collective commitment to the preservation of the Jewish people.

This reinvigorated Jewish identity, however, must be seen in much the same way as the resurgent ethnic identity among Italian Americans and other white ethnic groups: It is mostly symbolic. Jews continue to identify as Jews and to abide by some of the major rituals of Judaism such as attendance at synagogue during high holy days and having their sons (and increasingly their daughters) bar (bas) mitzvahed. But for most, the expression of their Jewishness does not extend much beyond these few practices. Put simply, a majority of American Jews have drifted away from religious participation (Wertheimer, 1989, 1993).

More specifically, Ritterband suggests a dichotomy in which the American Jewish community is becoming bimodal: "...the more committed Jews have increased their commitment while less committed Jews have more and more opened themselves up to the forces of assimilation" (1995:378). The latter appear to constitute the majority.

As observers have pointed out, however, the strength of Jewish communalism more than compensates for the weakness of Jewish religious associations (Lenski, 1963; Sandberg, 1986; Waxman, 1981). The commonality of social class and high educational achievement creates common lifestyles and, therefore, social bonds. Thus, even with a decline of religiosity, there remains a strong force for the continuation of Jewish identity. "The concentration of Jews in particular jobs and in college and post-graduate educational categories," notes Calvin Goldscheider, "link Jews to institutions, networks, families, neighbourhoods and political interests and therefore has become a powerful basis of ethnic continuity" (1995:136). Moreover, it should be noted that, despite their peripheral commitment to Jewish religious ritual and even belief, very few Jews convert to Christianity.

Contemporary Jewish identity is sustained, as well, by an awareness that, though their place in American society is probably as secure as it has been in any society in modern history, Jews still remain the potential object of differential treatment. The historical roots of antisemitism are deep and abiding. Ethnic identity, therefore, is a product of eternal vigilance, instilled by centuries of persecution. Jewish identity is reinforced also by the perception of exclusion. Despite their positions of prominence in academia, the arts, business, and politics, Jewish Americans still feel they are outsiders, and thus not entirely part of the centers of secure influence (Cohen, 1985).

The marginality of Jews in American society has been substantially reduced in recent decades, but there nonetheless remains a fundamental cultural difference from their Christian neighbors—religion. This has created a kind of collective schizophrenia for Jewish Americans. As one writer has put it, they live "in a state of tension between the two values of integration and survival, and, regardless of where they stand on a survival-integration continuum, they find themselves pulled in both directions" (Liebman, 1973:134). Jewish Americans will therefore continue to be unique in their path toward assimilation.

SUMMARY

In many ways, Jews collectively represent the prototypical American ethnic success story. Paralleling the immigration and settlement patterns of other groups of the new immigration, the East European Jews subsequently outpaced all others in upward social mobility.

The high skill level of Jewish immigrants entering an industrializing economy provided broad opportunities for them, but certain aspects of the Jewish cultural heritage also served to propel them quickly upward in occupational standing.

Anti-Jewish attitudes and actions (antisemitism) in the United States have included not only negative stereotyping but also widespread exclusion from certain educational institutions, business areas, and social organizations. Although certain forms of anti-Jewish prejudice and discrimination remain apparent, these are not organized and popular efforts, as they were in earlier periods.

As with Italian Americans, assimilation for Jewish Americans has proceeded with each successive generation though structural assimilation has been slower and less extensive than cultural assimilation. At the secondary level, Jews have begun to participate fully in almost all major institutions of the society and have also begun to penetrate the leadership echelons of these institutions. At the primary level, Jews continue to exhibit strong in-group cohesion. Sharply increasing rates of intermarriage, however, signify a growing change in this pattern. Although Jews continue to assimilate into American society, their religious uniqueness remains a source of ethnic identity and out-group recognition.

Suggested Readings

American Jewish Committee. (annually). *American Jewish Year Book.* A basic reference work for understanding the contemporary status of Jews in American society. Each edition contains up-to-date demographic statistics as well as research articles by social scientists.

Dinnerstein, Leonard. 1994. *Antisemitism in America.* New York: Oxford University Press. A history of antisemitism in the United States from the founding of the society to the present, concluding that it is a declining phenomenon, a trend not likely to reverse itself in the future.

Goldscheider, Calvin. 1986. *Jewish Continuity and Change: Emerging Patterns in America.* Bloomington, Ind.: Indiana University Press. A sociological analysis of the contemporary Jewish-American ethnic group, showing demographic and behavioral trends.

Hertzberg, Arthur. 1989. *The Jews in America: Four Centuries of an Uneasy Encounter: A History.* New York: Simon and Schuster. A basic history of Jewish Americans, concluding that they have assimilated thoroughly into all major institutional realms.

Howe, Irving. 1976. *World of Our Fathers.* New York: Simon & Schuster. Richly describes the immigration, urban settlement, and emergence of ethnic institutions among East European Jews of the late nineteenth and early twentieth centuries, focusing particularly on New York City.

Liebman, Charles S., and Steven M. Cohen. 1990. *Two Worlds of Judaism: The Israeli and American Experiences.* New Haven, Conn.: Yale University Press. Compares the meaning of Jewish life and identity in the United States and Israel.

Lipset, Seymour Martin, and Earl Raab. 1995. *Jews and the New American Scene.* Cambridge, Mass.: Harvard University Press. Describes the current status of Jews in American society and the social, economic, and political forces that have brought them to that place. The authors see strong indications of an eventual complete assimilation of Jewish Americans.

Prager, Dennis, and Joseph Telushkin. 1985. *Why the Jews?: The Reason for Antisemitism.* New York: Touchstone. Offers the thesis that antisemitism is a product of distinctive Jewish values that have made Jews outsiders.

Silberman, Charles E. 1985. *A Certain People: American Jews and Their Lives Today.* New York: Summit. Discusses the current position of Jews in American society, stressing their economic, political, and social achievements, as well as the declining potency of antisemitism.

Wistrich, Robert S. (ed.). 1995. *Terms of Survival: The Jewish World Since 1945.* London: Routledge. A collection of essays that discuss current economic, political, and religious issues of Jews in the United States and abroad.

THE EXPERIENCE OF African Americans[1] is unique among American ethnic minorities. No other group entered the society as involuntary immigrants, and no other group was subsequently victimized by two centuries of slavery. The vestiges of these social facts account for the uninterrupted, if vacillating, conflict between whites and blacks throughout American history and the agonizing nature of the adjustment of blacks to a predominantly white society.

Numbering 33 million (about 12 percent of the population), African Americans are the largest ethnic minority in the United States. They are also the most visible ethnic group, a fact that must be combined with the heritage of slavery in accounting for the extraordinary character of the black experience.

---

[1] In recent years, a debate has emerged concerning whether *African American* rather than *black* is a more appropriate and acceptable term for the American black population. A 1994 national poll of U.S. blacks (Gallup, 1995) indicated no strong preference for either. *African American* was preferred by 18 percent, *black* by 17 percent, and 60 percent answered "it doesn't matter." In this book, the terms are used interchangeably. Technically, however, whereas *black* stresses the racial distinction that sets off this group from others, *African American* emphasizes the group's ethnic identity, that is, its culture and historical origins.

The history of blacks in American society is complex and subject to a variety of interpretations. There is no disagreement, however, that it is a history marked above all by the continued stress of accommodation to the culture and institutions of a white-dominated society. Tension between whites and blacks has varied in intensity at different times, but it has never been far from eruption into overt and violent hostility. With the understanding that conflict has been a constant underlying theme of black–white relations, we can subdivide the history of those relations into three broad epochs, each characterized by a somewhat different pattern: slavery, the Jim Crow period, and the modern era.[2] During these three periods, relations were shaped most fundamentally by social and economic trends and structures of the larger society. As those trends and structures changed, black–white relations, in turn, took on new form and meaning.

Slavery: Paternalistic Domination

Blacks first entered American society when they landed in Virginia in 1619. Their legal status, however, was undetermined for at least forty years. Until the 1660s, the status of blacks as servants was not essentially different from that of many others, including some whites.[3] Various forms of bondage had been prevalent in all the colonies almost from the founding of the society (Handlin, 1957). Perhaps the most common form was voluntary servitude, or indenture, in which a person was bound by contract to serve a master for a certain length of time, usually four to seven years (Jordan, 1969). In exchange, the servant's passage was paid to the colonies. A great number of people, particularly from Scotland and Ireland, entered the country in this manner.

Although no specific date marks the establishment of chattel slavery, slave codes and statutes evolved piecemeal so that by the 1660s most southern states had enacted laws defining blacks as slaves rather than as indentured servants. Most of the features of American slavery were now firmly in place, and at least in the slaveholding states, the ambiguity of black status was resolved.

The Choice of Blacks Other groups in the colonies seemingly could have served as slaves; only blacks, however, were inevitably to qualify. Other groups were treated harshly, and even the enslavement of some was seriously consid-

---

[2]This broad historical subdivision of the black experience conforms generally to that delineated by William Julius Wilson (1980).

[3]Jordan (1969), however, maintains that some blacks were enslaved even before the 1660s.

ered. The Irish seem to have been dealt with particularly severely, and attempts to enslave Indians were also made. But why, in the end, were only blacks chosen for indeterminate bondage, that is, slave status? Historian Oscar Handlin gives us some clues. In the midseventeenth century, the colonies were in dire need of labor; immigration from Europe therefore had to be stimulated. As a result, the colonies adopted legislation that promised better conditions for servants and expanded the opportunities for release from servitude. To encourage immigration, then, the master–servant relationship gradually evolved into a contractual one. But blacks did not benefit from the liberalization of the bondage system. Because they were involuntary immigrants, the application of these policies could have no effect on their numbers. In short, there was nothing to be gained by placing blacks under this legislation. "To raise the status of Europeans by shortening their terms would ultimately increase the available hands by inducing their compatriots to emigrate; to reduce the Negro's term would produce an immediate loss and no ultimate gain" (Handlin, 1957:13).

The choice of blacks was dictated as well by white ethnocentrism. Most simply, their physical and cultural traits were more obviously distinct than those of other groups. Especially significant to the English settlers was what they interpreted as the heathenism of blacks. This served as a vital part of the rationalization for enslavement. Indeed, as historian Winthrop Jordan (1969) points out, the religious difference between whites and blacks was of greater importance, at least initially, than the physical distinction. As Christians, white indentured servants could not be dealt with as nonhumans. Yet the religious factor is not in itself a sufficient explanation for black enslavement because the colonists made no distinction between black nonbelievers and those who had been converted to Christianity; both groups qualified as slaves. Jordan concludes that it was an aggregate of qualities, the total of which set blacks sufficiently apart from others, that made them the most likely candidates for slavery. Their lack of Christianity and unique physical appearance were vital, but other traits that the English interpreted as savage and bestial "were major components in that sense of *difference* which provided the mental margin absolutely requisite for placing the European on the deck of the slave ship and the Negro in the hold" (1969:97). The prevalent white images of blacks, explains Jordan, can be traced to the first contacts between the English and Africans in the sixteenth century. From the first, "Englishmen found blackness in human beings a peculiar and important point of difference" (1969:20). Black skin became a key visible mark for identifying people who were believed to be defective in religion, savage in their behavior, and sexually wanton.

But what of Indians? Were not they, like blacks, nonbelievers and even physically distinct? On these bases, they were as subject to white ethnocentrism as were blacks and thus, logically, could have been eligible for slavery. But Indians maintained certain advantages. First, they were unused to settled agriculture and therefore not easily suited to plantation labor. Moreover, unlike blacks, Indians were familiar with the terrain and could more easily escape to their own people (Stampp, 1956). The most important difference between blacks and Indians as regards their ultimate status, however, was the power resources each could command. Put simply, Indians could offer greater resistance to enslavement. As William Wilson (1973) notes, blacks were forced to live in a foreign land, lacked organization, and were scattered about the countryside. In no way did Indians face a similar situation. Indians remained culturally and politically organized in nations, whereas blacks, following enslavement in Africa and the brutal middle passage, were dispersed and purged of their native cultures. As a result, they remained more dependent on white settlers in the way of culture and sheer survival and posed little collective threat (Jordan, 1969).

In sum, the cultural and physical differences of blacks were important rationales for their enslavement, but in the end, the most critical factor was differential power (Geschwender, 1978; Noel, 1968). Although black skin and "heathenism" gave rise to white ethnocentrism, this did not predetermine the choice of blacks. Rather, blacks simply lacked a viable community from which they might have mustered sufficient counterforce to resist slavery. Color became the visible symbol of slave status, and gradually a racist ideology was developed to buttress the system.

The American form of slavery that evolved was unique, having had no precedent in seventeenth-century England (Elkins, 1976; Handlin, 1957). Indeed, it was a system different from any earlier forms of slavery. The Portuguese and Spanish had already enslaved blacks in their colonies and had maintained slaves at home since the early fifteenth century. But the status of slaves in these societies differed somewhat from that of their American counterparts. As we will see in Chapter 13, in Brazil and other colonies of Portugal and Spain, slaves maintained certain property and family rights and were often freed. In the American form, however, the slave was essentially an object not to be afforded common human privileges.

In the final analysis, the evolution of slavery in the United States was primarily a consequence of economic rationality, prompted above all by the demand for cheap labor in the underpopulated colonies. "The use of slaves in southern agriculture," explains historian Kenneth Stampp, "was a deliberate choice (among several alternatives) made by men who sought greater returns than they could obtain from their own labor alone, and who found other types of labor more expensive" (1956:5). Although it had existed in all the colonies at one time or another, slavery by the middle of the eighteenth century was confined to the South, where plantation agriculture had become the foundation of the economy. In the North, political and moral factors had been influential in the decision to abandon slavery, but more important, the system was seen there as a hindrance to continued industrialization.

The Development of a Racist Ideology In Chapter 3, we noted that negative belief systems regarding ethnic groups are more likely to follow, rather than precede, the development of systems of discrimination. This is clearly seen in the case of American slavery. The pseudoscientific form of racism — the belief that blacks are innately and permanently inferior to whites — did not arise forcefully until well into the nineteenth century. Racial prejudice was evident before that time, of course, but black inferiority was seen as a product of slavery itself, not of inherent deficiencies (Fredrickson, 1971). Before the nineteenth century, beliefs in cultural, not biological, inferiority were in effect (Wilson, 1973). Blacks were seen as heathens and as savages, but not necessarily innately so.

During the 1830s, slavery came under serious attack from abolitionists. Drawing on the environmentalist thought of the eighteenth-century Enlightenment, they argued that blacks should be freed to develop their capacities to the fullest. During the three decades between 1830 and 1860, proslavery theorists, in response to abolitionist thought, developed a racist belief system replete with ideas pertaining to irreversible physiological differences, such as cranial size and shape, which allegedly explained mental and physical inferiority. Also stressed in these new racist theories were the failure of blacks to develop what Westerners considered a "civilized" life in Africa and the dangers of miscegenation, leading to racial degeneracy (Fredrickson, 1971).

Racist ideology was not limited to the South, however. Most northerners were likewise unprepared to accept blacks as equals and subscribed to basically similar ideas. Fredrickson notes that "a common Northern dream depicted an all-white America where the full promise of equality could be realized because there was no black population to be relegated to a special and anomalous status" (1971:323). Even antislavery forces included at least one school of thought (numbering among it Abraham Lincoln) that held that only by establishing a colony for blacks in the Caribbean or in Africa would the issue of slavery and the racist fears of white Americans be solved.

After the 1830s, almost all whites of whatever persuasion agreed that blacks were inferior to whites in certain fundamental qualities, especially intelligence and initiative, and that those differences were essentially unchangeable. Because of these differences, an integrated society, it was maintained, was not feasible. Barring their elimination, therefore, continued subordination of blacks was natural and necessary (Fredrickson, 1971). These ideas did not change fundamentally until well into the twentieth century.

The development of a racist ideology, then, is best seen as a response of southern slaveholders to the threat of abolition of the slave system, but a response that gained national acceptance. Perhaps most significant in contributing to the general acceptance of biological racism were the ideas of Social Darwinism, which gained widespread recognition in the late nineteenth and early twentieth centuries. Indeed, the fullest flowering of the ideology of racism did not occur until after slavery had been abolished (Woodward, 1974).

Master–Slave Relations The American brand of slavery was a system of paternalistic domination, creating, in some degree, a father–child relationship between master and slave (Elkins, 1976). The total control of the slave's existence by the plantation master was explicitly legitimized by the slave codes of the various slaveholding states. Slaves lacked virtually all legal rights. They could not own property, testify in court, inherit property, hire themselves out, or make contracts. Slave laws clearly provided that marriage between slaves held none of the rights of marriage between free people. As a result, families could be broken up in trade with no consideration given to keeping husband, wife, and children intact as a single unit. Finally, laws forbade teaching slaves to read and write.

Enforcement and interpretation of the slave laws were left almost entirely to the master, not the courts (Fogel and Engerman, 1974). With such wide discretion, utter cruelty was not uncommon, but it was felt that brutal treatment would be controlled through public opinion, by the master's sense of decency, and above all, by economic interest (Elkins, 1976). Slaves, representing a significant financial investment, would not be physically or emotionally abused to the point of incapacity. Historians Robert Fogel and Stanley Engerman (1974) explain, for example, that although slave marriages were not legally recognized, masters nonetheless encouraged their slaves to marry and to establish families in order to ensure an atmosphere of stability. Thus, slave families were not separated if this was avoidable (see also Gutman, 1976).

As noted in Chapter 4, systems of paternalistic ethnic relations do not require the maintenance of great physical distance between groups. Rather, it is the *social* distance that is of greatest importance. In the pre–Civil War South, a rigid stratification system was enforced through a well-understood racial etiquette, in which physical contact was not uncommon but the social places of whites and blacks were not violated. Social intimacy might often be seen between master and slave in the form of concubinage between white men and black women or the raising of white children by black "mammies." It was out of this apparent physical closeness that emerged the myth of racial harmony in the Old South. But such intimate personal contacts in a paternalistic system are always relationships between unequals. Moreover, the degree of physical intimacy between the two groups should not be overstated. On large plantations, where most black slaves were held, personal relations with masters were limited to household servants. The majority of slaves were field hands who rarely came into physical contact with the white master or his family.

Although most blacks in the South remained in rural slavery, a few hundred thousand free or quasi-free blacks, most concentrated in cities, lived within the slavery regime. But these free blacks were scarcely better protected by law than slaves, and the degree of discrimination mounted against them was not significantly less. They could make contracts and own property, but in most other respects, their civil rights were not much greater than slaves' (Stampp, 1956; Woodward, 1974). Further, blacks living in the North, though not subjected to slavery, were by no means accepted by whites into the mainstream society. As Wilson describes it, "Whites rejected slavery as an acceptable institution in the North but were unwilling to endorse the view that blacks should receive social, economic, and political equality" (1973:94).

The immediate effects of slavery on blacks were obvious, but the impact of the system on whites must be considered more carefully. Most whites in the South did not accrue direct benefits from the slave system. In fact, most whites were small farmers who owned no slaves. In 1860, there were 385,000 owners of slaves among 1,516,000 free families in the South. Moreover, the typical slaveowner held only a few slaves. In 1860, 88 percent held fewer than twenty, 72 percent held fewer than ten, and 50 percent held fewer than five (Stampp, 1956). Obviously, large slaveowners—the planter class—represented only a tiny fraction of the southern white population. Most of the slaves, however, were concentrated in these few large plantations.

For the elite planter class, slavery provided substantial economic and social rewards, and for smaller slaveholders, it offered correspondingly more modest benefits. But what did slavery provide for nonslaveholders, and why was it so universally accepted among them? Clearly, it could provide them no direct economic payoff, but it could serve as a means of controlling potential economic competition from blacks. In addition, it furnished to poor white farmers a kind of prestige that could be derived from the thought of belonging to a superior "race" (Stampp, 1956). In short, both fear of competition from blacks and social-psychological rewards account for the acceptability of slavery by nonslaveholders. Moreover, the power of the ruling planter class to impose its racist ideology on all whites, regardless of social class, was substantial. As explained in Chapter 2, nonruling groups ordinarily accept for the most part the prevailing social system in the belief that it works in their interests and is justifiable, even though it may work primarily in the interests of the powerful few.

In sum, the 200 years of slavery were a period in which the subservience of blacks to whites was legitimized in the South and informally recognized in the North. In addition, a racist ideology emerged out of the slave system that continued to influence the perception of blacks by whites well into the twentieth century.

Racial Caste: The Jim Crow Era

For a brief historical moment following the Civil War, it appeared that blacks might overcome the crippling social effects of slavery. Slavery had been abolished, and with the protection of the federal government, freed blacks began to acquire some of the power resources necessary to enter into the mainstream society with whites. They now exercised the vote and were even elected to high offices in state and federal governments. An educational system was also established that promised still greater potential opportunities. The decade of Reconstruction from 1865 to 1877, then, marked the end of paternalistic relations between blacks and whites and the beginning of competitive relations. But it was this competition that eventually led to the system of segregation as a means of controlling the black threat to white economic and social advantage. And to reinforce that control, a racist ideology was developed into a more sophisticated and generally acceptable belief system.

The process of denying blacks competitive power resources had begun well before 1877, but that year serves as a historical benchmark ending Reconstruction. The results of the presidential election of 1876 were disputed, and as a means of breaking the impasse, the Democrats and Republicans agreed to a compromise: Federal occupying troops, guaranteeing protection to blacks, would be removed from the South in exchange for the electoral votes of southern Democrats for the Republican candidate, Rutherford B. Hayes. This was not simply a symbolic action but enabled whites to establish further control over the black population without the fear of interference by the federal government. By that time, the planter class had already reasserted its traditional place of power, and the economic dependence of blacks on their former masters was revived with the advent of the sharecropping system. With the removal of the federal government's oversight, the drive to reinstitute white dominance was given greater impetus.

Perhaps the most significant mechanism in this process was the disfranchisement of blacks. When it became apparent that they might hold the balance of political power between competing white factions, efforts to deny voting rights to blacks began in earnest. This put an end to the incipient alliance that had begun to form within the Populist movement between poor blacks and poor whites. Through various constitutional provisions, blacks were effectively disfranchised throughout the South by the first decade of the twentieth century. In South Carolina, for example, a clause was adopted that called for two years' residence, a poll tax of $1, the ability to read and write any section of the U.S. Constitution or to understand it when read aloud, the owning of property worth $300, and the disqualification of convicts (Franklin, 1980). Such measures, of course, automatically disqualified most blacks from the registration rolls.

The Growth of Segregation Following disfranchisement, the path was cleared for the enactment of the entire array of Jim Crow measures designed to separate blacks and whites in almost all areas of social life: housing, work, education, health care, transportation, leisure, and religion. Trains, waiting rooms, drinking fountains, parks, theaters, and other public facilities were now segregated. Access to hotels, restaurants, and barbershops was limited to those serving either blacks or whites, and by 1885, laws requiring racially separate schools had been enacted in most southern states. With the 1896 Supreme Court decision in the case of *Plessy* v. *Ferguson,* the "separate but equal" doctrine was upheld, and the system of segregation in the South was securely in place.

Segregation as it now appeared was not the same as that which had been practiced during slavery. As van den Berghe (1978) explains, roles and institutions that had previously been racially complementary now became racially duplicatory. Rather than whites and blacks maintaining different statuses and roles within the biracial plantation, separate economic and social institutions now served each racial group in a castelike system. Although state laws requiring separate facilities for the two groups stipulated that these would be equal, the Supreme Court never defined *equality* while repeatedly upholding the "separate but equal" principle. As a result, the most consistent fact of this system was that separate facilities were inherently *un*equal; those reserved for blacks were grossly inferior or inadequate, usually by design.

In addition to physical separation, there was the need for a system of racial etiquette, for blacks and whites obviously could not be totally isolated from each other. The etiquette was intended to make clear the caste positions of dominant and subordinate groups. Whites could not shake hands with blacks, for example, nor would blacks be addressed as "mister" but rather as "boy" or by the person's first name. And most important, racial endogamy was rigidly enforced. Indeed, the prohibition of interracial marriage was one of the first formal measures enacted by southern states in the formation of the Jim Crow system.

Racial segregation was maintained by both force and ideology. As to the former, blacks in the South could be dealt with in an almost totally unrestrained fashion. Where legal techniques were found inadequate, extralegal measures were used. With the rise of vigilante groups such as the Ku Klux Klan, violence and physical intimidation became the order of the day. Lynching was the most extreme technique employed to maintain black subservience and was resorted to with increasing frequency. From 1884 to the outbreak of World War I, over 3,600 lynchings occurred, most in the South and with most of the victims black

(Franklin, 1980). Fear and humiliation were integral parts of everyday life for blacks in the southern states.[4]

Conditions in the North, though not supported by legal mechanisms as in the South, were not significantly different. Although black civil rights were acknowledged, segregation in virtually all areas of social life typified black–white relations. Indeed, as historian C. Vann Woodward (1974) points out, segregation actually emerged first in the North before the abolition of slavery. Despite the free status of blacks in the northern states, their inferiority and inability to assimilate had not been questioned in any quarter. Segregation was a means of assuring that blacks understood and kept "their place" and was well established in all the free states by 1860. Moreover, blacks in most of these states had been effectively disfranchised.

The system of segregation in both the South and North was abetted by the fullest development of a racist ideology. Ideas of racial superiority and inferiority were now supported by scientific thought and were applied not only to blacks but also, as we saw in the case of Italians and Jews, to all non–Anglo-Saxon groups. Woodward (1974) notes that a further contributing factor to the vehement racism of this era was American imperialistic ventures at the turn of the century, involving nonwhite peoples in the Philippines and the Caribbean. The negative stereotypes of blacks that emerged from slavery—lazy, childish, irresponsible, uncouth—became even more potent. In short, ideas of white, specifically, Anglo-Saxon, supremacy were at a zenith at the turn of the century and for two decades afterward.

Early during the period of caste, the black leadership strongly debated whether to adapt to or challenge the system of segregation and denial of black rights. One school of thought, reflected most clearly by Booker T. Washington, contended that the most realistic course was to accept a form of separatism based on the idea of self-improvement. In his famous "Atlanta Compromise" address in 1895, Washington deemphasized the effects of racial discrimination and called for blacks to learn industrial skills and trades that would help strengthen the black community internally. Opposed to this strategy of accommodation and withdrawal were those led by W. E. B. Du Bois, who argued for the complete acceptance of blacks into all areas of the society and advocated militant resistance to white racism. Du Bois contended that it was not solely the responsibility of blacks, nor was it in their capacity, to alter their collective place in American society but that it was primarily the responsibility of whites, who held the power to effect such change. His efforts were instrumental in the founding in

[4]A poignant description of the constant terror and degradation to which blacks were subjected is provided in the works of Richard Wright, one of the great black writers of the twentieth century. See especially *Black Boy* (1945) and *Uncle Tom's Children* (1938).

Table 8.1 Distribution of U.S. Black Population by Region

REGION	1790	1870	1910	1940	1960	1990	1994
South	91%	91%	89%	77%	60%	53%	55%
North	9	9	10	22	34	38	37
Northeast	9	4	5	11	16	19	17
North Central	—	6	6	11	18	19	20
West	—	—	1	1	6	9	8

SOURCE: U.S. Bureau of the Census, 1979, 1991a, 1995a.

1909 of the National Association for the Advancement of Colored People (NAACP), which eventually became the chief organization working for black civil rights.[5]

The alternative strategies of integration or separation were debated by succeeding generations of black leaders. The theme of separatism reemerged strongly in the 1920s when Marcus Garvey, a Jamaican immigrant, led a movement advocating the return of American blacks to Africa. Integrationist thought prevailed among most blacks, however, although the separatist issue, as we will see, surfaced again in the 1960s.

Northward Migration Relations between blacks and whites took on a more confrontational and violent tone during the early 1900s than at any previous time. This situation was brought about by demographic and economic changes in the society that placed the two groups in direct competition.

Industrial expansion in the North created an increased demand for labor, and with the cutoff of immigration from Europe, blacks from the South began to fill these places. If jobs were the key "pull" factor impelling black out-migration from the South, the declining cotton economy and the continued enforcement of Jim Crow were the chief "push" factors. The demographic changes involving the black populace were profound. In 1910, almost 90 percent of blacks were living in the South, but eighty years later, little more than half remained there (Table 8.1). Streams of blacks from the South to the North and West reached epic proportions from 1940 to 1970, with almost 1.5 million leaving in each of these three decades. Moreover, this migration was almost wholly to the cities, making blacks an increasingly urbanized population.

[5]Washington's and Du Bois's major ideas concerning the place of blacks in American society and their approach to change are contained in Rudwick and Meier (1969).

With their movement to the cities, black workers began to pose a real labor threat to white workers in the North, and the violence and intimidation that had characterized southern race relations became a national phenomenon. Adding to the tension created by job competition was the discontent expressed by the many black World War I veterans who had been exposed to more liberal social conditions in Europe. The upshot of the rising competition for work and housing and the growing impatience of blacks was an increase in the frequency and severity of black–white hostilities. Serious race riots occurred in numerous cities throughout the first two decades of the century (Rudwick, 1964; Rudwick and Meier, 1969; Tuttle, 1970).

In sum, black–white relations from the end of Reconstruction in 1877 to the 1940s involved a system of restrictive competition (Wilson, 1973). The forces of industrialization created the potential for significant changes in the structure of race relations, and blacks were now theoretically capable of competing with whites, in contrast to their previous slave status. But in fact, strict segregation in most areas of social life and the denial of political participation severely limited their accumulation of competitive resources—jobs, education, and housing. Wilson has noted that when groups lack the necessary political resources, whatever opportunities are created by industrialization are offset by new controls imposed by the dominant group. "Race relations become competitive, but only in the narrowest sense" (1973:64). Blacks could offer little resistance because of the strength and resolve of whites of all classes to maintain the discriminatory system through both legal and extralegal mechanisms. Furthermore, the system was buttressed by a now scientifically endorsed racist ideology, which posited the innate inferiority of black people.

Fluid Competition: The Modern Era

Whereas the Jim Crow period was one of restrictive competition between blacks and whites, World War II marked the emergence of more fluid competitive relations. In such a system, according to Wilson (1973), the minority group's power resources are increased to the point at which it can begin to challenge the dominant group's authority and invalidate racist stereotypes. Members of the minority begin to move into positions formerly occupied only by the dominant group, and in the process, institutional racism is exposed and challenged, and individual racism is seriously undermined. As these changes occur, some members of the dominant group begin to question the legitimacy of racial discrimination and either refrain from practicing racist norms or openly attack the racist system. As Wilson explains, "The greatest threat to any form of racism, then, is the significant entry of minority members into upper-status positions within the larger society" (1973:59).

The midtwentieth century was clearly that period in which the acquisition of power resources by blacks impelled the most forceful challenge to the American racist system. This challenge involved the efforts of organized pressure groups as well as unorganized forces. Together they made up what became the most compelling U.S. social movement of the twentieth century.

Several factors account for the emergence of the movement for black rights after World War II. Much of the impetus had been established in the 1930s, when the Roosevelt administration began to appoint blacks to nonpolicymaking government advisory positions (Franklin, 1980; Myrdal, 1944). Moreover, blacks benefited from the administration's relief programs even though these continued to be administered in a discriminatory fashion. Another significant inroad made by blacks in the 1930s was their entrance into the labor movement, specifically the Congress of Industrial Organizations (CIO). Unlike the American Federation of Labor (AFL), the CIO was relatively nonexclusionary in its organizing activities. Representing industries in which most black unskilled laborers were employed (steel, automobiles, mining, clothing), the CIO now afforded blacks labor union participation in significant numbers for the first time.

Wartime conditions in the 1940s opened new areas of work to blacks and provided an even greater push to the northward migration that had been occurring since the early part of the century. Jobs in the defense plants of cities like Detroit, Chicago, Philadelphia, Cleveland, and New York attracted a new surge of migrants from the South, whites as well as blacks. Blacks, however, continued to suffer the effects of discrimination in the distribution of these jobs. As a result, black labor leader A. Philip Randolph organized the March on Washington Movement and threatened to mobilize millions of blacks in protest. Hoping to squelch mass unrest, President Franklin D. Roosevelt in 1941 issued an executive order prohibiting racial discrimination in federal jobs and created the Fair Employment Practices Committee to monitor the order. Randolph's movement was an early sign of the effectiveness of black protest, which would reach much greater heights after the war.

In addition to the economic opportunities the war provided, it presented white political leaders with a dilemma: How could the United States support a racist system at home while it was at war abroad with the most flagrantly racist regime, Nazi Germany? This inconsistency was difficult to reconcile without a greater commitment to black rights. Moreover, following the war, racist policies proved an international liability in dealing with the emerging nations of nonwhite Africa and Asia. As a result, the federal government now seemed more amenable to efforts at ending patterns of discrimination. One of the first institutions to be desegregated was the armed forces.

Perhaps the most momentous single government action during the 1950s that stimulated the push for black advancement was the Supreme Court decision of 1954 in the case of *Brown* v. *Board of Education of Topeka.* In that case, the Court ruled invalid the separate but equal doctrine, which had been upheld in the famous *Plessy* v. *Ferguson* decision of 1896. The 1954 decision basically removed the legal foundation of segregation and thus marked the beginning of the end of the entire Jim Crow system. Influenced by the testimony of social scientists, the Court made clear in its ruling that separate schools were inherently unequal and imposed an inferior status on black children, causing irreparable psychological damage. School segregation laws, the Court declared, therefore violated the equal protection clause of the Fourteenth Amendment of the Constitution. This clause stipulates that no state shall deprive people of life, liberty, or property without due process of law, or deny them the equal protection of the law. The *Brown* ruling was subsequently used as the legal basis for court decisions that struck down other segregated institutions.

The significance of the *Brown* decision cannot be understated, for it provided the spark to ignite a surge of hope among blacks that had begun to build in the 1940s. Lewis Killian has described its psychological impact on blacks: "The authority of the law of the land was now cast on the side of change, not on the side of the status quo. Any Negro with even the vaguest idea of what the Supreme Court had said now had good reason to hope for an improvement in his life and, even more so, in the life of his children" (1975:43). The drive for integration and an end to societal discrimination may be said to have begun with *Brown.* One year after this decision was handed down, the civil rights movement was symbolically launched with the boycott by blacks of the segregated bus system of Montgomery, Alabama.

Nonviolent Protest From the late 1950s to about 1964, nonviolent protest and civil disobedience were the movement's chief tactics. The basic idea was that individuals had a moral duty to disobey the law when it was clearly unjust. This was a strategy that had been applied many times historically in a variety of social settings, its most notable success occurring in the 1940s in India, where Mahatma Gandhi led the movement for independence from Britain (Sibley, 1963). The technique was to actively oppose the law but in a peaceful fashion. The proponents of nonviolent protest, Martin Luther King, Jr., in particular, emphasized that it was necessary to win over the opposition through friendship and understanding, not through defeat or humiliation. Those participating in sit-ins at segregated lunch counters, freedom marches, or freedom rides were therefore prepared to accept suffering without retaliating in kind (King, 1964).

Hostile white retaliation against demonstrators and civil rights workers, much of it in the South carried out by civil authorities, included shootings, beatings, firebombings, and even killings. A pivotal year was 1963, when several key events occurred. Police brutally suppressed black demonstrators in Birmingham, Alabama, a Birmingham church was firebombed, killing four black children, and National Guardsmen were used to carry out the court-ordered desegregation of the University of Alabama. These events were given prominent media attention and galvanized public and governmental support for black civil rights. This supportive mood climaxed with the march on Washington of over 200,000 people in August of that year. One year later, the most comprehensive civil rights measure since the Civil War was enacted, prohibiting discrimination in voting, public facilities, schools, courts, and employment. And in 1965, the Voting Rights Act was passed, ending the systematic disfranchisement of southern blacks.

In essence, the civil rights acts of 1964 and 1965 signified the demise of official segregation and discrimination in the United States. But the end of de facto racial separation and inequality was not yet in sight. Moreover, black leaders became increasingly aware of the discrepancy between what had been accomplished in a legal sense and what had not been accomplished in the actual improvement of life conditions for most blacks. This was sharply evident in northern cities, where blacks were concentrated in depressed ghettos. The black movement, accordingly, shifted its focus from the South to the urban areas of the Northeast and Midwest. This shift signaled a move not only in the geographical locus of protest but also in the goals and tactics of the movement itself.

Securing legal rights was not the primary aim of black protest in the North, as it clearly had been in the southern states. Technically, those rights had been in place for many decades. Instead, emphasis was placed on economic issues—jobs, housing, education, and all other life chances—which, through discriminatory actions, were being denied or poorly provided to the majority of blacks. These issues, however, did not lend themselves to the kind of direct amelioration that had been the case with previously state-approved segregation and denial of civil rights. Legal maneuvering and sit-ins were inapplicable to socioeconomic problems.

As a result, the black movement passed temporarily through a more militant phase, in which the black leadership now proclaimed that ending racism required systemic changes in economic and political institutions, changes that purely nonviolent protest had failed to address. The shift away from defensive and nonviolent tactics produced not only more militant rhetoric but also serious urban violence throughout the late 1960s (Bergesen, 1980; National Advisory Commission on Civil Disorders, 1968).

Black Power Complementing the substantially changed strategy of black protest during this period was a reformulation of the movement's ideological goals. Until the late 1960s, the chief aims of black protest were the abolition of segregated public facilities, the achievement of legal equality, and the encouragement of eventual racial integration at all societal levels. It was the third of these objectives that now came under attack from black leaders like Malcolm X and Stokely Carmichael, who pronounced the need for independence of black action and the establishment of black control of black community institutions. This was the essence of the idea of black power. Integration was rejected in favor of a separatist position, coalitions with white liberal organizations were severed, and whites were generally purged from the movement. Racial prejudice and discrimination were seen not simply as the attitudes and actions of individuals but as structural features of the American social order. The notion of institutional racism now became popularized (Killian, 1975). The ideology of black power (or black nationalism) was espoused most vehemently by the Student Nonviolent Coordinating Committee (SNCC) and, later, by the Congress of Racial Equality (CORE). Less militant black protest organizations, such as the National Association for the Advancement of Colored People (NAACP) and King's Southern Christian Leadership Conference (SCLC), however, remained steadfastly committed to integration.

Black power was widely interpreted as the advocacy of militant tactics against the white power structure, and the stridency of its proponents represented what seemed to many whites a kind of racism in reverse. Black power, however, comprised a twofold campaign: to assert control over the development of the black community with minimal interference by whites and to instill self-pride among blacks, who, in the view of black power advocates, had for generations internalized the white stereotyped image of black people.

The black community in the United States was seen essentially as a colonial people dominated by white institutions—hence the rejection of integrationist objectives. Carmichael and Charles V. Hamilton stated this position: "The fact is that integration, as traditionally articulated, would abolish the black community. The fact is that what must be abolished is not the black community, but the dependent colonial status that has been inflicted upon it" (1967:55). Blacks were urged to build genuinely black communities by taking control of school boards, businesses, and political bodies in their areas of the central cities. Only by closing ranks and asserting power within their own communities, it was felt, could blacks emerge as a viable political force, able to compete within a pluralistic system.

The development of a black communal identity and pride that had long been stifled by white oppression prompted an acceptance of African cultural roots that had previously been consciously shrouded. As Isaacs contends, the slogan for which the black radicalism of this period is likely to be remembered is "Black is beautiful." "In both its literal and its symbolic meanings," he notes, "it became the password to a measure of self-acceptance by black Americans that generations of earlier leaders and tribunes of the people had sought in vain to achieve" (1989:85–86).

The militant phase of black advocacy was short-lived. In the late 1960s, public concern with black civil rights in general was increasingly displaced by preoccupation with the Vietnam War. Most important, with the advent of the Nixon administration, it became clear that the response of the federal government to black nationalist goals and tactics would be severely repressive (Button, 1978). The militant stage of the movement came to an end, according to Killian, not because the racial crisis had passed but because the white power structure effectively demonstrated the dangers of such measures by imprisoning, assassinating, or forcing into exile the most "dramatically defiant" black leaders (1975:155).

Though ephemeral and seemingly unsuccessful in achieving its long-range goals, the black power phase of the civil rights movement may, in retrospect, be judged more positively. First, black nationalism in the 1960s created a degree of self-pride among blacks that had not been achieved by previous nationalist movements such as Garvey's in the 1920s. Second, short-run profits in the form of increased government assistance followed the urban turbulence. Although it is commonly believed that violence is counterproductive to groups employing such techniques in political protest, historical evidence seems to reveal the contrary. When ruling elites make concessions to challenging groups or, conversely, when challenging movements are effectively stifled, violence or the threat of violence on the part of the successful group looms large in the picture (Gamson, 1975). Piven and Cloward (1971) maintain that protest movements like the one for black civil rights do not gain establishment concessions because of the successes of their organization and leadership but because of the social disturbances they create. Relief programs such as welfare and unemployment payments are set up not so much because of moral incentives as because of the need to contain the social disorder created by mass protests. This conclusion seems to be borne out by events of the 1960s. Following the riots of that decade, the federal government extended numerous programs to the cities designed to address those social conditions that were believed to be at the core of black discontent. As with the aid provided to the unemployed of the Great Depression, write Piven and Cloward, "The expansion of the welfare rolls was a political response to political disorder" (1971:198).

Of course, when concessions are granted under these circumstances, they may be only superficial, momentary, or symbolic; that is, their long-range effect may not be great. As Piven and Cloward (1971) explain, government relief to the poor and unemployed is usually withdrawn once social order has been restored. Such was clearly the case with aid to the urban ghettos. In his study of government responses to the riots of the 1960s, James W. Button (1978) shows that once the level of hostilities subsided and the areas were relatively pacified, the programs were either discontinued or emasculated by underfunding.

Assessing the Civil Rights Era By the early 1970s, the "Second Reconstruction" had ended, and blacks began to take stock of their advances and failures of the previous two decades. Most important, the legal infrastructure of the system of segregation in the South had been dismantled. No longer could state and local ordinances prevent blacks from entrance into hotels, theaters, restaurants, trains, buses, waiting rooms, and all other public facilities previously reserved for whites. Unlike the first Reconstruction following the Civil War, white vigilante groups like the Ku Klux Klan were limited in their influence and were not supported either materially or morally by the majority of whites. Most important, the federal government accepted the commitment to black civil rights and effectively countered state and local governments that sought to sustain segregation in some form. This was in sharp contrast to the abandonment of southern blacks by the federal government during the post–Civil War period.

A second major achievement of the civil rights movement of the post–World War II era was the expansion of black economic opportunities and the consequent emergence of a sizable black middle class. But it is also in the economic realm that the movement's most glaring failures were to be found. As we will see in the following section, blacks accumulated much greater competitive resources, such as white-collar jobs, middle-level income, and higher education, than at any previous time. These were not distributed equitably among the black population, however. Thus, although the black middle class expanded considerably, the conditions of a significant black underclass, submerged mainly in urban ghettos, were not seriously affected.

Finally, if the decline of biological racist ideas was clearly on the horizon as early as the 1930s, the postwar period of black–white relations, despite its volatility, seemed to further diminish the persuasiveness of this ideology. The notion of innate black inferiority was no longer given credence by the vast majority of white Americans.

AFRICAN AMERICANS IN THE STRATIFICATION SYSTEM

Where do African Americans stand in the society's class and power hierarchies, and how has that position changed during the group's sociohistorical development? Answers to these questions must be approached cautiously because there are various measures of stratification, and data can be marshaled to support a variety of often contradictory views. As with Italians and Jews, analyzed in the two previous chapters, we will concentrate on the effects of ethnicity on the place of African Americans in the economic and political hierarchies. As explained in Chapter 2, minority status does not necessarily indicate the position of a group at the time its members enter the society in large numbers but rather the patterns that develop to keep them in a disadvantaged position generation after generation. Such patterns are clearer for African Americans than for any other ethnic group in American society.

The Economic Status of African Americans

The fact of slavery weighs so heavily on the black experience that its lingering effects continue to make African Americans a special group in the American ethnic hierarchy. Obviously, blacks' collective status under slavery was at the bottom of the economic class system, even for those who were free in the North or South. Following the abolition of the slave system, this status changed little, as we have seen. Indeed, the economic standing of blacks was even more tenuous following slavery because they were denied the opportunity to compete on an equal basis with whites. Clearly, they did not fare well in resisting a system of blatant discrimination and repression that kept them in a subservient position.

Even after large-scale migration to northern cities starting in the second decade of the twentieth century, a system of direct and intentional discrimination effectively thwarted serious competition with the white population. For the most part, blacks were shut out of higher-skilled, more prestigious, and higher-paying jobs. Here we might recall the split labor market theory of discrimination, discussed in Chapter 3 (Bonacich, 1972, 1976). Not until the 1950s and 1960s did blacks begin to move out of the lowest-level jobs in significant numbers. In effect, then, meaningful comparisons with other ethnic groups can be made only for the post–World War II period.

African-American economic progress of the last five decades is debatable, and social scientists as well as policymakers, both black and white, have reached no consensus on either its extent or substance. There is, however, relatively broad agreement regarding several general patterns: (1) the 1960s was a decade of substantial improvement in the economic status of blacks collectively, as measured specifically by income, occupation, and education; (2) in the 1970s and 1980s, by comparison, the rate of progress was curtailed except for a few specific subsets of the African-American population; (3) despite black upgrading, the economic gap between blacks and whites collectively today remains wide; and (4) the economic progress of the modern period has created a substantial black middle class but has had little impact on blacks at the bottom of the economic scale.

Let us look at some specific variables of economic class, comparing the position of blacks and whites.

Income and Wealth Depending on the manner in which data are interpreted, there are grounds for both positive and negative views regarding black income.

Table 8.2 Median Family Income of Blacks and Whites

YEAR	BLACK	WHITE	RATIO B–W
1950	$ 1,869[a]	$ 3,445	.54
1960	3,230[a]	5,835	.55
1970	6,279	10,236	.64
1980	12,674	21,904	.58
1990	21,423	36,915	.58
1993	21,542	39,300	.55

SOURCE: U.S. Bureau of the Census, 1981, 1995b.
[a]Includes other nonwhites.

The real income of black families increased substantially after World War II, especially during the 1960s (Farley and Hermalin, 1972; Levitan, Johnston, and Taggart, 1975). This is the positive picture. But the negative view arises when we look at the gap between median income of blacks and whites. As is seen in Table 8.2, this gap has remained relatively constant since 1950. In 1993, median black family income was slightly more than one-half the median white family income, almost the same ratio as was evident in 1950.

Although the income gap between blacks and whites narrowed slightly in the late 1960s and early 1970s, in the 1980s and early 1990s, blacks seemed to fall further behind whites. Indeed, black family income in 1993 was, when measured in constant dollars, no different from what it had been in 1969. Although family income grew more slowly in the 1980s than in the 1970s for all families (and actually declined in the early 1990s), black family income grew more slowly than white family income. Moreover, income inequality between blacks and whites is evident throughout the class hierarchy. Thus, poor blacks are generally poorer than poor whites, and affluent blacks are less well-off than affluent whites (Swinton, 1992). In short, parity in black–white income is today extremely remote.

This general picture should not obscure the fact, however, that comparisons between black and white income are subject to a number of factors, including region and urban location. In a few communities, black median income has actually surpassed that of whites. More important, the black population is polarized between relatively stable middle- and working-class persons on the one hand and those in poverty on the other. The substantially greater percentage of blacks below the poverty line in comparison with whites accounts in large measure for the general differences between black and white income patterns. Among blacks with higher education, the racial gap has diminished and for black women specifically it has vanished (Farley, 1993).

Table 8.3 Persons Below the Poverty Level

YEAR	BLACK Number (1,000s)	BLACK Percent	WHITE Number (1,000s)	WHITE Percent
1959	9,927	55.1	28,336	18.1
1971	7,396	32.5	17,780	9.9
1980	8,050	31.0	17,214	9.0
1994	10,877	33.1	26,226	12.2

SOURCE: U.S. Bureau of the Census, 1979, 1983b, 1995b.

Another mixed indicator of the current economic status of African Americans is the percentage of blacks in poverty. Here, the negative view is stronger. In 1994, one-third of the African-American population lived below the poverty level, not essentially different from the proportion of blacks in poverty in 1971. Although the percentage of poor among both blacks and whites rose in the 1980s and early 1990s, the difference between the two groups remains striking (Table 8.3). Blacks are almost three times more likely to be poor than are whites.

An important factor in accounting for these statistics, both positive and negative, is changes in the African-American family structure. Those families that have experienced income increases have been dual-breadwinner families. Among such families there are only slight differences between blacks and whites. Similarly, much of the halt in the relative rise of black income from the late 1970s through the early 1990s is attributable to the increasing number of female-headed, single-parent families (Farley, 1993). In 1970, 28 percent of black families were headed by a woman; by 1989, that figure had risen to over 43 percent and by 1994 to 48 percent (U.S. Bureau of the Census, 1995a). Single-parent families earn considerably less than husband–wife families. In fact, over half of all families headed by women are below the poverty line. Moreover, a large percentage depend to some degree on government assistance.

Wealth, or property, refers to the economic assets that families possess—homes, bank accounts, stocks and bonds, automobiles, and so on. Wealth is extremely important as a determinant of class position because it can be invested and thus earn income or it can be accumulated and passed down to other family members. On this measure of socioeconomic status, differences between blacks and whites are significantly greater than differences in income. Blacks are less likely than whites to own any type of wealth, and the holdings of those blacks who do own wealth are far smaller than for white wealthowners. For example, less than 7 percent of black households own stocks compared to 23 percent of

Table 8.4 Occupational Distribution of Employed Persons Age 16 or Older

	OCCUPATION	BLACK	WHITE
Males	Managerial, professional	14.7%	27.5%
	Technical, sales	17.6	20.6
	Service	20.0	9.8
	Farming	2.0	4.3
	Skilled blue-collar	15.0	18.5
	Unskilled blue-collar	30.7	19.3
Females	Managerial, professional	20.1	29.9
	Technical, sales	39.4	43.2
	Service	26.9	16.8
	Farming	0.2	1.2
	Skilled blue-collar	2.5	2.1
	Unskilled blue-collar	10.8	6.8

SOURCE: U.S. Bureau of the Census, 1995a.

white households (Oliver and Shapiro, 1995). In fact, the typical white household has assets about ten times those of the typical black household. Most black income, then, even within the middle class, does not originate from property or self-employment.

Occupation If income is the chief determinant of life chances, work is, for most people, the medium through which income is acquired. Occupationally, African Americans as a group have improved their place over the past several decades, progressively moving into better-paying and higher-status jobs. In 1890, most male blacks were in agricultural jobs, and females were in domestic and personal service occupations. By 1930, the structure of black employment had changed substantially, with an enormous drop in agriculture and an increase in manufacturing. This was due, of course, mainly to the out-migration of rural southern blacks to the industrial cities of the North. But the most significant occupational changes occurred between 1940 and 1970. Whereas in 1940 about one-third of all employed blacks were farmworkers, by 1960 only 8 percent were, and by 1970, 3 percent. By contrast, in 1940 only 6 percent of all employed blacks were in white-collar occupations. By 1970, 24 percent were white-collar workers, and by the mid-1990s, over half were (Table 8.4).

Although it is clear that blacks have been increasingly integrated into the mainstream work force, black–white disparities remain evident. Blacks are un-

Table 8.5 Unemployment Rates for Persons Age 16 or Older

YEAR	BLACKS[a]	WHITES	RATIO B–W[b]
1950	9.0%	4.9%	1.8
1960	10.2	4.9	2.1
1970	8.2	4.5	1.8
1980	14.3	6.3	2.2
1990	11.3	4.7	2.4
1994	11.5	5.3	2.2

SOURCES: National Urban League, 1981; U.S. Bureau of the Census, 1979, 1995b.
[a]Includes other races from 1950 through 1980. Over 90 percent of this category is black.
[b]This means that in 1950, 1.8 black workers were unemployed for each unemployed white worker.

derrepresented in jobs at the top of the occupational hierarchy and overrepresented at the bottom. As can be seen in Table 8.4, a far higher percentage of whites are in managerial and professional occupations, while blacks predominate in service jobs, such as cleaners, cooks, security guards, and hospital attendants. Indeed, the only subset of the black population that has attained occupational parity with similar whites is college-educated women (Farley and Allen, 1987). Moreover, the apparent occupational advancement of blacks need not be seen only as a product of the decline of discrimination in employment. Much of it is due to changes in the occupational structure itself as the economy shifted from an agricultural to an industrial base and subsequently to a service base.

Further, the discrepancy between blacks and whites in rate of unemployment is wide (Table 8.5). Since the end of World War II, black unemployment has been, except for a few years, at least double that for whites. When one considers that government unemployment statistics do not take into account those discouraged persons who have given up the search for a job, the unofficial number and percentage of unemployed is much higher.

Education As with income and occupation, there is room for interpreting educational data for African Americans both positively and negatively. Measured by years of schooling, blacks have increased their educational achievement both absolutely and relative to whites during the last five decades. Especially noticeable gains were achieved in the number of high school graduates. As is seen in Table 8.6, among people 25–34 years of age, 84 percent of blacks in 1994 had completed high school or more, compared with 87 percent of whites. Thus, blacks have almost reached parity with whites in median years of schooling completed.

Table 8.6 Educational Attainment of Persons 25–34 Years Old, White and Black

EDUCATIONAL LEVEL	1940 W	1940 B	1960 W	1960 B	1970 W	1970 B	1980 W	1980 B	1994 W	1994 B
High school graduate	39%	11%	61%	33%	76%	53%	87%	74%	87%	84%
Four or more years of college	7	2	12	4	17	6	25	12	24	13

SOURCE: U.S. Bureau of the Census, 1979, 1995a.

Also, today there is little difference in the high school dropout rate for blacks and whites (U.S. Bureau of the Census, 1995b).

Substantial gains have also been evident in black college attendance and graduation. In 1960, there were 250,000 blacks attending college; by 1990, there were 1.2 million (*Chronicle of Higher Education,* 1992a; Wilson, 1989). Whereas in 1980, 8 percent of adult blacks had earned at least a bachelor's degree, by 1994, 13 percent had done so (U.S. Bureau of the Census, 1995a). Black college attendance, however, has vacillated in the past two decades. Starting in the late 1970s, the number of blacks attending college declined until the late 1980s, when it again rose. A similar trend was apparent at the graduate level, where the number of blacks earning master's degrees and doctorates dropped sharply and then rose again starting in the early 1990s. Nonetheless, the number of black Ph.D.s in 1992 was fewer than ten years earlier (U.S. Bureau of the Census, 1995b).

In any case, the quantity of education is not necessarily commensurate with its quality. Despite the narrowing differential in years of schooling, black achievement continues to lag behind white achievement at all school levels—elementary, secondary, and college (Coleman et al., 1966; Commission on Minority Participation, 1988; Farley and Allen, 1987; U.S. Department of Education, 1992). Lower educational achievements are, of course, translated into fewer opportunities for occupational advancement.

Explanations vary for racial differentials in educational achievement. At one extreme, some argue that much of the difference between blacks and whites is due to genetic factors (Herrnstein and Murray, 1994; Jensen, 1969, 1973a). This view remains highly controversial and is not accepted by the vast majority of social scientists and educators. At another extreme, many maintain that the schools themselves are inherently racist in both content and technique and thus guarantee the failure of blacks (Bowles and Gintis, 1976). A third position emphasizes social and environmental factors. The essential argument is that family background and individual and peer-group attitudes are more important determinants of academic performance than either genetic inheritance or the characteristics of the schools (Coleman et al., 1966; Jencks et al., 1972).

In sum, black educational advances in enrollment in the past four decades have not resulted in comparable academic attainment. Blacks have approximated equality with whites in years of schooling, but what that means beyond a mere time increase is difficult to evaluate.

Class or Race? It is clear that, on all economic measures, blacks as a group, despite their advances, continue to lag behind whites. The questions that such statistics prompt are whether the persistent gap between blacks and whites is attributable to lingering racism, whether there are deeper structural forces that impede the attainment of economic equality, or whether there is something unique about the black experience that contributes to the group's continued plight. We will look at the last-mentioned issue when blacks are compared with other American ethnic groups later in this chapter. The first two questions have stirred a lively debate among sociologists for several years. The issue is essentially this: Does racial discrimination continue to affect the economic status of blacks more than their socioeconomic background; that is, do blacks succeed or fail mainly on the basis of their skills and education, or does the racial factor still play a major role in their economic destiny?

The issue of race versus class is addressed most directly in the controversial work of sociologist William Julius Wilson (1980, 1981, 1987). His basic position is that although discrimination based on race accounted for the denial of necessary economic resources to blacks in earlier times, the traditional patterns of black–white interaction have been fundamentally altered so that class factors are today the dominant influence on the life chances of black people: "In short, whereas the old barriers portrayed the pervasive features of racial oppression, the new barriers indicate an important and emerging form of class subordination" (1980:2). Wilson does not maintain that prejudice and discrimination no longer exist in American society but that so far as access to power and privilege are concerned, they are no longer of paramount significance.

Wilson explains that to understand the current plight of the black populace, it is necessary to carefully subdivide it into at least three distinct classes: a black middle class, consisting of white-collar and skilled blue-collar workers, which has expanded measurably in the past two decades; a working class of semiskilled operatives; and a lower class made up of unskilled laborers and service workers. At the very bottom of this last group is an underclass that includes those lower-class workers whose income falls below the poverty level, the long-term unemployed, discouraged workers who have dropped out of the labor market, and the more or less permanent welfare recipients (Wilson, 1980). It is the disjunction between the black middle class and working class, on the one hand, and the lower class (particularly its most destitute segment), on the other, that is most important in explaining why economic statistics seem to indicate a halt to economic progress for the black populace.

During the post–World War II decades, writes Wilson, middle-class blacks were able to take advantage of the breakdown of racial barriers and as a result experienced substantial upward mobility. Those in the lower class, however, were unable to benefit from the expanded opportunities because of their lack of competitive skills and education. In fact, their position deteriorated. Thus, according to Wilson, a growing bifurcation has arisen within the black community between the "haves" and the "have nots." This growing gap is a product of changes in the society's economic structure that have made lower-class, unskilled blacks an excess labor force that is becoming a permanent underclass. The labor market today calls for jobs requiring technical and educational skills, for which such persons cannot qualify. Moreover, industry has moved to areas on the urban fringe so that those low-level jobs still available are outside the range of central-city blacks. The result is a semipermanent state of unemployment. The lack of job opportunities creates a vicious circle in which blacks are forced to remain in inner-city ghettos, attending inferior schools and thereby reinforcing their disadvantaged position in the labor market.

Wilson (1987) further explains that the problems of the black urban underclass—high rates of joblessness, school dropouts, crime, teenage motherhood, and so on—have been exacerbated by the exodus of middle- and working-class families from ghetto neighborhoods, removing vital role models and a support system of basic institutions. He points out that poverty in black urban areas in the past was not characterized by the degree of social isolation that is such a striking feature of contemporary inner-city ghettos. Residents of these areas become concentrated in disadvantaged areas of central cities and thus are increasingly cut off from the mainstream society. As Wilson puts it, "the communities of the underclass are plagued by massive joblessness, flagrant and open lawlessness, and low-achieving schools, and therefore tend to be avoided by outsiders. Consequently, the residents of these areas, whether women and children of welfare families or aggressive street criminals, have become increasingly socially isolated from mainstream patterns of behavior" (1987:58). Observing the situation of inner-city blacks twenty years after the Kerner Commission Report warned of increasing isolation, Gary Orfield also has noted that the scale of isolation is much greater than it was in the 1960s: "To a considerable extent the residents of city ghettos are now living in separate and deteriorating societies, with separate economies, diverging family structures and basic institutions, and even growing linguistic separation within the core ghettos" (1988:103). Sociologists

Douglas Massey and Nancy Denton (1993), in their study of residential segregation, reached a similar conclusion.

Wilson is not optimistic about the future of the black lower class. Significant improvement in their economic condition, he feels, is unlikely within the framework of current economic institutions. It is important to reiterate Wilson's key point, that it is not racial discrimination but the current occupational structure that sustains the black underclass. Programs aimed at alleviating traditional forms of racial exclusion and encouraging black competition with whites, such as affirmative action, are not relevant to the key problem: the surplus of unskilled labor in a modern industrial economy that calls for an increasingly better-trained work force. Catastrophically high unemployment rates among black youth and the burgeoning of low-income, female-headed families unable to support themselves independently threaten to perpetuate the black underclass, estimated at fully one-third of the black populace (Wilson, 1980).

Civil rights leaders as early as the 1960s recognized that the problems of the black underclass were not the same as those of the black middle class when they warned that, despite its political successes, the movement had not basically affected the life chances of ghetto blacks. What is controversial in Wilson's thesis is that race is no longer seen as the key factor in the perpetuation of these conditions. Although some agree fundamentally with his position (Gershman, 1980) or have largely corroborated his findings (Featherman and Hauser, 1976), others have criticized him for too easily dismissing the continued influence of racial discrimination in the economic sphere or, through their research, have offered findings that do not entirely support his conclusion (Clark, 1980; D'Amico and Maxwell, 1995; Landry, 1987; Pinkney, 1993; Willie, 1978).

Some, for example, have pointed out that unemployment rates are higher for blacks than for whites with the same educational attainment and sometimes even higher than for whites with less education (Hill, 1981; Kasarda, 1985). Critics view this as evidence of continued racial discrimination in employment. Moreover, criticism is made that Wilson and others who attribute the lower occupational status of blacks to lack of education and skills ignore the extent to which failure to attain those qualifications may itself be attributable to racial discrimination (Steinberg, 1995).

Another aspect of Wilson's thesis concerns the question of whether middle-class blacks have attained the equivalent status of middle-class whites or whether their middle-class status is marginal, making them "second-class" Americans despite their personal achievements (Landry, 1987). Between 1960 and 1970, the percentage of middle-class blacks doubled, from about one in eight to one in four black workers. This gain, notes Bart Landry (1987), exceeded their total increase during the previous fifty years. Landry explains, however, that the decade of the 1960s was unique to black upward mobility because it was a period that combined both newly enacted civil rights laws with a general economic prosperity, thereby giving not only legal incentive to blacks but the promise of real opportunities in middle-class occupations. This propitious combination had not occurred in the 1950s, nor was it to become apparent in the 1970s, thus creating a slowdown in occupational gains for blacks.

Landry further maintains that the black middle class remains significantly different from its white counterpart. Its path to middle-class status has been much more tortuous than for whites, its living standard is well below that of the white middle class, and its middle-class position remains far more tenuous than for whites. Moreover, Landry notes the continuation of subtle forms of discrimination against the new black middle class, including the application of timeworn stereotypes and unconscious expressions of racism. Oliver and Shapiro (1995) also show the insecure perch of the black middle class by comparison with its white counterpart. The differences in assets that middle-class blacks and whites hold are striking and are critical to understanding that "an accurate and realistic appraisal of the economic footing of the black middle class reveals its precariousness, marginality, and fragility" (1995:92–93).

Whether the black middle class will stabilize, expand, or even constrict will rest heavily on the state of the U.S. economy. The recessions of the 1970s, 1980s, and early 1990s took a greater toll on the living standards of the black middle class than on the white, demonstrating that economic vacillations continue to affect the black population more severely than the white.

African Americans and Societal Power

As we noted in Chapter 2, whatever changes occur in the distribution of the society's resources—education, income, jobs—necessarily depend on access to important decision-making positions in the polity and economy, those two institutions in which power lies most heavily. To what extent have African Americans begun to move into such positions?

Political Power The gains made by African Americans in the realm of politics in the past three decades have been largely a product of increased electoral participation. The black vote has become an important swing factor in many elections as well as a significant electoral force in its own right in certain localities. The post–World War II economic advances of blacks were very much tied to the changed role of the federal government in securing and protecting black rights, and this was even more clearly the case in the striking increase in black political participation during the 1960s and 1970s, especially in the South. Before that time, various measures kept blacks off the registration rolls in the southern states, and in the North, gerrymandering of black areas neutralized their vote.

Table 8.7 Cities over 100,000 in Population with Black Majorities

CITY	TOTAL POPULATION	BLACKS AS PERCENT OF TOTAL POPULATION
Gary, Indiana	117,000	80.6
Detroit, Michigan	1,028,000	75.7
Atlanta, Georgia	394,000	67.1
Washington, D.C.	607,000	65.8
Birmingham, Alabama	266,000	63.3
New Orleans, Louisiana	497,000	61.9
Baltimore, Maryland	736,000	59.2
Newark, New Jersey	275,000	58.5
Jackson, Mississippi	197,000	55.7
Richmond, Virginia	203,000	55.2
Memphis, Tennessee	610,000	54.8
Macon, Georgia	107,000	52.2
Inglewood, California	110,000	51.9
Savannah, Georgia	138,000	51.3

SOURCE: U.S. Bureau of the Census, 1995b.

Although an increasing number of blacks had begun to register and vote in the South during the 1940s, legal impediments as well as threats of violence still discouraged most from trying to exercise the ballot. As Gunnar Myrdal explained, “It is no test of the franchise that some Negroes are permitted to vote in a given community, for what is permitted to a few would never be permitted to the many” (1944:489). With the advent of the civil rights movement in the 1960s, however, enormous gains were achieved in black voter registration in the South, aided by the civil rights legislation of 1964 and 1965. Over two million blacks were newly registered in the southern states between 1964 and 1972.

In the North, the expansion of black electoral power was not a product of the civil rights movement but rather of the demographic changes that had occurred during the three decades from 1940 to 1970. It was the transformed racial composition of cities of the Northeast and Midwest that accounted for the rise in black electoral power more than any increase in black voting or concessions of power by whites. As the central cities became increasingly black and the suburbs increasingly white, blacks in the central cities found themselves with the numerical power to influence the outcome of local elections. Table 8.7 shows the largest central cities with black majority populations.

The upshot of increased political participation in the South and altered urban demographics in the North was a sharp rise in the number of black elected officials, particularly at the local and state government levels. Whereas in 1970, some 1,500 blacks had been elected to political office, that number had swelled to almost 8,000 by the early 1990s (U.S. Bureau of the Census, 1994a).

The substantial increase in the number of black elected officials in the past two decades is, on the surface, an encouraging indication of increasing black political power. This is particularly so when compared with the virtual absence of blacks in politics during the early part of the century. By the mid-1990s, the black presence in national politics was very apparent. Forty African Americans were in the U.S. Congress, and an African-American woman in the U.S. Senate. African Americans had served as governor of Virginia, as national chairman of the Democratic Party, and as chairman of the joint chiefs of staff, the nation's top military post. Perhaps most significant, four African Americans had been appointed to President Clinton's cabinet: secretaries of commerce, agriculture, energy, and veterans affairs. At the local level, the election of African-American mayors of major U.S. cities had become commonplace. And, for the first time, a black had been given serious consideration as a potential presidential candidate.

Closer analysis, however, indicates that there are mixed results of African-American efforts at political equity. To begin with, despite the increased number of African Americans elected to office, they do not represent the numerical proportion of the U.S. black population or that of any individual state. African Americans constitute less than 2 percent of all elected officials in the United States (Joint Center for Political and Economic Studies, 1994).

Second, the offices held by blacks are predominantly at the local level and in communities where black voters constitute a sufficient numerical majority. The general pattern of white support of black candidates is around 15 percent (Lusane, 1989). As Killian has explained, gains in political office for blacks have come, ironically, "from a combination of the increase in black voting and the persistence of residential segregation" (1981:46). Few of the black members of the U.S. Congress, for example, represent districts with white majorities. In fact, the increase in the number of African Americans elected to the U.S. Congress in 1992 from twenty-five to thirty-eight was the result of redistricting under the Voting Rights Act, which created new districts made up primarily of minority communities and thus designed specifically to elect minority members.[6] In short, black candidates, notes political scientist Charles Hamilton, "must rely heavily on

[6]A U.S. Supreme Court decision in 1995 questioned the legality of such racially designed congressional districts with its ruling that the race of voters could not be used as the "predominant" factor in drawing district boundaries.

black voters in order to contest seriously for office" (1989:115). That blacks have almost exhausted the political jurisdictions in which a black majority voting population exists would not seem to bode well for further gains in black political power, especially at the national level. Nonetheless, there are some positive signs that race may be declining as the primary factor in the election of African Americans to political office. An increasing number of blacks are being elected at national and state levels by predominantly white electorates (Swain, 1995).

Third, although the election of black mayors is symbolically significant (by the early 1990s, more than 300 American cities were led by black mayors), the power of mayors or any other elected city officials to affect income and jobs in their communities is severely limited. Local economic conditions are shaped mainly by corporate officials, not local political leaders, and mainly on the basis of short-term profit, not community needs. Moreover, the increasing black control of central cities has come at a time when political and economic power has shifted to the suburban areas. Industries as well as white middle-class residents have continued their flight to the suburbs, leaving the core cities with a declining tax base with which to finance public needs for those who remain—increasingly, the elderly, the poor, and disproportionately, the black poor. This has left the cities largely dependent on state and federal government funding. As they have become more dependent on state and federal governments to finance basic services, central cities have become captives of the decisions of government agencies and legislatures that may have no political interests or constituencies in these communities. Ironically, then, at a time when the difficulties of cities are vast—crime, poverty, unemployment, industrial flight, and deteriorating schools—they lack the financial means to deal with them and must appeal increasingly to often unresponsive and unsympathetic sources. Black control, therefore, may be a hollow victory.

Fourth, even in local areas where blacks constitute a significant potential electorate, their traditionally lower electoral participation has further diluted their political power. As with other American racial-ethnic minorities, this tendency is largely a function of the lower collective class position of African Americans. Those lower in occupation, income, and education generally participate in the electoral process to a lesser degree than those of the middle class and above. In 1994, for example, whereas 47 percent of whites voted in the congressional election, only 37 percent of blacks did so (U.S. Bureau of the Census, 1995b). Nonetheless, in certain states and at the national level as well, the black vote has proved critical and will continue to be given serious consideration by political candidates.[7]

---

[7]The fact that African Americans remain a numerical minority within a majoritarian political system, of course, may be an inherent limitation to their acquisition of great collective power.

Finally, and perhaps most important, an increase in numbers may mean little if the positions to which blacks are elected do not entail broad policymaking powers that provide opportunities for changing the distribution of economic and political resources. Despite their growing presence in political offices at the local and state levels of government and in the U.S. Congress, the black presence at the highest ranks of the executive branch of the federal government, precisely where the power to effect changes in the collective life chances of African Americans is most concentrated, has not kept pace. The appointment by President Clinton of four African Americans to cabinet positions, however, spelled significant change in this regard. Also, the serious attention given Colin Powell in 1995 as a possible presidential candidate must be seen as a very positive sign. Ironically, before he removed himself as a candidate, Powell's support had been far more solid among whites than among blacks.

Economic Power The assertion by political scientist Thomas R. Dye that "there are very few blacks in positions of power in America" (1995:175) is borne out in the extreme within the corporate world. Few African Americans serve on the board of directors or in a top-level executive post of any of the largest corporations, and these few are often token rather than bona fide decision-making appointments. Even those who have achieved middle-management positions are often assigned to race-related areas or to special marketing (that is, minority) groups. Rarely are they found in decision-making positions dealing with production, planning, or the supervision of large numbers of employees (Clark, 1980; Jones, 1986). This is an area of authority in American society, of course, that is not amenable to change through the electoral process; African Americans and other ethnic minorities, therefore, are dependent on the inclinations of the corporate elite to grant them a share of power. There is an encouraging pattern emerging, however, in which blacks continue to increase their presence in executive and managerial positions, small though their numbers are (Williams, 1995).

African Americans also control only a tiny fraction of the nation's income-producing assets, and they seldom own their own businesses (Swinton, 1992; U.S. Bureau of the Census, 1979). The meager share of business wealth controlled by blacks is reflected in a few comparative figures. In 1994, the 100 largest primarily black-owned industrial and service enterprises had combined sales of $6.7 billion. This was equivalent to the sales of the 173d largest American industrial corporation alone. As for banks, insurance companies, and other financial institutions, the marginal place of black enterprise is even more striking. For example, in 1994 the combined assets of the 25 largest primarily black-owned banks, $3.2 billion, were less than the yearly *profits* of the single largest American banking corporation (*Black Enterprise,* 1995a, 1995b; *Fortune,* 1995a, 1995b).

The modern period of black–white relations has seen profound changes in patterns of racial prejudice and discrimination. Whereas some form of biological racism was a solidly entrenched ideology in all parts of the nation well into the 1930s, by the 1960s a majority of Americans had rejected the notion of innate black inferiority. This shift in attitude accompanied the demolition of the legitimated system of racial segregation in which overt and direct forms of discrimination, both individual and institutional, were approved and defended by the state. But as these traditional forms of discrimination were abolished, covert and indirect forms replaced them, ones less amenable to alleviation.

Changes in Patterns of Discrimination

Before the civil rights movement of the 1960s, discrimination was de jure, that is, legitimized and enforced by local, state, and even federal governments. Today, however, prescribed discrimination has given way to de facto forms, which are the result not of direct antiblack measures but of spatial arrangements, demographic trends, and housing patterns (Pettigrew, 1979). Here we might recall the notion of indirect institutional discrimination, discussed in Chapter 3.

The most evident, persistent, and consequential forms of racial discrimination in American society are found in the area of housing. Moreover, the patterns of discrimination here are direct and deliberate to a far greater extent than in the areas of work and education. The effects of lingering racial discrimination in employment, for example, are, as we have seen, debatable. Employers may continue to discriminate against blacks, but the extent and impact of those actions as well as their intent are difficult to precisely ascertain. In education, too, although many urban school systems are as segregated today as they were before the 1954 Supreme Court decision barring segregated schools, these patterns are maintained more by residential segregation and the past drawing of racially distinct school districts than by the conscious design of whites. Discrimination in housing, however, is much clearer, and its intent is not easily disguised. It is well recognized that less progress in breaking down racial barriers has been made in this sphere of social life than in any other (Levitan, Johnston, and Taggart, 1975; Massey and Denton, 1993; Pettigrew, 1979).

Because the patterns are so clear-cut, focusing on housing is essential in looking at current forms of discrimination leveled at blacks. However, there are additional important reasons for concentrating more intently on residential discrimination. Its ripple effects touch every other area of social life (Farley, Bianchi, and Colasanto, 1979; Massey and Denton, 1993; Roof, 1972; Steger, 1973). Where people live determines in largest part the schools they attend (and thereby the quality of their education); the jobs they have access to (and thus their occupational destiny); the benefits they receive from public institutions like hospitals, recreational facilities, and transportation systems (and, as a result, their health and well-being); and the commercial establishments they have access to (and thus their style of life). Moreover, the society's residential arrangements determine the essential character of interethnic relations. With continued segregation in housing, whites and blacks are prevented from interacting at personal levels, thus reinforcing racial attitudes and perpetuating structural pluralism. Though it is difficult to select one element of the system of American ethnic relations and call it the "key" to the maintenance of conflict and misunderstanding, residential segregation is unquestionably among the most fundamental. It would hardly be an exaggeration to proclaim that the present pattern of housing in the United States is the chief obstacle to progress in all other aspects of race and ethnic relations. Pettigrew has put the matter succinctly: "The residential segregation of blacks and whites has emerged as a functional equivalent for the explicit state segregation laws of the past in that it effectively acts to limit the life chances and choices of black people generally" (1979:124).

The Formation of Black Ghettos Wide-scale residential segregation has been a continuous fact of urban life in the United States since blacks began to move to cities in large numbers in the early twentieth century. As with European immigrants before them, they were concentrated in urban ghettos, but unlike the European groups', black ghettos did not progressively break up with each succeeding generation.

Indeed, what is most significant regarding the formation of urban black ghettos is that they have remained so intractably persistent. Patterns of residential dispersion exhibited by white ethnic groups as they have experienced upward mobility (the Italians and Jews described in previous chapters, for example) simply do not hold for blacks (Hershberg et al., 1979; Kantrowitz, 1973, 1979; Massey and Denton, 1987, 1993; Roof, 1978, 1979; Spear, 1967; Taeuber, 1975; Taeuber and Taeuber, 1964). The black ghetto today is an integral facet of almost every large or medium-sized city in the United States. In their study of urban residential segregation, Karl and Alma Taeuber concluded that "a high degree of racial residential segregation is universal in American cities. Whether a city is a metropolitan center or a suburb; whether it is in the North or South; whether the Negro population is large or small—in every case, white and Negro households are highly segregated from each other" (1965:2). Their study showed that segregation in housing not only had not lessened since the early 1900s but had actually intensified in the 1940s and 1950s.

Subsequent studies have shown that these trends changed little throughout the 1960s and 1970s despite greater government activity in promoting desegregation, more liberal social attitudes, and more middle-class blacks with sufficient incomes to afford better housing (Farley, 1985; Levitan, Johnston, and Taggart, 1975; Massey and Denton, 1987, 1993; Orfield, 1985; Roof, 1979; Van Valey, Roof, and Wilcox, 1977; White, 1987). Table 8.8 shows the levels of black–white residential segregation in the largest metropolitan areas. Although these levels declined somewhat in the 1980s, blacks remain "a uniquely segregated group" (Farley and Frey, 1992:28). The simple fact is that most African Americans, regardless of their social class or the geographic area in which they live, continue to reside in predominantly black neighborhoods.

Migration to the suburbs, a movement followed heavily by whites since the end of World War II, has not been replicated to the same degree by blacks, who continue to occupy mainly central-city areas. In 1988, 57 percent of the black population lived in central cities of metropolitan areas, a rate more than double that of whites. Moreover, rising socioeconomic status has not measurably affected the overall level of black segregation (Massey and Denton, 1987, 1993). Hence, when blacks do move to suburban areas, they again find themselves, for the most part, in racially segregated communities (Berry and Kasarda, 1977; Dent, 1992; Logan, 1988; Massey and Denton, 1987, 1993; Reed, 1982; Siembieda, 1975).

Might these patterns be more a result of class than of race? That is, might the persistence of black ghettos merely reflect the lower economic status of blacks generally? If this were so, we could expect to find blacks and whites of similar social class residing in racially mixed neighborhoods. Studies have shown, however, that even where average rents paid by blacks and whites are equivalent, little racial integration is evident (Farley, 1985; Farley and Allen, 1987; Massey and Denton, 1993; Orfield, 1988; Taeuber, 1975). Blacks of every economic and educational rank are highly segregated from whites of the same economic and educational levels. Class factors, then, do not significantly account for black concentrations in central cities.

Might a part of the continuation of residential segregation be the preference of blacks themselves to remain in racially homogeneous areas? This view is also not supported by research evidence. For more than two decades, national samples have shown that blacks favor residential integration and prefer mixed neighborhoods by a wide majority (Farley, 1985; Farley, Bianchi, and Colasanto, 1979). A poll in 1989, for example, asked blacks whether they would prefer to live in a neighborhood with mostly whites, mostly blacks, or half whites and half blacks; only 10 percent answered "mostly blacks" (*Detroit News,* 1989). Surveys show that whites, on the other hand, are willing to accept residential integration only if the number of blacks is minimal.[8]

---

[8]Farley et al. (1994), however, have found that black preferences for racially mixed neighborhoods have weakened in recent years. Whether this trend is the result of a genuine preference to live in black communities, apprehensions about white hostility, or the view that integration yields few benefits is not clear.

Table 8.8 Black–White Residential Segregation for Metropolitan Areas with the Largest Black Populations[a]

REGION/ METROPOLITAN AREA	1990 BLACK POPULATION	BLACK PERCENT OF METROPOLITAN POPULATION	WHITE–BLACK RESIDENTIAL SEGREGATION SCORE[b]	
			1990	*Change 1980–1990*
Northeast				
New York	2,250,026	26.3	77.8	−0.6
Philadelphia	929,907	19.1	81.5	−1.1
Newark	422,802	23.2	82.8	−1.0
Boston	233,819	6.2	70.4	−5.9
Midwest				
Chicago	1,332,919	22.0	87.1	−3.6
Detroit	943,479	21.5	88.8	0.1
St. Louis	423,182	17.3	81.1	−3.9
Cleveland	355,619	19.4	86.1	−3.1
Kansas City	200,508	12.8	75.5	−5.1
South				
Washington, D.C.	1,041,934	26.6	67.5	−3.9
Atlanta	736,153	26.0	72.6	−6.4
Baltimore	616,065	25.9	75.4	−2.4
Houston	611,243	18.5	69.0	−9.2
New Orleans	430,470	34.7	73.8	−2.2
Dallas	410,766	16.1	66.2	−14.8

Studies indicate that Latinos and Asians, groups entering the society much later, have already begun to disperse throughout metropolitan areas to a far higher degree than African Americans (Farley, 1985; Farley and Allen, 1987; Farley and Frey, 1992; Langberg and Farley, 1985; Massey and Denton, 1987, 1989b, 1993; Taeuber and Taeuber, 1964; White, 1987). Segregation scores, or indexes of dissimilarity (see Table 8.8), indicate that residential segregation between Latinos and non-Hispanic whites as well as between Asians and whites is much lower than segregation between blacks and whites.

It is apparent, then, that the intensity and intractability of black–white residential segregation are unequaled in American society. Housing patterns between

REGION/ METROPOLITAN AREA	1990 BLACK POPULATION	BLACK PERCENT OF METROPOLITAN POPULATION	WHITE–BLACK RESIDENTIAL SEGREGATION SCORE[b]	
			1990	*Change 1980–1990*
South *(continued)*				
Memphis	399,011	40.6	75.6	−0.2
Norfolk	398,093	28.5	56.7	−8.5
Miami	397,993	10.5	74.6	−6.4
Richmond	252,340	29.2	64.1	−3.6
Birmingham	245,726	27.1	79.3	−0.3
Charlotte	231,654	19.9	64.6	−3.0
West				
Los Angeles	992,974	11.2	71.4	−8.7
Oakland	303,826	14.6	68.7	−6.0

SOURCE: Farley and Frey, 1992. Reprinted by permission.
[a]With 1990 black populations greater than 200,000.
[b]The *segregation score* or *index of dissimilarity* is a measure of the extent of residential segregation between two groups. It indicates the percentage of one group that would have to move in order to achieve total integration, that is, a population mix that corresponds to each group's proportion of the population of a designated area. For example, if a city were made up of 80 percent whites and 20 percent blacks, the two groups would be represented in each neighborhood in that same proportion. The index can range from 100, indicating total segregation, to 0, indicating total integration. Thus, the higher the index, the greater the degree of segregation. See Taeuber and Taeuber (1965).

blacks and whites are, as Pettigrew (1979) has noted, qualitatively different from those between other groups. Massey and Denton have described this uniquely disadvantaged black urban environment as "hypersegregation," that is, segregation across several dimensions. "Black Americans in these metropolitan areas live within large, contiguous settlements of densely inhabited neighborhoods that are packed tightly around the urban core. In plain terms, they live in ghettos" (1993:77). Roughly one-quarter of the African-American population, in ten large metropolitan areas, lives in such an urban area. No other racial-ethnic group, Massey and Denton explain, experiences such multidimensional segregation. This environment shuts out most contact with whites except through participation in the labor force. Moreover, these extraordinarily high levels of residential segregation almost assure their continuation into the foreseeable future. Farley and Frey (1992) note that if the average black–white segregation score continued to decrease at the rate of the 1980s, it would require fifty years for the segregation level of blacks to fall to the level now experienced by Latinos and Asians.

The Maintenance of Residential Segregation Despite the enactment of many state and federal antidiscrimination statutes since the 1940s, residential segregation is maintained by a set of discriminatory practices, both individual and institutional. Moreover, many are direct and intentional.

Past government policies created much of the framework for the current patterns of residential segregation. Beginning in the 1930s, the federal government encouraged home ownership among middle- and working-class people by establishing the Federal Housing Administration (FHA) as well as other housing-related agencies. The idea was to furnish cheap financing by providing government-backed mortgage insurance. From the outset, the FHA officially discouraged integrated residential areas by refusing to guarantee loans for homes that were not in racially homogeneous areas (Abrams, 1966; Feagin and Feagin, 1978; Massey and Denton, 1993). This policy was not changed until 1962, when nondiscriminatory pledges were required from loan applicants.

The real estate system has been another key element in the perpetuation of segregated housing. Until the 1950s, real estate boards, almost entirely white in membership, followed a strict code that restrained the renting or selling of property in white areas to blacks. Using various techniques, white real estate agents would steer blacks to black areas and whites to white areas in order to enforce this code. Banks and other lending institutions cooperated in the system by "redlining," that is, designating certain areas within which real estate loans would not be made. Another device used by real estate brokers was "blockbusting." By spreading word through a white neighborhood of an impending black influx, agents would frighten whites into selling their homes cheaply. These homes were subsequently sold to blacks at inflated prices. In the process, all-white areas were transformed quickly into all-black areas.

Where such tactics were unsuccessful, zoning regulations were established, specifying certain types of housing that could be erected in particular neighborhoods; these were designed to exclude low-income, mainly black, units. Restrictive covenants, agreements made by homeowners not to sell to members of particular groups, were also used to bar blacks and other minorities from white residential areas. Until the Civil Rights Act of 1968 prohibited discrimination in housing, such covenants had been widely applied and were even supported by law until 1948. When all else failed, violence and intimidation were frequently used to deter blacks from buying or renting in all-white neighborhoods.

Although some of these practices have been curtailed by more forceful antidiscriminatory legislation, many remain prevalent. Discriminatory measures still widely employed include resistance to public housing desegregation, zoning and annexation policies that foster racial segregation, redlining by lending institutions, and various informal practices of the real estate industry such as steering people toward predominantly white or black neighborhoods (Massey and Denton, 1993; Taeuber, 1975). Despite antidiscriminatory statutes, most of these techniques are difficult to detect and cannot be effectively policed. In addition to these institutional forms of residential discrimination, individual homeowners and landlords may discriminate in selling or renting their property with even greater impunity from legal restraints.

Farley et al. (1994) and Massey and Denton (1993) find that discriminatory real estate practices are often linked to the stereotypes of blacks by whites. When whites strongly hold racial stereotypes of blacks (crime, indolence, and so on), they are more inclined to resist residential integration and will hesitate to move into neighborhoods where there is more than a token number of blacks.

Essentially, then, there exist in the United States two housing markets: one white and one black. The real estate industry, banks and lending institutions, and government agencies at all levels have in the past directly laid the foundation for this dual system and continue to apply indirect and covert measures to sustain it. Government policies designed to mitigate discriminatory patterns have had only a limited effect.

Other racially related problems of American cities, including school segregation and chronic unemployment, are in large part directly traceable to the perpetuation of black ghettos. School segregation, for example, is sustained chiefly by the maintenance of segregated neighborhoods (Farley, 1984; Santiago, 1991). As a result, legal efforts to integrate schools have been nullified in countless cases (Abrams, 1966; Cataldo, Giles, and Gatlin, 1975; Farley and Taeuber, 1974; Kantrowitz, 1979; Taeuber, 1979). As previously noted, spatial segregation creates numerous types of "side effect discrimination" (Feagin and Feagin, 1978). Residential discrimination, which is most often overt, direct, and intentional, produces, naturally and often unintentionally, indirect discriminatory effects in other areas (see Chapter 3). In sum, although racial discrimination in other institutional realms of American society has broken down to a great extent, it remains tenaciously persistent in residential patterns.[9]

Following the urban riots of 1967, the Kerner Commission issued its report on civil disorders, warning that the United States was "moving toward two

[9]On the positive effects of residential segregation for blacks, see Marston and Van Valey (1979). They argue that residential segregation creates and sustains an institutional structure that caters to blacks, induces ethnic group consciousness, and in some cities, produces local political clout.

societies, one black, one white—separate and unequal" (National Advisory Commission on Civil Disorders, 1968:1). In no area of social life does this prognosis continue to be verified better than in housing.

Here we might consider Joel Kovel's distinction between "dominative" and "aversive" racism. In the former, actions are taken to oppress racial minorities and keep them subservient. In the latter, *in*action is characteristic. Racial minorities are simply ignored and avoided when possible; relations at best become "polite, correct and cold in whatever dealings are necessary between the races" (1970:54). Continued residential segregation provides the backdrop for such "aversive" discrimination in various areas of social life. While obvious and blatant forms of discrimination are no longer prevalent, more benign black–white relations are, in large measure, founded on intergroup avoidance (Pettigrew, 1980).

Changes in Racial Attitudes

As the nature of racial discrimination has changed in the past several decades, so have the attitudes of whites toward blacks. Although there is much room for optimism regarding these attitudinal shifts, there are also indications that racist beliefs are still firmly rooted in American thought, albeit in modified form. Moreover, there is ample evidence that blacks and whites do not see the same reality with regard to the current status of blacks in the society or the nature and causes of black problems.

Attitudinal changes during the past thirty years have been measured by extensive and frequent surveys among blacks and whites. At the outset, it should be noted that survey data regarding attitudes must be accepted cautiously as representative of people's real beliefs and feelings. At best, they comprise rough descriptions, not precise readings. Because they are designed to draw collective or general pictures, they will not necessarily reflect the views of particular individuals within the groups. Obviously, blacks and whites as collectivities are internally variegated, and views among individuals will differ on the basis of age, gender, education, and numerous other social variables. Moreover, as explained in Chapter 3, surveys, particularly those dealing with socially sensitive issues like race and ethnic relations, may reflect mainly what people believe interviewers want to hear or what they feel to be the most socially acceptable response, not necessarily their true feelings. Finally, attitudes may appear inconsistent from item to item or from survey to survey. For example, public opinion studies have shown a strong acceptance of integrated schools but an equally strong rejection of school busing (Gilbert, 1988; Greeley and Sheatsley, 1974). Interpretations of such inconsistencies are often problematic. In this case, the antibusing sentiment may be an indication of covert racism, or it may indicate other, nonracist attitudes such as commitment to the neighborhood school, fear of the dangers of busing,

Table 8.9 White Americans' Stereotypes of Blacks

STEREOTYPE	PERCENT OF WHITES WHO AGREED				
	1963	*1967*	*1971*	*1978*	*1995*
Blacks tend to have less ambition than whites.	66%	70%	52%	49%	29%
Blacks want to live off the handout.	41	52	39	36	22%
Blacks are more violent than whites.	—	42	36	34	38[a]
Blacks breed crime.	35	32	27	29	—
Blacks have less native intelligence than whites.	39	46	37	25	12
Blacks care less for the family than whites.	31	34	26	18	40[b]

SOURCE: Anti-Defamation League, 1993; National Conference, 1995; National Conference of Christians and Jews, 1978.

[a]1992

[b]Statement was "Blacks have less family unity."

or resentment over the inconvenience caused by busing (Greeley and Sheatsley, 1974). It may also reflect a particular view of what government should or should not do to accelerate racial equality (Sniderman and Piazza, 1993). With these caveats in mind, let us look at some specific patterns.

Stereotypes and Social Distance First, it is clear that negative stereotypes of blacks have diminished and that less support is given to blatant forms of discrimination (Brink and Harris, 1963, 1967; Campbell, 1971; Greeley and Sheatsley, 1974; Pettigrew, 1975, 1981; Schwartz, 1967; Sheatsley, 1966; Williams, 1977). As is seen in Table 8.9, a relatively steady decline in white notions of black inferiority has been measured since 1963. In the fifteen years between 1963 and 1978, the percentage of whites who believed blacks to be innately less intelligent fell from 39 to 25. By the mid-1990s, only 12 percent maintained that view (Sniderman and Piazza, 1993).

White attitudes toward racial integration have also changed somewhat in the past three decades (Table 8.10). Most whites now seem committed (at least verbally) to integration in most social areas entailing secondary relations (Greeley and Sheatsley, 1974; Schuman, 1974; Schuman, Steeh, and Bobo, 1985; Steeh and Schuman, 1992; Taylor, Sheatsley, and Greeley, 1978). Even the more negative policies and attitudes of the Reagan era of the 1980s did not seem to

Table 8.10 White Americans' Attitudes on Integration

ITEM	RESPONSE	PERCENT RESPONDING 1963	1970	1977	1982	1989
"Do you think white students and black students should go to the same schools or to separate schools?"	"Same schools."	63%	74%	86%	90%	93%
"How strongly would you object if a member of your family wanted to bring a black friend home to dinner?"	"Not at all."	50	63	71	78	77
"White people have a right to keep blacks out of their neighborhoods if they want to, and blacks should respect that right."	"Disagree slightly" or "disagree strongly."	45	51	56	71	78
"Do you think there should be laws against marriages between blacks and whites?"	"No."	37	48	72	66	77

SOURCES: *Public Opinion*, 1987; Schuman, Steeh, and Bobo, 1985; Smith, 1980; Wood, 1990.

reverse the steady liberalization of racial attitudes among whites (Steeh and Schuman, 1992). This rise in racial liberalism may be attributed to higher levels of education, an increasing liberal perspective on social issues generally, and increased contact between blacks and whites.

As for social distance, only a minority of whites nationwide object to contacts with blacks on the job or in public facilities. But the percentage rises dramatically when intimate contacts are involved. Few blacks today, regardless of social class, are likely to socialize with their white fellow employees outside the formal work setting. As Schuman, Steeh, and Bobo explain, "In all likelihood, the more impersonal and transient the interracial contact at issue, the more white support for integration; the more intimate and long-lasting the contact, the less the support" (1985:136). Hence, a majority of whites continue to object to interracial dating and marriage, and most, as we have seen, are also opposed to living in areas with a high percentage of blacks (Brink and Harris, 1963, 1967; Campbell, 1971; Farley et al., 1978; Gallup, 1995; Schuman, Steeh, and Bobo, 1985). It is this apparent fear of proximity at more intimate levels that looms as a significant impediment to black–white integration.

Attitudes and Actions Although white attitudes concerning discrimination and integration are apparently increasingly liberal, ambivalence arises when translation of these attitudes into concrete actions is suggested. This is obvious in various ambits of social life, including politics, jobs, housing, and education. Whites agree that blacks should have equal rights in all these areas but are reluctant to approve measures necessary to guarantee them (Pettigrew, 1979; Schuman, Steeh, and Bobo, 1985). For example, from 1964 to 1978, white attitudes favoring the principle of school integration rose by 22 percentage points, but support for federal intervention to implement such integration dropped by 17 percentage points (Schuman, Steeh, and Bobo, 1985).[10]

Similarly, though a majority of whites favor the principle of free choice in residence, only a minority support open housing legislation. In a classic work dealing with black–white relations in the United States, Swedish social scientist Gunnar Myrdal (1944) described the disjuncture between white attitudes and actions vis-à-vis blacks as "an American dilemma." The dilemma, according to Myrdal, is the conflict between, on the one hand, adherence to the "American Creed," entailing a belief in equality and Christian precepts, and on the other, group prejudice and discrimination based on material and status interests.

We have seen that in housing, especially, whites are adamant in maintaining separation from blacks. Opinion studies indicate that white Americans are willing to tolerate a limited amount of residential integration but are unwilling to formally recognize full black rights, without which housing markets cannot be meaningfully desegregated (Pettigrew, 1975; D. Taylor, 1979). Here also can be seen a more clear-cut relationship between attitude and action. In a sense, there is a circular movement in which prejudice feeds on segregation, and vice versa. As one observer has noted, "The most well-documented and well-replicated finding in research on the effect of interracial housing and neighborhood integration is that white prejudice declines with interracial contact, particularly when the interracial contact occurs in a situation where whites and blacks are of equal social status and are aware that they share similar values" (D. Taylor, 1979:35; see also Schuman, Steeh, and Bobo, 1985). Increased contact, in other words, seems to moderate racial prejudice, especially if the groups are of the same social class. With regard to the latter, it has been observed that a key reason for whites' fears of black penetration of their neighborhoods is the lower

---

[10]The issue of school busing is complicated. In 1994, 86 percent of whites and 94 percent of blacks agreed with the Supreme Court's 1954 decision barring school segregation. But most blacks, as well as whites, preferred that students attend their local schools, even if it meant most of the students would be of the same race. Only 9 percent of whites and 33 percent of blacks preferred to assign students to schools outside their local communities as a way to achieve more integration (McAneny and Saad, 1994).

social class of blacks, not necessarily negative racial attitudes alone (Farley, Bianchi, and Colasanto, 1979). Whites are concerned that their neighbors will be poorer and will have lifestyles very much at odds with their own. Studies have shown, however, that blacks moving into white neighborhoods are generally similar to their white neighbors in socioeconomic status (Duncan and Duncan, 1957; Taeuber and Taeuber, 1965). But fixed negative images of blacks (as "poor," "lazy," "squalid") often prevent whites from perceiving the class distinctions among them—hence the continued resistance to integrated residential areas. Without social contact, then, prejudicial attitudes and images persist, in turn leading to intransigence on matters of neighborhood desegregation.

Views of the Socioeconomic Gap Although whites generally recognize the disadvantaged status of a large proportion of the black population, there are striking differences in black and white explanations for that depressed condition. Most whites continue to view the problems as the result of blacks' own shortcomings, not social conditions or institutional forms of discrimination (Lipset, 1987; Schuman, 1982; Sniderman and Piazza, 1993). The majority, however, no longer attribute these deficiencies to biological inferiority (Schuman, 1982; Schwartz, 1967; Sniderman and Piazza, 1993). Most often, whites see black disadvantages as a product of "lack of ambition," "laziness," "irresponsibility," or "failure to take advantage of opportunities." The prevalent attitude is, "If we did it, so can they." Indeed, the image of blacks as lacking ambition and socially irresponsible is today the most widespread and common negative stereotype (Sniderman and Piazza, 1993). Blacks, by contrast, see their plight as deriving primarily from discrimination or a lack of the same opportunities as whites. For instance, a Detroit area survey reported that 61 percent of blacks but only 33 percent of whites believed that "a lot" of discrimination impedes blacks in getting good jobs (*Detroit Free Press,* 1992).[11]

Structural forces that strongly affect people's place in the stratification system (such as the society's changed occupational order or its basically altered demographic patterns) are not easily perceived. Thus, social success or failure is apt to be interpreted as the result primarily of individual effort. Moreover, the belief that each member of society is personally responsible for his or her social lot is a basic component of American ideology. The society's opportunity structure is pictured as open, providing equal chances for all to achieve material success or political power regardless of their class or ethnic origin. As a result, it seems only

---

[11]There is some ambivalence on the part of blacks, however. Although they are more likely to recognize structural factors as important in accounting for lower economic and social achievement, blacks no less than whites subscribe to the view that blacks fail to try hard (Sniderman and Piazza, 1993).

natural that whites would be more likely to attribute the lower social and economic achievement of blacks to individual and group characteristics than to the workings of a class and ethnic system that automatically favors success for the wellborn, especially those who are also white, and failure for the poor, especially those who are black. In addition, the more blatant forms of discrimination, which long held blacks in subservience — slavery, the Jim Crow system — are easier to perceive and understand by whites than are the indirect and largely covert forms of institutional discrimination that continue to render debilitating effects.

Problems and Solutions Whites and blacks do not share a common view of the problems facing African Americans. Whites today tend to view blacks as better off economically and socially than blacks view themselves and also see the extent and pace of racial change in the United States as far greater and more rapid than do blacks. A majority of whites today believe African Americans enjoy equal opportunities in education, jobs, housing, and in access to credit. African Americans have diametrically opposite views (National Conference, 1995).

Surveys indicate that whereas whites see problems as more of a "mopping up," with basic changes having been made, blacks perceive deeply rooted discrimination continuing to affect their chances in work and education. In a 1995 survey, middle-class whites and blacks were asked if they agreed with the statement "Past and present discrimination is the major reason for the economic and social ills blacks face." Eighty-four percent of blacks agreed, but only 30 percent of whites did so (*Washington Post,* 1995). Whites define racial progress mostly in political terms. Thus, if laws have been changed that previously denied blacks access to critical institutions, then it is assumed that an equal opportunity structure has been created. In this view, if blacks succeed or fail it is their own doing. Blacks, on the other hand, see not simply a few remnants of the past that need attention, but a system that continues to block their advancement. Moreover, even middle-class blacks, those who have seemingly "made it" in an economic sense, continue to experience what they perceive as indignities based on race (Cose, 1993) and express frustration regarding future opportunities (Hochschild, 1995).

Also, blacks see a justice system that treats blacks differently from whites (National Conference, 1995). A strikingly contrasting view of the O. J. Simpson case provided stark evidence of this. While most whites saw Simpson as guilty and as having evaded justice, a majority of blacks agreed with the "not guilty" verdict and were inclined to believe that the jury was fair and impartial (Whitaker, 1995). Many blacks saw the decision as a kind of symbolic retribution for what they viewed as a criminal justice system that more commonly victimizes blacks.

As to the pace of social change, a 1991 survey showed that only 39 percent of blacks as opposed to 64 percent of whites thought that job opportunities for blacks had improved in the last five years (*Detroit News,* 1991). A more optimistic view among blacks had been evident in 1969, when 70 percent thought the situation for blacks had improved in the previous five years. In short, black views of black progress are dramatically different from those held by whites. Whites are more likely to believe that racial injustices have been ameliorated and that discrimination therefore does not affect the pace and extent of black progress.

AFRICAN AMERICANS: ASSIMILATION OR PLURALISM?

In Chapter 4, we pointed out that assimilation and pluralism are processes and outcomes of interethnic relations that are not necessarily mutually exclusive for any particular group in a multiethnic society. Rather, a group may display aspects of both simultaneously. In the case of African Americans, certain patterns of assimilation seem apparent even though pluralism, specifically, the colonial form, has generally been more characteristic of their historical and even contemporary experience. African-American assimilation has been limited to *cultural* assimilation and *secondary structural* assimilation.

Cultural Assimilation

Because the African cultural heritage was largely destroyed under slavery, blacks had little choice about the adoption of the dominant group's major cultural traits, especially religion and language.[12] Some hold that, in light of this historical fact, cultural assimilation has been as fully accomplished for blacks as for any American ethnic group (Gordon, 1964; Pinkney, 1993). However, although it is true that blacks have adopted the dominant group's major cultural forms, it is equally evident that a black cultural variation has evolved in the United States, one that clearly sets blacks apart from Anglo-Americans and other ethnic groups (Hannerz, 1969; R. Taylor, 1979). African Americans have developed a linguistic style, have molded a unique version of Protestantism, and have created black art forms, especially in music.

As explained in Chapter 4, cultural assimilation is commonly a reciprocal process in which minority ethnic groups not only absorb the dominant culture but transform it to some degree with their own contributions. This has clearly been the case for African Americans in recent decades. Indeed, some, like sociologist Orlando Patterson, maintain that the black influence on the prevailing

[12]There is an unresolved debate among sociologists and historians concerning the degree to which blacks retained elements of their native African cultures under American slavery. Some, like Frazier (1949), assert that blacks were essentially stripped of African cultural ways., whereas others, like Herskovitz (1941), maintain that remnants remained intact.

American culture is today not simply evident, but pervasive. Blacks, he suggests, dominate American popular culture including music, dance, language, sports, and youths' fashion. "So powerful and unavoidable is the black popular influence," he writes, "that it is now not uncommon to find persons who, while remaining racists in personal relations and attitudes, nonetheless have surrendered their tastes, and their viewing and listening habits, to black entertainers, talk-show hosts and sit-com stars" (Patterson, 1995b:24).

Structural Assimilation

As our analysis demonstrated, structural assimilation at the secondary level has begun in earnest for African Americans as they increasingly acquire better jobs, higher incomes, and more education. For the black underclass, however, secondary structural assimilation has been stymied, and the prognosis for future movement in this direction is bleak. Even for the black middle class, of course, it is still only in the incipient stages.

Structural assimilation at the *primary* level — as measured by residential patterns, club memberships, friendship cliques, and intermarriage — remains minimal. Residential integration is a very basic prerequisite to other aspects of primary and even secondary assimilation, but in this area, as we have seen, African Americans have not followed the usual path of American ethnic groups. Though residential dispersion accompanied upward class mobility for other groups, this has not occurred for African Americans, who remain highly segregated despite advances in income, occupation, and education. As might be expected, interracial marriage also remains uncommon (Alba and Golden, 1986; Lieberson and Waters, 1988). Although the number of black–white marriages increased from 65,000 in 1970 to 296,000 in 1994, it is still an unusual phenomenon when one considers that in the latter year there were over 54 million married couples in the United States (U.S. Bureau of the Census, 1995b). Blacks and whites may increasingly work, shop, vote, and attend school together, but the formal interrelations in these areas of social life do not extend very far into more personal relations as neighbors, church members, close friends, or marital partners. Clearly, primary structural assimilation has progressed to a lesser extent for African Americans than for any other American ethnic group.

Assimilation: African Americans and Other Ethnic Groups

Are African Americans simply one ethnic group in American society, subject to basically the same patterns of ethnic relations as others? Or is their experience so exceptional that they cannot be compared using the same analytical models and criteria applied to other ethnic groups? This is an issue of some debate in sociology. The question is essentially this: Is the assimilation, or "ethnic," model—so clearly applicable to white ethnic groups, such as the Italians and Jews we examined in previous chapters—relevant to African Americans, or are they best viewed in terms of some form of inequalitarian pluralism? Let us briefly examine the two sides of this issue.

The Assimilation Model Social scientists who have applied the assimilation model to African Americans see their social evolution as generally comparable to that of white ethnic groups. In this view, African Americans, by virtue of their migration from the rural South to the cities of the North, face problems not unlike those faced by earlier urban immigrants. Just as Italians, Jews, and other southern and eastern Europeans arriving in the late nineteenth and early twentieth century had to compete for jobs and political power with the earlier-arriving Irish and others, so, too, African Americans must now compete with those who preceded them to the city. Changed urban conditions and group cultural differences are acknowledged, but the most important difference between African Americans and other urban ethnic groups, in this view, is simply the time at which they have entered the city: African Americans are latecomers (Glazer, 1971; Glazer and Moynihan, 1970; Handlin, 1962; Kristol, 1970; Sowell, 1981). The process, therefore, may be longer and more painstaking, but eventually, African Americans, like other ethnic groups, will be assimilated into the mainstream of American society.

In this view, the steady economic and political progress of the black middle class is evidence of continuing assimilation much in the fashion of other ethnic groups. The low level of societal integration of the African-American underclass, by contrast, is explained in large measure as a product of their failure to adopt conforming, middle-class social values (Banfield, 1968, 1990 [1974]; Sowell, 1978, 1981). Discrimination, it is argued, is not the major impediment to further assimilation for lower-status African Americans; instead, it is largely a matter of internal group norms and values. Only if and when those norms and values reflect the mainstream culture will the black underclass advance economically and socially and, subsequently, be integrated into the larger society. Thomas Sowell compared American blacks of slave ancestry with those whose families originally migrated from the West Indies and found striking occupational, income, and educational differences: West Indians ranked higher on all counts. Because the groups are racially the same, Sowell concluded, the development of different attitudinal traits, not racism, has been the major factor in the different rate of group advancement (1978, 1981).[13]

[13]Farley and Allen (1987), however, suggest that the socioeconomic differences between blacks of West Indian origin and other black Americans may not be so great as has often been assumed.

Assimilationists have found the contemporary black family to be a particularly critical hindrance to African Americans in following the adjustment patterns of earlier ethnic groups. Daniel P. Moynihan, in a controversial report issued in 1965, asserted that the black family's comparatively high instability, its disproportionate rate of illegitimacy, and its matriarchal structure were the chief sources of black social and economic problems. He traced these destabilizing family characteristics to slavery, in which, it was presumed, the black family was sapped of its vitality and a female-dominated unit was created (see Frazier, 1939). Moynihan's thesis stirred a great deal of controversy among social scientists and policymakers. Most criticism held that he had confused cause with effect: The unstable black family was not the cause of such circumstances as widespread unemployment and discrimination in the labor and housing markets but the result of those conditions (Billingsley, 1968; Farmer, 1967; Ryan, 1967). Some also pointed out that Moynihan and others who took his position had devoted most of their attention to problem-plagued lower-class families and had largely ignored stable middle- and working-class black families—hence the perception of the black family as "pathological" (Lewis, 1967; Staples, 1981). Moreover, the issue of slavery and its effects on the black family was opened to rigorous debate shortly after Moynihan's report was published. Some historians maintained that the slave system had not had the devastating effects traditionally attributed to it. Instead, they suggested, the slave family had been a haven of psychological security for its members and a social mechanism that aided in survival (Fogel and Engerman, 1974; Gutman, 1976). The prevalence of matriarchy and instability was therefore seen as a product not of slavery but of twentieth-century occupational and residential discrimination.

The role of the family structure in the perpetuation of the black underclass remains disputable. There is little question, however, about the deleterious impact of the female-headed family on income, occupation, and education. Moreover, the percentage of black female-headed families has risen sharply since the mid-1960s, when the Moynihan report was issued. Then, about a quarter of black families were headed by a female; by 1994, almost half were (U.S. Bureau of the Census, 1995a).

The Inequalitarian Pluralistic Model Some aspects of the assimilation model continue to be applied to African Americans. But a more prevalent view today regards the black experience as unique among American ethnic groups. Even exponents of the assimilation model have conceded that the path of societal integration followed by African Americans is exceptional vis-à-vis not only past immigrants, but even new immigrant groups (Glazer, 1993).

In explaining the noticeably different patterns of assimilation of African Americans compared with other ethnic groups, emphasis is placed on the fundamentally divergent African-American ethnic experience; no other group arrived so nearly completely as involuntary immigrants,[14] none was subjected to two centuries of slavery, none experienced the depth and persistence of racial prejudice and discrimination, none was so totally divested of its native culture, and none was so indelibly marked.

These social scientists contend that the African-American experience fulfills the basic components of what was referred to in Chapter 4 as the "colonization complex"—forced entry of the group, destruction of its indigenous culture, repression by the dominant group, and the application of a racist ideology (Blauner, 1969, 1972; Franklin and Resnik, 1973). Other ethnic groups did not face similar circumstances. In this view, then, African Americans clearly fit the internal colonial model, not the assimilation, or "ethnic," model.

Those who see African Americans as singular among American ethnic groups argue that the vestiges of slavery have immeasurably handicapped them in their efforts to compete with whites. The abolition of slavery did not change established relations of dominance and subordination. Blacks continued to be relegated to the lowest occupational levels and were denied fundamental political rights well into the 1960s. Furthermore, the economic and political structures of American society, they argue, have changed so basically from what they were for entering European immigrant groups that comparisons with them are rendered vacuous (Hershberg et al., 1979). African Americans simply cannot utilize economic and political institutions as European immigrants earlier did to secure better jobs, housing, and political clout and, thus, to effectuate assimilation.

Assimilation and Visibility Visibility, as we noted in Chapter 4, is usually a decisive factor in the rate and degree of structural assimilation for ethnic groups in multiethnic societies. The greater the visibility, the slower and less intense is the assimilation process. This would seem to be of overriding importance in the case of African Americans.

Despite its abatement in the past two decades, the racial stigma of African Americans remains an ever-present fact in the United States, making black–white interaction at all social levels the most obdurate problem of American ethnic relations. As Hughes and Hughes put it over four decades ago, "Few are the situations which both Negro and white Americans enter, side by side, unlabelled and undifferentiated. Even a monk or a hero must be named by his color unless he is white" (1952:28). Today, even those African Americans who have attained

---

[14]Though the overwhelming majority of American blacks trace their heritage to slavery, a minority are descendants of voluntary immigrants, most from the West Indies. On black Americans from the West Indies, see Farley and Allen (1987), Patterson (1995a), Reid (1939), and Sowell (1978).

middle-class status have not been able to reverse the order of their social identity as "black" first and "middle-class" second. Isaacs has poignantly expressed the significance of visibility as it affects the lives of black people in America:

> For black Americans, more clearly perhaps than for any other group, the element of color and physical characteristics lies at the very center of the cluster that makes up their basic group identity. It is the one element around which everything else in their lives has been made to revolve, the heart of the identity crisis that is with them every hour of every day and which they need more than anything else to resolve (1989:66).

In sum, although ethnicity may be declining in significance for most whites in American society, for blacks it continues to be the most crucial social characteristic, influencing almost all facets of social life. Clearly, the relations between blacks and whites will remain the major focus of interethnic relations in the United States for many generations.

Black Ethnic Diversity

It is becoming increasingly difficult to describe the American black population not only in broad class terms, but in ethnic terms as well. As we will see in Chapter 11, a significant element of the newest immigration to the United States has comprised people from Caribbean countries, many of whom are, by American standards, "black." Indeed, over 3 percent of the black population in the United States today is foreign-born. Haitians, Jamaicans, Trinidadians, Guyanese, and others bring cultural, and in some cases linguistic, features that are both varied and profound. While they may evince racial similarities, culturally they have little in common with African Americans. This is obviously the case as well for an increasing number of immigrants from sub-Saharan African countries.

As a result of the newest immigration, the American black population is evolving into a somewhat broad, multicultural ethnic category similar to that of Asian Americans and Hispanic Americans, though certainly to a lesser degree. As the American black population becomes more culturally diverse, one can no longer make assumptions about commonalities simply on the basis of racial designation.

SUMMARY

The black experience is unique among American ethnic groups. No other group entered the society so completely as involuntary immigrants, and no other group was subjected to such fully institutionalized degradation. The aftereffects of slavery continue to influence the patterns of black–white relations and the place of blacks in the social hierarchy.

Black–white relations can be divided into three major periods. Slavery, lasting from the 1600s until 1865, was characterized by paternalistic domination in which blacks were totally subservient, by law in the South and by informal processes in the North. Following the brief interlude of Reconstruction after the Civil War, black–white relations entered the Jim Crow era. Although no longer slaves, blacks were forced back into a subservient economic role and, through legal and extralegal measures, were denied the vote as well as other basic civil liberties. Moreover, a fully legitimized system of segregation was developed in the South, abetted by the application of a racist ideology. An era of fluid competition between blacks and whites commenced in the 1930s, and in the 1960s, the Jim Crow system was demolished in the wake of the black civil rights movement.

As a collectivity, blacks in the past three decades have made significant strides in income, occupation, and education. However, they continue to lag behind whites in overall employment and income, and the poverty gap between blacks and whites persists. Furthermore, the gains of the past thirty years have not been experienced in a balanced fashion within the African-American community. One segment has taken its economic place in the working and middle classes, but the other segment comprises an underclass, chronically unemployed, undereducated, and acutely indigent. A debate is being waged between those who see the economic problems of poor blacks mostly as a function of their lack of skills and education and those who see these problems as a function of continued racial discrimination.

The levels of prejudice and discrimination against blacks have historically been more severe and persistent than those against any other ethnic group in American society. In the past three decades, however, these patterns have changed measurably. White attitudes are more tolerant, and discrimination today is confined mostly to the institutional variety. Housing, however, is an area still largely unaffected by changes in black–white dynamics. Although African Americans have reached a high level of cultural assimilation and are in the incipient stages of secondary structural assimilation, primary structural assimilation remains minimal. Little black–white interaction occurs outside formal social settings such as work and education.

There is some debate concerning the relevance to African Americans of either the assimilation model or the internal colonial model. Some maintain that African Americans are simply latecomers to the urban environment and will eventually display assimilation patterns much like those of white ethnic groups. Others, however, see African Americans as a group whose historical experience and current conditions are so distinctive that comparisons with white ethnic groups are difficult at best.

Suggested Readings

Davis, James. 1991. *Who Is Black?: One Nation's Definition.* University Park: Pennsylvania State University Press. Deals with the issue of racial identification and the variable, often arbitrary, nature of racial classification in the United States as well as other societies.

Hacker, Andrew. 1992. *Two Nations.* New York: Scribners. Examines the continuing divisions and inequalities in various spheres of social and economic life between blacks and whites in the United States in the 1990s, citing current data.

Hill, Herbert, and James E. Jones, Jr. (eds.). 1993. *Race in America: The Struggle for Equality.* Madison: University of Wisconsin Press. Essays by prominent social scientists that examine the current socioeconomic and political status of African Americans and the nature of black–white relations.

Jencks, Christopher, and Paul E. Peterson (eds.). 1991. *The Urban Underclass.* Washington, D.C.: Brookings Institution. A collection of essays written by social scientists that deals with the contemporary urban underclass.

Killian, Lewis M. 1975. *The Impossible Revolution, Phase II: Black Power and the American Dream.* New York: Random House. One of the best accounts of the origins, strategies, and objectives of the black civil rights movement and the brief period of the black power movement. A rather pessimistic prognosis for American race relations that seems to have been borne out.

Landry, Bart. 1987. *The New Black Middle Class.* Berkeley: University of California Press. Describes the economic and social status of the contemporary black middle class, pointing out critical differences from the white middle class.

Lemann, Nicholas. 1991. *The Promised Land: The Great Black Migration and How it Changed America.* New York: Vintage. Chronicles the migration of blacks from the rural South to the urban North, interweaving personal narratives with an analysis of public policy.

Massey, Douglas S., and Nancy A. Denton. 1993. *American Apartheid: Segregation and the Making of the Underclass.* Cambridge, Mass.: Harvard University Press. Traces the development of residential segregation in U.S. urban areas. Argues that the maintenance of two housing markets, black and white, is the key to understanding the continuation of the economic gap and the social divisions between the two racial categories.

Myrdal, Gunnar. 1944. *An American Dilemma: The Negro Problem and Modern Democracy.* New York: Harper & Row. One of the most influential books written about race relations in the United States. Myrdal, a Swedish sociologist, challenged Americans to recognize the inconsistency between their society's commitment to social democracy and its institutional discrimination against blacks.

National Urban League. (annually). *The State of Black America.* New York: National Urban League. Presents up-to-date statistics on the economic, political, and social status of African Americans, as well as topical essays by social scientists.

Oliver, Melvin L., and Thomas M. Shapiro. 1995. *Black Wealth / White Wealth: A New Perspective on Racial Inequality.* New York: Routledge. Focuses on the differences in wealth held by whites and blacks, demonstrating the precarious position of the black middle class.

Stampp, Kenneth M. 1956. *The Peculiar Institution: Slavery in the Ante-Bellum South.* New York: Random House. Describes, in a very readable account, the nature of the American slave system as it developed and functioned in the eighteenth and nineteenth centuries.

Wilson, William J. 1987. *The Truly Disadvantaged: The Inner City, the Underclass, and Public Policy.* Chicago: University of Chicago Press. An update of Wilson's earlier work on the black underclass, *The Declining Significance of Race* (1980), which argues that the continuation of black inner-city poverty is the result not of racial discrimination but of a restructured urban economy.

THERE ARE MORE people of Mexican origin in Los Angeles than in all but one city in Mexico; in New York City, there are more Puerto Ricans than in San Juan; and only in Havana are there more Cubans than in Miami. Together, Hispanic Americans comprise 9 percent of the total American population, making them the country's second largest ethnic minority. And, given their rapid population growth, they will become the largest American ethnic minority in the next few years, surpassing African Americans.

Hispanic Americans[1] constitute several distinct ethnic groups, linked by a shared language and a cultural heritage derived, in the main, from Spanish colonialism. The most sizable groups among them are Mexican Americans — the largest — followed by Puerto Ricans and Cubans (Figure 9.1). The remainder are people from Central and South America as well as a small group whose origins are traced to Spain. Because they are the largest elements of the Hispanic-American population, our concern in this chapter is mainly with Mexican Americans, Puerto Ricans, and Cubans.

[1] Although the term *Hispanic* has been most frequently used in the past two decades, *Latino* is today also commonly used to describe those of various Latin-American origins. In this chapter, the two are used interchangeably. Mexican Americans, specifically, are often referred to as *Chicanos*. The U.S. Census has, at different times, used a variety of terms to describe the Hispanic-American population. The issue of which term, Hispanic or Latino, is today more acceptable has created some controversy (Gonzalez, 1992).

Figure 9.1 Hispanic-American Population

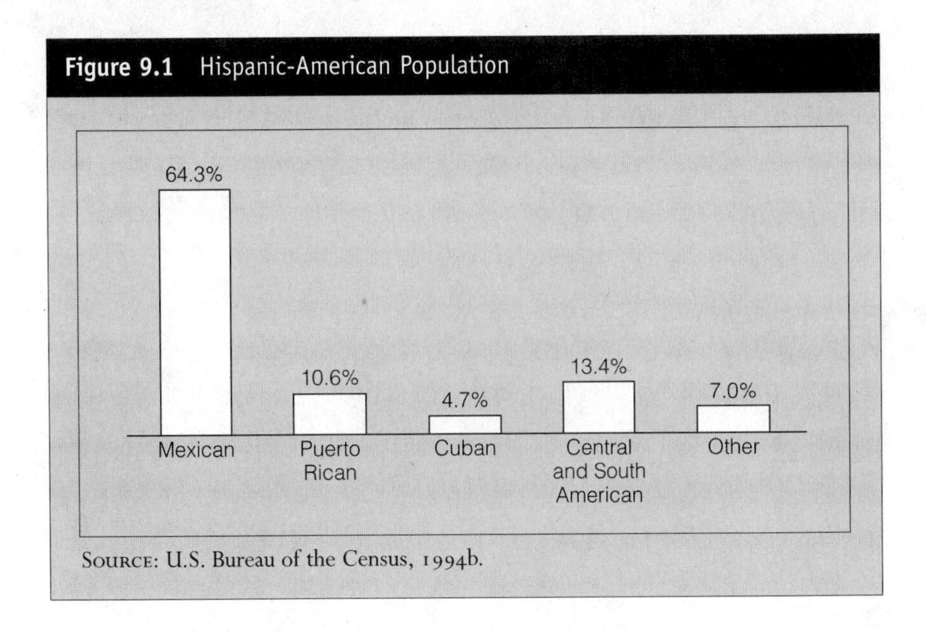

SOURCE: U.S. Bureau of the Census, 1994b.

THE "IN-BETWEEN" STATUS OF HISPANICS

If there is a common theme that runs through the unique histories and experiences of the several Hispanic groups in the United States, it is their intermediate ethnic status between Euro-American groups, on the one hand, and African Americans, on the other. In several important respects, Hispanics are an ethnic minority "in between."

Manner of Entry

Hispanic Americans have entered the society in a variety of ways. Although Mexicans have come to the United States in the twentieth century as voluntary immigrants, they retain a heritage of having been absorbed into American society originally as a conquered group. The area that today is the American Southwest was part of Mexico before being annexed by the United States in 1848 following the Mexican War. Mexicans living on those lands were subsequently incorporated into the United States much in the fashion of classic European colonialism. Unlike any other ethnic group, then, Mexican Americans today inhabit an area that was once part of their native society.

Puerto Ricans are a group not clearly part of either voluntary immigration or conquest. Puerto Rico became a territory of the United States in 1898 following the Spanish-American War, and in 1917, the inhabitants of the island were given the status of American citizenship. This enabled them to immigrate freely to the mainland with no restrictions. Puerto Ricans, then, are not technically immigrants even though they come to the mainland from a distinctly foreign culture.

The movement of Cubans to the United States in the past three-and-a-half decades has been a voluntary immigration. Yet it has been different from those of most past groups in that it was initially impelled mainly by political rather than economic motives. Moreover, the class characteristics of immigrant Cubans and the reception given them by the host society were unusual. As sociologist Alejandro Portes has put it, "Seldom has a foreign group come to the United States so well prepared educationally and occupationally and seldom has this country received one so well" (1969:508).

Hispanic Americans and Race

Perhaps the most obvious and consequential in-between feature of Hispanic Americans is their racial status. All three major groups include a range of physical types.

Mexico is racially amalgamated to a greater degree than perhaps any modern society (van den Berghe, 1978). Over the past two centuries, its population has been transformed into one dominated by mestizos, a physical type combining European and Indian traits. Although there are many who are either basically European or, at the other extreme, purely Indian, the vast majority of Mexicans today are racial hybrids. Most Mexican immigrants to the United States have been mestizo, making them physically distinct from the Anglo majority but still "the least divergent racially of all non-Caucasian groups" (Murguía, 1975:51). So ambiguous is the racial status of Mexican Americans that the U.S. Census Bureau has in the past classified them both as a separate "race" and as part of the white population. In any case, most Mexican Americans are physically distinct enough to be perceived by many Anglos in racial terms. (The term *Anglo* is used in the Southwest and in other areas with a large Hispanic population to distinguish any non-Hispanic white.)

Puerto Ricans' in-between racial status is even more problematic. Whereas Mexicans are mostly mestizo, Puerto Ricans inherit a racial background that is a combination primarily of European and African but also some Indian elements. The large African population originally brought to the island as slaves intermarried with whites over several generations, and today, although they cover the entire color spectrum, many Puerto Ricans are racially "intermediate" (Glazer and Moynihan, 1970). On the island, racial distinctions are not so acute, and one's color is seen along a scale or continuum (Fitzpatrick, 1987; Padilla, 1958; Rodriguez, 1989). But problems arise in the United States, where there is a racial dichotomy, allowing for no intermediate categories. Many who are "white" on the island may therefore find themselves classified as "black" when they come to the mainland. The differential treatment they subsequently receive on the basis of their darker skin (by American standards) is frustrating and problematic because interracial marriages are commonplace in Puerto Rico, and social treatment of darker groups is more equitable than in the United States.

In both Mexico and Puerto Rico, racial distinctions are more often a function of class and cultural differences than of physical type. Those who rise in socioeconomic status or who adopt Western cultural ways are accordingly "whitened." As we will see in Chapter 13, this is also the case in Brazil. In the United States, however, racial distinctions are based on skin color or other physical characteristics, in addition to ancestry, and are not modified by changes in class or culture.

Among the Hispanic groups in the United States, Cubans are the least racially heterogeneous, most falling into the American category of "white" (Diaz, 1980). Cuba itself is a racially variegated society, however, suggesting that the Cuban immigration has been racially selective (Portes, Clark, and Bach, 1977). The majority of early Cuban immigrants were of higher social status in Cuba and hence whiter in color, whereas many of the more recent immigrants have been of lower class origins and of darker color (Pedraza-Bailey, 1985).

THE DEVELOPMENT OF THE LATINO MINORITY

Despite broad similarities, the Mexicans, Puerto Ricans, and Cubans in the United States all display a unique history of immigration as well as unique patterns of settlement and adaptation.

Mexican Americans

As they have with blacks and American Indians, Anglos have come in conflict deeply and consistently with Mexicans throughout the history of the Southwest. Early in the contact between these groups, a pattern of Anglo domination and Mexican subordination emerged. And as with American Indians, the vehicle through which domination was initially established was physical conquest.

Conquest of Mexico Mexico gained its independence from Spain in 1810, and at that time its territory encompassed an area as far north as what is today Colorado. Most of this northern Mexican region, far removed from the administrative capital of Mexico City, was inhabited by Indians and a smaller number of Mexicans. As a means of populating the area, the Mexican government in 1821 granted permission for foreigners—mainly Americans—to settle in the area that today is Texas. By 1830, about 20,000 Anglos were living in the Texas region, constituting a numerical majority. Culturally and physically distinct from the indigenous Mexicans and Indians, the Anglo settlers from the outset translated their differences into contempt and hostility toward these groups.

The Mexican government could exercise only limited control over the Texas colony, and pressures from Anglos for independence grew steadily. In 1835–36, the colony revolted and established itself as an independent republic. Mexico never recognized the independence of Texas, and when it became clear that the United States favored annexation of the territory, much friction was created between the American and Mexican governments, leading eventually to war in 1848.

The Mexican War proved disastrous for Mexico, which lost more than half its territory. The United States acquired what are today the states of California, Colorado, New Mexico, Nevada, Utah, and most of Arizona in addition to Texas, which had been annexed previously. Under the Treaty of Guadalupe Hidalgo, ending the war, Mexicans living in what was now American territory were guaranteed the rights of American citizenship if they chose to remain. Most did, and it was from this nucleus that the Mexican-American ethnic group evolved.

The significance of the Texas rebellion and the Mexican War cannot be overstated as factors in the shaping of subsequent relations between Mexican Americans and Anglos in the Southwest. The borderlands remained an area of overt conflict long after the conclusion of the war, and this protracted hostility intensified the negative perceptions of Anglos and Mexican Americans of each other (Grebler, Moore, and Guzman, 1970). Anglo–Mexican conflict did not characterize all of the Southwest to the same degree. In New Mexico, for example, relations between the two groups were relatively peaceful by comparison with Texas, where the strife reached its most violent and bloody form. But in all cases, the war left a residue of mutually antagonistic feelings and group images among Anglos and Mexicans that continued to color interethnic relations for many generations.

Although their property rights and political rights were guaranteed by the Treaty of Guadalupe Hidalgo, Mexicans in the newly acquired American territory found themselves increasingly displaced by Anglos. Their property was taken through official and unofficial force and fraud, and they were gradually transformed into a colonized work force serving the area's labor needs. Mexicans were also dispossessed politically, as Anglos firmly established themselves in power. The scenario of heavy immigration of Anglos followed by the divesting of Mexicans' economic and political power was played out somewhat differently from area to area. In New Mexico, for example, the indigenous Mexican population, called Hispanos, did retain some political and economic power. These people were descendants of the original Spanish settlers dating back to the 1600s and did not identify themselves with the later immigrating Mexicans (Nostrand, 1973). In general, however, Anglos perceived Mexicans as a conquered—and therefore inferior—people who quite naturally were relegated to subordinate ethnic status (Alvarez, 1973).

The development of ethnic stratification in the Southwest seems clearly in accord with Noel's (1968) theory, discussed in Chapter 2. All three factors necessary to the emergence of an ethnic hierarchy—ethnocentrism, competition, and above all, differential power—came into play after 1821 in the confrontation of Anglos and Mexicans (McLemore, 1973). Anglos settling the region were most intent on securing land held by Mexicans—hence the ensuing competition. With them, they also brought their racial attitudes, which led them to see Mexicans as an inferior people whose exploitation could therefore be justified. Finally, with their numerical power and, following annexation by the United States, their political power, Anglos became the dominant group, displacing Mexicans both legally and by force.

Mexican Immigration As a result of the Texas and Mexican wars, the first Mexicans were incorporated into American society through conquest. But subsequent generations of Mexicans entered the United States as voluntary immigrants. Indeed, it must be remembered that the original Mexican Americans, those living on annexed lands, were a relatively small group, perhaps no more than 75,000 (McWilliams, 1968). There were over twice that number of Indians in the same territory. The overwhelming majority of Mexican Americans, then, historically entered as voluntary immigrants, motivated by the same push–pull factors as other immigrant groups.

The flow of Mexican immigration during the nineteenth and twentieth centuries was dictated primarily by the labor needs of the American Southwest (Camarillo, 1979; Tyler, 1975). Most simply, the movement across the border was heaviest during periods when the demand for cheap labor was high. Although immigration was evident during the remainder of the nineteenth century following the Mexican War, it reached enormous proportions during the twentieth century with the expansion of the American Southwest and the corresponding inadequacies of the Mexican economy. The pressures of a continually growing population and the political destabilization brought about by the Mexican Revolution in 1910 further impelled migration northward.

Between 1910 and 1930, the development of railroads, mining, and agriculture, all labor-intensive industries, spurred the first great wave of Mexican immigration into the United States (McWilliams, 1968). As the chief source of unskilled labor in the Southwest, Mexicans became a subordinate labor force. They were commonly kept in debt by their employers, paid less than others for the same work, concentrated in the least desirable occupations, often used as strikebreakers, and laid off most easily in distressful times (Barrera, 1979).

When they represented an excess labor supply, however, Mexicans were driven back across the border. During the 1930s, for example, not only were their jobs abolished, but they now served as convenient scapegoats for the area's unemployment and growing welfare rolls. Deportations were launched, in which illegal immigrants were rounded up and sent back to Mexico. Repatriations also took place, in which Mexicans, many of whom had been born in the United States, were induced to leave the country. This pattern was repeated in the 1940s and 1950s. During World War II, a severe labor shortage once again prompted the importation of Mexican workers. But by the early 1950s, the deportations and repatriations of the 1930s were grimly repeated in a policy called Operation Wetback, designed to stem the flow of illegal Mexican immigrants. In five years, 3.8 million persons were sent back to Mexico (Moore, 1976). Relatively few were actually deported in formal proceedings; most were simply rounded up and expelled. In the process, Mexicans were once again reminded of their tenuous status in the United States. Persons who simply "looked Mexican" were stopped and required to present evidence of their legal status. Although there is some question about the official number actually apprehended and deported (Garcia, 1980), Operation Wetback in any case created fear and suspicion among the Mexican-American populace.

Today, Mexicans are the country's largest immigrant group, and their numbers are swelled by a continual flow of illegal entrants. Between 1971 and 1991, Mexicans were almost one-quarter of all legal immigrants into the United States. As in past decades, economic conditions are the chief motive of this migration. This population movement will be discussed again in Chapter 11.

Puerto Ricans in the United States

The formation of the Puerto Rican minority in the United States differed considerably from the development of the Mexican-American community. To begin with, Puerto Ricans are a relatively recent immigrant group. Even though they had maintained a presence on the mainland for most of the twentieth century, their numbers were quite small until after World War II. The greatest influx occurred during the 1950s, when nearly 20 percent of the island's population moved to the mainland. Whereas in 1940, only 1,000 Puerto Ricans migrated to the mainland, in 1953, almost 75,000 did so (U.S. Commission on Civil Rights, 1976). In the 1970s, as the U.S. economy stagnated and low-skilled jobs disappeared, the migration to the mainland ceased and, in fact, a net return migration was sustained throughout the decade.[2]

---

[2]Fitzpatrick points out that immigration statistics involving Puerto Ricans are subject to serious difficulties in interpretation. "There is no way of telling the characteristics of the person traveling, that is, how many are Puerto Ricans coming to the mainland for the first time and how many are coming with the intention of returning to the island" (1987:17). Calculating net migration as the difference between the number arriving on the mainland and the number leaving Puerto Rico (as has customarily been done) may therefore tell us little about the actual size of the Puerto Rican migration.

The push–pull factors impelling Puerto Ricans to migrate have been similar to those for other voluntary immigrant groups: A surplus population in an economically depressed society seeks economic betterment in a society that promises improved conditions. Puerto Rican migration has therefore fluctuated, depending on economic conditions both on the island and on the mainland. An unusually high birthrate coupled with a seriously underdeveloped economy created great pressures for out-migration from the island after World War II. Two factors increased the appeal of the mainland United States. First, with citizenship rights, Puerto Ricans could enter the country with no restrictions, quotas, or other legal steps. Second, with the establishment of airline service, transportation between the mainland and the island became relatively cheap and rapid. The migration process for Puerto Ricans, then, was considerably less complicated than that for other groups.

Although the absolute size of the Puerto Rican migration to the mainland is, compared with past groups, small, it is numerically significant in relation to the population of the island. By 1960, there were one-third as many Puerto Ricans on the mainland as in Puerto Rico, and by 1970 one-half as many (Bahr, Chadwick, and Stauss, 1979). Moreover, migration statistics do not accurately indicate the frequent back-and-forth movement between island and mainland, so that it is quite reasonable to assume that a much larger number of islanders have, at one time or another, migrated. Indeed, during the three decades between 1940 and 1970, it is unlikely that a single Puerto Rican family was not affected by this movement (U.S. Commission on Civil Rights, 1976).

Today, one-third of all Puerto Ricans live on the mainland, almost half having been born here (Fitzpatrick, 1987). Thus, the Puerto Rican population is increasingly a second- and third-generation ethnic group. This is especially evident in New York, where the largest number of Puerto Ricans continue to reside (Fitzpatrick, 1995).

Cubans in the United States

The conditions of immigration for Cubans and their subsequent settlement patterns have been quite different from those experienced by either Mexicans or Puerto Ricans.

Given its geographical proximity to Cuba, the United States often served as a haven for Cuban exiles during times of political turmoil before and after the Cuban movement for independence from Spain in the late nineteenth and early twentieth centuries. In fact, one of the key figures of Cuban political history, José Martí, sometimes called Cuba's George Washington, spent much of his exile during the 1880s in the United States. Small colonies of Cubans had for many years been present in Key West, Tampa, Miami, and New York. But only in the 1960s, following the revolution led by Fidel Castro, did Cubans begin to come to the United States in significant numbers.

Whereas the pull factors of migration were stronger for Mexicans and Puerto Ricans, the push factors seemed more crucial for early Cuban immigrants. Their chief motivation was political refuge rather than economic betterment. Nonetheless, the underlying economic objectives of the émigrés should not be overlooked. Because the revolution imposed a socialist system in Cuba, the power and economic opportunities for most of the early immigrants—the majority of them middle-, upper-middle-, or upper-class people—had been severely constricted. As Portes has explained, "The wealthy, the successful, the powerful, the educated saw their status challenged, their influence radically curtailed, and their economic position continuously menaced" (1969:506). Departure from Cuba, then, was not simply politically inspired but was calculated in the hope of reestablishing a relatively high socioeconomic status in the United States and returning to Cuba when political conditions were favorable. For later Cuban immigrants, economic factors in the decision to migrate were even more clearly apparent (Portes, Clark, and Bach, 1977; Wong, 1974).

Cuban immigration from 1960 onward was aided immeasurably by American policies that, in effect, permitted the entrance of Cuban refugees without restriction. The political objectives of this policy are not difficult to understand. The chilled relations between the Cuban and American governments soon after the Cuban Revolution were followed by strongly belligerent actions on the part of the United States, including an American-supported invasion of Cuba at the Bay of Pigs in 1961 and the imposing of a trade embargo against Cuba in the same year. The Cuban missile crisis of 1962 firmly established the Castro regime as a basic element in the Cold War, and from that point forward, Cuba became a *bête noire* for successive American administrations. Cuba was portrayed in the United States as a "communist stronghold" and the Cuban people as "politically enslaved." The emigration of large numbers of Cubans, therefore, was seen as confirmation of this perspective, and the refugees were welcomed enthusiastically.

Large-scale Cuban immigration ceased in 1973 but resumed for a brief period in 1980, when more than 125,000 people came to the United States. The conditions under which this second immigrant wave left Cuba were graphically described by the American media, and the manner in which they arrived—most in small boats—further dramatized their entrance. These immigrants, however, were generally lower in social class than the original Cuban refugees, and many were black or mulatto (a mixture of black and white). As a result, they did not find the adjustment to American society as smooth or their reception as enthusiastic. They experienced higher rates of unemployment, low-paid work, and dependence on charity and public welfare (Portes, Clark, and Manning, 1985). Moreover, the presence of a criminal element among them served to taint the entire group.

With the exception of part of the second wave, most Cuban immigrants have entered the United States with a strong family and kin support system that has aided their adjustment considerably (Portes, Clark, and Bach, 1977). Indeed, following the initial immigration in the early 1960s, succeeding immigrants were encouraged to come to the United States by relatives or friends who could offer them physical and economic assistance (Fagen, Brody, and O'Leary, 1968). Although the presence of friends and relatives has also characterized the Mexican and Puerto Rican migrations, the support system of Cubans has been stronger and more complete. Moreover, Cubans have experienced substantial economic success and have therefore been enabled to offer greater assistance to those who have followed them. Mexicans and Puerto Ricans, by contrast, have remained collectively in a far more insecure economic position.

Hispanic-American Demographics Today

Unlike African Americans, each of the Hispanic-American groups is heavily concentrated regionally. Mexican Americans live mainly in the five southwestern states of California, Texas, Arizona, New Mexico, and Colorado. Almost 80 percent are in California and Texas alone. In recent decades, there has been a dispersion of Mexicans throughout the United States, particularly to the industrial cities of the Midwest (especially Chicago), but the vast majority remain in the Southwest, where they are numerically the most significant ethnic minority. Within the Southwest, however, there are regional differences in the size and importance of the Mexican-origin population. In some southern Texas towns, for example, Mexicans are a numerical majority, whereas in other areas of Texas, they may be a minuscule element.

Throughout the 1950s and 1960s, the Puerto Rican population was heavily concentrated in New York City. In 1950, over 81 percent of all Puerto Ricans on the mainland resided there, and in 1970, 62 percent. During the 1970s and 1980s, Puerto Ricans began to disperse, though almost half remained in New York City. As Joseph Fitzpatrick explains, "the Puerto Rican experience is no longer a predominantly New York experience, as it once was" (1987:10). Other significant Puerto Rican communities are found in Chicago and northern New Jersey, especially Newark and Jersey City. The remainder are primarily in cities of the Northeast.

Table 9.1 States with the Largest Hispanic Populations

STATE	NUMBER OF HISPANICS (1,000S)	PERCENT OF STATE POPULATION
California	7,688	25.8
Texas	4,340	25.5
New York	2,214	12.3
Florida	1,574	12.2
Illinois	904	7.9
New Jersey	740	9.6
Arizona	688	18.8
New Mexico	579	38.2
Colorado	424	12.9

SOURCE: U.S. Bureau of the Census, 1991c.

Cubans are the most concentrated of the three major Hispanic groups, with about two-thirds residing in the Miami area.[3] Another relatively sizable Cuban community is in northern New Jersey.

Like African Americans, Latinos are primarily urban dwellers, whether Mexicans in the Southwest, Puerto Ricans in the East, or Cubans in Florida. In fact, all three groups are more urbanized than the general U.S. population. Eighty percent of Mexican Americans and virtually all Puerto Ricans and Cubans live in urban areas (Bean and Tienda, 1987). Tables 9.1 and 9.2 show the states and urban areas with the largest Latino populations.

In the past two decades the Hispanic-American population has been infused with newer groups from Central and South America and from the Caribbean whose numbers in the United States previously were very tiny. Among the largest of these are Dominicans (who today make up the largest immigrant group in New York City), Nicaraguans (mostly in Miami), Colombians (mainly in New York and Miami), Guatemalans (mainly in Los Angeles), and Salvadorans.

[3] While Cubans are overwhelmingly the predominant group with 650,000, Miami in recent years has become a magnet for other Hispanic groups. There are 75,000 Nicaraguans, 65,000 Colombians, and tens of thousands from virtually every country in Central and South America and the Caribbean (Nordheimer, 1996).

Table 9.2 Hispanic Population in Metropolitan Areas with More Than 100,000 Hispanic Persons (1,000s)

METROPOLITAN AREA	HISPANIC POPULATION	PERCENTAGE OF TOTAL POPULATION
Los Angeles–Anaheim–Riverside	4,779	32.9
New York–Northern New Jersey–Long Island	2,778	15.4
Miami–Fort Lauderdale	1,062	33.3
San Francisco–Oakland–San Jose	970	15.5
Chicago–Gary–Lake County	893	11.1
Houston–Galveston–Brazoria	772	20.8
San Antonio	620	47.6
Dallas–Fort Worth	519	13.4
San Diego	511	20.4
El Paso	412	69.6
Phoenix	345	16.3
McAllen–Edinburg–Mission, Tex.	327	85.2
Fresno	237	35.5
Denver–Boulder	226	12.2
Philadelphia–Wilmington–Trenton	226	3.8
Washington, D.C.	225	5.7
Brownsville–Harlingen, Tex.	213	81.9
Boston–Lawrence–Salem	193	4.6
Corpus Christi, Tex.	182	52.0
Albuquerque	178	37.1
Sacramento	172	11.6
Tucson	163	24.5
Austin	160	20.5
Bakersfield, Calif.	152	28.0
Tampa–St. Petersburg–Clearwater	139	6.7
Laredo, Tex.	125	93.9
Visalia–Tulare–Porterville, Calif.	121	38.8
Salinas–Seaside–Monterey, Calif.	120	33.6
Stockton, Calif.	113	23.4
Detroit	105	2.0
Orlando	101	8.2

SOURCE: U.S. Bureau of the Census, 1995b.

Hispanic Americans as a whole are also a relatively young and, by comparison with others, rapidly growing minority. Since 1980, the Hispanic population has increased by more than 50 percent, five times as fast as the U.S. population in total and eight times as fast as non-Hispanics. About half of the Hispanic growth resulted from migration and half from natural increase.

LATINOS IN THE STRATIFICATION SYSTEM

Economic Status

As in other societal dimensions, Hispanic Americans seem to occupy an intermediate position, below Euro-Americans but, in most ways, above African Americans and American Indians in the degree of economic success they have achieved and in the paths to upward mobility they are following. However, we need to bear in mind that "Hispanic American" is an umbrella category encompassing several distinct groups. Thus, important differences will be evident among them. Specifically, there is a class range among the three major Latino groups, with Cubans at the top, Mexicans in the middle, and Puerto Ricans at the bottom.

Income As is seen in Table 9.3, the discrepancy between Hispanic and non-Hispanic white median family income is substantial: Hispanic income is about 60 percent of non-Hispanic white income. Moreover, while 37.5 percent of non-Hispanic white families earned more than $50,000 in 1993, only 16.7 percent of Hispanic families did. In short, despite moderate gains during the past two decades, there remains a sharp income gap between Latinos and the majority of Americans. Also evident in Table 9.3, however, is a range within the Hispanic category. Whereas the median income of Puerto Rican families in 1993 was $20,301, for Mexicans it was almost $24,000, and for Cubans, over $31,000. All three, however, were well behind the median income of non-Hispanic whites.

With incomes well below average, Latinos, especially Mexicans and Puerto Ricans, are, predictably, disproportionately among America's poor. Over one-quarter of all Hispanic families are below the poverty line, a rate almost four times as high as for non-Hispanic white families (U.S. Bureau of the Census, 1995b). As with income, however, there is a poverty range within the Hispanic-American category. As is seen in Table 9.3, Puerto Ricans below the poverty line are more numerous than either Mexicans or Cubans. Indeed, among a significant part of the Puerto Rican population in New York, economic conditions are much akin to those of the black underclass (Fitzpatrick, 1995; Lemann, 1991).

Table 9.3 Hispanic-American Median Family Income

GROUP	MEDIAN INCOME	PERCENT IN POVERTY
Hispanic	$23,912	26.2
Mexican	23,714	26.4
Puerto Rican	20,301	32.5
Cuban	31,015	15.4
Central and South American	23,649	27.0
Other Hispanic	28,562	21.7
Non-Hispanic White	40,420	7.3

SOURCE: U.S. Bureau of the Census, 1995b.

Occupation The low incomes and high poverty rates of Latinos, particularly Mexicans and Puerto Ricans, reflect their generally lower occupational levels. Except for Cubans, Latinos are underrepresented in the higher-status occupational categories and overrepresented in the lower ones. They also exhibit high unemployment rates (Table 9.4). And, as with African Americans, official statistics understate the extent of this problem.

How are we to account for the relatively low levels of income and occupation for Hispanic Americans? While employment discrimination against Latinos has historically been a factor for particular groups, especially Mexicans in the Southwest and Puerto Ricans in New York, today low levels of education and their relatively recent entry into the labor force seem most critical. Large proportions of all Latino groups have immigrated to the United States since 1950 (Massey, 1981a). As latecomers to the American post-industrial economy, they have necessarily occupied the least skilled and lowest-paying jobs. Like African Americans, they have become disproportionately part of the most expendable element of the work force, generally lacking the requirements for positions above the semiskilled or unskilled levels. Moreover, as an unusually young population, Mexican Americans and Puerto Ricans are most likely to lack marketable skills and work experience. Education, however, appears to be an even more important factor in explaining the lower income and occupational positions of the Hispanic-American population. As the gap grows between those with higher education and those without, Latinos have been especially hard hit, given their relatively lower educational levels generally and the low proportion with college education specifically (Carnoy et al., 1993).

The effects of lingering discrimination can, in some cases, present additional difficulties even after age, education, and lack of skills are accounted for (Melén-

Table 9.4 Unemployment Rate of Hispanics

GROUP	PERCENT UNEMPLOYED
Hispanic	11.9
Mexican	11.7
Puerto Rican	14.4
Cuban	7.3
Central and South American	13.2
Other Hispanic	10.8
Non-Hispanic White	6.1

SOURCE: U.S. Bureau of the Census, 1994b.

dez, 1993). Here it is important to consider differences among specific Latino groups and the region in which they live. While discrimination against Mexicans in the Southwest or Puerto Ricans or Dominicans in New York may be a continuing problem in attaining higher-level jobs, Cubans in Miami, by contrast, have faced no such obstacles. Indeed, regarding employment, for them ethnicity may serve as a positive rather than negative factor.

Education As noted, perhaps the most serious factor in accounting for the lower occupational status and earnings of Hispanic Americans in general is their lower educational level. As shown in Table 9.5, only 53 percent of Latinos 25 years of age and over have completed high school, compared with 82 percent of non-Hispanic whites. Dropout rates among Latinos also are extraordinarily high, more than double those of both whites and blacks. Advances in recent years have begun to close this educational gap, however. The proportion of younger Latinos who have completed high school is substantially higher compared to older Latinos, and the proportion of Latinos over age 25 who completed four years or more of college doubled between 1970 and 1990. Also, as with income, it is important to consider the differences in educational attainment within the Latino category. As is seen in Table 9.5, the discrepancy between Mexicans and Cubans, for example, is very great.

Despite improvements in the past two decades, Latinos, especially Mexican Americans and Puerto Ricans, continue to rank lower than other ethnic groups on all measures of education. Reasons for their lower educational attainment have been debated. Some contend that inappropriate cultural values fail to encourage successful school performance; others maintain that schools themselves, Anglo-dominated as they are, do not address the unique needs of

Table 9.5 Hispanic-American Educational Level, Persons 25 Years Old and Over

GROUP	HIGH SCHOOL GRADUATE	BACHELOR'S DEGREE OR HIGHER
Hispanic	53.1%	9.0%
Mexican	46.2	5.9
Puerto Rican	59.8	8.0
Cuban	62.1	16.5
Central and South American	62.9	15.1
Other Hispanic	68.9	15.1
Non-Hispanic White	82.0	22.9

SOURCE: U.S. Bureau of the Census, 1995b.

Hispanic students. One point about which there is general agreement is that language difficulties intensify the academic problems of these students (Celis, 1992; Nava, 1975; U.S. Commission on Civil Rights, 1972, 1974, 1976). Most Mexican-American students in the Southwest are from families with non-English-speaking backgrounds, making their school experience trying, at best. Language problems create or exacerbate much of the high dropout rate and poor academic performance among Puerto Ricans as well. In New York in 1990, 66 percent of Puerto Ricans born on the mainland had finished high school, but only 37 percent of the first generation (Puerto Rican–born) had done so. Dropout rates were still high, however, even for the second generation, due in large measure to poor language proficiency (Fitzpatrick, 1995).

In sum, Mexican Americans and Puerto Ricans are disproportionately absent from the higher levels of the mainstream economy and overrepresented among the poor. In looking at the socioeconomic levels of these two groups, it should be noted that these are aggregate figures that reflect the general position of the groups in the American stratification system. There will obviously be an entire range of occupational and income positions within each group despite their disproportionate concentration at the lower levels. Moreover, for Mexicans there are regional differences of some importance. In the Southwest, where they have traditionally played a subservient economic role, they clearly make up the bulk of the poor. But in northern industrial cities, they are a more stable and well-entrenched part of the blue-collar work force (Cardenas, 1976; Estrada, 1976;

Nodín Valdés, 1989; Samora and Lamanna, 1967). For Puerto Ricans, too, analysis of New York City data reveals that there has been a gradual increase over the last two decades in professional and managerial positions (Fitzpatrick, 1995). In general, however, Mexican Americans and Puerto Ricans have entered the U.S. work force at an inopportune time. Because they face the same increasingly technical and high-skilled job market as African Americans with a correspondingly low skill level, they will continue to make up a large segment of the poor in the foreseeable future. And, like the black underclass, many will continue to face the intractable problems of chronic unemployment and a large percentage of female-headed families.

Cubans: A Special Case Cubans are unlike Mexican Americans and Puerto Ricans in their adaptation to the American economic system. Indeed, few groups in American ethnic history have equaled the phenomenal socioeconomic rise of the contemporary Cuban-American community in so short a time. On all measures of socioeconomic status, Cubans are clearly well in advance of Mexicans and Puerto Ricans despite their more recent arrival. Their income and educational levels are higher, their unemployment rates are lower, and they are well represented across the occupational spectrum. What accounts for these striking differences?

Cuban immigrants to the United States brought with them a class background markedly different from that of other Latinos. A disproportionate number derived from the middle and upper strata of prerevolutionary Cuban society. In their study of early Cuban refugees, Fagan, Brody, and O'Leary (1968) found that professional and semiprofessional occupations were heavily overrepresented among this group, whereas fishing and agricultural jobs—those predominating in Cuba—were hardly represented at all. The occupational skills Cubans brought with them enabled them to work themselves more easily into the mainstream economy and climb rapidly even though most of them experienced downward mobility when they first entered the United States. Indeed, research has shown that among younger Cubans in the United States, the rate of upward mobility has been similar to that of immigrants from English-speaking countries (Massey, 1981b). In addition to their high occupational status in Cuba, the refugees were far better educated than most Cubans, and the majority were urban residents. Succeeding immigrant cohorts displayed a somewhat lower class background but still represented a privileged sector of Cuban society (Portes, Clark, and Bach, 1977).

Cubans in the United States, then, differ radically from other Hispanics in their social origins. Unlike Mexicans and Puerto Ricans, the Cuban immigrants were likely to have been occupationally successful and well educated before emigrating. These higher skill and educational levels have translated into rapid upward mobility in the United States. Also, most immigrant Cubans have been white and therefore have not been exposed to the added handicap of racial discrimination in the labor market, with which many Mexicans and Puerto Ricans have had to deal.

Another advantage that Cubans have had over other Hispanic groups has been the creation of an institutionally complete community in Miami, where the majority of Cuban Americans reside. As described by Wilson and Portes (1980), an ethnic enclave economy was established by early Cuban immigrants in Miami, comprised of numerous small businesses serving the ethnic community. In addition, a small manufacturing sector emerged that subsequently employed other Cuban immigrants who arrived with fewer skills and financial capital. The presence of an ethnic subeconomy not only created jobs for these immigrants but also enabled them to avoid the secondary labor market of the mainstream economy, where jobs ordinarily are temporary and do not provide for occupational advancement. Many of the Cuban immigrant workers who were employed by Cuban employers eventually acquired the skills and capital needed to open their own establishments, which created an expanding Cuban subeconomy (Portes and Bach, 1985). Neither Mexicans nor Puerto Ricans have been as demographically concentrated as Cubans, thus making the development of similar economic arrangements impracticable. Moreover, neither group contained a substantial entrepreneurial class, possessing both business capital and skills, that could re-create in the United States the position it held in its country of origin and thereby serve as the nucleus of an ethnic subeconomy. Pedraza-Bailey (1985) further asserts that Cuban economic success in the United States has been very much a product of the fact that Cubans were received as political migrants, that is, as refugees, rather than as economic migrants. As a result, the U.S. government provided aid to early-arriving Cuban immigrants that helped them in reestablishing their class position. Mexicans, by contrast, have traditionally been received as economic migrants and have never been given similar governmental assistance.

Despite the extraordinary achievement of Cubans relative to other Hispanic Americans, their economic success in the United States should not be overstated. Though they are certainly in a very different category from Mexicans and Puerto Ricans, their median income is still below Anglos', and their poverty rate and unemployment rate are higher. Also, even though they are well represented in the higher-level white-collar occupations, most are still in clerical, semiskilled, or service jobs (Pérez-Stable and Uriarte, 1993). Moreover, the economic gap between Cubans and other Hispanics is closing (Betancur et al., 1993; Diaz, 1980; Perez, 1986).

Latinos and Political Power

Latino political power in the United States is still in its infancy. Because Latinos are regionally concentrated, their political gains have necessarily been limited mostly to the local and state levels. Even at these levels, however, Latinos have not yet mobilized as effectively as have African Americans in recent years.

Mexican Americans As the largest and geographically most widespread group, Mexicans constitute potentially the most significant political force among Hispanic Americans. Yet this potential has hardly been realized. Mexican Americans have maintained a notoriously low rate of political participation (de la Garza, 1977; Garcia and Arce, 1988; Griffen, 1992; McCleskey and Merrill, 1973; Welch, Comer, and Steinman, 1973). Several factors may account for this. First, and perhaps most important, a large percentage of Mexican Americans are poor, and as we previously noted, lower-income people in general display lower rates of political activity. Second, the fluid nature of Mexican immigration, in which people constantly move back and forth across the border, has militated against the development of long-term political loyalties and interests. Third, a vast number of Mexicans in the United States either are not naturalized citizens or are illegal immigrants, further diluting the Mexican constituency (Kirschten, 1992). Finally, in the Southwest, tactics designed to prevent Mexican Americans from voting, such as the poll tax, gerrymandering of ethnic districts, literacy tests, intimidation, and even violence, were, in the past, commonly applied (Pachon and Moore, 1981; U.S. Commission on Civil Rights, 1970). Though these impediments have been removed as a result of the voting rights legislation of recent decades, their aftereffects are still felt in many areas.

Low political participation and discriminatory practices have resulted in political underrepresentation at local and state levels. In Los Angeles, for example, with the largest Mexican-American community in the country, only two Chicanos sit on the fifteen-member city council (Griffin, 1992). And not until 1991 was a Mexican American elected to the Los Angeles County Board of Supervisors, an extremely powerful local governing body. By contrast, San Antonio, also with a very large Mexican population, achieved proportional ethnic representation on its city council by 1979, and in 1981 a Chicano was elected mayor. This is still an exceptional case, however, though it may augur growing Mexican political power in the Southwest.

Only in New Mexico have Mexican Americans (Hispanos) attained a relatively equitable share of political power with Anglos. Indeed, the Anglo–Hispano political balance there has endured since the original annexation of the territory in the nineteenth century (Garcia and de la Garza, 1977; Knowlton,

1975; Padilla and Ramirez, 1974). The American conquest of New Mexico was not violent, as it had been in Texas and other parts of the Southwest, resulting in less historical division between the two groups. Moreover, after the Mexican War, Hispanos in New Mexico were a numerical majority. By contrast, in Texas, even before annexation Mexicans were outnumbered five to one, and Anglo political and economic dominance were almost complete. The upshot was that Mexicans there had to struggle for local and state political power (Moore, 1970).

Mexican Americans have engaged in other, nonelectoral, forms of political activity in recent decades. Perhaps the two most dramatic episodes of Mexican-American political activism were the movements led by Cesar Chavez and Reies Lopez Tijerina. Chavez's objective was to organize migrant farmworkers, and Tijerina focused on regaining communal lands that he charged had been illegally taken from Hispanos in New Mexico.

Chavez's well-publicized movement in the 1960s and 1970s consisted of efforts to force growers in California and the Southwest to recognize his United Farm Workers Union (UFW) as the representative of thousands of migrant laborers, most of them Mexican Americans. Past efforts had been made to organize farmworkers in the Southwest, some even dating back to the early years of the twentieth century, but Chavez was the first to succeed. Through his organizing activities, Chavez became something of a folk hero to Mexican Americans, and his charisma thrust him into the forefront of the larger Chicano movement (Solis-Garza, 1972; Steiner, 1970; Stoddard, 1973).

The movement led by Tijerina and his organization, the Alianza de las Mercedes, was more openly militant, and its objectives were less pragmatic than Chavez's. Tijerina argued that several million acres of land in New Mexico had been fraudulently taken from Hispanos following annexation of the territory at the conclusion of the Mexican War, and his aim was to recover them. To draw attention to his cause, he provoked several confrontations with federal and state officials, leading to his arrest and subsequent imprisonment (Knowlton, 1972). Although the Alianza helped in creating a national awareness of the Chicano movement, its goals were too visionary to yield concrete benefits.

Chavez's and Tijerina's movements were the two most prominent elements of the larger Chicano movement of the 1960s and early 1970s, which extended well beyond specific personalities and organizations. *Chicanismo* was very similar to the black nationalist movement and was spawned in the same general atmosphere of minority activism in the 1960s. Its major goal was to galvanize Mexican-American political action in order to assert group rights and expand economic and social opportunities. Many of the same tactics employed by the black movement, including legal cases, protests, boycotts, and direct encounters with authorities, were adopted.

Like its black nationalist counterpart, the Chicano movement also represented an assertion of militant cultural pluralism. Rather than assimilation, the idea was to affirm the unique Mexican-American culture and to instill in Mexican Americans a sense of ethnic pride and awareness (Macías, 1972; Steiner, 1970). Younger activists began to refer to themselves as Chicanos rather than Mexican Americans and to proclaim a more militant and antiestablishment position. The term *Chicano* had in the past referred to lower-class Mexicans, but it now began to represent an ethnic viewpoint (Macías, 1972; Moore and Pachon, 1975). Just as *black* had been substituted for *Negro, Chicano* came to symbolize an ethnic identity that repudiated the negative group image that had been imposed for generations by Anglos.

That *Chicanismo* seemed to coincide with black nationalism was not accidental; indirectly, the latter contributed to its emergence. Black power demonstrated that securing civil rights did not in itself assure meaningful social and economic change. Mexican Americans observed aid programs to black inner cities and began to petition for their own "fair share" of assistance. Their problems, they felt, were equally pressing, and Chicano ideology, like black nationalism, placed the onus for minority deprivation on the larger society.

Just as black militancy subsided in the 1970s, so, too, did the Chicano movement. But as with black power, the long-range effects of *Chicanismo* were more substantial than they first appeared. Undeniably, concrete gains were not broad; governmental programs aimed at depressed Mexican-American communities were short-lived. Intangible benefits, however, were more meaningful and lasting. A positive group identity emerged among Mexican Americans, and ethnic unity was instilled through the idea of *La Raza.*[4] Most important, national attention was drawn for the first time to Mexican Americans as a sizable minority group.

Today, Mexican Americans are a growing political force in those areas where they make up a significant portion of the population, namely southern California and south Texas. In 1996, twelve members of Congress, mostly from those locales, were Mexican-American. Moreover, two Mexican Americans were serving in President Clinton's cabinet—Henry Cisneros, former mayor of San Antonio, as secretary of housing and urban development, and Federico Peña, former mayor of Denver, as secretary of transportation.

Puerto Ricans To a limited extent, Puerto Ricans participated in the ethnic militancy of the 1960s and early 1970s. Several groups emerged at that time advocating control by Puerto Ricans of their communities and emphasizing ethnic identity and pride. The Young Lords, for example, a militant group in Chicago

---

[4] *La Raza* ("the race") is not similar in meaning to the North American idea of race, but is a somewhat vague concept denoting a common heritage or peoplehood.

and New York, imitated the activities of the Black Panthers in organizing Puerto Rican ghettos and calling for a redress of squalid conditions (Lopez, 1980). As with black and Chicano political activism, Puerto Rican militancy resulted in drawing the serious attention of national authorities to the Puerto Rican community for the first time. But just as the attention paid other minority ghettos dwindled with the passing of militancy, so, too, most of the programs of federal and community agencies directed at Puerto Ricans disappeared in the 1970s.

Today, the political activity and influence of Puerto Ricans in the United States are limited primarily to New York City because in no other area do they represent a group large enough to make a serious impact on local politics. Political participation among Puerto Ricans, however, as among other Latinos, is lower than average. The same factors at work for Mexicans—heavy back-and-forth movement between the United States and the society of origin, language difficulties, a young population, and above all, lower-class status—have militated against significant Puerto Rican political participation. Moreover, Puerto Ricans in the New York metropolitan area are relatively dispersed residentially so that in few districts are they alone able to carry an election (Fitzpatrick, 1987).

Despite their relatively low political profile in New York City, Puerto Ricans have become more politically active in recent years and have recorded some electoral successes. By the mid-1990s, three Puerto Ricans had been elected to the U.S. Congress and several to state and city offices.

Cuban Americans As more Cubans have become American citizens, they have emerged as a political force in Miami with its large Cuban population. Their political power has begun to match their economic power, as they increasingly occupy important positions in local government. They are rapidly emerging as the most significant single bloc of voters in Miami (Silva, 1984; Warren and Stack, 1986). In 1985, a Cuban American was elected mayor of the city, and by 1992, two Cuban Americans had been elected to the U.S. Congress from the Miami area and a third from northern New Jersey.

The political leanings of Cubans in general tend to be more conservative than those of either Mexican Americans or Puerto Ricans, which may be a reflection of their higher-class position. When they do participate in the electoral process, they tend toward the Republican party, in sharp contrast to the overwhelmingly Democratic leanings of Mexican Americans and Puerto Ricans. Cuban conservatism is also the product of a very strong anticommunist sentiment, particularly among the first-generation refugees (Portes and Mozo, 1985). Within the Cuban community, much attention remains focused on the political scene in Cuba, and political issues are often heavily tinged with ideology (Rieff, 1993). Younger Cuban Americans, however, are likely to see politics as they relate to Cuba and its relations with the United States in less passionate terms than their parents (de los Angeles Torres, 1988; Perez, 1980). Moreover, politics at the local level among Cuban Americans is increasingly concerned with local, not foreign policy, issues (Portes and Truelove, 1987).

The Outlook for Hispanic Political Power Unlike African Americans, Latinos have not yet been able to exert significant political influence at the national level. This is due primarily to two factors. One is the relative youth of the Hispanic-American populace. Thirty-five percent are under age 18 and thus not eligible to vote. Another substantial portion of the population is among the age categories (18 to 24 years) that are least active politically. A second, and more important, factor is the relatively low percentage of citizenship among Hispanic Americans. Among the Latinos who did not vote in 1992, for example, more than 50 percent were ineligible due to noncitizenship (*National Roster,* 1994). Thus, although the Hispanic-American population has increased dramatically, its political clout has not followed suit.

Despite the fact that Latino political power has not yet realized its potential, there are growing signs of change. If a good measure of a group's position in the political system is the presence of that group's members in elective offices, Latinos have made significant progress in the past decade. By 1994, almost 5,500 Latinos held publicly elected offices at various government levels—local, state, and federal—an increase of almost 50 percent from 1984 (*National Roster,* 1994). Although most of these officeholders were elected by majority Latino constituencies, they represented a variety of geographic areas. The voting Hispanic electorate has grown 50 percent in the last decade, and this has had a significant impact. Although Hispanic officeholders were only about 1 percent of all elected officials, this represented a severalfold increase over the previous twenty years. Perhaps most significant, by 1994, Latinos, aided by redistricting designed to create several largely ethnic minority districts, had increased their representation in the U.S. Congress to seventeen.

Despite these indications of increasing Latino political power, Hispanic-American politics will, in the near future, remain largely a local phenomenon, limited to those particular states and communities where Latinos constitute a numerically substantial part of the population. In California, for example, where the largest part of the Hispanic-American population resides, the number of Latino elected officials doubled in the seven years between 1988 and 1995, and the number of Latino members of the State Assembly increased from four to ten (Mydans, 1995). Nationally, however, while Latinos can be a swing factor in certain elections, alone they cannot yet play a role of major influence.

The long-range outlook for Hispanic political power, then, is mixed. Although Hispanic voting rates will lag behind those of whites and blacks for some time to come, continued assimilation will probably stimulate increased political activism. Moreover, that Latinos are a growing element of the population has not been ignored by the two major parties at both local and national levels. Today, it is possible to speak of a Latino vote, and as the Hispanic-American population increases, its potential impact can only expand.

Economic Power As with African Americans, the role of Latinos in leadership positions of the American economy is limited. The 1,000 largest U.S. corporations have only a few dozen Hispanics on their boards of directors and less than 200 at the vice presidential level or above (*Hispanic Business,* 1995). Although there is an increasing number of Hispanic-owned firms that are growing in economic significance, most enterprises owned by Mexican Americans are small, family-operated businesses, and the same is true of Puerto Rican businesses (Fitzpatrick, 1980; *Hispanic Business,* 1995; O'Hare, 1987).

Cubans have displayed more substantial business success than other Latinos, the result primarily of their generally higher-class background and the entrepreneurial skills and resources many brought with them from Cuba. For example, they are well represented at the executive level of the Miami banking industry (Grenier and Stepick, 1992; Neil, 1982). The increasingly significant economic power exercised by Cubans in Miami, however, remains a local phenomenon and does not translate into national influence within the corporate economy. Moreover, even in Miami, most Cuban-owned businesses are small enterprises.

PREJUDICE AND DISCRIMINATION

Because of their "in-between" minority status, Latinos have not been subjected to the dogged prejudice and discrimination aimed at African Americans, but neither have they been dealt with as European immigrant groups were. John R. Howard refers to such intermediate groups as "partial minorities" and calls attention to their paradoxical position: "By being somewhat below the threshold of public consciousness with regard to 'minority problems' they escape some of the pervasive indignities visited upon blacks, but, precisely because of this it becomes difficult to arouse public indignation at the depredations they do suffer" (1970:8–9).

More precisely, an entire range of attitudes and actions has characterized the relations between Anglos and the several Hispanic groups. Variations have turned on each group's historical context, its geographical location, and its racial makeup. Those close to the dominant Anglo group, like most Cubans, have experienced little prejudice and discrimination; at the other extreme, dark-skinned Puerto Ricans have often encountered ethnic antagonism equal in form and degree to that suffered by African Americans.

Hispanic Stereotypes

For as long as they have been part of the American ethnic mosaic, Mexicans and Puerto Ricans have labored under strongly negative stereotypes. These have arisen in response to the manner in which both groups became part of the society (more so in the case of Mexicans than Puerto Ricans) and as manifestations of American racist ideology.

Mexican Americans in the Southwest Feelings of racial superiority on the part of Anglos and policies reflecting those beliefs have been obvious since the American conquest of the Southwest in the nineteenth century. More than anything, the historical circumstances that originally brought Mexican and Anglo into confrontation fashioned a set of group images that have remained relatively fixed since that time.

As we noted in Chapter 3, negative group images are ordinarily formed as a result of some form of interethnic competition, usually economic. In the Southwest, land and labor were the critical economic components in the development of mutual antagonism between Anglos and Mexicans. "To the early American settlers," writes Carey McWilliams, "the Mexicans were lazy, shiftless, jealous, cowardly, bigoted, superstitious, backward, and immoral" (1968:99). These images subsequently served to rationalize the dispossession of Mexicans' lands and the exploitation of their labor. The negative stereotyping of Mexicans was abetted by the fact that they were a conquered people. To the Anglos this only confirmed even more positively their inferior nature. Correspondingly negative stereotypes of Anglos naturally arose in this situation. To the Mexicans, Anglos were "arrogant, overbearing, aggressive, conniving, rude, unreliable, and dishonest" (McWilliams, 1968:99).

Also critical in the development of negative stereotypes of Mexicans were the racial attitudes Anglo settlers brought with them to Texas. From the beginning, Anglos viewed Mexicans as innately inferior on the basis of their racial characteristics (Moore and Pachon, 1975). Mexicans were different in language, culture, and religion, but they were also physically distinct from the Anglos. As a mestizo people, they were not blacks, but neither were they whites. This in-between racial status led to the classification of Mexicans as a separate racial group, one perhaps a notch above blacks but clearly inferior to the Anglos.

The poverty and illiteracy of those Mexicans with whom Anglo settlers had the most dealings further strengthened these negative views. McWilliams notes that most of the Anglos who settled in the Southwest borderlands in the mid-nineteenth century were middle-class people who failed to find a Mexican middle-class counterpart. "If a larger middle-class element had existed," he explains, "the adjustment between the two cultures might have been facilitated and the amount of intermarriage might have been greater" (1968:75). Although upper-class Hispanos were initially seen in a different light, they, too, were eventually absorbed into the adverse image.

The negative stereotypes of Mexicans that developed during and after the period of conquest were maintained and even strengthened not only in the Southwest but also in the society generally. Mexicans were included among the "undesirable" groups denoted by the Dillingham Commission in the early 1900s, along with southern and eastern Europeans, falling somewhere between blacks, on the one hand, and "white" Americans, on the other. It is instructive to note that despite their undesirable status, Mexicans were so vital as a cheap labor supply in the Southwest that they were exempted from the restrictive immigration laws that were drafted at that time largely on the basis of the commission's report.

The stereotypes developed in the nineteenth century have continued to influence the views of Anglos in the Southwest and elsewhere. In his study of a south Texas community in the early 1970s, Simmons (1971) found that Anglos believed Mexicans to be unclean, prone to drunkenness and criminality, deceitful in dealings with Anglos, hostile, and unpredictable. Among favorable images of Mexicans were their "romantic" rather than "realistic" attitude and their fun-loving nature. But even the positive traits implied a childlike and irresponsible character, making the Mexicans appear deserving of their subordinate status.

One of the well-entrenched mental pictures of the Mexican American is the lazy peasant, usually propped against a wall, sombrero over his face, asleep. This dovetails with the image of unskilled farmworker or migrant laborer that many Americans still believe fits Mexican Americans. That they perform a wide range of occupations and that the vast majority of them are city dwellers are overlooked.

Television and movies have played a large role in perpetuating negative stereotypes of Mexican Americans, much as they have for African Americans and American Indians. In Hollywood films, Mexicans have traditionally been portrayed as villains in confrontation with Anglos, and even when they are allied with the Anglo heroes, their roles are ordinarily subservient and often ludicrous (Petit, 1980; Wilson and Gutiérrez, 1995). Television commercials have also perpetrated the villainous image of Mexicans (Martinez, 1972; Wagner and Haug, 1971). One of the better known of these ads, the Frito Bandito commercial, provoked a strong protest by Mexican Americans, and eventually the Frito-Lay Company was forced to remove it. The ad featured a cartoon character who appeared as a Mexican bandit sneaking Fritos corn chips from the kitchens of unsuspecting housewives. Because he was cunning and clever, they were advised to buy two bags (Westerman, 1989; Wilson and Gutiérrez, 1995).

Latino representation in general on American television continues to be slight, and when Latino characters do appear, they tend to be typecast as gang leaders, drug dealers, or clownish domestics (*Newsweek*, 1989). As TV producers and mass media advertisers have become more attuned to a growing Latino market, however, these stereotypical images are subject to change. Ten years after it was forced to remove its Frito Bandito ad, for example, the Frito-Lay Company advertised Tostitos, another corn chip product, this time with a character who spoke with a lilting Spanish accent and presented a romantic image of a distinguished, cultured Latino. No protests were forthcoming (Wilson and Gutiérrez, 1995).

Puerto Ricans Although they have not experienced a history of confrontation and conflict with Anglos, as Mexicans have in the Southwest, Puerto Ricans have nonetheless commonly been stereotyped negatively. This is mainly a product of their racial distinction and their generally lower-class position. We have already noted that much of the Puerto Rican population in the United States is dark-skinned and thus categorized as black. This means, of course, that Puerto Ricans will often experience the same forms of prejudice as non–Puerto Rican blacks. As a group whose members are disproportionately among the poor, Puerto Ricans are also often seen as lazy, irresponsible, and "on the dole." This is no better illustrated than in the remark of businessman J. Peter Grace, who, as head of the Reagan administration's committee on governmental waste, derisively described Puerto Ricans living in New York City as "all on food stamps," thus making the federal food stamp program "basically a Puerto Rican program" (*New York Times,* 1982).

Cubans As a group heavily concentrated in a single city and having arrived more recently, Cubans have not had the kind of national exposure that Mexican Americans or even Puerto Ricans have had. Moreover, as a group that, for the most part, is physically indistinguishable from the Euro-American population, Cubans automatically enjoy a status that precludes racially based stereotypes. Further, their generally higher class position has made for a more positive image among Anglos.

Although the attitudes of Anglos toward Cubans in Miami were initially benign, certain events in recent years have created an increasing bifurcation between the two groups. One area that has fueled Anglo resentment is language. Increasingly, Miami has become a bilingual city, and this has met with Anglo resistance. In 1980, a referendum was held to bar Dade County (Miami) from using public funds to conduct official business in any language but English and from promoting any but "American" culture. The measure, which narrowly passed, polarized the community along ethnic lines (Burkholz, 1980; Castro, 1992).

The referendum on bilingualism disclosed the depth of the anti-Cuban sentiment among Anglos. A survey conducted immediately following the vote revealed that nearly half of those who had voted for the ordinance did so as an opportunity to express their protest, not because they saw it as a good idea. Furthermore, more than half the Anglos who supported the measure said they would be pleased if it "would make Miami a less attractive place for Cubans and other Spanish-speaking people" (Tasker, 1980). Resentment among Anglos toward Cubans is also based on the belief that inflated crime rates in Miami reflect Cuban activities, particularly in drug trafficking. This view was fueled by the presence of a large number of Cuban criminals among the 1980 immigrant wave. An additional source of animosity has been Cubans' rabid and sometimes violent anticommunism.

This negative image has apparently extended to the society as a whole in recent years. A national survey conducted in 1982 inquired about attitudes toward ethnic groups and their contributions to American society. Cubans, as is seen in Table 9.6, were judged below all other American ethnic groups. It is significant that all three major Hispanic-American groups ranked considerably lower than all European groups and, surprisingly, blacks, Chinese, and Japanese. Why Cubans, concentrated as they are in a single locale, should have elicited such a negative societywide response can be explained in some part as a result of negative stereotyping. Films and television dramas like *Scarface* and *Miami Vice* contributed to an image of Cubans as drug dealers and criminals of one sort or another. To some extent, this image was shared by other Latino groups, like Colombians. "In the eighties," notes one film researcher, "the easily identifiable darkskinned Hispanic drug dealer became the ethnic enemy" (Richard, 1994:xxxii).

Discriminatory Actions

At the outset, we noted that Latinos in the United States have occupied an intermediate ethnic status between Euro-American ethnic groups, on the one hand, and African Americans, on the other. This in-between status is quite evident in looking at patterns of discrimination. Latinos have lacked the heritage of a legal system that, in the case of African Americans, made discrimination legitimate. As a result, such actions have ordinarily been extralegal rather than formally endorsed by the state. Moreover, although discrimination against Latinos has been widespread and often intense, it has not been so profound and intransigent as that experienced by African Americans.

This is illustrated by Pinkney's study (1963) of one community with sizable Mexican-American and black populations. General patterns of discrimination for both groups, he found, were similar but were much stronger in degree for blacks. For example, greater acceptance of Mexican Americans was evident in public accommodations like hotels and restaurants as well as in residential areas and social clubs. Anglos also expressed greater approval for granting equal rights in employment and residence to Mexican Americans. Pinkney attributed these

Table 9.6 Public Attitudes Toward American Ethnic Groups

GROUP	HAS BEEN GOOD FOR COUNTRY	HAS BEEN BAD FOR COUNTRY	MIXED FEELINGS	DON'T KNOW
English	66%	6%	21%	8%
Irish	62	7	22	9
Jews	59	9	24	8
Germans	57	11	23	8
Italians	56	10	25	9
Poles	53	12	25	11
Japanese	47	18	26	9
Blacks	46	16	31	7
Chinese	44	19	27	10
Mexicans	25	34	32	10
Koreans	24	30	31	15
Vietnamese	20	38	31	11
Puerto Ricans	17	43	29	11
Haitians	10	39	26	26
Cubans	9	59	22	10

SOURCE: *Public Opinion*, 1982.

differences in part to differences in skin color: Mexican Americans, closer to the dominant Anglo group, were more acceptable.

That it has not been as rigid and intense as that aimed at African Americans should not disguise the severity of discrimination to which Latinos in some circumstances have been exposed, particularly Mexican Americans in the Southwest. Many of the actions against them in areas such as southern Texas have been similar in degree and scope to those experienced by blacks, including segregated restaurants, churches, schools, and other public facilities (Burma, 1954; Grebler, Moore, and Guzman, 1970). Physical violence against Mexicans in the Southwest also has a long heritage. McWilliams notes that in the mining camps of the nineteenth century, any crime committed was immediately attributed to Mexicans, for whom lynching was the accepted penalty. "Throughout the 1860's," he writes, "the lynching of Mexicans was such a common occurrence in Los Angeles that the newspapers scarcely bothered to report the details" (1968:130).

Police Relations A particularly harsh and often violent relationship has traditionally existed between Mexican Americans and various law enforcement agencies in the Southwest. The police and courts have seemed to consistently mete out a different standard of justice to Mexican Americans (Moore and Pachon, 1975; Morales, 1972; U.S. Commission on Civil Rights, 1970). Although the level of discrimination in the criminal justice system has historically been unusually high in Texas, several of the most serious large-scale incidents of the modern era have taken place in Los Angeles. These include the notorious *pachuco,* or zoot suit, riots of 1943 and the East Los Angeles riot of 1970. In both cases, the police played major roles in either precipitating or sustaining the hostilities.

The events surrounding the zoot suit riots have been well documented (Garcia and de la Garza, 1977; McWilliams, 1968). During the 1940s, the Los Angeles press waged a virulent anti-Mexican campaign, creating a rancorous mood among Anglos. Sensational accounts were presented of Mexican-American youth gangs whose members took to wearing bizarre outfits called zoot suits. Following an incident involving Mexican-American youths and sailors stationed in Los Angeles, 200 servicemen began a four-day rampage, in which they drove through the Mexican area in taxis, randomly attacking those on the streets. The Los Angeles police did virtually nothing to halt the violence and in fact arrested many of the victims. Serious rioting continued for almost a week, ending finally with the intervention of the military police. This incident left permanent scars on Mexican–Anglo relations in Los Angeles, and confirmed the ever-precarious status of the Mexican-American community in an Anglo-dominated environment. Although other such incidents have rarely garnered as much public attention, commonplace police brutality (or indifference, as the case may be) in dealing with Mexican Americans has been repeatedly affirmed.

Residential Segregation As we saw in Chapter 8, residential segregation has been a particularly critical problem for African Americans and remains a most obdurate area of discrimination. For Latinos, however, discrimination in housing is generally less severe (Grebler, Moore, and Guzman, 1970; Massey, 1981a, 1981b; Massey and Denton, 1993; U.S. Commission on Civil Rights, 1976). Recall the index of dissimilarity, a measure of residential segregation, explained in Chapter 8. Studies using this measure have shown that Hispanic–Anglo segregation is markedly lower than black–white (Anglo) segregation (Massey, 1979; Massey and Denton, 1989a, 1989b, 1993; White, 1987). Moreover, whereas black–white patterns of residential segregation do not seem to be affected significantly by social class, this is not the case for Latinos. Rather, Hispanic–Anglo segregation seems to decline with increasing socioeconomic status. That is, as Latinos experience upward class mobility, there is a tendency toward residential dispersion.

There are specific differences, however, in spatial patterns among the three major Hispanic-American groups. Whereas both Mexicans and Cubans are highly segregated from blacks, Puerto Ricans are less segregated from blacks and more segregated from whites (Bean and Tienda, 1987; Massey and Bitterman, 1985; Massey and Denton, 1989a). In housing patterns, therefore, Mexicans and Cubans are increasingly similar to the general white population, whereas Puerto Ricans are much like African Americans. Massey and Denton suggest that, given their continued spatial isolation, Puerto Ricans are, like many African Americans, in danger of "becoming part of a permanent urban underclass" (1989a:75).

In sum, discrimination against Latinos has been ongoing throughout American history, but it has differed in scope and intensity for each Latino group and has differed as well by region. In any case, its severity has not been similar to that experienced by African Americans and it appears to be declining. It is of note that a majority of Mexicans, Puerto Ricans, and Cubans reported in a recent national survey that they had not been discriminated against because of their ethnicity (de la Garza et al., 1992). Today, anti-Latino prejudice and discrimination seem focused mostly on large immigrant populations in specific locations like southern California and south Florida, particularly on those who have entered the United States illegally. At times, of course, this has had a spillover effect on all Hispanic Americans.

HISPANIC-AMERICAN ASSIMILATION

In Chapters 6 and 7 we saw that Italians, Jews, and other European ethnic groups in American society are best analyzed within the framework of the assimilation model of ethnic relations, wherein ethnicity becomes less intense with each succeeding generation. African Americans, by comparison, were seen in Chapter 8 as a case that more closely fits the inequalitarian pluralistic model; rather than decreasing in significance, ethnicity continues to shape intergroup relations and to affect in a basic manner most life activities of group members. Where do Hispanic Americans fit on the continuum of ethnic relations, and which model is best suited to their experience in American society? As with African Americans, there is some disagreement over whether they are basically like previous immigrant groups who have gone through stages of assimilation or whether they are essentially a semicolonial people. Here we must be careful to distinguish among the three major Latino groups. Patterns of assimilation and pluralism for Mexicans, Puerto Ricans, and Cubans have not been entirely similar.

Patterns of Assimilation

Those who see Hispanics as simply the latest among a continuous stream of American immigrant groups emphasize the slow but steady upward mobility of Mexicans, Puerto Ricans, and Cubans; the resemblance of their social adjustment to that of previous groups; and their increasing social interaction with non-Hispanics (Chavez, 1991; Handlin, 1962; Jaffee, Cullen, and Boswell, 1980). There is evidence to support the view that Latinos in general have experienced a relatively low but increasing level of both cultural and structural assimilation. More specifically, however, Hispanic Americans exhibit a wide range of assimilation, extending from Hispanos in New Mexico, who have strongly assimilated both culturally and structurally into the dominant group, to black Puerto Ricans, who have not displayed more than minimal structural assimilation.

Cultural Assimilation In some ways, all three major Hispanic-American groups have displayed a lower degree of cultural assimilation than other ethnic groups. Unlike African Americans, they were not stripped of their native cultures on entering the society and have retained many ethnic traits, especially the Spanish language. Today, Spanish is the most widely used non-English language in the United States.

It is important to understand, however, that only a minority of Hispanic Americans speak only Spanish, and most of these are first-generation immigrants. As Latinos move to the second and third generations, English becomes their primary language (Cafferty, 1985; Crawford, 1992; Veltman, 1983). Moreover, even among those who continue to use Spanish in the home, the majority use English in other social contexts, like work and friendship groups.

A recent survey of the Latino population found strong evidence that, in terms of language, Latinos are clearly moving along an assimilation path (de la Garza et al., 1992). Not only did a majority of U.S. citizens from each of the three major groups use English or a mixture of English and Spanish at home, but English-language media were preferred to either bilingual or Spanish-language media. Further, more than 90 percent of Latinos supported the proposition that citizens and residents of the United States should learn English. Clearly, then, contrary to a common public perception about the maintenance of Spanish, most Latinos prefer English and support its usage.

The persistence of Spanish in some degree, however, has contributed to difficulties in educational and occupational adaptation (Bean and Tienda, 1987; Portes and Truelove, 1987). Moreover, the tendency for Latinos to concentrate in particular cities and regions and in some cases to maintain relatively tightly bound ethnic enclaves (Cubans in Miami, for example) retards language assimilation (Portes and Truelove, 1987).

Another factor contributing to slower cultural assimilation among Latinos is the proximity of their society of origin. This is especially evident for Mexican Americans. The nearness of Mexico and the relatively porous border with the United States tend to make Mexicans different from European immigrants, who, if they came as permanent residents, were forced to sever themselves more completely from their native society and culture. Nearly three decades ago, McWilliams observed that "In migrating to the borderlands, Mexicans have not found immigrant colonies so much as they have 'moved in with their relatives'" (1968:38). Today that observation seems even more compelling. In a sense, the Mexican immigrant in the Southwest never really leaves Mexico. The frequency and ease of travel between the United States and Latin America have created a similar phenomenon for other Latino groups. Immigrant networks have made back-and-forth movement so commonplace that family units may be said to reside simultaneously in two societies.

Mexican Americans who have moved to more cosmopolitan urban areas like Los Angeles, Houston, or Chicago, however, do take on more of the characteristics of traditional urban ethnic groups. Occupational and residential patterns in such environments resemble somewhat those of earlier groups, and the process of assimilation is correspondingly similar (Cardenas, 1976; Padilla, 1987; Samora and Lamanna, 1967). Here, Mexican Americans are forced to interact with Anglos in schools, work, and other institutional areas to a greater degree than in the smaller towns or rural areas of the Southwest.

Language and closeness of the native culture have also affected the assimilation patterns of Puerto Ricans and, less so, of Cubans. In the case of Puerto Ricans, the easy access to and from the island acts as a disincentive to assimilate culturally into dominant American ways (Garcia-Passalacqua, 1994; Lewis, 1965). In addition, for this group there is a racial factor that complicates matters. For those darker in skin color, the incentive to retain the Spanish language and culture is a product of their perceived need to separate themselves from the African-American population. It is more advantageous, they feel, to be identified with the Latino populace than to be lumped with non-Hispanic blacks. For those who are neither clearly white nor black by American racial standards, the problem is even more acute for they find themselves marginal to both categories (Mills, Senior, and Goldsen, 1950). For Cubans, a relatively substantial portion who arrived as part of the 1980 refugee wave are, in the American sense, racially defined as black. They, too, find themselves in a marginal position, peripheral to the larger Cuban community and unable to assimilate into the African-American community (Nordheimer, 1987).

As for most Cubans, because the majority are still first-generation immigrants, only minimal cultural assimilation is evident. Moreover, because they originally came not expecting to stay and take on the life of the new society,

Cubans have perhaps clung to the old culture more strongly than most other ethnic groups. This pattern is not uncommon in the case of political exiles, whose commitment to their adopted country is never complete. It will no doubt change with the second and third generations, for whom there are no thoughts of return. Further, political realities have prevented frequent back-and-forth movement between Cuba and the United States, which has had a stabilizing effect on the Cuban-American community.

Another factor that has contributed to a retention of the ethnic culture among Cubans is their concentration in a single community. Unlike Mexicans and Puerto Ricans, Cubans in the United States, especially those in Miami, have established a well-knit institutional structure that caters exclusively to the unique social and cultural needs of the group. Raymond Breton (1964) has suggested the idea of "institutional completeness" to describe the extent to which the needs of an ethnic group are met by institutions within the ethnic community — businesses, churches, newspapers, schools, and so on. Ethnic communities may range from those that are institutionally complete, in which members need to make no use of the host society's institutions, to those that are institutionally incomplete, in which the network of interpersonal relations is almost totally within the context of the host society. Cubans in Miami appear to be at the extreme of institutional completeness. There is not even a need to learn English because Miami is very much a bilingual city. Television stations broadcast and newspapers publish in Spanish, and food stores, restaurants, and businesses of all kinds cater to the unique ethnic needs of the Cuban community.

Despite the problems of language and proximity to the origin societies, Latinos do appear to be moving toward cultural assimilation, though more slowly than Euro-American groups of the past. Peñalosa, three decades ago, maintained that Mexican Americans were "clearly moving further away from lower-class Mexican traditional culture and toward Anglo-American middle-class culture" (1967:43). He believed that Mexican Americans were moving increasingly into the mainstream of American life just as previous immigrant groups had done. *The Latino National Political Survey* revealed much evidence that Latinos in general, not only Mexican Americans, are, in fact, moving in that direction (de la Garza et al., 1992). Regarding social networks and social distance, for example, although members of each specific national-origin group felt closer to and interacted more frequently with co-ethnics (that is, Mexicans with Mexicans, Cubans with Cubans, and so on), a majority felt closer to Anglos than to any of the other Hispanic national-origin groups.

The survey found that most Latinos do not identify themselves as members of a pan-ethnic group — Hispanic or Latino — but as members of their specific national group (Mexicans, Puerto Ricans, Cubans) or simply as Americans. Although there were issues and attitudes on which members of the three groups expressed common views, there were also many on which there were marked differences. Hence, "to the extent that the Hispanic political community exists, there is scant evidence that it is rooted in alleged distinctive cultural traditions such as Spanish-language maintenance, religiosity, or shared identity" (de la Garza et al., 1992:13–14). In sum, the survey found that Latinos were well within the mainstream of American political thought and action. "There is no evidence here," the study concluded, "of values, demands, or behaviors that threaten the nation's cultural or political identity" (de la Garza et al., 1992:16).

Moreover, as with African Americans, cultural assimilation is not entirely a one-way process. Rather, there is a growing interchange of Latino and American core cultures. Latino cultural influences are increasingly evident in American music, language, and cuisine: salsa and Latin jazz are now standard musical genres; Spanish is today the most popular foreign language, taught nationwide in primary and secondary schools and in every U.S. college; and Americans prefer salsas and picantes as often as ketchup (Fabricant, 1993).

Secondary Structural Assimilation As we have already seen, all three major Hispanic-American groups continue to lag behind Anglos in income, occupation, education, and political power. Despite these shortcomings, steady, if slow, progress has been made in closing the gap. Each succeeding generation has improved its status over that of the previous one, and it appears that Latinos are moving into greater participation in the society's mainstream economic and political systems. The pace and extent of secondary structural assimilation, however, have been uneven for these three groups. Puerto Ricans have proceeded more slowly and less effectively than Mexicans, who, in turn, have not progressed as rapidly as Cubans.

Entering the urban environment later than other ethnic groups, Mexicans and Puerto Ricans have not been able to use the economic and political systems as leverage in their pursuit of upward mobility. By the time they began to enter the cities in great numbers, the unskilled factory jobs that prior groups had used as economic springboards were already dwindling. And the local political machines, which had doled out many benefits to immigrant groups, were also no longer of much impact. The economic and political conditions faced by these groups, then, are closely parallel to those encountered by African Americans.

The factor of race in the case of Puerto Ricans and, less so, of Mexican Americans may also account for some of the lag in secondary structural assimilation. As we noted, prejudice and discrimination, though not so well rooted and intense as those experienced by African Americans, are continuing facts for Latinos, particularly in areas where they constitute a numerically significant part of the population.

Cubans in the United States have clearly exceeded Mexicans and Puerto Ricans in both the pace and degree of economic success. However, their upward economic mobility in the past thirty-five years has been not so much within the mainstream economic system as within a parallel Cuban economy, particularly in the Miami area. Whether they will begin to move into economic and political institutions of the larger society in significant numbers can only be speculated because most Cubans are still first-generation immigrants. Given their class advantage and, for most, their racial indistinctness, however, assimilation at the secondary level should be a less painstaking and lengthy process than it has been for either Mexican Americans or Puerto Ricans.

The Culture of Poverty Some have explained the slower pace and lower degree of secondary structural assimilation among Mexican Americans and Puerto Ricans as a product of the value systems of these groups. Essentially, the argument is this: Most Mexicans and Puerto Ricans retain the values of a peasant society when they enter the United States, and failure to discard these values results in an inability to satisfactorily adjust to a modern urban environment. Hence, assimilation is obstructed by these groups' cultural traits as much as by the opportunity structure they encounter in the new society. Much of this view is incorporated in the idea of the "culture of poverty." As originally explained by anthropologist Oscar Lewis (1961, 1965, 1966), some of the key features of the culture of poverty are a present orientation rather than a future orientation, a fatalistic view of the world and one's place in it, a tendency toward female-headed families, authoritarianism within the family, a high rate of abandonment of wives and children, frequent use of violence in settling disputes and in disciplining children, a high rate of alcoholism, a belief in male superiority, and a martyr complex among women.

Much of Lewis's research dealt with poor Mexican and Puerto Rican families, the latter in both Puerto Rico and New York. He maintained that these values and behaviors are used as a means of adapting to the conditions of poverty but that in the process they become well-entrenched cultural traits that are passed on from one generation to the next. "By the time slum children are age six or seven," writes Lewis, "they have usually absorbed the basic values and attitudes of their subculture and are not psychologically geared to take full advantage of changing conditions or increased opportunities which may occur in their lifetime" (1965:xlv). Poverty, therefore, is perpetuated indefinitely.

Critics of this view have pointed out that the basic argument of the culture of poverty blames the poor themselves for being poor. That is, their condition is essentially a product of their failure to adopt middle-class norms and values rather than a restructured economy or ethnic discrimination (or both) that makes escape from poverty, for most, difficult at best (Ryan, 1975; Valentine, 1971). Moreover, much debate has centered on the culture-of-poverty concept itself. Are the behaviors and values of the poor a culture in the sense of a way of life passed on from one generation to the next, or are these traits simply adaptive mechanisms that would be discarded once the material conditions of poverty were removed (Gans, 1968; Lewis, 1967)? This debate relates closely to that discussed in Chapter 8 concerning the slow pace of black structural assimilation and the role of the black family in sustaining poverty.

Primary Structural Assimilation At the primary level of structural assimilation, Latinos have clearly advanced further than African Americans. We have already seen that on one important measure of primary structural assimilation, residential integration, Mexican Americans, in particular, have begun to show strong signs of dispersion. There are similar indications in other areas.

As explained earlier, the rate at which members of an ethnic minority group marry those of the dominant group is perhaps the best evidence of how far the assimilation process has advanced. We have seen that marriage across ethnic lines has occurred increasingly among Euro-American ethnic groups, like Italians and Jews, but has not occurred significantly among African Americans. As in other aspects of comparison, Latinos seem to fall somewhere between the two.

In 1994, there were almost 1.3 million couples in the United States in which one spouse was of Hispanic origin and one not. These constituted 26 percent of the 5 million couples in which the husband or wife were of Hispanic origin (U.S. Bureau of the Census, 1995b). This indicates significant marital assimilation. Moreover, studies have shown increasing rates for each of the three major Hispanic groups (Fitzpatrick, 1987; Grebler, Moore, and Guzman, 1970; Jaffee, Cullen, and Boswell, 1980; Mittelbach and Moore, 1968; Murguía and Frisbie, 1977; U.S. Commission on Civil Rights, 1976; Valdez, 1983).

Internal Colonialism and Hispanic Americans

Although there is evidence that Latinos are assimilating into the mainstream society, though at a slow and uneven pace, there are nonetheless aspects of group history and interrelations with Anglos that conform to the internal colonial model.

The Mexican-American experience, in particular, seems to closely fulfill several features of internal colonialism: forced entry of the dominant group, alteration of the indigenous culture, administration by the dominant group, and the application of a racist ideology. Following the U.S. conquest, as we noted,

Mexican culture was denigrated by Anglos, indigenous Mexicans were stripped of political and economic power, and the system of dominant–subordinate relations that emerged was justified by a racist ideology: Mexicans, it was believed, were innately inferior to Anglos.

Some have maintained that the conditions established at the initial encounter of Mexicans and Anglos in the Southwest have been upheld over the years so that Mexicans, like African Americans, remain a colonized people who lack the fundamental political and economic power to control their destiny. Schools, police, businesses, and other vital institutions, it is argued, are controlled by Anglos and do not serve the needs and interests of Latino communities (Acuña, 1972; Alvarez, 1973; Garcia and de la Garza, 1977; Murguía, 1975). Unlike European immigrants who were able to utilize unskilled jobs to sponsor either their own or their children's upward mobility, Latinos, in this view, have been held at the lowest occupational levels not only because of the society's changing work structure but also because of discrimination. The depressed barrios of southern Texas and other parts of the Southwest are seen as similar to urban black ghettos, and in such cases, the internal colonial model is deemed relevant.

Although they have entered the society as voluntary immigrants, Puerto Ricans also seem to evince several aspects of the internal colonial model, notably minimal control of their communities on the mainland. Puerto Rico itself has been an American colony for most of the twentieth century, and though its official status was changed in 1952 to "commonwealth," the colonial-like relationship between island and mainland has remained essentially unchanged (Christopulos, 1980). The status of Puerto Ricans in this situation is somewhat in limbo. Although they are officially American citizens, they are not permitted to vote in presidential elections and do not have voting representatives in the U.S. Congress. The issue of commonwealth status, statehood, or full independence for the island has been an overriding and often emotional issue in Puerto Rican politics and remains unsettled among both Puerto Ricans and American policymakers (Rodríguez-Vecchini, 1994).

Most important, Puerto Ricans have traditionally been viewed with a colonial mentality, a product in large part of their racial background, which, in terms of the American racial system, places them among the darker-skinned categories. As we noted, racial factors have seriously impeded Puerto Rican assimilation. The dilemma of Puerto Ricans regarding their racial status in the United States is expressed dramatically by Piri Thomas, a dark-skinned Puerto Rican who has written of his New York experience. Thomas tells of an early incident that thrust on him the American race ethic. After applying for a door-to-door sales job, he is told by the employment manager that he will be called when "the new territory is opened up." His white friend applies immediately afterward and is told to start "on Monday." Realizing what has happened, Thomas relates the story to his black buddy, who tells him that "a Negro faces that all the time." "I know that," Thomas replies, "but I wasn't a Negro then. I was still only a Puerto Rican" (1967:104).[5]

Assimilation Versus Pluralism: An Assessment

Clearly, neither the assimilation model nor the internal colonial model of ethnic relations can alone account for the patterns of adjustment of Hispanic Americans and their relations with Anglos. Again, we must return to the theme of intermediacy, or in-betweenness, of Latinos. Unlike Euro-Americans, who obviously fit the assimilation model, or African Americans, who in large measure do not, Hispanic groups fall somewhere between these two extremes.

Murguía has written that Mexican Americans "seem to lie in that area between those groups that are so culturally and racially similar to the dominant society that they will assimilate completely in time, and those groups that are so racially and culturally divergent from the majority society that, given present attitudes and behavior, they will never assimilate with the dominant society, but instead will proceed with a decolonization movement of cultural nationalism and separatism" (1975:57). This seems equally true in the case of Puerto Ricans. At present, only for Cubans among the three major Hispanic-American groups does the traditional ethnic, or assimilation, model seem largely relevant. For Mexicans and Puerto Ricans, however, the nature of their entry into the society, their racial characteristics, and their class handicap make the assimilation model less clear-cut and impose at least some aspects of the internal colonial model.

Some analysts maintain that Hispanics, for the most part, may be following the general path of assimilation followed earlier by Euro-American groups, though their slower and more tortuous movement in that direction is occurring with little notice. Linda Chavez (1991), for example, argues that ongoing Hispanic assimilation is concealed by the continued entrance of new Hispanic immigrants in great numbers who remain more attuned to the ethnic culture and language. She also contends that Hispanic leaders, strongly committed to multiculturalism, are responsible for generating heightened ethnic consciousness in Latino communities in their attempts to retard assimilation. We will revisit the issues of immigration and multiculturalism in Chapter 11.

---

[5]Although the problem of racial identity remains unsolved, an increasing number of New York Puerto Ricans identify themselves as "New Yoricans" or "Ricans," terms implying recognition of and pride in the synthesis of the two cultures. These names also imply an acceptance of nonwhite and ghetto status (Rodriguez, 1974).

SUMMARY

As part of the American ethnic system, Hispanic Americans occupy an intermediate status between Euro-Americans and African Americans. Three major groups make up the Hispanic-American minority—Mexicans, Puerto Ricans, and Cubans. Each group is concentrated regionally, and each is primarily urban. Although Mexicans, the largest of the three, have entered American society mostly as voluntary immigrants, their roots in the United States reach back to Spanish colonial times before the American conquest of the Southwest. Puerto Ricans are technically American citizens yet have arrived from a foreign culture, much as other immigrant groups have. Cubans are more recent arrivals and differ from Mexicans and Puerto Ricans in class origins and racial features.

Hispanic Americans generally are below average on all measures of socioeconomic status and are a large element of the American poor. There are differences among the three major groups, however, with Cubans in a relatively higher position than Mexicans, who are somewhat higher than Puerto Ricans. Latinos are in a less developed stage of political participation than are African Americans, and like the latter, are underrepresented in top positions in all institutional areas.

The level of prejudice and discrimination directed at Hispanic Americans has not been as intense as that experienced by African Americans but has been more severe than that suffered by Euro-American ethnic groups. Mexican Americans have commonly been the chief targets of ethnic hostility in the Southwest, where traditionally most have lived.

As with African Americans, there is some disagreement over whether Hispanic Americans more appropriately fit the assimilation model or the internal colonial model. Although Cubans seem to conform to the immigrant experience of previous white ethnic groups, Mexicans and Puerto Ricans exhibit certain features of internal colonialism. Compared with both African American and Euro-American ethnic groups, Hispanics have displayed a lower degree of cultural assimilation. The speed and extent of secondary structural assimilation for Latinos have generally resembled African American patterns, but at the primary level, Latinos have outpaced African Americans.

Suggested Readings

Bean, Frank D., and Marta Tienda. 1987. *The Hispanic Population of the United States.* New York: Russell Sage Foundation. A compilation of social and economic data on Hispanic Americans, drawn largely from the U.S. Census.

Chavez, Linda, 1991. *Out of the Barrio: Toward a New Politics of Hispanic Assimilation.* New York: Basic Books. Argues that the economic and social assimilation of Hispanic Americans is concealed by the flood of new Hispanic immigrants and by Hispanic leaders with an interest in fueling separatism.

Connor, Walker (ed.). 1985. *Mexican-Americans in Comparative Perspective.* Washington, D.C.: Urban Institute Press. Essays by social scientists that deal with social, economic, and political issues of the largest Hispanic-American group.

Fitzpatrick, Joseph P. 1987. *Puerto Rican Americans: The Meaning of Migration to the Mainland.* 2d ed. Englewood Cliffs, N.J.: Prentice-Hall. A comprehensive study of the Puerto Rican experience in the United States, especially New York, where the bulk of the Puerto Rican population has traditionally resided.

Grasmuck, Sherri, and Patricia R. Pessar. 1991. *Between Two Islands: Dominican International Migration.* Berkeley: University of California Press. Examines one of the newer, and growing, Hispanic groups in the United States, particularly the process of their immigration from the Dominican Republic.

Grenier, Guillermo J., and Alex Stepick III (eds.). 1992. *Miami Now!: Immigration, Ethnicity, and Social Change.* Gainesville, Fla.: University Press of Florida. A collection of articles describing the social composition of the most thoroughly Hispanic city in the United States and the political, economic, and social issues that center on its various ethnic groups.

Massey, Douglas, Rafael Alarcón, Jorge Durand, and Humberto González. 1987. *Return to Aztlan: The Social Process of International Migration from Western Mexico.* Berkeley: University of California Press. An analysis of the factors that motivate Mexicans to immigrate to the United States. Shows that immigration is driven not only by economic circumstances but also by the establishment of social networks.

Portes, Alejandro, and Robert L. Bach. 1985. *Latin Journey: Cuban and Mexican Immigrants in the United States.* Berkeley: University of California Press. Deals with the economic and social adaptation of immigrants of two Hispanic groups in the United States. Also contains an important theoretical discussion of immigration and of different modes of immigrant adaptation.

Rieff, David. 1993. *The Exile: Cuba in the Heart of Miami.* New York: Touchstone. Describes the Cuban community of Miami—how it has impacted on the Anglo community and, conversely, how it has increasingly been transformed by American culture.

ASIANS HAVE BEEN in the United States since the early part of the nineteenth century. Only in recent decades, however, have they emerged as a sizable element of the American ethnic configuration. Like other ethnic groups, Asians have displayed variant patterns of settlement and adaptation and have been received by the dominant group in different ways, depending on social, political, and economic circumstances. As it has been with other groups, our objective is to trace the historical development of Asians in the United States and then to examine their place in the stratification system, the forms of prejudice and discrimination they have encountered, and their patterns of assimilation into the larger society.

At the outset, it is important to stress that Asian Americans constitute an extremely diverse ethnic category. In some ways, it is even more disparate than the Hispanic-American category, most of whose groups share a common language and religious tradition. To lump together those from numerous Asian societies as "Asian Americans" is, therefore, a serious oversimplification. In fact, Asian Americans consist of at least a dozen distinct groups, with extremely diverse cultural, and even physical, features. Currently the largest groups are Chinese, Filipino, and Korean. But substantial populations of Japanese, Asian Indians, and Vietnamese, as well as small numbers from other Asian and Pacific Island societies, are included. The term *Asian American* must be understood as nothing more than a convenient category that enables us to look at the general characteristics of the various Asian groups together, in comparison with other ethnic populations in the United States.

IMMIGRATION AND SETTLEMENT

The Asian experience in the United States can be divided into two distinct and divergent parts. The first consists of the old Asian immigration, occurring roughly from the middle of the nineteenth to the early years of the twentieth centuries. The Chinese were the first to arrive, followed by the Japanese and, in much smaller numbers, Koreans and Filipinos. These first Asian immigrants were mostly unskilled laborers, recruited for construction or agricultural work. Following restrictive measures, very few additional Asians entered the society until the revision of immigration laws in 1965. That marked the onset of the second period of immigration to the United States from Asia, which continues to the present.

The second, and current, immigration has been markedly different from the earlier movement in that most of the new immigrants have been noticeably higher in class origin, and many have been well educated and occupationally skilled. Moreover, the new Asian immigration has been considerably more diverse in national origin, made up of people from almost every contemporary Asian society. Today, except for the Japanese, Asians in the United States are predominantly foreign-born. Over 60 percent of the Chinese and Filipino populations, 80 percent of the Koreans, 70 percent of the Asian Indians, and 90 percent of the Vietnamese are first-generation immigrants (U.S. Commission on Civil Rights, 1992).

Chinese Americans

The Asian-American experience may be said to begin in the midnineteenth century. Unfavorable social and economic conditions and political unrest at that time in China, combined with the lure of labor opportunities in Spanish, Portuguese, Dutch, and British colonies, created a swell of Chinese emigration, mostly from the country's southeastern coastal provinces.[1] The United States was also among the destinations of these emigrants, mostly single males, some

[1] Chinese immigration over many generations has constituted one of the great diasporas. Today Chinese communities can be found in almost every country of the world (see Poston and Yu, 1990). Ethnic Chinese minorities have played an especially important economic role in most Southeast Asian societies (Minority Rights Group, 1992).

of whom were lured by the discovery of gold in California in the 1840s (Dinnerstein and Reimers, 1988; Lyman, 1970). In the 1860s, the construction of the transcontinental railroads drew many additional Chinese laborers to the United States. Like other immigrants before them, most intended to return after bettering themselves economically.

In a very short period of time, the influx of Chinese produced one of the most hostile movements in American ethnic history. The cast of an anti-Chinese movement had been set even before the Chinese landed in the United States, notes Stanford Lyman, having been "preceded by a richly embellished but almost entirely negative stereotype" (1974:56). White workers perceived the Chinese as a labor threat and, as a result, they were forced out of their jobs and commonly driven from one town to another. In addition to the antagonism they encountered from hostile white workers, the Chinese were severely restricted by legal measures used to limit their occupational and residential movement. These culminated in the Chinese Exclusion Act, passed by the U.S. Congress in 1882, designed to preclude Chinese immigration for a ten-year period. The act was extended for another ten-year period in 1892, however, and the ban against Chinese was made permanent in 1907. These measures marked a watershed in American ethnic history: For the first time, a specific group was formally barred from entrance. In fact, the Chinese Exclusion Act was the precursor of numerous anti-immigration laws that would be enacted in the following decades.

Although the exclusionary laws never fully suspended Chinese immigration, they effectively checked any significant population growth of the Chinese in America. At the time of the passage of the initial act in 1882, there were around 125,000 Chinese in the United States. For the next six decades, the Chinese population not only failed to grow but actually declined. By 1910, it had dropped to around 70,000, a figure that did not change radically until after 1965. This decline was the result not only of the restrictive legislation itself but also of the overwhelmingly male composition of the Chinese population that had made up the midnineteenth-century immigration. Before 1882, more than 100,000 men but fewer than 9,000 women had immigrated to the United States from China (Lyman, 1974).

Because they were almost totally male, there existed no possibility of a natural population increase and thus no creation of a stable and thriving ethnic community. The severely unbalanced sex ratio, in addition to the relentless discrimination to which they were subject, forced the Chinese at the turn of the century into urban ghettos, the familiar "Chinatowns," from which they did not begin to disperse until well after World War II.

By the early 1960s, the Chinese population had begun to grow somewhat as a result of the repeal of the Chinese Exclusion Act in 1943 as well as the entry of war brides, refugees, and some scientific personnel after World War II. This also helped correct the unbalanced sex ratio. As with other non-European groups, however, it was the revised immigration legislation of 1965 that created a large-scale Chinese population increase in the United States. Not only were more immigrants admitted, especially from Hong Kong and Taiwan, but the Chinese population was augmented as well by a large number of Vietnamese refugees who, though having lived in Vietnam, were ethnically Chinese. We look more closely at these new immigrants, along with others from Asia, in succeeding sections.

Japanese Americans

The formal exclusion of the Chinese provided for the entrance of Japanese immigrants beginning in the latter years of the nineteenth century. In their early immigration and settlement patterns, the Japanese resembled in some ways the Chinese who preceded them. Most of the first Japanese immigrants were males who were confined to lower-status occupational positions. And, like the Chinese, they soon came to be seen as a labor threat in California, where most had settled. The result was a movement of severe antagonism, culminating in formal legislation that prohibited their further entrance. In 1907–8, the United States and Japan entered into a Gentlemen's Agreement, which effectively restricted further Japanese immigration to the United States. Under the terms of the agreement, only nonlaborers and relatives of resident Japanese would be permitted to enter. The Oriental Exclusion Act of 1924 carried the restriction a step further, barring *all* subsequent Japanese immigration.

Despite some similarities, in a number of important ways the early Japanese immigrant experience differed from that of the Chinese. To begin with, although they worked initially in many of the same jobs as the Chinese, a large percentage of the early Japanese immigrants gradually turned to farming, where they carved out a niche for themselves in the California economy. More important, the sex ratio of the Japanese immigrant population, though heavily male in the early years, eventually was balanced with the introduction of a substantial female immigration. The Gentlemen's Agreement of 1907–8 had permitted wives and families of men already in the United States to join them. As a result, the social difficulties created by the demographic imbalance in the case of the Chinese did not materialize. With families intact, the base for a second generation was provided and thus a natural increase in the Japanese-American population. By 1940, nearly two-thirds of the Japanese in the United States were native-born (Kitano and Daniels, 1995). Today, unlike other Asian-American groups, the Japanese have not increased their numbers through large-scale immigration. As a result, the Japanese-American population has stabilized and is rapidly being surpassed by other, more recent, Asian groups.

The demographic and historical characteristics of the Japanese-American group today set it apart in some ways from others that make up the Asian-American population. Perhaps most important, because most (over two-thirds) are native-born, Japanese Americans have a longer history in the United States and are generationally divergent from others.[2] As a result, Japanese-American assimilation is, as we will see, far more advanced than for other Asian groups. Also, there is a strong anti-Japanese sentiment among many of the recent Asian immigrants, stemming from the harsh experiences of their home countries under Japanese colonization, particularly during World War II.

Other Asian-American Groups

The Chinese and Japanese were the first and largest of the Asian groups to enter the United States. Not until the liberalized immigration law of 1965 did a new stream of Asian immigrants augment these two communities. The number and variety of Asian societies represented by the post-1965 stream were considerably greater than any previous immigration, and it is these groups that today make up the major components of the Asian population in the United States. Although small numbers of all these groups had been in the society since the late nineteenth century, none could be said to represent a sizable community comparable to those of the Chinese and Japanese. Other than the Japanese, no more than half of the total population of any Asian-American group is native-born. The history of these new Asian-American groups, therefore, is essentially of the last three decades.

Koreans In the early years of this century, a few thousand Koreans were recruited to work the sugar plantations in Hawaii. Ironically, they had been needed to replace Chinese workers who had been barred by the 1882 Chinese Exclusion Act. Further Korean migration to the United States in any significant numbers did not occur again until the late 1950s, following the Korean War, when many came either as refugees or war brides. Prior to this time, Koreans on the U.S. mainland never exceeded 2,000 (Light and Bonacich, 1988).

It was in the post-1965 period, however, that the formidable growth of a Korean ethnic group in the United States occurred. Between 1970 and 1980, the Korean-American population increased five times, from 70,000 to 355,000. By 1990, Korean Americans numbered almost 800,000. Like other Asian groups, Koreans have concentrated in California, where almost one-third of the Korean population is found, 17 percent in Los Angeles alone (Light and Bonacich,

---

[2] Although the Chinese have an even longer history in the United States, most of the current Chinese-American population is foreign-born.

1988). Contemporary Korean immigrants have tended to be from the urban middle class of Korean society, including many college-trained professionals. What's more, most have come as family units, and most are Christians (Kitano and Daniels, 1995; Min, 1991). Koreans in the United States have been especially prominent as owners of small businesses, usually operated as family enterprises (Light and Bonacich, 1988; Kim, 1981; Min, 1988, 1995).

The strong economy of Korea in recent years has begun to impact the Korean population in the United States. The flow of immigration between 1970 and 1990, which had created such a substantial increase in the Korean-American group, has declined considerably and there appears to be a nascent return migration occurring. Economic opportunities in Korea are now seen by many as more promising, and political instability, which had characterized Korea in the past, has been reduced. These factors have diminished the appeal of immigration to the United States. They have also created an incentive for some to return to Korea. For those who have been successful in the United States, having earned a university degree or having run a profitable business, return represents an opportunity to enjoy even greater success in their country of origin, where they may feel more culturally at ease. Return is also appealing for those who have experienced increasing difficulties in adapting to American culture or who find that operating a small business, so common among Korean Americans, is simply too risky in terms of physical safety as well as economic security (Belluck, 1995a; Min, 1995). The image of Korean shops being burned and looted during the Los Angeles riots of 1992 and the frequent harassment of Korean shopowners in other cities have made Korean Americans more conscious of ethnic conflict, a condition not encountered in a homogeneous Korea.

Filipinos Few people think of Filipinos[3] when they consider the makeup of America's Asian population. Yet, Filipinos are now the second-largest Asian-American group. As with other Asian groups, the liberalization of immigration laws in 1965 accounts for their rapid and substantial increase. During the past two decades, only Mexicans have outnumbered Filipinos as immigrants to the United States.

The Philippines, before becoming an American possession in 1899, was a colony of Spain for over three centuries, and as a result, most Filipinos are Roman Catholics and have Spanish surnames. After the Spanish-American War, the United States took possession of the Philippines and the American cultural influence became dominant. Following independence in 1946, political ties between

---

[3]There is no "F" sound in Tagalog, the language, other than English, that is most widely spoken in the Philippines. Hence, rather than *Filipino, Pilipino* is sometimes used as a preferred term of reference.

the two countries remained close and American cultural and economic influences continued to be strong. English is one of the country's major languages and is spoken by most educated people. Historically and culturally, then, the Philippines has had little in common with other Asian countries. Geographically, however, it is part of the Pacific rim, and thus Filipinos are considered Asian people.

No group better illustrates the link between immigration and labor needs than do the Filipinos. Like the early Koreans, a few thousand Filipinos came to Hawaii in the early 1900s, recruited as agricultural workers to replace the excluded Chinese. Many more immigrated to the mainland in the 1920s as California farm producers faced a labor shortage created by newly restrictive quotas limiting cheap labor from Mexico. By 1930, almost 50,000 Filipinos were living on the mainland United States, most in California (Mangiafico, 1988). Because the Philippines at this time was a U.S. territory, Filipino immigrants were actually considered American nationals though not U.S. citizens. Because of this, they were not subject to the same kind of quota restrictions as other Asians until 1935, when the Philippines was granted deferred independence.

Like the Chinese before them, the early Filipino immigrants were almost entirely single males. And with the general anti-Asian social and political climate on the West Coast, where most settled, Filipinos found themselves subjected to the same kinds of discriminatory actions as the Chinese and Japanese. Following the outbreak of World War II, the Philippines was a critical ally in the Pacific war against Japan, and as a result, the social climate in the United States reflected a more tolerant view of Filipinos.

Along with discrimination, the low level of education of Filipino immigrants produced a restricted range of occupational opportunities. Most worked as seasonal agricultural laborers and in low-status positions in the service sector such as restaurant workers, hospital attendants, and hotel workers.

Between 1960 and 1970, the number of Filipinos in the United States nearly doubled. Cultural compatibility, especially language, and political ties account for the attraction of the United States for Filipinos. Also the strong U.S. military presence in the Philippines for over four decades following World War II enabled many Filipino women to immigrate as wives of U.S. servicemen. It is important to consider that the social characteristics of the new immigrants were considerably different from those of the earlier period. Most were now families with children, seeking the promise of better economic opportunities. Recent Filipino immigrants have also been better educated than those of earlier decades, and many come with professional credentials, enabling them to move into highly skilled occupational fields like medicine and education (Cariño, 1987). Their generally higher occupational positions, along with English proficiency, have dictated against the formation of cohesive and distinguishable ethnic enclaves comparable to Chinatowns (Agbayani-Siewert and Revilla, 1995).

Vietnamese The entrance of Vietnamese into the United States has been unlike that of any other Asian group. Most have come as political refugees since the end of the Vietnam War. Prior to that time, very few Vietnamese had been resident in the United States.

Between 1966 and 1975, some 20,000 Vietnamese arrived in the United States, but the Vietnamese immigration reached epic proportions following the fall of the South Vietnamese government in 1975. In the nine years between 1975 and 1984, over 700,000 Southeast Asian refugees came to the United States, most of them Vietnamese (Gardner, Robey, and Smith, 1985). Perhaps the most dramatic entrance of these refugees occurred in 1979 and 1980 when thousands fled Vietnam, Laos, and Cambodia in crude boats. The United States absorbed over 200,000 of these so-called "boat people" in a period of fifteen months. Most of the refugees coming in the second wave were unlike the earlier Vietnamese in that they were relatively unskilled occupationally, less educated, and spoke little English (Wong, 1986). Regarding differences in education, for example, almost 80 percent coming to the United States between 1965 and 1969 were college graduates, as opposed to less than 16 percent of those coming between 1975 and 1980 (U.S. Commission on Civil Rights, 1988). Their adjustment to American society, as a result, was far more difficult, and today they remain the most economically depressed element of the Asian-American population. The Vietnamese, then, display a bipolar class breakdown—those who have successfully adapted to American society and are at a comfortable economic level, and those at the opposite end of the class hierarchy, disproportionately unemployed, among the poor, and socially alienated (Weiss, 1994).[4] Other Asian-American groups show a somewhat similar class division.

Though the heaviest concentration of Vietnamese is in California, the group has dispersed widely throughout the United States. Substantial Vietnamese populations can be found today in Houston, New Orleans, and Arlington, Virginia, among others. The largest single community, however, is in Orange County, California. There, in the town of Westminster a thriving population of 80,000 has created a Little Saigon, made up of about 1,000 stores, offices, and restaurants serving the ethnic enclave.

Other Southeast Asians In addition to the Vietnamese, several hundred thousand refugees from other Southeast Asian countries arrived in the United States following the end of the Vietnam War in 1975. The largest among these were Laotians (mostly Hmong people) and Cambodians.

---

[4]As a result of subsequent immigration waves, the Vietnamese in America today constitute not only a class-diverse group, but one that reflects a cross section of Vietnamese society in terms of language, religion, and region of origin.

These groups, like many Vietnamese, have not easily adapted to American society. The intention of U.S. policy when they entered as refugees was to resettle them in various regions and cities. The expectation was that this would accelerate their adaptation and would avoid any negative impact on the labor market in communities that might absorb a large concentration of the refugees.

The resettlement of the Hmong, in particular, has been fraught with serious problems (Beck, 1994). Culturally, the group had not been exposed to Western ways before emigrating from a relatively isolated region of Laos. The class and educational characteristics of most of the refugees have been significantly lower than those of other Asian groups, further contributing to their problems of adaptation.

Asian Indians Like the Chinese, Indians have a long heritage of immigration to societies in all parts of the world. Until recently, however, the United States had not been one of their principal destinations. In the nineteenth century, most had gone as indentured servants to various parts of the British colonial empire, including South Africa and a number of Caribbean societies. In the early years of this century, a few thousand workers from India were recruited as agricultural laborers in California and Washington, but because of immigration restrictions, the Indian population did not expand. Like other Asian immigrants at this time, they experienced extremely virulent forms of discrimination. In this racially hostile atmosphere, the relatively small Asian-Indian population remained isolated in agricultural communities, mainly in California.

As with other Asian groups, the liberalized immigration measure of 1965 represented the point at which Indians began to enter the United States in significant numbers. And like the others, the new immigrants were sociologically quite distinct from their earlier counterparts. Those who have come in recent decades represent, for the most part, a highly educated and occupationally very skilled population. Indeed, their educational level, compared with that of other ethnic groups, is quite remarkable. Almost 60 percent of the adult Asian-Indian population are college graduates, and fully one-third have earned graduate or professional degrees (U.S. Bureau of the Census, 1993a). Clearly, Asian Indians are among the most highly educated segment of the American population. What's more, their income levels are far above average.

Most Indians coming to the United States, then, represent a very select subset of the Indian population—more educated, more skilled, and more generally privileged. They are part of the worldwide migration of those with educational and occupational skills leaving developing societies in favor of the industrialized societies of the West, where economic and professional opportunities are more plentiful. In fact, many of the Asian-Indian immigrants of the recent period have been students at American universities who, after completing their studies, find employment and remain. The relatively high occupational status and high level of education of Indian immigrants in the past two decades have enhanced the speed and extent of their economic and social adaptation to American society. Indians have also benefited from the fact that on arrival most are fluent in English and have already been exposed to Western values.

In the past decade, the Asian-Indian community in the United States has become more diverse, with a larger commercial class augmenting the predominantly professional and technical character of the first post-1965 immigrants.[5] Hence, in addition to their place in skilled occupational areas, Asian Indians have played a significant role as entrepreneurs in the small-business sector. This is in some measure comparable to the economic path followed by the recent generation of Korean immigrants.

It is important to understand that those classified as "Asian Indian" represent a group that is internally quite diverse culturally. India itself is an ethnic potpourri, with virtually dozens of languages and cultural communities. The religious division among Hindus, Muslims, and Sikhs is only the most apparent and profound variance. Moreover, the national and geographical origins of Asian Indians in the United States are varied. While most have come from India, sizable numbers have come from Pakistan, Bangladesh, and Sri Lanka, as well as other countries where East Indians had settled earlier in the nineteenth century.

Asian-American Demographics Today

Today, Asian Americans remain a relatively small part of the total American population—3 percent—but one that is rapidly growing not only in numbers but in economic and cultural significance. In fact, the Asian-American population is today the fastest-growing ethnic minority in the United States. Between 1970 and 1980, it grew by 128 percent, and between 1980 and 1990, by 108 percent. It is projected that the current Asian-American population of 7.3 million will grow by the turn of the century to 12 million, and by 2050, to 41 million (Pear, 1992). This rapid and sizable population growth is attributable mainly to the fact that Asians now account for the largest element of immigration to the United States, more than 45 percent. Figure 10.1 shows the breakdown of the Asian-American population today.

In 1970, the Chinese and Japanese composed the bulk of the Asian-American population, but today these two groups are well under one-half, and by 2000, it is expected that they will be only one-quarter (Bouvier and Agresta, 1987). The Chinese category remains the largest component of the Asian-American

---

[5]It is interesting to consider that the post–World War II Indian immigration to Britain, by contrast, comprised not only a much larger but also a more heterogeneous population, with many unskilled and poorly educated among them (see Watson, 1977).

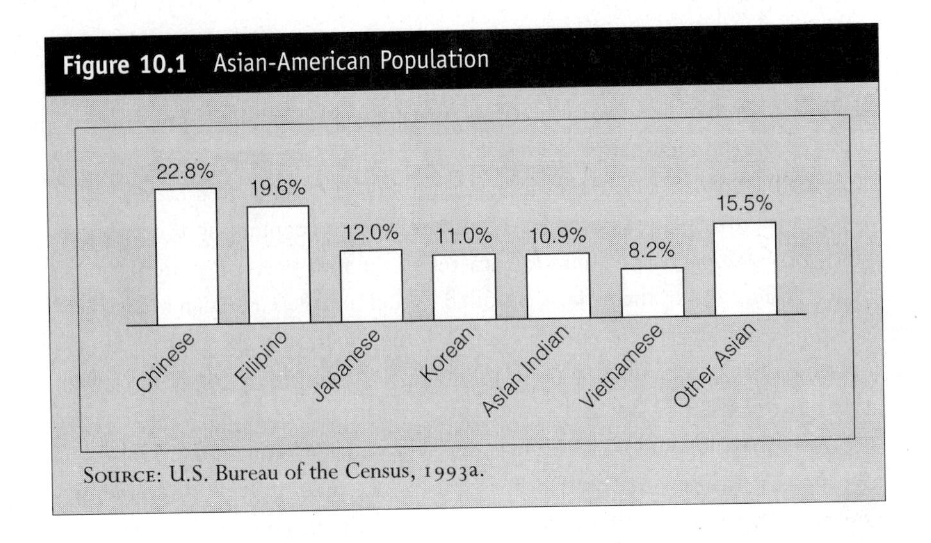

SOURCE: U.S. Bureau of the Census, 1993a.

population (1.6 million), but Filipinos are almost as large. The Asian-Indian and Korean populations have also continued to grow steadily, each with around 800,000. The Japanese-American population (about 850,000) is growing much more slowly because of very limited immigration from Japan and a low fertility rate; by the turn of the century, it will become one of the smallest Asian-American groups.

During the 1980s, the Asian-American population expanded not only in size but in diversity as well. Hmong tribespeople from the mountains of Laos, for example, increased from 5,200 to 90,000, and Cambodians increased from 16,000 to 147,000 (Worsnop, 1992).

The rapid increase in the Asian-American population is evident in the numbers who arrived in the 1980s. Of all Filipinos in the United States, 32 percent entered in the ten years between 1980 and 1990, as did 39 percent of Chinese, 41 percent of Koreans, 44 percent of Asian Indians, and 62 percent of Vietnamese (U.S. Bureau of the Census, 1993a).

Asian Americans are most heavily concentrated in five states: California, Hawaii, New York, Illinois, and Washington. More than one-third reside in California alone. Among Asian Americans, only Asian Indians have not concentrated in California, having dispersed more widely. Also, like other ethnic populations who have arrived primarily in the post-1965 era, Asian Americans dwell mostly in large metropolitan areas. The largest communities are found in just four urban areas: Los Angeles, San Francisco, New York, and Honolulu (Table 10.1). In Los Angeles, Asians make up more than 9 percent of the total metropolitan population, and in San Francisco, they make up almost 15 percent (U.S. Bureau of the Census, 1993d).

One of the more intriguing demographic patterns exhibited by Asians, compared to other nonwhite groups, is their rapid rate of suburbanization. Unlike

Table 10.1 Metropolitan Areas with the Largest Asian Populations

METROPOLITAN AREA	ASIAN POPULATION	PERCENT TOTAL
Los Angeles–Anaheim–Riverside	1,339,000	9.2
San Francisco–Oakland–San Jose	927,000	14.8
New York–Northern New Jersey–Long Island	873,000	4.8
Honolulu	526,000	63.0
Chicago–Gary–Lake County	256,000	3.2
Washington, D.C.	202,000	5.2
San Diego	198,000	7.9
Seattle–Tacoma	164,000	6.4
Houston–Galveston–Brazoria	132,000	3.6
Philadelphia–Wilmington–Trenton	123,000	2.1
Boston–Lawrence–Salem	121,000	2.9
Sacramento	115,000	7.7

SOURCE: U.S. Bureau of the Census, 1993d.

blacks and Hispanics, Asians, especially those coming since 1970, have been leaving the central cities in ever greater numbers. In the Los Angeles metropolitan area, many of the Asian enclaves are now found in suburbs rather than the central city. Monterey Park is an interesting example. In the early 1970s, it was a Los Angeles suburban town of mainly Anglos and Hispanics. Now the population is about 50 percent Asian, most of it middle class, and a majority of its businesses are Asian-owned (Horton, 1995; Lemann, 1988).

ASIAN AMERICANS IN THE STRATIFICATION SYSTEM

Economic Status

What is perhaps most striking about the Asian-American population today is its unusually high ranking on most measures of socioeconomic status. Put simply, Asian Americans rank higher than most ethnic groups in family income, occupational prestige, and level of education.

The socioeconomic success of Asians in the United States, however, must be looked at with some caution. There are differences among specific groups as well as among members of the same group. For example, although nearly half of employed Asian Indians are professionals or managers, less than 20 percent of

Table 10.2 Asian-American Median Family Income

	ALL	NATIVE-BORN	FOREIGN-BORN	PERCENT IN POVERTY
All Asians	$41,583	$50,513	$40,011	11.4
Chinese	41,316	56,762	37,966	11.1
Filipino	46,698	42,114	47,595	5.2
Japanese	51,550	52,728	47,836	3.4
Asian Indian	49,309	38,256	49,567	7.2
Korean	33,909	42,427	33,401	14.7
Vietnamese	30,550	23,819	30,621	23.8
Cambodian	18,126	21,974	18,063	42.1
Hmong	14,327	22,171	14,272	61.8
Laotian	23,101	20,795	23,144	32.2
Non-Asians	35,108	35,451	30,354	9.9

SOURCE: U.S. Bureau of the Census, 1993a.

Vietnamese are in such occupations (U.S. Bureau of the Census, 1993a). The income and educational levels of the latter are also considerably lower than those for other Asian-American groups. Another internal difference we need to consider is the tendency toward economic polarization. That is, members of specific Asian-American groups tend to cluster at the two extremes of the economic hierarchy, either at or near the top (such as professionals and business owners) or at or near the bottom (service workers and other low-status job classifications).

Income The family income of Asian Americans far exceeds that of most other minority ethnic categories and, with the exception of Koreans and Vietnamese, surpasses most Euro-American groups' as well (Table 10.2). Median family income for Asian Americans in 1990 was $41,583, compared to $35,108 for non-Asians (U.S. Bureau of the Census, 1993a).

Here again we need to carefully consider the internal differences within the Asian category. As is seen in Table 10.2, the median income for Laotians, for example, is less than half of what it is for Filipinos or Asian Indians. Moreover, although most Asian groups have higher family incomes than whites, they also have more family members in the labor force than whites.

Occupation The rate of labor-force participation for Asian Americans is higher than that of any other ethnic category, and the range of occupations held by Asian Americans is broad, extending from working-class jobs to professions such as medicine and engineering.

Here the split between those of previous generations and the more recent immigrants is very evident. In striking contrast to the earlier Asian immigrants, most of whom were unskilled laborers, Asians today are disproportionately among the most highly skilled. One of the key features of the immigration reform of 1965 was the preferential treatment given to family members and to those with valuable occupational training. Because Asians had been virtually excluded for several decades, most of the early post-1965 immigrants entered not as family members but as skilled workers. As a result, many of the recent arrivals are found in professional, technical, and managerial jobs. Between 1976 and 1980, almost half of working Asian immigrants reported professional occupations, whereas no more than 10 percent of any Asian group reported laborer occupations (U.S. Commission on Civil Rights, 1988). Asian immigrants to the United States are helping to fill shortages in a number of professional and technical areas.

High occupational rank is characteristic of native-born and foreign-born Asian Americans alike. Except for Filipinos and Vietnamese, Asian-American men are more heavily represented in white-collar occupations and less represented in blue-collar occupations than are non-Hispanic white men. The percentage within the managerial and professional category is especially noteworthy. More than one-third of Asian-American males are in managerial and professional occupations compared with 27 percent of non-Asian males (U.S. Bureau of the Census, 1993a).

Asian Americans are also rising faster in the corporate world than any other minority group and are increasingly occupying high-powered jobs in finance and advertising. One should not misread such trends as evidence of the arrival of Asian Americans into the corporate world on a par with the white population, however. They are underrepresented in managerial positions despite their comparatively large numbers in professional and technical positions. And at the very top of the corporate ladder, their numbers are as small as for African Americans and Hispanic Americans. In 1990, 0.3 percent of senior corporate executives were Asian Americans (U.S. Commission on Civil Rights, 1992). Asian Americans have often spoken of a glass ceiling, referring to the limits to which they are able to rise in the corporate world. For example, in California's Silicon Valley, the major center of the American computer industry, Asians make up perhaps a third of the engineering work force. But they are not well represented in upper management at the larger companies. Some, as a result, have begun to start up high-technology enterprises of their own (Pollack, 1992). This was a route followed earlier, in the 1950s, by the late An Wang, Chinese-American founder of Wang Laboratories, for many years one of the country's largest computer companies. Moreover, despite their scant numbers at the executive level, the extraordinarily high level of education of most Asian Americans will likely accelerate their movement into top management positions at a faster rate than other minorities. Furthermore, as the number of Asian Americans in the general population grows, corporations will find it in their interests to seek out and place more Asian Americans into higher ranking positions, much as they have done with African Americans and Hispanic Americans.

One of the salient features of the economic role played by Asian Americans today is the unusually large number who choose to own small businesses. Indeed, Asian Americans have the highest rate of business ownership of all U.S. minority groups. In 1982, for every thousand Asians in the population, almost fifty-five owned a business, a rate that far exceeded that for blacks—12.5 per thousand—and for Hispanics—seventeen per thousand (Manning and O'Hare, 1988).

Specifically among Asian-American groups, the rate of business ownership among Koreans, Asian Indians, Japanese, and Chinese exceeds the rate for all Americans. Koreans have been especially prominent in small business (Kim, 1987; Light and Bonacich, 1988; Min, 1988, 1991). Indeed, a greater proportion of the Korean ethnic population is active in small business enterprises than of any other ethnic group (Hoffman and Marger, 1991). In the New York metropolitan area, for example, where they number about 100,000, Koreans run some 9,000 small business enterprises, including most of the city's green groceries (Kim, 1981, 1987). In Los Angeles, they operate about 6,000 retail businesses.

Several factors seem to account for the high rate of participation of Asians in the small-business sector. Most of these businesses are family operated, in which several or even all members of the family participate. As a result, labor costs are minimal, and relatively small amounts of capital are needed. Many immigrants thus find readily available opportunities in small enterprises. Many also fear that opportunities for advancement within the mainstream sector of the economy, dominated by big corporations, is difficult for nonwhites, especially those who do not speak English fluently. Having a small business of one's own thereby provides the immigrant with some independence, and enables one to avoid having to deal with what are seen as unreceptive dominant economic institutions.

Education The differences between Asian Americans and other minority ethnic groups regarding education are striking. All Asian-American groups are either close to or exceed the average level of schooling of the white population. Eighty-two percent of white adults in the United States are high school graduates, compared to 85 percent of Asians. The rate of college graduation is even more dramatic. More than 41 percent of adult Asian Americans in 1994 had graduated from college, compared to 23 percent of whites (U.S. Bureau of the Census, 1995b). Many Asian immigrants, of course, are highly educated before

Table 10.3 Asian-American Educational Level, Persons 25 Years Old and Over

GROUP	PERCENT HIGH SCHOOL GRADUATES		PERCENT COLLEGE GRADUATES	
	Foreign-Born	*Native-Born*	*Foreign-Born*	*Native-Born*
All Asians	74.8	86.4	37.0	32.0
Chinese	70.1	91.1	38.7	51.0
Filipino	82.2	85.1	42.3	22.3
Japanese	86.0	88.3	35.1	34.2
Asian Indian	84.8	82.2	58.5	44.2
Korean	80.0	82.7	34.4	36.5
Vietnamese	61.0	67.7	17.4	16.9
Cambodian	34.6	55.4	5.5	12.4
Hmong	31.0	41.3	4.8	9.4
Laotian	39.9	43.5	5.2	13.1
Non-Asians	54.1	76.9	15.4	20.3

SOURCE: U.S. Bureau of the Census, 1993a.

they enter the United States, making them sharply different from most contemporary Hispanic immigrants as well as Asian immigrants of past eras. Table 10.3 shows the percentage of college graduates among native-born and foreign-born Asians. Only the Southeast Asian groups (Vietnamese, Cambodian, Hmong, Laotian) do not have higher rates than non-Asians.

In terms of academic achievement, Asian-American students are much above average on almost every measure. For example, Asian-American high school students score higher than students from any other ethnic group in mathematics scholastic aptitude tests and are second only to whites on verbal aptitude tests (U.S. Department of Education, 1992). Predictably, Asian students enter top universities far out of proportion to their numbers. At Harvard and MIT, for example, Asians constitute about 20 percent of the student body and almost 40 percent at the University of California at Berkeley and UCLA.[6]

Science, medicine, and other highly technical fields are academic areas in which the excellence of Asian-American students is especially apparent. But they are entering the humanities and the arts in increasingly greater numbers

[6]It is noteworthy that in 1990, an Asian American was named chancellor for the prestigious Berkeley campus. This represented the first appointment of an Asian American to head a major research university on the U.S. mainland.

and are excelling in those fields as well. A quarter of the student body of the Juilliard School of Music in New York, perhaps the best-known and most prestigious institution of its kind, is Asian and Asian American. The representation of Asian Americans on the faculties and in the administration of American universities is also growing (Tien, 1995).

Ethnic and Class Factors in Upward Mobility

The economic and educational record of Asian Americans is truly extraordinary compared with averages for other ethnic groups. Indeed, in recent years, Asian Americans have been referred to as a "model minority," implying that they have adapted to their new society better than other minority ethnic groups and are well on the way to assimilation. The now well-recognized success of Asian Americans has prompted a number of questions. Why have they moved up the economic ladder more quickly than other ethnic groups, and why do they seem to outdistance others in the realm of education? Is it because of a unique cultural heritage, or are class factors more important? Might Asian Americans even be innately different from other racial and ethnic groups?

In looking at the Asian-American experience, two caveats are in order. First, we need to remember the diversity of this ethnic category. Not only is the Asian-American population made up of many different groups, but each group itself is internally varied, stretching over an entire range of social class. Second, the Asian-American population consists of two historical components, those who came as early as the midnineteenth century, and those who have come since 1965. As we have now seen, many of the more recent Asian Americans have come with class and educational backgrounds very different from those of earlier ones.

Attempts to explain the unusually high rates of Asian-American mobility, educational achievement, and social adaptation have employed a number of different lines of thought. First, some believe Asians are innately superior. An eminent pediatrician at Harvard, for example, suggests that Chinese and Japanese babies are more alert and sensitive at birth, making them faster learners (Butterfield, 1986). As explained in Chapter 1, most social scientists do not recognize the validity of such biological arguments, whether applied to seemingly high-achieving or low-achieving groups. Moreover, this only prompts the nature–nurture debate, which, as we suggested, is seemingly irresolvable. The intelligence factor is one that nonetheless confounds observers because there is no denying that on certain measures of intelligence, Asian Americans consistently outperform other ethnic groups.

As noted in Chapter 7, social scientists have emphasized either structural opportunities or cultural factors in explaining why some ethnic groups seem to display greater collective achievement in American society than others. In the Asian-American case, many sociologists and psychologists point to cultural factors in accounting for high achievement. Stressed in this explanation are values that emphasize the importance of education and the cohesiveness of the family. Almost all studies that have addressed this issue seem to converge on one point: Asian Americans are prepared to work arduously and, in turn, are able to instill in their children a great motivation to work hard. Hence, Asian-American business owners, it is often observed, work longer hours than their competitors, and time spent on homework or other school activities by Asian-American students exceeds considerably that spent by non-Asian students (Butterfield, 1986; Kim, 1981; Light and Bonacich, 1988). In his study of Koreans in New York, Illsoo Kim traces their business success largely to cultural factors. "An emphasis on economic motivation and mobility," he writes, "has contributed to acculturation in the new land because it was already a deep part of contemporary Korean culture and values" (Kim, 1981:295). He claims that values such as industriousness and the control of impulses and emotions, which correspond closely to dominant American values, were carried by Korean immigrants to the United States. A similar argument has been advanced regarding the economic success of Japanese Americans, who, it is claimed, brought values from Japan highly compatible with white, Protestant, middle-class American values (Caudill and De Vos, 1956).

That cultural factors are of key importance in explaining Asian-American educational and economic success is suggested by a study, conducted by University of Michigan researchers, investigating the adaptation of Vietnamese refugees. At the time of their arrival in the United States, most of the subjects of this study (part of the boat people of the late 1970s) had completed little formal education, few had any marketable skills, and only one in a hundred spoke fluent English. The researchers found that, despite these handicaps, many had achieved economic independence after only a few years. Furthermore, whatever success they had realized was almost entirely a product of their own efforts; the assistance of outside agencies had played a very limited role (Caplan, Whitmore, and Choy, 1989).

The academic achievement of school-aged children from these families was more remarkable still. In spite of their severe language deficiencies and family poverty, after an average of only three years, they were experiencing success in American schools. On national standardized tests of academic achievement, 27 percent of the refugee children scored in the ninetieth percentile on math achievement—almost three times better than the national average. And despite their somewhat lower scores in English language proficiency, they earned higher grade point averages than their peers (Caplan, Choy, and Whitmore, 1991, 1992). The researchers found that this rapid educational success was closely linked to a cluster of core values deriving from the Confucian and Buddhist traditions of East and Southeast Asia: education and achievement, a cohesive family, and hard work. These, they found, "are, if not identical, congruent with what have been viewed as mainstream middle-class American beliefs about the role of education in getting ahead" (1991:172).[7]

Other researchers have also emphasized the importance of Confucian values in explaining the rigid work ethic that Asian Americans and their children seem to exhibit. Kim (1981), for example, views Confucianism as a significant factor in accounting for Korean business success. In addition to the sanctified place of education and the family, Confucian thought stresses that people can always be improved with effort and instruction, values well suited to the competitive worlds of American business and schooling.

Although most observers today recognize the importance of unique cultural factors in explaining Asian-American success, many also see a fortuitous fit between the opportunity structure of contemporary American society and the class background of these immigrants. We have already noted the disproportionate number of highly skilled and educated persons who make up the Asian immigrant population. Hence, the existence of many doctors, engineers, and other highly trained professionals among them is largely a function of their preimmigration class position. In some ways, then, today's Asian immigrants represent a select subset of the population of their origin societies. What has been referred to as a "brain drain" consists of the out-migration of highly skilled and educated persons from developing societies. Such persons recognize the more abundant opportunities to employ their skills and training in the United States and also the possibility of earning substantially greater income. Hence, many are not poor, underprivileged immigrants but are already imbued with achievement-oriented values along with the class qualifications to implement those values. The importance of a propitious opportunity structure, therefore, cannot be understated in explaining the rapid and substantial upward mobility of Asian Americans, especially the more recent arrivals. In brief, many are people who bring their skills and training with them, enabling them to adapt more easily than others with fewer skills and lower educational backgrounds. Structural opportunities, then, in addition to a cultural heritage compatible with American achievement values, are critical in accounting for the relatively rapid success of so many Asian Americans.

---

[7]The researchers did find a critical difference between the two value systems, however, in their orientation to achievement: "American mores encourage independence and individual achievement, whereas Indochinese values foster interdependence and a family-based orientation to achievement" (1992:41).

Asian Americans and Political Power

Even though their accomplishments in the economic and educational spheres of American society are increasingly evident, Asians have not exhibited a similar prominence in the society's political institutions. With the exception of Hawaii, this is true even at the community level in areas where Asians are found in significant numbers. In California, for example, where they are 10 percent of the population, Asian Americans hold only one seat in the state legislature. In San Francisco, where almost one-third of the population is Asian, only one member of the eleven-member county board of supervisors is Asian American. And in Los Angeles, about 12 percent Asian, only one Asian American sits on the fifteen-member city council (Gross, 1989). There are, however, four Asian Americans from California in the U.S. Congress, and all of Hawaii's congressional and senatorial seats are held by Asian Americans.

The relatively low level of participation of Asian Americans in political life has been attributed to a number of factors. The nationally diverse and factionalized nature of the Asian-American population seems to be a critical hindrance to effective political power. Asians also are not as concentrated spatially as Latinos or African Americans, making it difficult to form cohesive voting blocs. Moreover, because such a large proportion of the Asian-American population is foreign-born, many are not U.S. citizens and thus not eligible to vote. Also, among older Asian Americans, especially those who have been victimized by past discrimination, the tendency to remain as invisible as possible continues to militate against strong political activism (Gross, 1989). Finally, many of the newest Asian immigrants come from societies where citizen participation in politics is not institutionalized and where there exists a profound distrust of government generally.

As their population base widens and as the newer immigrants experience progressive assimilation, Asian Americans are likely to play a more active political role, especially in those areas where they constitute a sizable proportion of the population or where they are a significant minority. Even if Asian Americans do not rapidly enter the political arena by putting forth candidates of their own for office, their increasing visibility, numbers, and financial power will necessarily force politicians to take notice of them. Already, for example, Asian Americans have established themselves as significant campaign donors (Kirschten, 1992; Stokes, 1988). Perhaps most important, future issues may arise that will have a mobilizing effect on the Asian-American population. Widespread occurrences of discrimination, for example, real or perceived, are apt to solidify the disparate Asian-American ethnic groups and to create a more potent political base. Interethnic competition and hostilities, such as the boycotts against Korean shops by blacks or the attacks on Korean businesses during the Los Angeles riots of 1992, may also stimulate Asian-American political participation. Moreover, the growing importance of U.S. economic and political relations with Asian nations may create issues that will redound to the Asian-American population, pushing them increasingly into the domestic political arena.

PREJUDICE AND DISCRIMINATION

Among voluntary immigrants to American society, none suffered more severe forms of prejudice and discrimination in their settlement experiences than Asians. Although they were not enslaved, like African Americans, nor were genocidal measures taken against them, as with Native Americans, they were constant targets of all forms of social and physical abuse. Indeed, if there is a common thread running through the early history of Asian-American groups, it is the experience of rampant prejudice and discrimination of the most vehement and often violent nature.

The Anti-Asian Heritage

Anti-Asian sentiments have lengthy historical roots, reaching back, as we have seen, to the entrance of the Chinese into the American West in the midnineteenth century. The intensity of early anti-Asian actions traditionally seemed to hinge on the perceived threat of Asian immigrants to the jobs of native workers. The anti-Chinese movement, which began in the mines of California, spread to other West Coast areas and eastward to other parts of the country (Lyman, 1974). Everywhere, Chinese were met with derision and hostility. They enjoyed hardly any legal rights and were common targets of assault and even murder. "The bewildered Chinese," note tenBroek, Barnhart, and Matson, "speaking little or no English and ignorant of Western customs, became the victims of every variety of fraud and chicanery, abuses encouraged by the absence of an active public opinion which might have alerted the police and the courts, and by the law prohibiting the testimony of Chinese in cases involving whites" (1968:15).

The "Yellow Peril" The standard panoply of negative ethnic stereotypes—dirty, immoral, unassimilable—was applied to the Chinese. In addition, they were seen as sly, untrustworthy, and inscrutable. Much of the negative stereotyping of the Chinese, and later the Japanese, stemmed from their labor role and their alleged impact on the job market. The term *coolie labor,* for example, denoting undignified work, derives from this period. That the Chinese were inferior to whites was beyond question in most people's minds. Perhaps the most devastating aspect of Chinese stereotyping, however, involved the notion of a "yellow peril." Chinese immigration, in this view, was seen as an invasion of people who were loyal only to their country of origin and who, if not stopped, would eventually take over the United States (tenBroek, Barnhart, and Matson, 1968).

This sinophobia, fueled especially by labor leaders, culminated in the previously mentioned Chinese Exclusion Act of 1882. By 1910, the anti-Chinese movement had achieved its primary objectives: Chinese workers had been all but eliminated from the labor market, and restrictive legislation was in place that effectively barred further Chinese immigration. Moreover, the ban on future immigration, particularly of women, seemed to assure that the Chinese-American population would, in the long run, disappear.

Many of the same attitudes and actions were cast on the later Japanese immigration. At the time of their first substantial entrance in the 1880s, the Japanese were received enthusiastically by large farm owners and manufacturers who saw them as a natural substitute for the Chinese laborers who had been excluded by the restrictive measure of 1882. So long as they occupied only menial or unskilled jobs, for which there was a shortage of workers, they were seen as filling a need. But like the Chinese before them, they soon found themselves the targets of labor groups who viewed them as potential competitors. Anti-Chinese agitation easily shifted to the Japanese, with many of the same stereotypes now applied to them, including the yellow peril (tenBroek, Barnhart, and Matson, 1968). Despite the Gentlemen's Agreement between the United States and Japan in 1907–8 limiting Japanese immigration, as well as a California measure restricting Japanese ownership of land, anti-Japanese ferment never fully subsided.

In many ways, the anti-Japanese movement of the 1910s and 1920s was merely a continuation of the earlier anti-Chinese movement. Despite their modest group size (in California, where most were concentrated, they numbered but 70,000 in 1919, little more than 2 percent of the state's population), the Japanese were seen by many as a threat to white dominance. As already noted, the Japanese in California had become successful farmers, and although they controlled only about 1 percent of California's farmland, their highly efficient agricultural practices produced more than 10 percent of the value of the state's crops (Kitano and Daniels, 1995). Many whites were alarmed at the growing Japanese economic influence and began to agitate for restrictions not only on further immigration but on the right of the Japanese to own land in the state. California newspaper publishers William Randolph Hearst and V. S. McClatchy emerged as two of the leading figures in the anti-Japanese movement, outlining in their newspapers the themes of the Japanese "threat." Their propagandistic attacks claimed that the Japanese refused to assimilate, that their birthrate was so great they would eventually outnumber whites, and that their "low standard of living" presented an unfair advantage in economic competition with whites (McClatchy, 1978). All these charges, of course, had been applied to southern and eastern European immigrants as well, but the anti-Japanese campaign brandished a unique fear of an immigrant "invasion." Following World War I, a full-blown exclusionist movement developed in California, designed to isolate those Japanese already in the United States and, as one California politician described it, "to influence Congress and the administration at Washington to enact such legislation, even if the amendment of the Constitution be necessary, as will protect the white race against the economic menace of the unassimilable Japanese" (quoted in Daniels, 1977:84).

The vehemence of the anti-Japanese movement in California was unmatched against any other American ethnic group outside the South. Much of this hostility was the result of the racial distinctiveness of the Japanese. Southern and eastern European groups, though the targets of virulent racist attacks, were not so indelibly marked. Perhaps of greater importance, however, was that the Japanese were economically successful, in both agriculture and business, and therefore were resented and easily targeted as competitive threats to the majority population. The Japanese fear was further fueled by the growing disrepute of Japan as an aggressive, militaristic international power during the early decades of the century. The Japanese military threat, covered sensationally by the press, served as cause for suspicion of Japanese Americans. Finally, that the Japanese were concentrated in California provided an opportune setting for a movement aimed at an Asian group. As one historian describes it, "California, by virtue of its anti-Chinese tradition and frontier psychology, was already conditioned to anti-Orientalism before the Japanese arrived" (Daniels, 1977:106).

Japanese Internment Fueled by the shock of Pearl Harbor and reinforced by the stereotype of Japanese treacherousness, the anti-Japanese movement culminated after the start of World War II with the internment of the Japanese population living on the West Coast. No other series of events in American ethnic history, other than the enslavement of African Americans and the genocidal measures employed against Native Americans, were comparable to this action. Two months after the Japanese attack on Pearl Harbor in 1941, President Franklin D. Roosevelt, acting on the recommendation of Secretary of War Henry L. Stimson, issued Executive Order 9066, authorizing the U.S. Army to remove any group viewed as a security risk. Almost all those of Japanese ancestry living in California, Oregon, Washington, and Arizona— 120,000—were affected. They were subsequently rounded up and sent to internment camps (called "relocation centers") in several Rocky Mountain states and in Arkansas. Having no time to prepare, most were forced to quickly liquidate their businesses, and many abandoned their homes and possessions, losing their life savings. They were permitted to carry with them only a single suitcase of personal belongings.

The Japanese Americans were held in the detention camps almost until the end of the war. In an ironic twist, thousands of incarcerated Japanese men were permitted to enter the U.S. Army as a special unit; in the European theater, where they fought the Nazis, they became one of the most highly decorated battle units in American military history.

Several aspects of the Japanese incarceration stand out. To begin with, most of the detained had been born in the United States and were American citizens. Hence, people were denied their fundamental civil rights solely on the basis of their ethnicity. In his final report, the commanding U.S. Army general responsible for carrying out the Japanese removal put the matter bluntly: "The Japanese race is an enemy race and while many second and third generation Japanese born on United States soil, possessed of United States citizenship, have become 'Americanized,' the racial strains are undiluted" (quoted in tenBroek, Barnhart, and Matson, 1968). Second, that the removal of the Japanese was racially motivated was demonstrated by the fact that Americans of neither German nor Italian ancestry were similarly treated despite the war against Germany and Italy. Moreover, after the war, it was acknowledged that the Japanese on the West Coast had never presented a military threat, and, in fact, their incarceration may have actually retarded the war effort. The emptiness of the security issue is reflected in the fact that the larger Japanese population in Hawaii was never removed despite its more strategic location. In the end, the exclusionist movement that had begun in California in the earlier years of the century had accomplished its objective—removal of the Japanese. The fear, suspicion, misperception, and envy, bred by decades of negative stereotypes, were released in a grievous deed, responsibility for which was shared by nativist groups, farmers, entrepreneurs, the military, the courts, and state and federal politicians.

Not until 1983 did a congressional committee formally acknowledge the political injustice that had been dealt the Japanese Americans, recommending a formal apology and compensation of $20,000 in damages to each survivor of the incarceration. Legislation to accomplish this was enacted in 1988.[8] The long-term effects of the internment camps, of course, could not be erased so easily, particularly for first- and second-generation Japanese Americans. For them, American democracy had failed to provide its promise of protection for all

---

[8]A similar internment was carried out by the Canadian government at the start of World War II against the 23,000 people of Japanese origin living in the West Coast province of British Columbia, 75 percent of whom were Canadian citizens (see Adachi, 1976; Daniels, 1981; Sunahara, 1980). In much the same fashion as the U.S. Congress, the Canadian parliament in 1988 moved to formally apologize to Japanese Canadians and to make a reparation payment of $19,325 to each survivor (Burns, 1988).

citizens. Even for the third and fourth generations, the World War II internment remains the paramount event of the history of the Japanese in America.

Although the internment of the Japanese is the most infamous case of direct institutional discrimination, all Asian groups have been subjected to various legal forms of discrimination at one time or another. Indeed, during the early years of the century, anti-Chinese and anti-Japanese measures were usually designed to apply to all Asians. In California, intermarriage of whites and Asians, for example, was barred, and residential segregation was enforced by restrictive covenants aimed specifically at Asians. Even federal laws were designed to handicap Asians. Perhaps most revealing is the fact that not until 1952 was federal legislation passed making all Asian immigrants eligible for U.S. citizenship (U.S. Commission on Civil Rights, 1988).

Contemporary Forms of Prejudice and Discrimination

Though in no way comparable in scope and vehemence to earlier hostility, anti-Asian incidents have become increasingly evident in recent years. Many of these are simple cases of verbal harassment or of indirect forms of institutional discrimination, but a growing number have been more overt and even violent (Noble, 1995; U.S. Commission on Civil Rights, 1986b, 1992; Zinsmeister, 1987). As with Jews in past eras, the very success of Asian Americans in education, business, and the professions has created a potential backlash of resentment from majority whites, who see themselves being surpassed by a nonwhite minority, and by other racial and ethnic minorities, who see themselves being leapfrogged in competition for the society's rewards. One of the most serious charges of discrimination emanating from Asian-American success has been leveled against some of the most prestigious colleges and universities in the country, who have been accused of limiting Asian enrollment (Biemiller, 1986; Johnson, 1989; Salholy, 1987; U.S. Commission on Civil Rights, 1992). Such measures are markedly similar to the quotas used by Ivy League schools in the 1920s to restrict the admission of Jews, who qualified in disproportionate numbers.

In addition to the success of Asian Americans relative to other ethnic groups, recent anti-Asian sentiments have been fueled by the economic prowess of Asian societies, especially Japan, and their ability to compete with the United States in the American as well as the global market. Business leaders, labor officials, and politicians have accused Japan of unfair trade practices, which, they claim, have created many of the ills of American industry. Advertisements encouraging people to buy American-made products have been tinged with anti-Asian sentiments, and many politicians have used economic nationalism as a primary component of their campaigns. These actions have redounded to Asian Americans (Mydans, 1991a).

Perhaps the most compelling illustration of the impact of economic nationalism on Asian Americans, and surely the most dramatic anti-Asian incident of recent years, was the killing in Detroit of Vincent Chin in 1982. While celebrating his impending wedding with friends at a bar, Chin, a young Chinese-American engineer, was verbally accosted by two white autoworkers, one of them unemployed at the time. Assuming that Chin was Japanese, they told him that he and one of his friends, also an Asian American, were "the reason we're all laid off." A fight ensued, after which the two chased Chin and clubbed him to death on the street. The attackers were fined $3,780 and given three years probation after one pleaded guilty and the other pleaded no contest to charges of manslaughter. The leniency of the sentence outraged the Asian-American community, and the U.S. Justice Department subsequently filed civil rights charges against one of the attackers. He was convicted by a federal jury of violating Chin's civil rights and sentenced to twenty-five years in prison. That conviction was later overturned on appeal, however, and he was acquitted of all charges, ending any further criminal prosecution (Moore, 1987).

In addition to the Chin case, which strongly aroused the ire of Asian Americans, two other incidents in the 1980s provoked similar reactions. One involved an altercation between students at a Davis, California, high school, which resulted in the fatal stabbing of a Vietnamese student by a white student. Like the Chin attackers, the defendant was convicted of the less serious charge of voluntary manslaughter and sentenced to a six-year prison term in a youth offenders' program. A second case involved a racially motivated assault on a Laotian immigrant by a white in Fort Dodge, Iowa. The attacker was sentenced to six months in jail, but the sentence was suspended. He was ordered to pay the victim's medical expenses and to write an essay of no fewer than twenty-five words (U.S. Commission on Civil Rights, 1986b).

Although these cases received much publicity, other less prominent but no less severe anti-Asian actions, many motivated by unadulterated racist sentiments, have been reported in recent years (Noble, 1995; U.S. Commission on Civil Rights, 1992). In Jersey City, New Jersey, for example, a community of several thousand Asian Indians had been subjected to constant harassment by white, black, and Hispanic youths. In 1987, the violence took on a far more serious cast as one Indian was beaten to death and another found beaten unconscious on a busy street corner (Marriott, 1987). Anti-Indian incidents continued to occur frequently in the Jersey City area for at least a year after the killing. In another incident, on Christmas Eve 1986 in Revere, Massachusetts, a working-class Boston suburb, arsonists burned down a house sheltering twenty-eight Cambodian immigrants. This was only one in a long series of anti-Asian attacks in Revere, including brick throwing, beatings, and vandalism (Zinsmeister, 1987). Similar incidents occurred elsewhere in Massachusetts, such as a fire set by arsonists that left 31 Cambodians homeless in Lynn in December 1988 (U.S. Commission on Civil Rights, 1992). In another incident, a young Vietnamese student was fatally beaten by five men in Coral Springs, Florida, in 1992. The student objected when one of the five made a disparaging remark about his Vietnamese ancestry, and in the ensuing beating, he was repeatedly punched and kicked in the head (*New York Times,* 1992).

These kinds of incidents have become increasingly common. Harassment of Asian Americans in recent years has been constant and has frequently involved physical attacks. A 1995 study indicated a 35 percent increase in anti-Asian hate crimes nationwide over the previous year, involving a range of acts from racist graffiti to beatings. In one case, the home of a Chinese city councilman in Sacramento, California, was firebombed (Noble, 1995). In fact, the extent of anti-Asian hate crimes is probably much greater than is actually reported. Asian Americans, especially recent immigrants, are reluctant to report crimes for a variety of reasons, including language problems, distrust of police, the desire not to cause problems, and shame at becoming a victim of crime (U.S. Commission on Civil Rights, 1992).

Asians and Other Ethnic Minorities In addition to harassment and hostilities issuing from the white majority, Asian Americans increasingly have become targets of violence by inner-city blacks and Latinos. As previously noted, many Asian immigrants turn to small business, and they often operate groceries, restaurants, liquor stores, and laundries in minority areas of large cities. In this, they are playing a middleman minority role. And as we saw in Chapter 2, such middleman business owners frequently become the targets of hostility from their customers. Tensions between Asian business owners and blacks in inner-city areas have been especially high in recent years. Animosities have arisen from the perception of blacks that Asian shopkeepers often deal with them disrespectfully and have no interest in the black communities in which they operate other than earning a profit. Hence, they see Asians as exploiters. Shopkeepers, on the other hand, complain of the frequency of harassment, theft, and vandalism of their businesses. In a widespread pattern, black boycotts of Asian-owned businesses have occurred in several cities in recent years, including New York, Philadelphia, Washington, and Los Angeles. During the Los Angeles riots of May 1992, shops owned by Koreans in the South Central district, where most of the disturbances occurred, seemed to be particular targets of looters and arsonists. Before the riots, relations between Korean shopkeepers and their black customers had been volatile for years and seemed to reach a high point when a Korean-born grocer shot to death a fifteen-year-old black girl whom the merchant had accused of stealing a bottle of orange juice. Black resentment grew even stronger when the woman grocer, who was convicted of voluntary manslaughter, was given only a probationary sentence (Mydans, 1992b). While discordant relations between Koreans and blacks remain most dramatic in Los Angeles as a result of the 1992 riots, a similar undercurrent is evident in the inner cities of other urban areas where Koreans operate small businesses serving minority residents (Chesley and Gilchrist, 1993; Goldberg, 1995; Wilkerson, 1993).

Friction between Koreans and blacks in business relationships should not be seen as unique, however. Similar animosities have arisen among blacks and merchants of other ethnic groups operating their enterprises in inner-city areas. In Detroit, for example, most inner-city neighborhood groceries and liquor stores are owned by Arab Americans or Chaldean Americans, who are seen in much the same way as Koreans in other cities (Dawsey, 1992; Gerdes, 1993). Moreover, conflict between neighborhood residents and shopowners of another ethnic group has a long tradition in many countries and has been evident in diverse parts of the world in recent times (White, 1993).

Moderating Levels of Prejudice and Discrimination Despite evidence of anti-Asian sentiments as well as disturbing acts of violence in recent years, there seems little chance of a return to the extreme levels of prejudice and discrimination that characterized the Asian-American experience of earlier decades. One of the more apparent differences today is that negative stereotypes of the past have given way to a generally positive collective image. Indeed, as noted earlier, Asian Americans are increasingly seen as groups to which others should look as role models. As sociologist Peter Rose has put it, "The pariahs have become paragons, lauded for their ingenuity and industry and for embodying the truest fulfillment of the 'American Dream'" (1985:182). Moreover, despite cases of blatant and direct discrimination, such as those described, most instances of discrimination today are more subtle and indirect.

One key indication of a lower rate of discrimination against Asian Americans than against other nonwhite groups today is their more rapid rate of residential integration, exceeding that for both blacks and Latinos. Even though their presence in American society is more recent than either of these two groups, the level of Asian–white residential segregation is comparable to that of Hispanics from whites and much lower than the segregation of blacks from whites (Langberg and Farley, 1985). The Vietnamese are an exception, however, in that they display a very high degree of residential segregation (Jiobu, 1988).[9]

---

[9]Eighteen percent of the respondents in a 1989 national survey indicated their aversion to having Vietnamese as neighbors. This was a higher negative response than for any other ethnic group though not significantly different from those given for Hispanics (16 percent), Indians or Pakistanis (15 percent), or Koreans (14 percent) (*Gallup Report,* 1989).

Chinese and Japanese Patterns

As with economic stratification, in a number of ways, assimilation for Asians in the United States takes on a dual pattern. On the one hand, within all Asian-American groups, those who are highly educated and who hold prestigious occupations are increasingly moving toward both cultural and structural assimilation. On the other hand, those less affluent and educated retain much of their traditional ways and have had more difficulty in moving beyond the ethnic enclave.

In examining assimilation among Asian Americans, it is important to consider that the bulk of the Asian-American population today is foreign-born. It is thus too early to marshal solid evidence regarding the rate and extent of assimilation for those who have not gone beyond the first generation. Most evidence must necessarily come from Chinese and Japanese Americans, both groups well into the third and even fourth generations. With these limitations in mind, let us look at the patterns and prospects of Asian-American assimilation.

Chinese-American Assimilation Given their longer history in the United States, we might first consider assimilation patterns among Chinese Americans. As in other societies to which they have migrated, the Chinese in America traditionally exhibited a resistance to assimilation. The characteristic features of Chinese ethnic communities worldwide seem to be social and cultural exclusiveness and a low level of absorption into the larger society (Lyman, 1968b, 1974; Purcell, 1980). Much of this self-imposed isolation is the product of Chinese community organizations, which in the new society reflect extended family patterns, known as lineages. Traditionally, much of Chinese social life in the United States was organized around various kin, clan, and secret societies.

Additional reasons have been suggested for the historical lag in assimilation among Chinese Americans. For one, the early demographics of this group—almost all males without wives and families—prohibited a second generation from establishing stable families and moving along an assimilation path comparable to that of other immigrant groups (Kitano and Daniels, 1995; Lyman, 1974). Moreover, the Chinese were prohibited from acquiring citizenship and thus remained legally marginal. Finally, the levels of prejudice and discrimination were unusually harsh, barring Chinese from closer relationships with members of the dominant society or from entering into mainstream institutions. For the first 100 years, then, the Chinese were a relatively isolated ethnic group in American society.

Despite past patterns of insularity, it appears that Chinese Americans are now moving rapidly in the direction of assimilation, both cultural and structural. This is evident at the secondary level in occupational and residential patterns (Jiobu, 1988) and at the primary level in the substantial rise in Chinese intermarriage. One study (Wong, 1989) indicates that currently over 30 percent of all Chinese marriages involve non-Chinese, most of them whites. This may be seen as solid evidence of the diminishing of social boundaries between the Chinese and the larger society.

Japanese-American Assimilation Among Asian-American groups, the Japanese have experienced a more profound assimilation than others. This is partly because they have resided in the United States for several generations and their numbers have not been swelled in recent years by new immigrants. Patterns of assimilation among Japanese Americans, however, differ on the basis of generation. Japanese Americans refer to each generation with a particular name. The first, or immigrant, generation is called *Issei;* the children of the Issei, or second generation, are referred to as *Nisei;* and their children, the third generation, are called *Sansei.* Japanese Americans are now entering the fourth generation, who are referred to as *Yonsei.*

The Issei represent a generation that did not assimilate. In this, they did not differ essentially from other immigrant groups. However, the much more severe prejudice and discrimination directed against them produced a greater disincentive to venture out of the ethnic milieu and attempt to interact with the larger society.

By contrast, Nisei, and especially Sansei and Yonsei, are highly assimilated culturally, as indicated by the very thorough adoption of dominant cultural traits (Levine and Rhodes, 1981; Onishi, 1995). This is not surprising in light of the high level of education among Japanese Americans of the second generation and beyond and the great strides each generation has made in socioeconomic advancement.

Regarding secondary structural assimilation, we have already seen the marked success of Japanese Americans in economic and educational realms of the society and their increasing movement into various institutional areas. Residential patterns of Japanese Americans also suggest a relatively fluid movement into the mainstream society (Kitano, 1976; Levine and Rhodes, 1981).

At the primary level of structural assimilation, there is strong evidence that Japanese Americans have begun to leave the ethnic community and forge close associations with non-Japanese. This is particularly apparent among those beyond the second generation (Levine and Rhodes, 1981; Onishi, 1995). The most telling indicator of primary structural assimilation is intermarriage, and in this, several studies have found significant increases, especially for the Sansei (Kikumura and Kitano, 1973; Kitano and Daniels, 1988; Levine and Rhodes, 1981; Montero, 1980; Tinker, 1973). Among his sample of Japanese Americans, Montero (1980) found that 40 percent among the Sansei had married a non-Japanese. This compared to only 10 percent of the Nisei and 1 percent of the Issei. Kikumura and Kitano (1973) found a similar pattern and concluded that Japanese Americans now choose marital partners as frequently from outside the ethnic group as from within. Data on out-marriage of Asian Americans in Los Angeles between 1975 and 1989 show a rate of over 50 percent for Japanese Americans (Kitano and Daniels, 1995).

In general, these studies indicate a progressive trend toward primary structural assimilation among Japanese Americans, and an increasing dilution of the ethnic community with each succeeding generation. In light of their rapid and extensive assimilation, some have even begun to question the ability of Japanese Americans to maintain a viable ethnic community into the next (Yonsei) generation (Montero, 1980; Onishi, 1995).[10] Others, however, maintain that despite very strong assimilation tendencies, Japanese Americans do not appear to be losing all cultural distinctiveness and identification (Kitano, 1976; Woodrum, 1981). One recent study, in fact, concludes that Japanese Americans appear to have been able to successfully combine high levels of structural assimilation with a strong retention of ethnic identity and group cohesion (Fugita and O'Brien, 1991).

The Future Path of Asian-American Assimilation

In examining issues of stratification, we emphasized the internal variability of the Asian-American population, and in looking ahead to future assimilation patterns, these differences once again must be kept in mind. As with stratification, the rate of assimilation among Asian Americans will differ from group to group. Vietnamese and other Southeast Asian groups, most recent and least prosperous, at the present time appear to be that part of the Asian-American population experiencing the most tortuous path toward assimilation. At the other extreme, Japanese Americans, one of the oldest and most prosperous among Asian-American groups, continue to move most directly and rapidly in that direction (Jiobu, 1988). For the new immigrants the occupational and educational achievements of many will enhance the prospects for both cultural and structural assimilation, particularly for those beyond the first generation.

We noted in Chapter 7 that the assimilation of Jews in the United States had proceeded as far as in any society where Jews had lived in large numbers. Yet, the religious divergence between Jews and non-Jews, we suggested, remains a

[10]Montero (1980) notes that, ironically, the remarkable upward economic mobility of the Japanese in America may lead to a diminution of the traditional ethnic values that helped to stimulate that advancement in the first place; hence, the loss of those values may create a leveling off of the achievement of future generations.

factor that continues to impede full assimilation of Jewish Americans. The same can be said for Asians, whose physical distinctiveness makes them even more indelibly conspicuous within the larger society. Because of this visibility, Americans continue to see Asian minorities as somehow not fully American, as "outsiders," even though their ancestors may have been in the United States for several generations, they may speak only English, and they have no more ties to Asian societies than Swedish Americans have to Sweden or Italian Americans to Italy. Because of their racial visibility, as Peter Rose explains, "No matter how adaptive in values and aspirations, no matter how similar to whites in mannerisms and actions, Asian Americans cannot be members of the majority" (1985:212). Ethnic identity and consciousness among Asian Americans, therefore, regardless of the extent of their assimilation, is not likely to fully erode.

Asian Americans as a "Model Minority"

The clear movement toward assimilation, demonstrated most powerfully by the extraordinary success they have achieved in the realms of education and the economy and by their comparatively problem-free adaptation to American society, has led in recent years to the popular characterization of Asian Americans as a model minority. Although the label is seemingly complimentary, many Asian Americans today see it as condescending and problematic in a number of ways. Some have pointed out that this new, positive image tends to gloss over the shabby treatment of Asians in the past and disguises the serious social and economic problems that continue to confront some elements of the Asian-American population. Obscured is the fact that some, particularly Southeast Asian refugees, remain poor, uneducated, and relatively unsuccessful.

In addition, the model minority image has become so well entrenched that it places tremendous social and psychological pressures on Asian Americans, especially young people. Students, for example, must do exceedingly well in their studies and in other pursuits in order to uphold this image and may even feel that they must outperform their non-Asian peers. This becomes especially problematic for recent Asian immigrants whose command of English may be minimal and who therefore suffer added burdens in an educational environment that is not attuned to the unique cultural needs of Asian students (Bernstein, 1988).

The model minority image has also seemed to create a new set of stereotypes, which though generally positive, nonetheless create the same kinds of group effects as other, negative, stereotypes: They induce the perception of commonality of traits among all individual ethnic group members. A Chinese-American schoolteacher expresses the frustration produced by the new stereotypes: "Years ago they used to think you were Fu Manchu or Charlie Chan. Then they thought you must own a laundry or restaurant. Now they think all we know how to do is sit in front of a computer" (quoted in *Time,* 1987). Moreover, as explained in Chapter 3, positive traits attributed to ethnic groups may easily be manipulated so that they become negative in content and thus serve as the rationales of prejudicial thought and discriminatory behavior. If Asian Americans today are seen as "brainier" and "more ambitious" than others, such seemingly positive traits may easily be transformed into negative ones like "shrewd" and "pushy," should Asians come to be seen as outcompeting the dominant group. Recall the similar circumstances regarding the rise of antisemitism in the United States in the late nineteenth and early twentieth centuries.

The characterization of Asian Americans as a model minority may also work to deflect attention from structural problems of other minority ethnic groups in the society. If Asian Americans can attain success despite their racial visibility, it is argued, even race cannot be viewed as an overwhelming handicap in American society. As Asian Americans are seen as hardworking, persevering, and highly motivated, the failure of other minority racial and ethnic groups to reach comparable levels of achievement is attributed to a lack of determination and hard work rather than to structural problems created by changing economic and social institutions.

Asian Americans in the Future

The term *Asian Americans,* as we have seen, is something of a misnomer given the cultural and physical variety of the population to which it refers. Yet, a social and political coming together of the various elements of this broad and extremely diverse ethnic category may become increasingly apparent in the future. The continued use of the collective *Asian American* by government, academic, and media institutions already reflects the perception of a more coherent ethnic population than actually exists. This is not unlike the use of the term *Hispanic American* or *Latino* to collectively describe the various groups of Latin American origins. And it is not unlike the process of ethnic group formation that characterized the American experience of many European groups of the earlier immigration periods. Only in the American context were Italians from various provinces in Italy, for example, designated and dealt with collectively as "Italians." Through this common treatment by Americans, an Italian-American group consciousness was engendered, and eventually Italians in America identified themselves primarily as "Italian," rather than "Calabrian," "Sicilian," or "Neapolitan." Though surely a long-term process for the various Asian-American groups, such a collective identity may emerge more clearly with each successive generation. Moreover, Asian Americans of diverse ethnic origins themselves may increasingly recognize this collective designation as a useful political strategy, just as Latinos are beginning to do (Espiritu, 1992). As separate and disperse groups, none can exert the kind of political leverage at the national or even local level that a more consolidated position may produce.

SUMMARY

Asian Americans are a diverse collection of distinct ethnic groups, the largest among them Chinese, Filipinos, and Japanese. The Asian immigration to the United States consists of two periods. The first began in the midnineteenth century with the immigration of Chinese manual laborers. Exclusionary measures beginning in 1882 halted further Chinese immigration and provided an opening for Japanese workers to replace them in the early part of the twentieth century. Discriminatory laws brought further Asian immigration to a virtual halt in the 1920s. The second period extends from 1965, when immigration laws were liberalized, to the present. The vast majority of Asian Americans today are first-generation immigrants who stem from this second wave. Many of those of the second wave have come to the United States with much higher occupational skills and more education than earlier immigrants. Both immigrant waves settled primarily on the West Coast, especially California, where Asian Americans remain heavily concentrated.

Asian Americans, as a whole, are at or near the top of the stratification system in terms of income, education, and occupation. There are important differences among the various Asian groups, however. Whereas Japanese and Asian Indians rank among the highest income earners, the Vietnamese and other Southeast Asian groups, most of whom have come as political refugees, are among the most impoverished groups in the United States. There are also internal differences in class within each Asian-American group.

High levels of prejudice and discrimination have marked the Asian-American experience from the outset of immigration. Exclusionary measures designed to bar the entry of Asians into the United States were enacted in the nineteenth and twentieth centuries, and basic civil rights were denied Asians in Western states until well into the twentieth century. In recent years, Asian Americans have been portrayed as a model minority because of their exceptional economic and educational attainments. Nonetheless, remnants of anti-Asian prejudice and discrimination remain evident. More serious is the potential backlash of anti-Asian sentiment that may emerge from these groups' economic success.

It is still too early to gauge the extent of assimilation among Asian Americans because most are first-generation immigrants. However, high levels of cultural assimilation characterize those groups who have entered the third and fourth generations, specifically, the Japanese and Chinese, and there is also evidence of increasing rates of both primary and secondary structural assimilation among them.

Suggested Readings

Bonacich, Edna, and John Modell. 1980. *The Economic Basis of Ethnic Solidarity: Small Business in the Japanese American Community.* Berkeley: University of California Press. A good account of the way in which Japanese Americans carved out a place for themselves in the California economy. Also contains a valuable explication of the role of small business among ethnic minorities.

Caplan, Nathan, John K. Whitmore, and Marcella H. Choy. 1989. *The Boat People and Achievement in America.* Ann Arbor: University of Michigan Press. Along with its companion work, *Children of the Boat People: A Study of Educational Success* (1991), explores why Southeast Asian refugees have excelled educationally and occupationally despite their language and other handicaps on arrival in the United States.

Espiritu, Yen Le. 1992. *Asian American Panethnicity: Bridging Institutions and Identities.* Philadelphia: Temple University Press. Describes the process by which an inclusive Asian-American ethnic group may be forming, one that transcends its members' specific national origins.

Kim, Illsoo. 1981. *New Urban Immigrants: The Korean Community in New York.* Princeton, N.J.: Princeton University Press. Examines the social and economic institutions of the Korean ethnic group in the United States and the entrepreneurial role Koreans play in many large cities.

Kwong, Peter. 1987. *The New Chinatown.* New York: Noonday Press. The author describes the changing makeup and social institutions of New York's Chinatown by drawing on his observations as a long-time resident.

Lyman, Stanford M. 1974. *Chinese Americans.* New York: Random House. An examination of the Chinese-American experience, with good discussions of the development of Chinese community organizations and the historical patterns of discrimination, both formal and informal, suffered by the Chinese in the United States.

Min, Pyong Gap (ed.). 1995. *Asian Americans: Contemporary Trends and Issues.* Thousand Oaks, Calif.: Sage. Describes the history, current demography, sociological features, and patterns of assimilation of the major Asian-American groups.

Pido, Antonio J. A. 1986. *The Pilipinos in America: Macro/Micro Dimensions of Immigration and Integration.* New York: Center for Migration Studies. Analyzes the immigration experience of Filipinos to the United States and the integration process that follows.

Takaki, Ronald. 1989. *Strangers from a Different Shore: A History of Asian Americans.* New York: Penguin. Traces the history of Asian Americans beginning with the first immigration of workers in the nineteenth century up to the most recent wave of immigrants.

CHAPTER

11

CURRENT AND FUTURE ISSUES OF RACE AND ETHNICITY IN THE UNITED STATES

AMONG MULTIETHNIC SOCIETIES of the modern world, the United States seems to fall at a midpoint along a scale of ethnic conflict and inequality. As we will see in Part III, in some societies, ethnic divisions and stratification are far more severe. But understanding that other societies may suffer even greater problems that derive from ethnicity should not cause us to lose sight of the fact that the United States has been and remains a society in which racial and ethnic inequality is a fundamental characteristic. That inequality, in turn, produces ethnic conflict, ordinarily subdued and contained, but at times blatant and explosive.

Today, three interrelated issues seem to clearly stand out regarding American race and ethnic relations: (1) How should public policies address the continued economic gap between Euro-American and most Asian-American groups on the one hand and African-American, Hispanic-American, and Native-American groups on the other? (2) How is the influx of large numbers of new immigrants and the creation of a more pluralistic ethnic mix affecting the society today, and how will it affect the future of race and ethnic relations? (3) Should the end product of interethnic relations be some form of assimilation, in which diverse groups become more culturally alike and socially integrated, or some form of pluralism, in which they maintain or perhaps increase their cultural differences and social boundaries?

THE CONTINUING GAP BETWEEN EURO-AMERICANS AND RACIAL-ETHNIC MINORITIES

Let us briefly review the American ethnic hierarchy that we have looked at in the previous five chapters. Roughly, the hierarchy was seen as divided into three comprehensive ranges: a top range of white Protestants, essentially the dominant ethnic group; a second, intermediate range made up of white Catholics of various national origins, Jews, and many Asians, for whom ethnicity remains of significance but decreasingly so; and a bottom range, comprised of racial-ethnic minorities, for whom ethnicity continues to render great effect in the distribution of societal rewards and in shaping patterns of social interaction. The most important division within this hierarchy is between the bottom range and the other two.

The third range of groups continues to maintain a collective place at the bottom of the society's economic and political hierarchies. This does not mean that many individuals in each of these groups have not demonstrated substantial upward mobility, particularly in the past three decades; indeed, our discussions have shown that the resistance of whites to nonwhite economic, political, and social advancement has greatly diminished in the past three decades, and as a result, the economic and political status of large numbers of racial-ethnic minorities has risen significantly. But in looking at their collective status vis-à-vis Euro-Americans or many segments of the Asian-American population, members of these groups remain disproportionately among the poor and the powerless. Moreover, the social boundaries separating the third range of groups from the first and second give little indication of dissolving. Members of third-range groups continue to encounter higher levels of prejudice and discrimination, and the extent to which they have been afforded entry into the full range of societal institutions remains lower.

Efforts to deal with this persistent economic gap and the social distance between majority and minority ethnic groups have, in recent decades, given rise to a number of controversial public policies.

Compensatory Policies

Out of the atmosphere of the civil rights movement and the new pluralism of the 1960s and 1970s, the federal government undertook sweeping measures to foster political and economic equity among ethnic groups. Specifically, these policies were aimed at improving the social standing of the long-term victims of discrimination—blacks, Latinos, and American Indians. Later, women were added as a target group. These new public policies, designed as they were to raise group positions, marked a significant departure from the traditionally understood role of government in the area of civil rights.

In the past, government's function had been to ensure that everyone was afforded equal opportunities in work and education, regardless of ethnicity. Because this guarantee had been violated for nonwhites, legislation was enacted beginning in the 1960s that was designed to protect members of minority groups from discrimination in schools, workplaces, and other societal institutions. But the question now raised was whether eliminating discriminatory practices alone could counteract the effects of past discrimination. Given the generations of denied opportunities, was it fair to expect blacks and other racial-ethnic minorities to compete on an equal basis with the white majority? How could these groups ever catch up if they entered the competition burdened by decades of imposed disabilities?

The Use of Affirmative Action In response to this dilemma, a series of compensatory measures and programs evolved, grouped under the rubric of "affirmative action," that was intended to advance the economic and educational achievement of the minorities that had been most severely and consistently victimized by past discrimination. That government was now acting to guarantee not simply equality of opportunity but also equality of result evoked a strong public outcry and continues to generate much controversy.

As affirmative action policies unfolded in the 1960s, they stipulated that those doing business with the federal government (universities as well as businesses) were required to take steps to increase their minority representation and to establish goals and timetables to meet that objective. The aim was basically to affect minority employment and student admissions. With more widespread and stringent application in the 1970s, affirmative action policies grew increasingly unpopular and created a backlash of displeasure among majority whites.

Proponents of affirmative action have argued that advancing the position of minorities through the use of goals and preferential hiring practices is necessary if the victims of past discrimination are eventually to attain equity with the majority (Kennedy, 1994; Livingston, 1979; W. Taylor, 1995). Even if direct forms of discrimination no longer prevail, they point out, the indirect and institutional forms continue to perpetuate nonwhite (and female) disadvantages in the labor market and in higher education. Simply protecting minority individuals against ethnic or sex discrimination, therefore, is inadequate by itself. Hiring employees or admitting students without regard to ethnicity or sex will automatically preserve the disproportionate representation of white males because they enter the competition with background advantages accumulated over many generations. Artificial incentives for minorities are needed temporarily, therefore, until the opportunity structure is made more truly equitable.

Those who oppose these programs or who strongly criticize the way they have been carried out argue that, in effect, affirmative action has become a kind of reverse discrimination in which those previously discriminated against are given preference over others merely on the basis of ethnicity (Glazer, 1975; Yates, 1994). Hence, they maintain, the very objective intended—reducing ethnic discrimination—has been undermined, the victims now being white males. Government is seen as re-creating racial and ethnic categories that had rightfully been abandoned. The effect of affirmative action in many cases has been to create quotas favoring minorities, a particularly sensitive issue for certain groups, such as Jews, who themselves were the past victims of quotas limiting their entrance into prestigious colleges and professional schools.

Opponents of affirmative action have pointed out that preferential measures shifting emphasis from equality of opportunity to equality of result create aspects of corporate pluralism in which social benefits are distributed on the basis of group membership, not individual merit.[1] Hence, less qualified minority persons may be promoted over better qualified majority persons. Moreover, the targeted groups of affirmative action, they argue, are stigmatized because of their special treatment, thus producing negative social and psychological effects on those who are supposedly the beneficiaries of these programs (Sowell, 1990). Shelby Steele (1990), for example, has suggested that affirmative action programs create a kind of implied inferiority. Whites commonly view blacks as having acquired their positions on the basis of special preference and thus as less than qualified or competent. This not only fuels negative stereotypes by whites, argues Steele, but also creates self-doubt among blacks and other minorities who are the beneficiaries of affirmative action.

In addition to questions of reverse discrimination and effects on employment and educational qualifications, affirmative action programs have been criticized as too sweeping in application and therefore unable to distinguish from among the various targeted minorities those who are truly the past or present victims of discrimination. For instance, although Latinos are covered under the principles of affirmative action, we have seen in Chapter 9 that this broad ethnic category comprises several disparate components, each with different American experiences. Should Cubans, who have not encountered discrimination in work and education, be entitled to the benefits of these programs in the same way as Mexicans in the Southwest, who have been severely discriminated against? Such problems suggest that minority group membership is no longer unambiguous. Moreover, many have pointed out that the major recipients of affirmative action benefits have been middle-class racial and ethnic minorities, as well as women,

[1]Ethnic pluralism as either "liberal pluralism," in which ethnicity does not enter into legal or political matters and does not define a person's rights or duties, or as "corporate pluralism," in which group affiliation is very much a determinant of rights and benefits, is discussed fully by Gordon (1975, 1981).

not the "truly needy," that is, those in disadvantaged class positions. Filtering out those truly deserving of compensatory benefits has therefore become more complicated not only between various groups, but within them as well.

The issue of affirmative action continues to generate heated debate, with advocates of these programs countering their detractors on various points. As to the lack of qualifications of those employed or admitted through preferential policies, for example, proponents argue that "qualifications" can be variously interpreted, no matter how seemingly objective and valid tests or other sorting mechanisms may seem. Placement tests are often biased in favor of those with a white middle-class background and in any case are never wholly adequate in measuring one's potential on the job or in the classroom. Moreover, merit, they point out, has never really been the sole criterion used in filling occupational and educational positions. On the matter of quotas, it is argued that the objective of affirmative action is to facilitate the entrance of minorities into various institutions, not to keep them out, as was the purpose of earlier discriminatory quotas. Furthermore, they point out, preferential treatment is already given certain groups, such as veterans or athletes, in employment or education. And, while opponents see racial minorities being stigmatized by affirmative action, defenders argue that their absence in jobs and schools would create an even greater stigma (Kennedy, 1994).

The Legal Issues of Affirmative Action The legal questions of affirmative action programs have involved their scope and intent as well as whether they are racially discriminatory and therefore in violation of the Fourteenth Amendment of the Constitution. They have been tested in the courts on numerous occasions, but judicial decisions have been inconsistent and have not fully clarified these issues.

As to their scope, an early case of great significance was *Griggs* v. *Duke Power,* heard by the Supreme Court in 1971. The specific question addressed in this case was whether tests given by companies to potential employees could be required if it was believed that they had adverse effects on minority applicants. This went beyond the Fifth and Fourteenth Amendments stating that only *intentional* harmful actions against people on the basis of race or ethnicity violated their constitutional rights. The Court ruled that company tests as a qualification for employment could be invalidated unless they could be shown to be directly related to job performance or were a business necessity. That is, the use of practices such as testing was now severely limited if their *effect*—not simply their intent—was to eliminate minority applicants. If the number of minorities in a company's work force, for example, did not reflect the proportion of the minority population in the area where the company was located, that fact might be sufficient to demonstrate the effect of such indirect discrimination or, as the term came to be known, "disparate impact." Thus, in this case the Court extended affirmative action to address instances not only of overt, intentional discrimination, but institutional discrimination as well. In 1989 in another ruling, however, the Court weakened the business-necessity requirement and shifted the burden of proof for demonstrating such indirect discrimination from the employer to the applicant (Ezorsky, 1991).

As to whether affirmative action programs are themselves discriminatory, the *Bakke* case is perhaps the most significant, particularly as affirmative action relates to higher education. Allan Bakke was denied admission to the medical school of the University of California at Davis even though his entrance qualifications exceeded those of minority applicants who had been admitted under a special program. As a means of increasing the number of minority students, 16 of 100 places in each entering class had been reserved for minority applicants. Bakke maintained that he had been discriminated against on the basis of ethnicity and sued the university for admission. The case eventually reached the U.S. Supreme Court, and in 1978, by a five-to-four decision, the justices ruled in favor of Bakke. Their decision was ambivalent, however, leaving the way open for schools to establish goals for meeting an ethnic balance. In effect, the Court ruled that quotas (which, in this case, the university was found to have used) were illegal but that the use of race as a criterion of preference was legitimate as long as it was one among many criteria and that its purpose was to create a more ethnically balanced student body.

In another case, *Weber* v. *Kaiser Aluminum,* affirmative action policies that affected employment were tested in the Supreme Court one year after the *Bakke* decision. Brian Weber, a white worker, contended that he had been passed over in a company training program in favor of blacks with less seniority. As in Bakke's case, a fixed number of places had been reserved for underrepresented blacks. Here the Court ruled with less ambiguity than in *Bakke,* declaring that employers seeking to increase minority representation in certain occupational areas not only could consider race as a criterion of hiring and promotion but also could do so even if such efforts involved the use of quotas. A similar decision was rendered in 1984 in a case involving the promotion of blacks as officers in the Detroit Police Department. Both of these decisions were read as victories for affirmative action. In another 1984 case, however, the Court seemed to move in the opposite direction, ruling that employer layoffs had to be made according to applicable seniority rules, even if increases in minority employment from past affirmative action were wiped out in the process.

The Politics of Affirmative Action The Reagan administration, unlike previous administrations, both Republican and Democrat, declared its opposition to the use of affirmative action measures and throughout the 1980s sought to subvert them through actions (or inactions) of the Justice Department. "Opposition to race-based affirmative action," notes Thomas Edsall, "became for the Reagan regime a matter not only of principle and of policy, but of partisan strategy" (Edsall and Edsall, 1992:187). During the early 1980s, the Supreme Court, despite its inconsistent rulings, generally resisted these destructive efforts, upholding affirmative action programs in a number of decisions. With the appointment of additional conservative justices by Reagan, however, the Court by 1989 was now philosophically poised to reduce the impact of affirmative action and to weaken civil rights measures. In several decisions, it clearly moved in that direction.

The Reagan administration's position on affirmative action seemed indicative of a general shift in presidential policy regarding ethnic minorities and civil rights issues (Edsall and Edsall, 1992; Miller, 1984). Though its attempts to undermine federal policy were not entirely upheld either in the courts or by Congress, the administration seemed intent on rescinding the federal government's role in advancing and protecting minority rights. George Bush's presidency seemed to sustain the efforts of Ronald Reagan to pull back from the application of affirmative action. Although not driven by the ideological fervor of the Reagan presidency, Bush, like Reagan, employed opposition to affirmative action as part of a strategy through which Republicans could galvanize white electoral support.

Since the inception of affirmative action in the 1960s, Democratic and Republican parties had generally aligned themselves on opposite sides of the issue (Edsall and Edsall, 1992; Hacker, 1992). The election of a Democratic president in 1992 therefore augured a more sympathetic presidential view of minority concerns. While Bill Clinton did, in fact, seem to uphold a commitment to the principle of affirmative action, a growing public disaffection with these policies and the election in 1994 of an extremely conservative Republican Congress, most of whose members strongly opposed affirmative action, combined to create the most serious movement to significantly modify or entirely abolish these policies since their inception in the late 1960s.

Public sentiment regarding affirmative action measures had always been ambivalent. In the 1990s, however, the antipathy to affirmative action among whites grew increasingly popular and vocal. A 1995 national poll revealed that 79 percent of white Americans opposed racial preferences in employment or college admissions (Fineman, 1995).[2] Quite simply, whites had come to see such measures as "reverse racism" and in violation of the American creed. So negatively had they come to see affirmative action that, as political scientists Paul Sniderman and Thomas Piazza concluded, it had "led some whites to dislike blacks—an ironic example of a policy meant to put the divide of race behind us in fact further

[2]The poll revealed that although racial minorities were not as strongly opposed to such preferences (46 percent disapproved), they were by no means overwhelmingly supportive either.

widening it" (1993:109). Furthermore, a widespread public perception, particularly on the part of whites, was that by the 1990s, an equal opportunity structure had been created, obviating the need for compensatory measures.

Public officials in the mid-1990s voiced uncertainty about how, and whether, affirmative action could be sustained. A Supreme Court decision of 1995 reflected that continued uncertainty. In June of that year, the Court issued an important ruling in the case of *Adarand Constructors* v. *Peña* that placed tight limits on affirmative action. Though not going so far as declaring them unconstitutional (two of the justices did, in fact, support that position), it struck down federal contracting programs that favored firms owned by racial minorities (so-called "set asides"), ruling that such programs had to be narrowly tailored to address identifiable past discrimination.

A number of other legal rulings dealt further blows to affirmative action. A federal appeals court in 1994 found that the University of Maryland's scholarship program for black students violated the rights of students of other races (a ruling that the U.S. Supreme Court later upheld), and in 1995, the governor of California declared an end to the use of preferential measures in admissions to the University of California system. In 1996, a U.S. Court of Appeals rendered a decision that appeared to rescind the *Bakke* decision allowing public universities to consider race as one criterion in choosing students. The judges ruled that the University of Texas Law School could no longer use race as a factor in its admissions decisions simply to increase racial diversity. Legal experts and educators agreed that, if upheld, the eventual impact of this decision would be far-reaching. All of these measures seemed indicative of a growing trend toward severely limiting or perhaps even abolishing altogether affirmative action policies.

The Future of Affirmative Action As with racial attitudes generally (discussed in Chapter 8), although whites today seem to generally accept the appropriateness of compensatory policies for racial minorities *in principle,* their support is not as unambiguous when efforts at applying those principles, in the form of concrete measures and programs, are made. It is important to consider, however, that the public response to affirmative action is not so clear-cut as one might expect. To see the issue as a fundamental split between whites and racial minorities — whites oppose affirmative action, racial minorities support it — is too simple. National surveys have consistently indicated that a majority of Americans, regardless of race or ethnicity, support preferential policies as long as they are not perceived as "rigid quotas" (Harris, 1991a, 1991b; Morin, 1995; *USA Today,* 1995a). Even a significant percentage of blacks are opposed to affirmative action when it takes such form. As Sniderman and Piazza found in their surveys, substantial numbers of whites are prepared to support policies that improve the status of blacks, but "The idea of quotas and preferential treatment is the reef on which affirmative action founders" (1993:130). Racial attitudes, then, are not necessarily the critical, much less the only, factor comprising this issue. People may oppose or feel resentment toward affirmative action policies not because they are antiblack or wish to keep things as they are (surely many *do* oppose them for these reasons), but because they see them as fundamentally in violation of the value of fairness.

How they are to be made more fair has been subject to similar debate. Some maintain that instead of race or ethnicity, social class should be the major criterion by which people are deemed eligible for compensatory advantages (Kahlenberg, 1995) or that policies should be applied in a "race neutral" fashion (Wilson, 1994). Others, however, maintain that such measures bring with them the same kinds of inherent problems as those based on race and ethnicity, that is, issues of determining who should be eligible and how they are to be chosen (Hacker, 1994; Kinsley, 1991).

Another confounding aspect of affirmative action concerns the very objective of such policies. Originally, they were intended to provide a temporary advantage to those racial minorities who had been the historical victims of systematic discrimination. As they evolved, however, their purpose was transformed into an effort to create schools and workplaces that resembled proportionately the racial and ethnic makeup of the society, state, or community. The question, then, is whether their purpose is to help bring about greater equity among a society's diverse groups or to create more ethnic diversity within the society's major institutions. Efforts at attaining one of these objectives do not necessarily correspond to attainment of the other.

Whether affirmative action policies will continue to be applied by governments, corporations, schools, and the media is no longer certain. It is unlikely that preferential measures in some form will be completely abandoned in the near future, whether dictated by government policy or not. Corporations, for example, recognize that recruiting more minority managers is simply good business in an increasingly ethnically diverse society (Glater and Hamilton, 1995). But rising political and social pressures seem to make it even more likely that the use of affirmative action will be reduced sharply and its shape reformed in a basic way.

Affirmative action, in sum, has emerged as one of the most vexing issues of race and ethnicity in America. Yet, it is important to consider the widespread use of such measures in other multiethnic societies. Preferential policies designed to bring about more equity among diverse groups have been adopted in countries as divergent as Australia, India, Canada, and Malaysia. Thus, we should not think that the problems revolving around this issue are unique to the United States. As countries everywhere are transformed into multiethnic societies, ameliorating the inequitable distribution of societal resources among various ethnic groups has become an inescapable predicament.

The prominence of ethnic issues in the United States today stems not only from the continuing socioeconomic gap between whites and nonwhites but also from the large-scale immigration of diverse peoples in the past three decades. These latest immigrants have begun to have a significant impact on the ethnic flavor of the society and have introduced new problems and questions of ethnic relations. The United States, it should be recalled, experienced two major waves of voluntary immigration: the old immigration of the midnineteenth century and the new immigration of the late nineteenth and early twentieth centuries. The latest, third, wave, beginning in the mid-1960s, might be called the "newest immigration."

The New Ethnic Groups

The newest immigration has been one of the most consequential for American society from the standpoint of numbers and group composition. So great have the numbers been, in fact, that the flow has reached proportions that resemble the classic, or new, immigration period. A few statistics illustrate its magnitude. During the three decades between 1960 and 1990, 15 million people either came to the United States as legal immigrants or were granted permanent residence. During the 1980s alone, 8.3 million immigrants, both legal and illegal, entered the society. This approached the 8.8 million figure that was reached during the first decade of the 1900s, when, as can be seen in Figure 11.1, immigration to the United States was at its peak.[3] By 1994, the foreign-born made up almost 9 percent of the society's total population, a proportion higher than at any time since World War II and nearly double the percentage in 1970.

During the 1980s, net immigration (a figure computed by subtracting the number leaving from the number entering) averaged around 600,000, representing almost one-third of U.S. population growth. A revision in the immigration laws in 1990 raised the legal level of immigration to 700,000 annually. A large portion of this increase was the result of a provision for raising the number of employment-based immigrants, to be admitted on the basis of needed skills. Moreover, additional immigrants may be admitted as political refugees. These figures also do not take into account illegal immigration, the extent of

[3]The proportion of immigrants to total population, however, was much greater in the 1900s than it is today. The average annual number of immigrants admitted from 1981 to 1990 was 3.1 immigrants per 1,000 U.S. residents; from 1901 to 1910, the annual rate was three times as great. Further, the percentage of foreign-born as part of the total population was much greater in the earlier part of the century.

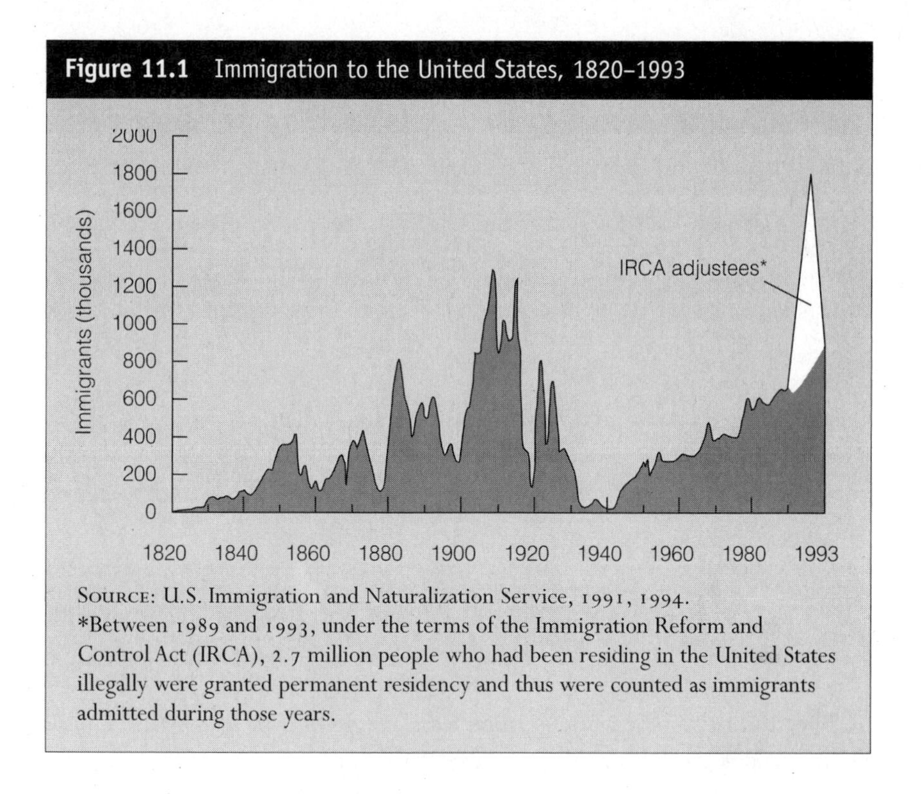

SOURCE: U.S. Immigration and Naturalization Service, 1991, 1994.
*Between 1989 and 1993, under the terms of the Immigration Reform and Control Act (IRCA), 2.7 million people who had been residing in the United States illegally were granted permanent residency and thus were counted as immigrants admitted during those years.

which can only be guessed. Estimates range from 100,000 to 300,000 illegal immigrants entering annually and anywhere from two to four million currently residing in the United States (Bean et al., 1990; Heer, 1990; U.S. Immigration and Naturalization Service, 1994). Most are from Mexico, but illegal immigrants come from virtually all countries. Estimates of the number of undocumented Irish immigrants in the United States, for example, range from 40,000 to 110,000 (Egan, 1990). Immigration in all forms, then, will surely continue to play a large role in U.S. population growth (Bouvier, 1992).

But numbers alone do not tell the whole story of the current period of American immigration. Perhaps more important, the character of immigration is radically different from that of past eras. The vast majority of immigrants throughout the nineteenth and early twentieth centuries were from European societies. Today, most are from Latin America and Asia (Table 11.1). This has been the result of a fundamentally changed immigration policy. In 1965, legislation was enacted abolishing the discriminatory quota system that since the late 1920s had favored northwestern European groups and severely restricted the entrance of others. The passage of this revised immigration law may be as significant in its long-range impact on ethnic issues as the legislative and judicial acts of the 1950s and 1960s removing discriminatory barriers against blacks.

Table 11.1 Legal Immigrants Admitted to the United States, by Region of Birth

REGION	1820–1860	1861–1900	1901–1920	1921–1960	1961–1970	1971–1980	1981–1990
Northern and Western Europe	95%	68%	41%	38%	18%	7%	5%
North America	3	7	6	19	12	4	2
Southern and Eastern Europe	—	22	44	20	15	11	5
Asia	—	2	4	4	13	35	37
Latin America	—	—	4	18	39	40	47
Other	2	1	1	1	3	3	4
Total	100	100	100	100	100	100	100

SOURCE: U.S. Immigration and Naturalization Service, 1991.

Since the 1970s, the leading countries of origin of legal immigrants to the United States have been Mexico, the Philippines, Korea, and four small Caribbean nations—the Dominican Republic, Jamaica, Haiti, and Guyana (a South American country that considers itself Caribbean in its economic and social structure). But the sheer diversity of non-European immigrants is quite astonishing. Among the Asians, for example, are Filipinos, Koreans, Chinese, Vietnamese, Cambodians, Laotians, Hmong, Indians, and Pakistanis. From the Middle East have come Palestinians, Iraqis, Iranians, Lebanese, Syrians, Egyptians, and Israelis. Those from Africa include Somalis, Ethiopians, and Nigerians. And from Latin America, the national origins of the immigrants cover the entire Central and South American regions, in addition to the Caribbean. Table 11.2 shows the leading countries of origin of immigrants admitted in 1993.

What has prompted this newest large-scale immigration to the United States? Essentially the same push–pull factors motivating past immigration to America have been operative for the newest groups. Although many have come as political refugees (mostly from Southeast Asia, Cuba, and Central America), the majority are responding to the population and economic pressures of their societies of origin. Most, in other words, seek economic betterment and improved social conditions. A national survey asked foreign-born adults living in the United States what they preferred about America in comparison with their homeland. Some answered political freedom, fair treatment under laws, safety

Table 11.2 Immigrants Admitted to the United States from Leading Countries of Birth, 1993

COUNTRY	NUMBER	PERCENT
Mexico	125,561	13.9
China	65,578	7.3
Philippines	63,457	7.0
Vietnam	59,614	6.6
Soviet Union	58,571	6.5
Dominican Republic	45,420	5.0
India	40,121	4.4
Poland	27,846	3.1
Other countries	418,124	46.2
Total	904,292	100.0

SOURCE: U.S. Bureau of the Census, 1995b.

from crime, or moral values. But the most common answer was job opportunities (Saad, 1995).

Contemporary immigration to the United States, as well as to other industrialized societies of North America, Europe, and Australia, then, accelerated starting in the 1960s as a consequence of the widening gap between rich and poor nations. The economies of the latter, many of them newly independent states, were generally incapable of supporting rapidly growing populations, thereby creating a migration push. Most of the world's population growth has occurred since the end of World War II, and the overwhelming majority of that growth has taken place in the less developed countries. Moreover, the gains in economic development made by many of these countries during the 1960s and 1970s slowed or reversed in the 1980s due to economic recession, growing debt, and internal political conflicts. These conditions induced further pressures to migrate to the wealthier countries, like the United States, whose industrialized economies promised better employment opportunities and a superior quality of life. Ironically, much of the migration to the United States from such societies has been induced by American political and economic influences. As American-based multinational corporations bring capital and technology to these developing economies, new wealth is created, but it remains highly maldistributed. Thus, the expectations of work and consumption opportunities are frustrated, leading to a natural gravitation to the United States, which promises to deliver these opportunities. "It is only natural," notes Alejandro Portes, "that many trek North in search of the means to acquire what transnational firms and the mass media have so insistently advertised for years" (1979:434). The newest immigration to the United States, then, must be seen as part of the internationalization of capitalist economies and the resultant globalization of the work force (Castles and Miller, 1993; Chaney, 1979).

Like so many American immigrants of past generations, many today come with few expectations of remaining permanently in the United States. A great number do leave, and many (particularly from Mexico) engage in a continual back-and-forth movement. But also like their predecessors, many who at first see their immigration as temporary become firmly ensconced in their new society. Thus, as in the past eras of European immigration, ethnic enclaves have flourished in cities that attract large numbers of new immigrants. Much as New York, Chicago, and other cities of the East and Midwest served in the past as meccas of immigration, cities of the Sun Belt—Los Angeles, Miami, Houston—are the focal points of the newest immigration. The immigrant population in Los Angeles County, for example, is almost a third of the total population, equivalent numerically to the foreign-born population of New York City during the 1920s (Muller, 1993). Reflecting the ethnic diversity of the metropolitan area is the fact that students in the Los Angeles School District speak more than eighty languages.

Sun Belt cities, however, have by no means monopolized the newest immigration. New York remains a traditional port of entry for new ethnic groups. The city has absorbed especially large numbers of newcomers from the Caribbean basin, including English-speaking West Indians from Jamaica, Trinidad and Tobago, Barbados, and Guyana; French- and Creole-speaking Haitians; and Spanish-speaking groups from the Dominican Republic, Cuba, and Colombia. Hispanics from virtually every Caribbean, Central American, and South American country comprise the largest component of New York's newest immigrants, but large numbers of Asians have also settled in the city, including Chinese, Koreans, Filipinos, and South Asians from India, Pakistan, and Bangladesh. Even Europeans entered in significant numbers in the late 1970s and early 1980s, most notably thousands of Soviet Jews (Foner, 1987; Muller, 1993; Winnick, 1990). So great has been the immigrant influx that Euro-Americans now constitute less than half of the city's total population (U.S. Bureau of the Census, 1995b). Other eastern and midwestern cities have also attracted large new immigrant populations. Detroit, for example, is home to an estimated 200,000 Arabs; in suburban Dearborn, Arabs now make up over 20 percent of the city's population, and an estimated 45 percent of the students in its schools are of Arab origin (Cook, 1989; Hamada, 1990).

Table 11.3 Ethnic Composition of the Largest U.S. Metropolitan Areas

METROPOLITAN AREA	WHITE	BLACK	HISPANIC	ASIAN	AMERICAN INDIAN
New York	62.9%	17.7%	14.6%	4.6%	0.2%
Los Angeles	48.8	8.5	32.9	9.2	0.6
Chicago	66.8	19.0	10.9	3.1	0.2
Washington–Baltimore	66.9	25.2	3.9	3.7	0.3
San Francisco	62.4	6.6	15.5	14.8	0.7
Philadelphia	75.6	18.4	3.8	2.0	0.2
Boston	88.1	4.8	4.4	2.5	0.2
Detroit	75.7	20.5	2.0	1.4	0.4
Dallas–Fort Worth	70.1	14.0	13.0	2.4	0.5
Houston	57.6	17.9	20.7	3.5	0.3
Miami-Ft. Lauderdale	46.6	18.5	33.3	1.4	0.2
Seattle	85.1	4.5	3.0	6.1	1.3
Atlanta	70.8	25.2	2.0	1.8	0.2
Cleveland	81.3	15.6	1.9	1.0	0.2
Minneapolis–St. Paul	91.4	3.5	1.5	2.6	1.0
San Diego	64.5	6.4	20.4	7.9	0.8
St. Louis	80.7	17.0	1.1	1.0	0.2
Pittsburgh	91.1	7.5	0.6	0.7	0.1
Phoenix	75.7	3.5	17.0	1.6	2.2
Tampa–St. Petersburg	82.9	9.0	6.7	1.1	0.3

SOURCE: U.S. Bureau of the Census, 1995b.

Table 11.3 shows the growing ethnic diversity of the largest U.S. metropolitan areas, most of which has been the result of increasing levels of immigration.

Issues of the Newest Immigration

The large-scale entrance of both legal and illegal immigrants has created problems of economic and social absorption and has stirred much public controversy, reminiscent of earlier immigration periods.

Economic Issues The economic impact of the newest immigration is widely debated. Whereas some argue that these people constitute an added burden to an already swollen labor pool, others contend that the jobs they typically hold are those that native workers shun and, furthermore, that they create as many jobs as they take. Similarly, while some maintain that the new immigrants overburden the social welfare system, others hold that they pay in taxes far more than they collect in benefits (Borjas, 1994; Simon, 1991).[4]

The newest immigration appears to favor some sectors of the economy and harm others. Most of the immigrants from Mexico (the largest single group) and the Caribbean are unskilled and take their place at the lowest employment levels. They benefit employers in labor-intensive industries, such as clothing manufacturing or other work areas calling for cheap labor, but they depress the job opportunities of native low-status workers, especially blacks. Not all the new immigrants are impoverished and unskilled, however. A large segment of some new groups, particularly Asian Indians, Chinese, and Filipinos, are highly trained professionals and managers whose economic impact is far different from that of those who enter with few occupational resources. Many immigrant doctors, for example, now staff large city hospitals, which would find it difficult to operate without them. The United States has also lagged in producing engineers and other highly trained scientific workers, and many of the newest immigrants are filling these needs (Muller, 1993; Suro, 1989; Szabo, 1989).

Regarding the economic impact of the newest immigration, the debate rages on, with each side pressing its case with government officials and the public generally. One reason why immigration has emerged as such a controversial issue today and why it has resonated so loudly is that immigrants are entering at a time of economic dislocation. Workers are fearful of losing their jobs; immigrants, from this perspective, are seen as a potential threat, prepared to work at lower wages and under poorer conditions. Indeed, this has been a historically consistent pattern, with labor generally opposing continued large-scale immigration.

Social Issues The current immigrants are highly visible, bringing to American society cultural and physical features distant from those of the white majority. This makes them considerably different from previous immigrants, who, though always viewed ethnocentrically by those already present, were nonetheless primarily European. Today's issues, therefore, concern not simply the new immigrants' economic integration but their social assimilation as well. These issues have a familiar ring, recalling those that raged at the turn of the century with regard to southern and eastern European immigrants.

---

[4]Another argument for restricting the number of immigrants to American society is that they constitute a growing environmental problem. The United States, it is held, has limits on what it can provide while still maintaining a high standard of living. Greater numbers of newcomers strain the society's resources and create environmental problems (Beck, 1996; Bouvier, 1992).

These questions and attitudes have come together most clearly around the issue of bilingualism. In Canada, as we will see in Chapter 14, the nucleus of ethnic conflict is the preservation of one group's language. A somewhat similar situation has begun to surface in certain regions of the United States, where large numbers of immigrants, particularly the Spanish speaking, use their native tongue. That these immigrants are able to maintain close links with their origin societies through easy communication and travel and through a constant influx of fellow immigrants from the homeland means that there is no pressing need to quickly relinquish the native language (Chavez, 1991; Heath, 1985; Muller, 1993). This makes their situation somewhat different from that of past immigrants. Although the latter also came speaking languages other than English, it was assumed that through the use of English in the school and other institutions, those languages would gradually be abandoned, if not by the first then by the second generation. Ethnic groups, of course, were not prohibited from maintaining their own institutions (private schools, newspapers, and so on), which often served as preservers of the native language for a time. Today, however, the assumption of language assimilation is being challenged. A greater tolerance of ethnic pluralism has led to efforts to provide public educational and other services in the language of the new groups, especially Spanish. This has created controversy among those who favor or oppose such measures (Baron, 1990; Cafferty, 1983; Chavez, 1991; Crawford, 1992).

Bilingual education has emerged as a particularly divisive political issue, especially in states with large Hispanic populations like California, Texas, New York, and Florida. In New York City, classes are now taught not only in Spanish, but in Chinese, Haitian Creole, Russian, Korean, Arabic, Vietnamese, Polish, Bengali, and French. And in Los Angeles, almost half of all students have limited English proficiency. Forty-three percent of U.S. school districts have at least some non-English-speaking children (Hornblower, 1995). The controversy swirls around whether such students should be instructed in their native languages or in English and in what proportion.

Some opposing greater language diversity have sought to have English legally declared the "official" American language. By 1995, more than twenty states had passed legislation to that effect. Proponents see such legislation as a means to reverse a perceived decline in English language use and to force immigrants to more quickly adopt the dominant culture. They advocate strict limits on bilingual education, the elimination of voting ballots in languages other than English, and raised language-proficiency standards for prospective citizens. Supporters of the movement to make English the country's official language maintain that without a common language, new immigrants will resist assimilation and the United States risks becoming linguistically divided, like Canada. In response, critics of the English-only movement charge that efforts to declare English the official language of the United States are a backlash against the new immigrants and essentially a mask for ethnic antagonism. Moreover, they claim that such legislation is useless because it does nothing to help promote the learning of English (Braverman, 1988; Crawford, 1992). By the mid-1990s, public opinion seemed to favor the "official English" side of the controversy. A national poll revealed that almost two-thirds of Americans favored a law that would make English the country's official language (*Time*, 1995).

The issue may be moot in any case because studies indicate that, as in the past, the children of recent immigrants adopt English as their major language and gradually drift away from the use of the native language outside the home (Portes and Rumbaut, 1990; Saad, 1995). This shift to the English language, notes sociologist Calvin Veltman (1983), has been characteristic of every previous immigrant group and is evident among the newer groups today. "There is no language group in any region," Veltman concludes, "which possesses the retentive characteristics which would remotely sustain the theory that continued immigration is laying the groundwork for linguistic nationalism" (1983:217). Research among Latinos, for example, shows that first- and second-generation immigrants are usually bilingual but that by the third generation, English has become the primary—and for most, the only—language spoken (Fishman, 1987). *The Latino National Political Survey* revealed that, among U.S.-born Latinos, 62 percent of Mexicans, 50 percent of Puerto Ricans, and 31 percent of Cubans used English predominantly or exclusively at home. More than two-thirds, regardless of national origin, were better at English or spoke no Spanish (de la Garza et al., 1992). An earlier study carried out in Miami revealed that 98 percent of Latino parents felt it was essential for their children to become competent in English (Braverman, 1988). And, as we saw in Chapter 9, the vast majority of Latinos believe that citizens and residents of the United States should learn English (de la Garza et al., 1992). What keeps Spanish alive in the United States, therefore, is not a determination of Latinos to retain their native language but the continuation of large-scale Hispanic immigration.

Resurgent Nativism The social issues of the newest immigration go well beyond language. The public backlash against immigration of recent years stems in some measure from the fact that the overwhelming majority of immigrants are nonwhite. Some whites have seen the entrance of large numbers of non-Europeans as a threat to what, in their view, is the essence of the American social fabric (Brimelow, 1995). The new groups are seen as unassimilable and thus a continual drag on the society. This has created a revitalized nativism, similar to the movement in the earlier part of the twentieth century that called for a halt to the large-scale immigration of southern and eastern European groups, who were viewed in much the same fashion.

But the surfacing of economic and social issues concerning the newest immigrants has provoked negative reactions not only from whites, but from nonwhites as well, who perceive the new groups as a threat to either their jobs or their language or as an increasing pool of welfare recipients. This has been especially apparent in areas that have absorbed large numbers of Hispanic immigrants like Miami, Los Angeles, and New York. Moreover, African Americans in these as well as other cities with large immigrant populations have expressed resentment over the adverse effects of the immigrant influx on their traditional segment of the labor market (Beck, 1996; *Business Week,* 1992; Portes and Stepick, 1993; Warren and Stack, 1986). As we saw in Chapter 10, anti-Asian incidents also have become more common with the continuation of Asian immigration.

National surveys in the 1990s revealed some of the misgivings of Americans toward the newest immigration (*The American Enterprise,* 1994; *Business Week*/Harris Poll, 1992; *Newsweek,* 1993, 1995b; *USA Today,* 1995b). They indicated, for example, that though most believed that immigration throughout American history had been good for the country, most thought that at the present time it was bad. What's more, an overwhelming majority felt that immigration levels should either stay the same or be decreased; few felt they should be increased.[5] Even though the newest immigrants were seen by most as very hardworking, they were also viewed by most as taking jobs away from American workers, driving down wages, and using too many government services.

Whether immigrants are a positive or a negative social influence remains a highly debated issue, particularly among politicians, who ultimately will decide the fate of immigration policy. Some believe that current policy should be continued or made even more generous in allowing for additional immigrants. Their view is that immigrants not only create economic activity, but are a fresh cultural influence on the society. Others, however, support highly restrictionist measures, maintaining that the new immigrants are causing a radical and unprecedented change in the social and cultural makeup of the society, which can only lead to more racial and ethnic conflict. Some see this change as undesirable in itself, suggesting that the United States remain a primarily European-origin society (Brimelow, 1995).

Immigration Reform: Closing Doors? Immigration has, in recent years, become one of the most virulent issues of national politics. Public opinion has seemed to favor more restrictionist measures. And, reacting to this public fervor,

[5]This was true of ethnic minorities as well as Euro-Americans. *The Latino National Political Survey* showed, for example, that more than 65 percent of U.S.-born Mexicans, Puerto Ricans, and Cubans believed that there were currently too many immigrants coming to the United States (de la Garza et al., 1992).

politicians at the national level and in states and locales where the impact of large-scale immigration is most acute have exploited this issue repeatedly. To date, the most consequential measure that has been enacted is California's Proposition 187. In 1994, voters approved this measure, which denies illegal immigrants access to welfare benefits, schools, and health care.

Given the growing public backlash against further immigration, it is almost certain that serious changes in current immigration laws, affecting the numbers and socioeconomic characteristics of immigrants, will be enacted in the near future. The Commission on Immigration Reform has called on Congress to cut annual immigration from 750,000 to 550,000. The Commission has also recommended that immigration priority go to immediate family members and that sisters and brothers and adult children no longer be granted automatic entry, as in the past.

It is unlikely, however, that future immigration measures will bring to an end the debate regarding the number and character of immigrants that should be permitted to enter the United States. Those seeking more stringent limits will continue to argue that immigrants have a negative economic and social influence, whereas those advocating more liberal immigration laws will maintain that immigrants ultimately have a positive impact, both economically and culturally. The absorption of immigrants has been a persistent issue of American social history. In one sense, the United States has always been regarded as a "golden door," open to all seeking political refuge or economic opportunity. But the acceptance of new groups has been countered with a tradition of protectionism, which has manifested itself repeatedly in efforts to limit or exclude newcomers for both economic and ethnocentric motives. The current public controversy is, therefore, only the latest in a long tradition.

CULTURAL ASSIMILATION OR PLURALISM?: COMPETING GOALS

The current high levels of immigration to the United States and the extremely diverse national origins of the immigrants have created a more ethnically diverse society than at any time in American history. And that diversity will, in all likelihood, continue to expand. "The United States," demographer Leon Bouvier has suggested, "is inexorably on its way to becoming a society with no one predominant group" (1992:148). The continued ethnic dominance of Euro-Americans, therefore, is no longer a given. As seen in Table 11.4, if demographic trends continue, non-European ethnic groups will make up almost half the American population by the middle of the twenty-first century.

As we have now seen, this ongoing population shift has already provoked fundamental economic, political, and social issues. But it also prompts a cultural issue: Given America's increasing ethnic diversity, should the end product of in-

Table 11.4 Population of the United States by Ethnicity

ETHNIC GROUP	1980	1995	2030[a]	2050[a]
Euro-American	80.0%	73.6%	60.5%	52.8%
African American	11.5	12.0	13.1	13.6
Hispanic American	6.4	10.2	18.9	24.5
Asian American	1.5	3.3	6.6	8.2

SOURCE: U.S. Bureau of the Census, 1996.
[a]Projected.

terethnic relations be some form of cultural assimilation, in which diverse groups become more alike, or some form of pluralism, in which they maintain or perhaps increase their differences? The question of whether the proper course for disparate ethnic groups in American society should be aimed at eventual assimilation and, concomitantly, whether ethnic policies should be directed toward assimilationist goals, has arisen many times throughout American history. The political and social conditions of the last three decades, however, have given this issue renewed substance and made it especially salient.

In the past, ethnic issues in the United States were generally reduced, by sociologists, policymakers, and laypeople alike, to problems of assimilation (Metzger, 1971). How, it was asked, could diverse peoples achieve a maximum of social integration with a minimum of social travail? Put differently, the direction of ethnic relations was rarely questioned; it was assumed that groups would move toward some form of assimilation. Today, that assumption is very much in dispute. In the 1960s, the black movement, especially its nationalistic phase, served as a catalyst in prompting feelings of pluralism across the entire spectrum of ethnic groups. Heightened collective consciousness among blacks and black political activism stirred similar movements among Latinos, Asians, American Indians, and even white ethnics. A new ideology seemed to materialize in which ethnic differences were not only to be tolerated and respected but perhaps even to be encouraged. In the past three decades, the failure of group boundaries, particularly between whites and blacks, to dissipate, despite higher levels of racial tolerance and the end of blatant forms of discrimination, along with the influx of millions of new immigrants, radically different in culture from the European norm, has given added incentive to stronger forms of cultural pluralism.

Consciousness and tolerance of cultural diversity seem greater today than at any previous time in American history. These attitudes are reflected in almost all major societal institutions, including education, business, government, and the mass media. The impact has been especially great in education, where non-European cultures are increasingly acknowledged in school and university curricula, generally labeled "multiculturalism." The essential objective is to give greater recognition to the contributions of non-European groups in U.S. and global historical development and to study those groups' literature and art. Multiculturalism has stirred a heated debate among educators and policymakers, with one side viewing it as an attempt to demean European culture and, in the process, to create sharper group divisions, and the other side viewing it as necessary to counterbalance the traditional emphasis on "white" (that is, European) cultural and historical traditions (Goldberg, 1994; Schlesinger, 1992).

Some suggest that in this new multicultural environment, even WASP Americans are evolving into simply one more ethnic group among many. William Greenbaum may have overstated their declining influence only somewhat when he declared that "after having *been* the society, the Protestants have been relegated to a *place within* the society, and increasingly they experience a bewildering sense of themselves as a new minority" (1974:412).

That the United States today is more ethnically diverse and more tolerant of that diversity seems unarguable. The trend is clearly toward some form of cultural pluralism. Not all, of course, see this pluralistic tendency and the accommodative responses to it by the society's major institutions in a positive light. In a larger sense, however, this issue may be hollow. Despite the accentuated cultural diversity of the United States, acculturative forces are at work that seem to have homogenizing effects on all groups. Mass forms of communication and transportation increasingly break down cultural singularities and compress them into common forms, spiced with only slight ethnic variations. Billboards that in Des Moines picture white people and are written in English may in Detroit feature blacks or in Miami be written in Spanish, but they advertise the same products for all. Moreover, as the society becomes more ethnically varied, cultural and social influences begin to cross in bewildering combinations. Journalist David Rieff describes, for example, the ethnic fusion occurring in contemporary Los Angeles:

> Just as the Irish, Poles, Jews and Italians, who had rubbed shoulders and more in the wake of the European immigration of 1900, had, by the 1950s and '60s, begun to intermarry en masse, so the process is beginning to take place among the recent arrivals in L.A. One can find every sort of nonwhite combination in the city now: Hmong and Salvadoran, Ethiopian and Taiwanese, Mexican and Filipino (1991:20).

It is in this sociocultural sense that, ironically, the "melting pot"—such an inaccurate metaphor of American ethnic relations in previous generations—may be closer than ever to realization today. Steinberg (1989) suggests that although pluralist principles have become much stronger in recent years, the fact remains that ethnic differences have been diminishing. Thus, though groups extol the need to retain an ethnic culture and encounter declining resistance to its retention from the dominant group, societal trends continue to erode those cultural differences. Indeed, some maintain that the resurgent ethnic awareness of recent years has been, particularly among white ethnics, largely a response to this homogenizing process. As their European ties increasingly disintegrate, they feel uneasy, wrote Irving Howe, "before the prospect of becoming 'just Americans'" (1977:18).

This homogenized culture is brought home strongly when Americans of any ethnic origin travel to the society of their forebears. Most quickly realize how little they have in common with the people of Ireland, if they are Irish Americans, of Poland, if they are Polish Americans, or of African nations, if they are African Americans. It is in such foreign contexts that a unique "Americanness" is most fully revealed. The same acculturative process will no doubt have a similar effect on the newest immigrants to the United States.[6]

Looking Ahead

Although ethnic strife has been a historically constant theme in the United States, there appears to be a direness about the state of ethnic relations today given the disturbing regularity of occurrences of interethnic hostility, the tenaciousness of ethnic inequality, and the emergence of a population more sharply divided by race and culture. Moreover, national surveys in recent years have consistently revealed public pessimism regarding current and future ethnic relations. Following the Los Angeles riots in mid-1992, for example, only 25 percent of Americans thought that race relations were "good," whereas 68 percent thought they were "bad" (*Newsweek,* 1992b). But even before the riots, surveys had revealed a similar public perception of the deterioration of ethnic relations in the United States (*Black Enterprise,* 1990; Hugick, 1992; *Newsweek,* 1992a). Three years after the riots, the same bleak picture seemed evident: only 23 percent of whites and 12 percent of blacks said that race relations were "good" or "excellent" (*Newsweek,* 1995a).

In all likelihood, then, ethnic discord will continue at a high level well into the twenty-first century. Still, there is cause for at least some cautious optimism.

---

[6]Contrary to conventional thought, immigrants themselves seem most committed to assimilation, not retention of their ethnic cultures. A national survey in 1995 found that only 27 percent of immigrants favored the so-called "multiculturalism" effort to preserve ethnic identity, whereas 59 percent favored assimilation (Saad, 1995).

Table 11.5 Changes in the Social Standing of American Ethnic Groups

ETHNIC GROUP	1964	1989	CHANGE
Native white Americans	7.25	7.03	−0.22
People of my own ethnic background	6.16	6.57	0.41
British	6.37	6.46	0.09
Protestants	6.59	6.39	−0.20
Catholics	6.36	6.33	−0.03
French	5.73	6.07	0.34
Irish	5.94	6.05	0.11
Swiss	5.50	6.03	0.53
Swedes	5.41	5.99	0.58
Austrians	5.06	5.94	0.88
Dutch	5.60	5.90	0.30
Norwegians	5.48	5.87	0.39
Scotch	5.73	5.85	0.12
Germans	5.63	5.78	0.15
Southerners	5.25	5.77	0.52
Italians	5.03	5.69	0.66
Danes	5.20	5.63	0.43
French Canadians	5.08	5.62	0.54
Japanese	3.95	5.56	1.61
Jews	4.71	5.55	0.84
People of foreign ancestry	4.84	5.38	0.54
Finns	5.08	5.34	0.26
Greeks	4.31	5.09	0.78
Lithuanians	4.42	4.96	0.54
Spanish-Americans	4.81	4.79	−0.02
Chinese	3.44	4.76	1.32
Hungarians	4.57	4.70	0.13
Czechs	4.40	4.64	0.24
Poles	4.54	4.63	0.09
Russians	3.88	4.58	0.70
Latin Americans	4.27	4.42	0.15
American Indians	4.04	4.27	0.23
Negroes[a]	2.75	4.17	1.42
"Wisians"		4.12	
Mexicans	3.00	3.52	0.52
Puerto Ricans	2.91	3.32	0.41
Gypsies	2.29	2.65	0.36

[a]Blacks were referred to as Negroes in the 1989 survey to conform with the wording in the 1964 survey.

First, we should remember that there have been few times in American history when ethnic issues have not been in the forefront of public consciousness and the focus of public policy. We should therefore not think of ours as a unique era.

Second, there is some evidence of a growing ethnic tolerance in the society as a whole. Table 11.5 shows the results of a poll in which a national sample of Americans in 1964 and 1989 was asked to rank different American ethnic groups. On a scale of 1 to 9, 1 was the lowest standing, and 9 was the highest. The figures shown are averages. (Note the inclusion of a fictitious group, "Wisians," in 1989.) As can be seen, the rank order of the groups corresponds closely to the social distance hierarchy that emerged from the earlier Bogardus studies of social distance (described in Chapter 3), indicating much continuity in the social standing of American ethnic groups. Perhaps equally significant, however, is that the score for almost all groups rose during the twenty-five years between 1964 and 1989. Especially large increases are noticeable for blacks and Asian-American groups.

Finally, as we will see in the next five chapters, the United States is not alone in its ethnic problems. Indeed, few societies of the contemporary world do not face equally daunting problems stemming from an expanding ethnic diversity and its attendant intergroup conflict.

SUMMARY

Three interrelated issues of race and ethnic relations in the United States are today of pressing importance. One concerns the continuing socioeconomic gap between Euro-American groups and racial-ethnic groups, specifically, African Americans, Latinos, and Native Americans. To address this problem, preferential policies have been used by government and other institutions during the past three decades aimed at creating greater ethnic equity. These programs have been the subject of much controversy among politicians, scholars, and the general public.

A second major issue concerns the large-scale immigration to the United States that has occurred since the mid-1960s. This has been a period during which more immigrants have entered the society than at any time since the classic period of immigration in the early years of the twentieth century. Most important, the vast majority of the current immigrants are non-European, coming mostly from Asia and Latin America. Their presence has prompted much social debate regarding their economic impact and their social assimilation.

The final issue of American race and ethnic relations concerns the question of whether the society is — or should be — moving toward greater cultural assimilation or pluralism. Unquestionably, the United States today is a more ethnically diverse society than at any time in its history. Moreover, there appears to be greater tolerance for that diversity, with various societal institutions acknowledging and even sanctioning the expression of ethnic differences. Whether in the long term these differences will dissipate or be sustained is not yet clear.

Suggested Readings

Beck, Roy. 1996. *The Case Against Immigration: The Moral, Economic, Social, and Environmental Reasons for Reducing U.S. Immigration Back to Traditional Levels.* New York: Norton. Argues that large-scale immigration to the United States during the past three decades has contributed significantly to American economic, social, and environmental problems.

Crawford, James. 1992. *Hold Your Tongue: Bilingualism and the Politics of "English Only."* Reading, Mass.: Addison-Wesley. Traces the current controversies surrounding the issue of bilingualism, prompted by the surge of new immigrants, especially Latinos, whose native language is not English.

Edsall, Thomas Byrne, with Mary D. Edsall. 1992. *Chain Reaction: The Impact of Race, Rights, and Taxes on American Politics.* New York: Norton. An engrossing analysis of the politics of race in the United States. Maintains that race has been the defining factor in American elections of the past two decades and dominates the domestic political agenda.

Ezorsky, Gertrude. 1991. *Racism and Justice: The Case for Affirmative Action.* Ithaca, N.Y.: Cornell University Press. Presents moral and practical arguments for affirmative action policies and assesses their impact, particularly on African Americans.

Glazer, Nathan. 1975. *Affirmative Discrimination: Ethnic Identity and Public Policy.* New York: Basic Books. Questions the manner in which affirmative action policies have been applied, arguing that they have brought about a shift from individual rights to group rights.

Kasinitz, Philip. 1992. *Caribbean New York: Black Immigrants and the Politics of Race.* Ithaca, N.Y.: Cornell University Press. Describes the economic, political, and social institutions of America's newest black ethnics, West Indians.

Muller, Thomas. 1993. *Immigrants and the American City.* New York: New York University Press. Explores issues pertaining to the newest immigrants to American cities. Maintains that they have had a generally positive, sometimes transformative, economic and social impact on urban areas.

Portes, Alejandro, and Rubén Rumbaut. 1990. *Immigrant America: A Portrait.* Berkeley: University of California Press. Examines the sociological issues of America's newest immigrants, such as occupational adaptation, urban settlement patterns, political participation, and language.

Schlesinger, Arthur M., Jr. 1992. *The Disuniting of America: Reflections on a Multicultural Society.* New York: Norton. A prominent historian analyzes the current upsurge in ethnic consciousness and its impact on education, especially the teaching of history as part of a multicultural curriculum.

Sowell, Thomas. 1990. *Preferential Policies: An International Perspective.* New York: William Morrow. Compares preferential policies instituted in a number of societies including India, Malaysia, Sri Lanka, Nigeria, and the United States. Concludes that rationales for such policies rarely conform to actual results.

Steele, Shelby. 1991. *The Content of Our Character: A New Vision of Race in America.* New York: HarperPerennial. A black scholar argues that affirmative action policies no longer function beneficially for African Americans but in fact act as a group handicap and accentuate perceived racial differences.

Ungar, Sanford J. 1995. *Fresh Blood: The New American Immigrants.* New York: Simon & Schuster. Drawing on extended interviews, describes many of the new ethnic communities arising in different parts of America. The author sees the newest immigration as having generally positive societal effects.

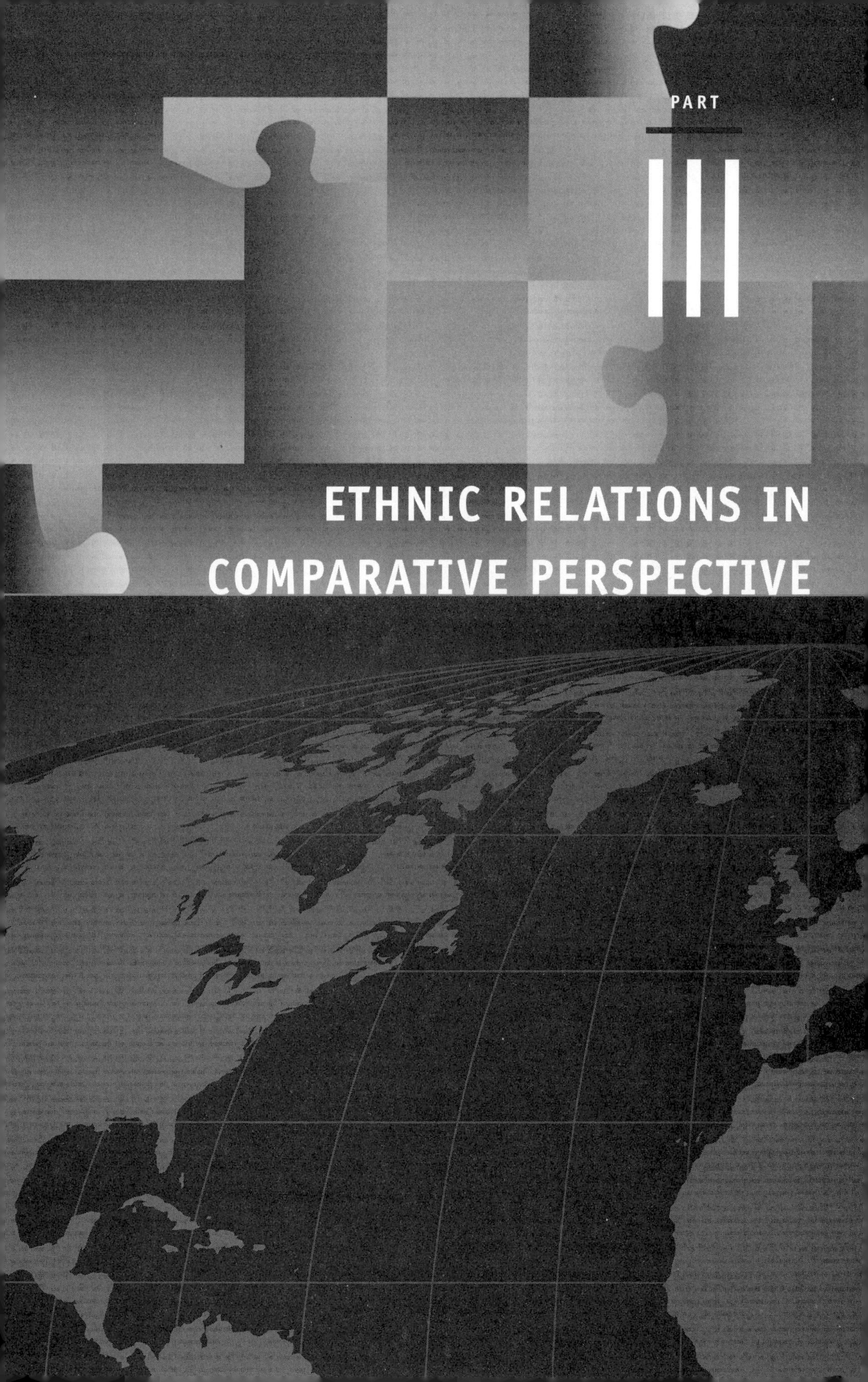

PART

III

ETHNIC RELATIONS IN COMPARATIVE PERSPECTIVE

Having investigated patterns of race and ethnic relations in the United States, we now examine some non-American cases as a basis for comparison. People often assume that what occurs in their own society is unique and that others are simply not like them. Conversely, they sometimes assume that what happens in their own society is characteristic of human societies in general. In part, both of these suppositions are valid. To ascertain in what ways U.S. society is similar to and in what ways it is different from others requires that Americans transcend the limited perspective imposed by the social system with which they are most familiar and comfortable. That is, they need to adopt a broader, less parochial approach to human affairs.

The societies examined in this part lend themselves readily to comparison with the United States. It will not be difficult to detect similarities, nor will the differences be obscure. Each is part of a different geographic region, is a product of a distinct historical tradition, and displays quite different bases of ethnic stratification and conflict.

The patterns of race and ethnic relations in each of these societies represent one of the major types described in Chapter 4. South Africa, the focus of Chapter 12, resembles in its recent history the *colonial* (or extreme inequalitarian pluralistic) type and Brazil, examined in Chapter 13, is a clear example of the *assimilationist* type. In Chapter 14, Canada will be seen as a society with many features of a *corporate pluralistic* system. Northern Ireland, described in Chapter 15, exhibits aspects of both colonial and corporate pluralistic systems. These societies also seem to fall roughly on a scale, or continuum, based on the severity of ethnic stratification and conflict in each. South Africa is at one extreme, with gross ethnic inequality and institutional separation. At the opposite pole is Brazil, where the blending of groups culturally, structurally, and even physically is officially endorsed. As in the United States, it is the disparity between groups of different color that represents the thrust of ethnic relations in both South Africa and Brazil. In Canada and Northern Ireland, however, ethnic relations revolve primarily around the juxtaposition of two cultural groups, English and French in Canada and Protestant and Catholic in Northern Ireland.

In Chapter 16, we explore the rising significance of ethnicity across the globe as nations everywhere, through immigration and political turmoil, become more ethnically diverse. We look first at the impact of this new heterogeneity on western European countries. This will provide us some opportunity to compare their relatively fresh encounter with ethnic conflict and change to the American experience, in which ethnic issues have been a wellspring of social friction and strife for many generations. Focus then falls on the case of the former Yugoslavia. There, as we will see, conflict among ethnic groups in the 1990s sank to a level of brutality and hate unseen in Europe since World War II. Although comparison with the United States is dubious, the Yugoslav case nonetheless forces us to consider what can happen to a society in which fear and distrust of ethnic outgroups become the preeminent characteristics of interethnic relations.

UNTIL VERY RECENTLY, South Africa, in its system of ethnic relations, was unique in the modern world. As an inequalitarian pluralistic society, it had had no peer. Whereas almost all contemporary multiethnic societies, despite persistent intergroup conflict and hostility, made attempts at ethnic unification, South Africa, for most of the past forty years, moved in the opposite direction, compelling greater divisions among its ethnic populations. What had made South Africa most singular among modern societies was that its racist system was fully legitimized and enforced by the state. South Africa, as we will see, has been a congeries of societal inconsistencies and contradictions. Throughout most of its history, however, one thing was clear and unambiguous—the dominance by whites, the numerical minority, over blacks, the numerical majority.[1] No area of South African life was untouched by the racial policies and ideology of the dominant white group.

---

[1]The ethnic terminology used in describing South Africa's population has not been consistent. Traditionally, black Africans were called *Bantu,* a name corresponding to their languages. During the apartheid regime, they were officially called *blacks* but referred to themselves as *Africans,* a term used by most scholars as well. Some have used the term *blacks* to describe the entire nonwhite population, including Coloreds and Asians, as well as Africans. Today *black* is used most commonly to refer specifically to Africans, and that usage is adopted in this chapter. As the new South Africa evolves, ethnic terms will likely take on new meanings (see Adam, 1995; Horowitz, 1991; Lever, 1978; Thompson, 1990; and van den Berghe, 1967).

SOUTH AFRICA

Today, South Africa has embarked on what appears to be an irreversible course of change, moving toward a more equalitarian system. Indeed, the scope and depth of change this society has experienced since 1989 would have been unimaginable just a few years earlier. South Africa therefore provides an instructive case of the fluidity of ethnic relations and how, no matter how rigid and seemingly implacable, they can change profoundly in a relatively short period of time. Despite the magnitude of change, however, the ultimate shape of the new South Africa as a multiethnic society remains very much uncertain. Its racial and ethnic problems, therefore, will engage the attention of the world for many years. Furthermore, although its system of ethnic separatism has been dismantled officially, enough of it endures informally to make this society still an extraordinary case in its extremes of ethnic division and inequality.

As we look at South Africa it is important to keep in mind that events of change have moved so swiftly in recent years that any description of its prevailing institutions risks almost immediate obsolescence. South Africa's significance as a case study of race and ethnic relations, however, lies as much in its history as in its current affairs. Furthermore, its system of extreme ethnic separation and stratification is not something of the distant past but has only recently been successfully challenged. It is fresh enough, therefore, to warrant a careful examination.

THE DEVELOPMENT OF ETHNIC INEQUALITY

Few who read newspapers or watch television are unaware of at least the basic outlines of the ethnic drama that has unfolded in South Africa. The extraordinary social and political changes that continue to take place and the new South Africa that is emerging as a result, however, can be understood only if they are positioned alongside this society's long heritage of intergroup conflict. As in other multiethnic societies, ethnic stratification in South Africa has been the result of a historical sequence of diverse groups coming together, viewing one another ethnocentrically, competing for scarce resources, and ultimately forming a hierarchy in which one imposes its superior power on others.

White Settlement

Historical evidence indicates that what is today South Africa was inhabited by Bantu-speaking peoples, the society's major African linguistic group, as early as the sixteenth century and probably much earlier (Wilson and Thompson, 1969). Modern South African history, however, involving the confrontation of European and native African peoples, begins with Dutch settlement in 1652. At that time, the Dutch East India Company established a colony in Cape Town as a refreshment station for its ships traveling to and from India. The indigenous people of the area, the Khoikhoi (Hottentots) and the San (Bushmen), were unable to satisfy the labor needs of the colonists, who therefore began to import slaves from other parts of Africa and from the Dutch East Indies only a few years after their original settlement. The negative view of these slaves by the colonists did not prevent widespread miscegenation; hence, a mixed racial group was created whose descendants are today called Coloreds.

As the colony grew, many of the Dutch, German, and French Huguenot colonists chose permanent settlement and started at the beginning of the eighteenth century to leave Cape Town, establishing farms on land farther into the interior. Thus began the penetration of the northern and eastern frontiers, a movement called trekking. These settlers became known as Boers, a Dutch word meaning "farmers." The trekking movement lasted 150 years, with the Boers continually pushing farther and farther inland. As they moved into the interior, the major patterns developed that would basically affect the shape of South African society: conflict with the indigenous peoples and the emergence of a unique white South African ethnic group.

The Boers came into conflict not only with native African peoples occupying these lands but also with migrating African tribes moving across the continent, westward and southward. These were Bantu-speaking tribes who were well organized and who presented serious resistance, unlike the Khoikhoi and the San, who had been more easily subdued. The clashes over land and cattle that developed between the Boers and the Bantu, called Kaffir wars, lasted for many decades before the dominance of the Boers was finally established. The disdain for the natives based on their racial features was exacerbated by the bitterness that resulted from these encounters. Additionally, because they greatly outnumbered the Boers, the Bantu were seen as a constant threat whose suppression therefore had to be ensured. Many of the racial attitudes of white South Africans in recent times can be traced to these initial contacts with Africans in the interior.

The second major consequence of the trekking movement was that the Boers increasingly distanced themselves from their European roots. As the settlers pushed inland, Dutch control over them waned. The farther they removed themselves from Cape Town, the more they became "a law unto themselves" (Thompson, 1964:186). As subsistence farmers, they disengaged themselves even further from intercourse with the economic and political institutions of the colony and its civil administration.

With their economic and political isolation, the Boers developed a culture thoroughly distinct from their European heritage. Differences were most apparent in language and religion. A unique Dutch dialect evolved over several generations, producing eventually the Afrikaans language, today only remotely related to the original Dutch. Divergent religious beliefs also contributed to the cultural distance between the Boers and their European ancestors. The Boers resisted the enlightened Protestantism that was emerging in Europe in the late seventeenth and early eighteenth centuries and clung instead to a primitive Calvinism whose ideas of predestination and rugged individualism seemed to complement well the pioneering conditions of life in rural South Africa. They increasingly saw themselves in a biblical light as "chosen people," destined to conquer the frontier against the numerically superior Bantu and, later, the colonial British (Fredrickson, 1981). The chosen people theme continued to run strongly through the ideology of the Boers (today, called Afrikaners) and was reflected in their stubborn resistance to changes that threatened to reduce their power.

By the late eighteenth century, the Boers were a unique social entity with a culture and language distinct from any other. They were geographically and intellectually isolated. Their European roots, notes historian Leonard Thompson, "were almost completely severed" (1964:187). Thus was born the Afrikaner ethnic group of South Africa. The Afrikaners' knowledge of and ties to Europe were so minimal that they had become, in effect, "white Africans." Unlike other European settlers on the African continent, then, the Afrikaners did not see themselves, nor did they even function, as colonials representing and maintaining strong cultural and political ties to a motherland.

This sense of isolation and permanence explains much about the determination of Afrikaners in recent decades to defend their way of life and what they came to see as their territory against the threat of black rule. "The clue to South African history," wrote Douglas Brown over thirty years ago, "is that the Boer, though early forced into an intimate local relationship with other races, brown, white, and black, has since the first decade of the eighteenth century had no backing, no allies, no means of reinforcement, no possibility of retreat. His nationalism soon became absolute, and so it remains in the jet age, to the discomfiture of the rest of the world" (1966:19). Not until the late 1980s were the Afrikaners compelled to alter that view. As we will see, Afrikaner nationalism has been tempered and appears to have given way to a recognition that only by sharing societal power with those groups it dominated so completely for most of South Africa's history could it survive in the modern world.

The British Entrance

Whereas the first century and a half of modern South African history is largely a chronicle of confrontation between white and black—Afrikaner and African—the following 150 years are marked principally by the political rivalry between the Afrikaners and the colonial British. The British entered the South African scene at the start of the nineteenth century when, in a move designed to protect their colonial interests in the East, they took possession of the Cape colony. The next fifty years saw the development of a competitive and often bitterly hostile relationship between the British and the Afrikaners that has been a continuing element of South Africa's social dynamics.

The British were not a large settler group like the Boers, and they brought a strikingly different political and social temperament to South Africa, more liberal and cosmopolitan. Their values clashed with the stern parochialism of the Afrikaners, and a wide gulf opened between them. This gap has remained relatively broad to the present time.

Soon after their arrival in South Africa, the British instituted liberalized policies regarding race relations, which exacerbated the already strained relations with the Afrikaners. The British attitude toward the native Africans was not different from the Afrikaners' in its substance: white dominance. But the British saw black Africans in a more paternalistic frame, whereas the Afrikaners saw them as a threat to their ultimate survival as a people. "The abstract difference between British imperial and Afrikaner republican conceptions of how to rule Africans," explains historian George Fredrickson, "might be described as a conflict between the trusteeship ideal and what the Boers called *baaskap*—which in essence meant direct domination in the interest of white settlers without any pretense that the subordinate race was being shielded from exploitation or guided toward civilization" (1981:193). The practice of white supremacy by the Afrikaners was thus consistent with their fundamental belief that force and exploitation were perfectly legitimate means to protect their group interests.

In 1834, slavery was abolished throughout the British empire, and this act was followed during the next two decades by the lifting of other legal restrictions against nonwhites. These radically new racial policies acted as a major incentive for the second and far more substantial migration of Boers into the interior. In 1836, a sweeping movement was launched in which one-quarter of the entire Afrikaans-speaking population would eventually set out on their own into the interior (Robertson and Whitten, 1978). This was the commencement of the Great Trek, comparable in many ways to the westward movement of Americans in the nineteenth century. As the Western frontier had served as a kind of social safety valve in the United States, the Boer trek northward and eastward after 1836 provided a means of escape for those who found the British presence a constant source of anguish.

These Voortrekkers, as they were called, were doggedly pursued by British colonial advances, however. After fierce battles with the Zulu tribe in 1838, Boers founded the republic of Natal, which was subsequently annexed by Britain in 1843. Pushing farther northward, the Voortrekkers founded the Orange Free State and the Transvaal, which were eventually recognized as independent states by the British. These states became the core of the Afrikaner culture, and within them the sense of autonomy and group solidarity was developed to its fullest. In both, the notion of white supremacy was a fundamental principle.

The discovery of diamonds around Kimberley in 1867 and of gold in the Transvaal in 1886 once again spurred British advances. Britain had earlier attempted to extend its influence into the Transvaal, resulting in the first Boer War. Although the Boers successfully resisted at that time, the influx of British railroads, mining interests, and workers following the gold discovery led to a more serious challenge to Boer autonomy. The British finally intervened with military force, provoking the second Boer War in 1899, a conflict that would leave permanent scars of bitterness between the two groups. The Boers used guerrilla tactics against the larger and superior British forces, and in response, the British interned the Boers in concentration camps where 26,000, including women and children, died of disease. This is an event of enormous significance in Afrikaner history, symbolizing the martyrdom of the Boer people and their determination to survive against all odds. This second encounter with the British was a key factor in the subsequent development of militant Afrikaner nationalism. As we will see, though they were defeated in war, the Afrikaners eventually asserted political ascendancy and emerged as South Africa's dominant ethnic group.

Unlike the Afrikaners, the British never coalesced into a nationlike unit in South Africa. Their loyalties were fragmented between South Africa and the British Commonwealth (Thompson, 1964). As a result, Afrikaner dominance, though repeatedly diluted by British incursions in education, immigration, and, especially, the economy throughout the nineteenth century and part of the twentieth, was eventually entrenched and solidified. Although they had been militarily successful in the second Boer War, the British began to lose influence afterward and reached a tenuous political compromise with the numerically superior Afrikaners.

In 1910, the four territories of South Africa—the Cape, Natal, the Orange Free State, and the Transvaal—were merged into a union. This marked the beginning of a period in which the British and moderate Afrikaner political forces were sufficiently consolidated to frustrate the efforts of the more extreme Afrikaner-dominated National party to excise the British influence. Following the formal union of South Africa, the foundation of the society's racial laws was also secured. Although the subordination of blacks had always been informally understood and enforced, during this period a succession of government policies institutionalized many segregationist and inequalitarian practices, particularly those regarding participation of Africans in government and the economy.

The Nationalist Ascendancy

The modern period of South African history can be said to have begun in 1948 with the political victory of the National Party. This ended the almost half-century of delicate balance between British and Afrikaner interests and established the Afrikaner political ascendancy. The National Party represented the most extreme element of Afrikaner nationalism, viewing itself as the spearhead of a God-ordained mission to shape South Africa in the interests of Afrikaners and to throw off finally the yoke of British influence. The authoritarian character of political rule, so evident until recently, took shape at this time, and Afrikaner dominance was established in all phases of the political system. This development culminated in 1961 with the withdrawal of the Union of South Africa from the British Commonwealth, severing once and for all the society's British ties.

Most important, the stage was now set for the formal establishment of an institutionalized system of white supremacy, called *apartheid,* the dynamics of which we will look at shortly. While other African nations of the post–World

War II era were moving toward national unification and a breakdown of tribalism, South Africa was moving in the opposite direction, creating a social order of strict separation between whites and blacks and encouraging the retention of tribal cultures among the African populace. In the modern period of South African history, the Afrikaner-led whites presented to the world a system of formalized ethnic separatism unmatched in rigidity and scope.

ETHNIC STRATIFICATION

In no other modern multiethnic society has ethnic stratification been so clear-cut as in South Africa, and in no other has it been so inflexible. Until very recently, the boundaries between ethnic strata were, by law, relatively impermeable. South Africa's racial system was castelike, sustained by endogamy and by the maintenance of separate institutional systems for each ethnic group. Only in the larger economic system did the different ethnic groups come together, and there, as in all other societal areas, the intergroup relations were grossly inequitable. Indeed, along with its rigidity, the other major feature of the South African system of ethnic stratification was the enormous gap between the dominant white group and the several subordinate black groups, a gap that, as we will see, remains very much in evidence today. South Africa, then, was the prototypical inequalitarian pluralistic society, in which only the economic system served as a kind of social glue holding the various groups together.

Undergirding and enforcing the system of ethnic inequality was the South African state, dominated by whites, specifically, the Afrikaners. So complete was white control of the society's political apparatus that blacks were allowed no meaningful participation and were accorded few legal rights. Pierre van den Berghe (1978) referred to the South African sociopolitical system as a *Herrenvolk* democracy, that is, a state that provides most democratic features of political rule to whites while ruling blacks dictatorially. Others described the system as a "race-oligarchy" (Adam, 1971b) or a "pigmentocracy" (Thompson and Prior, 1982). All these appellations denote the fact that skin color was the single overriding criterion of societal power in South Africa.

In a basic sense, the South African ethnic hierarchy in recent times has been reducible to two parts: whites at the top and blacks at the bottom. More accurately, however, the society has been divided into four officially designated categories: whites, Coloreds, Asians (mostly Indians), and Africans. Each of these major ethnic categories is further divided into subethnic groups. In almost all multiethnic societies, ethnic status is a significant factor in the distribution of life chances. The allocation of justice, health, occupation, education, living quarters, and so on depend, to some degree or another, on one's ethnic classification. But in no case has that classification been so critical in determining one's social place as it has in South Africa. So closely does the ethnic hierarchy overlap with the class hierarchy that class stratification in South Africa is almost wholly a function of ethnic status. Whites receive the bulk of the society's resources, Africans receive the least, and Coloreds and Indians occupy a position somewhat above the Africans. The distance between the whites and the other three ethnic categories is so great, however, that the hierarchy is essentially a dichotomy—whites over nonwhites. In short, one's skin color fundamentally has determined most aspects of social life.

The South African castelike system of apartheid was traditionally maintained by an elaborate arrangement of law and custom ensuring white dominance. Before investigating the dynamics of apartheid, let us look more closely at each of South Africa's major ethnic categories.

The Whites

Whites (or Europeans) constitute less than 15 percent of the population (Table 12.1) but, until 1994, completely dominated government. As the only group with effective political rights, whites for decades filtered all social rewards, first siphoning off their disproportionate share before distributing the remainder to the three subordinate groups. Although they no longer dominate the political system, whites continue to monopolize positions of power in the economy and own most of the society's wealth. In an economic sense, then, they remain the dominant group, despite the fact that they have been displaced as the political ruling group.

As already noted, however, whites are by no means a culturally or even politically unified group. Rather, they comprise two major elements, Afrikaners and the English-speaking, most of the latter of British origin. The two major white subgroups differ not only in language and culture but in class status and political power as well. Afrikaners traditionally held most high-ranking government posts and generally dominated the state bureaucracy. This political dominance was established with the victory of the National Party in 1948. For almost four decades, Afrikaner political power was not seriously threatened because the Afrikaners outnumbered the English-speaking whites by almost three to two, and nonwhites played virtually no effective role in the formal political system.

Although the Afrikaners were politically dominant, the English-speaking whites traditionally maintained the strongest influence in the society's economic institutions: trade, finance, mining, and manufacturing. The English-speaking whites, compared with the Afrikaners, traditionally occupied higher-ranking jobs, were better educated, and were, in general, wealthier (Adam, 1971b). These discrepancies, however, have diminished in recent years.

Table 12.1 Ethnic Population of South Africa[a]

YEAR	TOTAL (1,000S)	AFRICANS (1,000S)	%	COLOREDS (1,000S)	%	ASIANS (1,000S)	%	WHITES (1,000S)	%
1911	6,044	4,091	67.7	525	8.7	152	2.5	1,276	21.1
1936	9,591	6,597	68.8	770	8.0	220	2.3	2,004	20.9
1951	12,672	8,560	67.5	1,103	8.7	367	2.9	2,642	20.8
1970	21,794	15,340	70.4	2,051	9.4	630	2.9	3,773	17.3
1993	40,716	31,089	76.4	3,480	8.5	1,013	2.5	5,134	12.6
2000[b]	47,592	37,260	78.3	3,783	7.9	1,122	2.4	5,428	11.4

SOURCE: South African Institute of Race Relations, 1992, 1993, 1994.
[a]Includes all ten homelands.
[b]Estimated.

The English-speaking whites, having South African roots that do not extend as far back as the Afrikaners', are less cohesive and do not display the same sense of ethnic solidarity. Moreover, though most are of British origin, among them is a community of 115,000 Jews and smaller communities of other groups of European origin.[2]

The political and class rivalry between the two major white ethnic groups extends back to the early nineteenth century, when, as we have seen, the British first entered South Africa as a colonial power. The Afrikaner-British conflicts of the nineteenth century left a residue of intergroup animosity that continues to manifest itself today. On the one hand, Afrikaners have resented the economic role of the English, and anti-British sentiment has been a staple of Afrikaner nationalism for many generations. On the other hand, the English-speaking whites see the Afrikaners as crude parvenus whose worldview is anti-intellectual and parochial. Except for business and politics, contacts between the two groups are minimal. They attend different schools and churches, reside in different neighborhoods, and generally maintain separate communities (Brown, 1966; Lemon, 1987; Thompson and Prior, 1982).

Language is a factor that is particularly important in separating the two white groups. Afrikaans, the language of the Afrikaners, traditionally was established throughout the state school system and, when Afrikaners dominated South Africa politically, was the language of government and politics. However, in business and industry, dominated as they were by English-speaking whites, English

[2]On the Jewish community in South Africa, see Della Pergola and Dubb (1988) and Shimoni (1988).

was the predominant language. The language division, however, goes beyond even politics and commerce. For Afrikaners, it is a vital component of Afrikaner nationalism. As Lemon has observed, "In the minds of the Afrikaners themselves, the language is closely associated with their wider historical struggle against the British, and what they see as their present struggle to survive as a distinct nation or *volk*" (1987:63).[3]

The class, status, attitudinal, political, and linguistic divisions between the Afrikaners and the English, however, should not be overstated. First, with the ascendancy of the Nationalists in 1948, a significant closing of the gap in wealth and economic influence between the two groups gradually occurred (van zyl Slabbert, 1975). Class differences, though real, have been minimized by the primary importance of one's racial classification. Second, both groups benefited from the castelike system and were united in their desire to preserve white rule. This common interest was strong enough to override the political differences between them. Though the English-speaking whites traditionally displayed greater liberalism regarding ethnic stratification, their differences with the Afrikaners revolved mainly around the most efficient and politically feasible means of ensuring white dominance, not necessarily changing South Africa into a fully democratic society. Thus, their opposition to apartheid stemmed mainly from the view that it was economically impractical and inefficient, not that it was morally objectionable (Banton, 1967; Lewin, 1963; van den Berghe, 1967). Although the social gap between the English and Afrikaners remains evident, the political differences between them have diminished considerably (Horowitz, 1991).

The Coloreds

Out of the sexual and sometimes marital unions of whites and blacks during the early settlement of the Cape was created an intermediate racial group called Coloreds. During the nineteenth century, the Cape continued to evolve along different social and political lines from the frontier settled by the Voortrekkers. As a result, for a long period following the abolition of slavery in 1834, the legal system of the Cape was essentially color-blind, allowing for a great deal of racial intermixing (Fredrickson, 1981; Mason, 1970). Today, almost all of South Africa's three million Coloreds live in the Western Cape province, the majority in or around Cape Town.

The Coloreds are perhaps the most peculiar element within the South African ethnic system because in all ways except their physical features, they are thor-

[3]With the replacement of Afrikaners as the dominant political group by blacks in recent years, many fear that Afrikaans will increasingly be displaced by English in all areas of social and political life (McLarin, 1995).

oughly Europeanized. Most speak Afrikaans, are Christian (mostly Protestant), and are in other cultural ways barely distinguishable from the Afrikaners. The sole, and obviously critical, distinction between the two groups is skin color. Many Coloreds, of course, are even physically indistinguishable from the Afrikaners.

The position of the Coloreds has been marginal to both whites and Africans not only physically but also politically, economically, and psychologically. Coloreds maintained a few political privileges during the apartheid regime, which, limited though they were, placed them apart from the Africans. The economic position of the Coloreds has also been somewhat privileged by comparison with that of Africans, though well below the white standard. Most are part of the unskilled proletariat, both in agriculture and industry, though there is a substantial middle class. With the increasing dichotomizing of the populace in recent decades, however, and with the legal favoritism afforded whites in job competition with nonwhites, the economic mobility of the Coloreds was deliberately thwarted (Curry, 1972; Hunt and Walker, 1974).

Finally, the Coloreds are psychologically marginal to both the white and African groups. To some extent, they identify with the Afrikaners, hoping for eventual acceptance into the white society. That they are thoroughly Europeanized and, under apartheid, held at least a few token privileges also created feelings of superiority to and distance from the Africans. Their attitudes toward apartheid, therefore, were somewhat ambivalent. As Dickie-Clark (1972) has noted, through apartheid they lost their earlier legal equality with whites but gained a somewhat more advantageous position over Africans and Indians. This accounts in some measure for their reluctance to identify or interact more closely with the Africans (Gevisser, 1994).

The Asians

The bulk of South Africa's Asian population is made up of Indians, most of whom are descendants of indentured workers brought from India in the late nineteenth century to work in the sugar-cane fields of Natal. Following their term of service, many remained and became traders, shopkeepers, and workers at various skill levels. Most still live today in Natal, in and around Durban, the province's largest city. Like the Coloreds, they are politically, socially, and economically a marginal group, part of neither the privileged white nor the disadvantaged African population.

Indians within the South African ethnic system are the most complex group in both culture and class. They are of several linguistic origins and are further divided along religious lines. They have retained their ethnic culture more tenaciously than the other nonwhite groups, but the process of Westernization is clearly evident among them. Surprisingly, despite this cultural diversity, there is a much stronger cohesiveness within the Indian community than within the other ethnic populations (Moodley, 1975, 1980).

Within the Indian population, there is an entire range of classes, from wealthy merchants to unskilled agricultural and industrial workers. All, however, exist within the context of an institutionally separate community. Although Indians generally occupy a more privileged economic position than either Coloreds or Africans (in most occupational sectors, they earn more than Coloreds and far more than Africans), the vast majority are poor relative to the whites (Ginwala, 1977; Lemon, 1987; Moodley, 1980).

South Africa's Indians have functioned as a classic middleman minority. In Chapter 2, such minorities were described as occupying an intermediate economic status between dominant and subordinate ethnic groups. Anthony Lemon has written that "European attitudes to Indians have been characteristically ambivalent; employers found their labour useful, but traders and businessmen feared competition" (1987:245). For a long time the permanence of Indians in South Africa was questioned by the white power structure. Moreover, prohibitions on their movement in South Africa as well as against further immigration limited their size. At the same time that Indians have been viewed suspiciously and with derision by the whites, they have been seen by Africans as exploiters. One scholar, in fact, asserts that in general "Indians are more resented by most Africans than are Whites" (Moodley, 1980:226).

The Africans

As part of the South African ethnic hierarchy, the African population is by far the largest (over 75 percent) and traditionally had been the most subordinate. Though until recently Africans shared a common societal powerlessness, they are not a single culturally unified group. Rather, they derive from several Bantu tribes, the largest among them the Zulu, Xhosa, and Sotho. Even though almost all Africans still speak a native tongue and retain certain other uniquely Bantu traits, the tribal culture for most is residual. Except for those who remain in rural areas, they are assimilated culturally into Western ways. They are at least nominally Christian and in urban areas speak either English or Afrikaans in addition to their Bantu language. Within the urban environment, tribal or subethnic boundaries become indistinct, and the common political and economic oppression they have suffered has impelled the formation of an encompassing African ethnic group.[4]

---

[4]A particularly sharp political division within the African population has emerged in the last several years, however, which will be discussed in a later section of this chapter.

A crude economic hierarchy is evident among Africans, just as it is among the other three ethnic populations. Occupational differences are particularly acute between those living in the urban areas and those living in the African "homelands" (under the apartheid regime, officially designated tribal areas) or rural white areas. The latter are mainly subsistence farmers or agricultural workers while those in the urban areas are industrial workers. In addition, a substantial number of migrant laborers from neighboring African nation-states work in the gold and diamond mines.

Despite their occupational differences, most Africans have occupied the lowest-ranking jobs in the labor structure, those requiring little or no skill. With the end of the official policy of racial discrimination, however, this traditional pattern is beginning to change. Job opportunities for blacks have begun to open in significant numbers and blacks have been afforded increasing access to higher education. As a result, the black middle class, growing even before the end of apartheid, is expanding rapidly.

These marked improvements in the status of blacks, however, should not disguise the fact that the economic differences between blacks and whites remain vast. As one observer of South Africa has described it, "perhaps no country on earth has cultivated such a wide gulf between the country-club, luxury-car, boutique life style of white privilege and the wretchedness of black denial" (Keller, 1995:A6). The rate at which the socioeconomic differences between blacks and whites can be reduced will determine whether South Africa will move through its social transformation relatively peacefully or with renewed conflict.

PREJUDICE AND DISCRIMINATION

In Chapter 4, inequalitarian pluralistic systems were seen as characterized by several key traits: separation and duplication of institutions among the society's various ethnic groups, extreme polarization between dominant and subordinate groups, high levels of prejudice and discrimination, and a vastly inequitable distribution of wealth. The coordinating mechanism of such systems is the absolute authority and power of the dominant group. Under apartheid, the South African system was characterized by all these traits. A system was created by which white supremacy was assured and separation between the racial categories rigorously maintained.

As they took shape starting in 1948, political policies designed to create and sustain inequality among South Africa's four ethnic populations were grounded in two overriding principles: (1) the subordination of nonwhites in all areas of social, political, and economic life and (2) the maximum physical and social separation of the four groups, particularly the two major groups, white and African.

The latter principle was justified by the premise that peace and order could be ensured "only if people of the various colours mix as little as possible" (Wiechers, 1989:8). So great was the incompatibility of black and white ethnic groups, it was felt, that contact between them could lead only to conflict; hence, minimizing contact would be the logical solution to this perceived problem.

The Dynamics of Apartheid

Apartheid provided specific legislation mandating the social separation of the racial groups, the maintenance of separate social institutions for each, and, eventually, the formal division of South Africa into separate and independent white and black nations. Indeed, so blatant and uncompromising was the South African system that the very term *apartheid* has, in recent times, come to be applied commonly outside the South African context, used to signify any rigid and durable aspect of ethnic separation and inequality.

In 1994, the Nationalist regime relinquished power in a peaceful transfer to a democratically elected black-dominated government. This officially ended apartheid. Although it is no longer formally in effect, examining this system of ethnic separation today is instructive in that we may see how extreme inequalitarian pluralism may operate. Based on an ideology of ethnic incompatibility, this was a complete system of ethnic separatism that shaped all aspects of social life. There have been few comparable cases in the contemporary world. Moreover, although, as we will see, with the advent of black political dominance apartheid was officially dismantled, its basic principles and underlying framework continue, in large measure, to guide interethnic relations unofficially and informally. Let us look more closely at several of the most important measures that shaped the apartheid system.

Racial Classification Where physical distinctions among people are not clear-cut, ethnic boundaries are difficult to define and enforce. In South Africa, the physical dissimilarities between whites and blacks are usually apparent, but in some cases they are not. To establish and enforce a classification system that could be used to distinguish the racially dominant and subordinate groups, the Nationalist government soon after it came to power in 1948 installed several formal measures.

The first cornerstone of apartheid was the Population Registration Act of 1950, providing for the official racial categorization of all persons. Questionable cases were decided by race classification boards, made up of whites, who ruled on the identity of such persons. Although the registration system was technically applied to everyone in South Africa, it was designed primarily to monitor the movement of Africans and was enforced mainly against them. A "passbook," comprising detailed identification papers, was required of all Africans, and failure to produce it on demand was an automatic assumption of criminal status. Over the years, eighteen million arrests were made for passbook violations (Lipton, 1987). The passbook system was a particularly odious element of apartheid for Africans and was a constant reminder of their almost total subordination (Klaaste, 1984). The pass laws were abolished in 1986. Most important, the Population Registration Act was repealed in 1991, removing in effect one of the most fundamental legal props of the apartheid system.

Petty Apartheid Physical contact between blacks and whites was minimized through a system of petty apartheid, or what van den Berghe (1967, 1979a) called "micro segregation," consisting of dozens of laws and mandates created to maintain ethnic separation in almost every conceivable area of social relations. Restaurants, hotels, buses, trains, public toilets, waiting rooms, hospitals, schools, parks, beaches, theaters, and the like were, for the most part, segregated. Moreover, these separate facilities were invariably unequal in quality and quantity. Crowning the rules of petty apartheid was a policy of strictly enforced endogamy, prohibiting interracial marriage and even sexual intercourse between whites and blacks. In the realms of social life where people come into physical contact with each other, then, the South African apartheid system was strikingly similar to the Jim Crow system of segregation enforced in the pre-1960s American South. Unlike the Jim Crow system, however, there were no pretensions of "separate but equal."

Most of the measures of petty apartheid were officially repealed beginning in the 1980s. Neighborhoods, hospitals, public transportation, parks, beaches, and other public facilities were desegregated. But these official changes have not been implemented uniformly, nor have social attitudes, cast by decades of socialization to apartheid, been easily reversed. As a result, petty apartheid continues to guide many areas of social life. "To the casual visitor," noted a study group to South Africa in 1986, "apartheid may appear to be on the way out. In its essential elements, it remains very much intact" (Commonwealth Group, 1986:33). Almost a decade later, the observation still seemed valid. A former Canadian member of parliament visiting South Africa in 1995 related an incident that poignantly illustrates apartheid's legacy. Riding on the airport bus, he struck up a conversation with a black politician whom he recognized. Two other whites on the bus immediately stopped speaking. Although he didn't think much of it at the time, later the black politician chatted with him about the shock that many South African whites still feel about "normal conversations between a black man and a white man" (Langdon, 1995).

Residential Segregation A more fundamental and significant level of apartheid concerned the maintenance of separate living areas for whites and nonwhites.

Officially mandated residential segregation, then, became the second cornerstone of apartheid. Through the Group Areas Act of 1950, each racial group was assigned specifically demarcated living areas. Any area might be proclaimed "reserved" for a particular group, after which residents not belonging to that group had to move elsewhere. As a result, depending on their skin color, people might be arbitrarily removed and sent to their "appropriate" area. Because this measure was designed to assure racial integrity of white areas and to displace nonwhites from them, over the years its effects were felt primarily by the latter.

The designation of specific areas for each ethnic category was closely related to the South African labor system. With the industrialization of the society, Africans functioned as the unskilled industrial proletariat. South African industry was located in traditionally white areas, and this spurred the migration of African workers into these areas. Because the industrial economy could not function without them, some provision had to be made to accommodate Africans without upsetting the ethnic purity of white neighborhoods. The solution was a migrant laboring system in which Africans were permitted to enter white areas to work but would be housed in separate communities, called townships, adjacent to the white cities. These townships, then, served essentially as dormitories for Africans who labored in white areas. Africans who lived in the townships were officially only "temporary" residents or migratory workers, even if they were born in the urban areas and their ancestors had lived there for generations.

The township of Soweto, adjacent to Johannesburg, is perhaps the epitome of this purposefully designed system of residential segregation. Constructed to house Johannesburg's African migrant laborers, Soweto is actually larger than the white population of Johannesburg itself! Because it was seen by the government as simply a housing area for temporary workers, the physical conditions often have been squalid and extremely dense.

Years of restricted movement of Africans created severe social disruption among the urban African populace. In a classic illustration of the self-fulfilling prophecy, social conditions were produced among urban Africans that inevitably led to the decadent behavior commonly attributed to them. Forced to migrate to the cities to find employment, African men were forbidden to bring their families with them. Demoralization and alienation consequently thrived among them, in turn leading to high rates of crime, violence, and alcoholism in the townships.

Restricting the movement of blacks into white areas was referred to by the Nationalist government as "influx control," and the passbook system was a key device in enforcing this policy. The system, however, failed to stem the urban flow of Africans, and the policy was scrapped in 1986. Nonetheless, residential segregation today remains the norm, in large measure due to the fact that blacks in general lack the means to acquire housing in what have been traditional white areas (Sachs, 1992).

The Homelands As the Nationalist government shaped apartheid, it envisioned the culmination of this racially divided system as the complete separation of the ethnic populations in independent nations. This policy of ultimate exclusion, or what van den Berghe (1967, 1979a) called "macro segregation," entailed the division of the society into geographic areas corresponding to race and ethnicity. As a result, certain areas, variously called "homelands," "Bantustans," or "black states," were set aside for Africans, based on the Nationalist version of historic tribal areas. Each homeland roughly corresponded, in this view, to an area originally inhabited by a particular Bantu tribe. The areas set aside for Africans were less than 14 percent of the land area of South Africa, leaving more than 85 percent for whites. (Here it should be recalled that Africans make up more than 75 percent of the South African population.) All Africans, regardless of where they resided or where they were born, were declared "citizens" of one of the homelands, based on their ethnic or tribal origins. This included those living in officially designated white areas, the majority of the African populace.

When the homelands policy was formulated, the Nationalists expected that eventually all Africans in South Africa would live in one of these areas, each of which would be granted independence. Once this grand design was fully realized, it would be possible for the South African government to declare that there were no Africans in South Africa, only "guest workers" temporarily living in the remaining (white) areas of the country. In short, until all Africans could be "repatriated to their homelands," those remaining in the white areas were to be considered transient laborers, migrants entitled to few legal rights or privileges. Sebastian Mallaby refers to the homelands as "the most ambitious injustice that apartheid ever devised" (1992:26).

From the outset, the homelands were fictional creations, constructed essentially to provide South Africa with a continued flow of cheap, exploitable labor and to serve as a convenient dumping ground for unemployed, aged, and infirm Africans (Carter, 1980). None of these areas was economically viable, capable of sustaining itself independently of the South African economy. Moreover, Africans living in South Africa's cities had no cultural or even social ties to their official homeland. The Nationalist government acknowledged these facts by encouraging and even subsidizing industries to locate in adjacent border areas or in the homelands themselves so that jobs could be created for Africans living there (Stultz, 1980). None was ever recognized by any international body or nation outside South Africa. With the advent of a black-dominated government in 1994, the homelands as political entities were officially abolished.

Apartheid and Economic Inequality

One of the cruel ironies of apartheid is that it was financed primarily by its victims. All nonwhites—Africans, Coloreds, and Indians—were discriminated

Table 12.2 Share of Personal Income Received by Each Racial Category

YEAR	AFRICAN	ASIAN	COLORED	WHITE
1985	29%	4%	8%	59%
1990	33%	4%	9%	54%

SOURCE: South African Institute of Race Relations, 1992.

against, yet they had to underwrite the oppressive system by accepting artificially low wages and seriously deprived working and living arrangements. Plainly, apartheid enabled whites to live at a level of comfort they would not otherwise have enjoyed.

The class of poor whites, a substantial part of the Afrikaner population before World War II, was essentially eliminated following the advent of the Nationalist regime. This was accomplished through discriminatory occupational and wage policies that guaranteed to whites the higher-ranking jobs and dictated grossly discrepant wages between blacks and whites. As a means of ensuring the most desirable occupations for whites, a system of job reservation was established whereby no black could advance above a white in the same occupational area. In certain areas of work, blacks were totally excluded. To complete the system of labor exploitation, trade union activity and strikes by African workers were forbidden.

The requirements of South African industry forced the circumvention of these measures, however, and dictated official policy changes. Whites were unable to fill all higher-skilled jobs, and the movement of blacks into these positions therefore was necessitated. In 1969, for example, only 3 percent of professional and managerial positions were occupied by blacks, but by 1989, blacks accounted for 14 percent of them (South African Institute of Race Relations, 1992). With the end of apartheid, many companies have adopted affirmative action policies and are making efforts to employ more blacks in managerial positions.

In recent years, the economic gap between whites and blacks has declined, but huge discrepancies remain. Table 12.2 shows the share of personal income received by each racial category. As can be seen, whites, though less than 15 percent of the population, receive a majority of personal income. In all occupational areas, the wage gap between whites and blacks remains wide though it continues to narrow. Whereas in 1985, Africans earned a little more than one-quarter of what whites earned, by 1990 this had increased to about one-third (South African Institute of Race Relations, 1993). Unemployment is also extremely high among Africans, with estimates ranging from 33 to 50 percent.

Table 12.3 Proportion of White Expenditure for Education for Nonwhite Groups

YEAR	AFRICAN	COLORED	INDIAN
1969–1970	5%	20%	27%
1979–1980	8%	20%	33%
1989–1990	25%	44%	72%
1991–1992	28%	61%	N/A

SOURCE: South African Institute of Race Relations, 1992, 1993.

The economic disparities among the racial groups are duplicated in other areas of social life. In education, for example, whites receive the bulk of resources. Per capita government spending on African education in 1991–1992 was 28 percent of white expenditure. In other words, the state spent almost four times as much on a white student as on an African student. Nonetheless, this represented a significant improvement since 1969–1970 when black educational expenditure was only 5 percent of white expenditure. Similar improvements are evident for Colored and Asian groups (Table 12.3). The results of the huge discrepancy in resources allocated to blacks and whites during the apartheid era are staggering. The illiteracy rate among blacks is four times higher than among whites (P. Taylor, 1995). And, dropout rates for blacks are extraordinarily high; of 10,000 who start school, only 113 pass matriculation examinations (Lemon, 1994).

At the university level, institutions that previously had barred blacks now are increasingly racially integrated. A majority of the 1995 freshman class at the University of Witwatersrand in Johannesburg, one of South Africa's most prestigious universities, was black (Keller, 1995). Below the university level, however, schools remain de facto segregated. Nonwhites make up fewer than 1 percent of students in traditionally white schools, and these are mainly Colored children in the more liberal Cape Town area (Berger, 1992; Keller, 1992). Schools in the black townships have also been disrupted in recent years by the sporadic violence that has gripped these communities. Racial parity in education, then, remains a distant goal that the new South African regime will have to address.

Enforcement of Apartheid: Force

We noted earlier that all societies are characterized by elements of order and conflict. Social control is therefore always a combination of consent and coercion. In a society in which the majority of the population does not accept the legitimacy of the prevailing system, however, coercion becomes the more common of the two. In South Africa, where four-fifths of the population rejected the system of white domination, "the element of consent," wrote South African legal expert Albie Sachs, "must be extremely low and the element of coercion extremely high" (1975:224).

The South African criminal justice and legal institutions traditionally provided for suppression of any challenges to white rule. Indeed, the use of terror against the African population was routine. Writer Mark Mathabane recalls his own experience: "I remembered the brutal midnight police raids launched into the ghetto to enforce apartheid; the searing images of my father's emasculation as he was repeatedly arrested for the crime of being unemployed; my parents constantly fleeing their own home in the dead of night to escape arrest for living together as husband and wife under the same roof" (1994:38). Police violence against Africans was so pervasive and routine, explains political scientist Gwendolyn Carter, as to be almost unnoticeable except when a well-known person was affected or the circumstances "so unusual as to attract attention" (1980:12).

The preparedness of the Nationalist government to use force in averting change in the distribution of power was demonstrated repeatedly. At Sharpeville in 1960, a crowd of Africans peacefully protesting the pass laws was fired on by the police, and sixty-nine were killed. Sachs has pointed out that the Sharpeville massacre, though the best-reported incident of its kind, was "merely one in a series of characteristically South African episodes in which police fired on crowds of blacks that were either totally unarmed or armed only with sticks and stones" (1975:227). A second well-reported incident occurred in Soweto in 1976. What began as a student protest against inequalities in the educational system culminated in the slaughter of over 600 Africans. Thousands were injured in the disturbances that followed, and schools and universities were closed and industries shut down by strikes. This event represented a spontaneous protest against the entire apartheid system and has since taken on great symbolic importance among the African populace (Dugard, 1980).

Socialization to Apartheid: Ideology

South Africa under apartheid illustrates vividly the power of a dominant group to uphold a system of ethnic supremacy using not only the power of raw force and political manipulation but also material, psychological, and ideological commitment. As in any multiethnic society, most members of the dominant group in South Africa were effectively socialized to accept and justify the prevailing system of ethnic inequality. Socialization to apartheid was especially pronounced among Afrikaners. Traditionally, in every institution—family, school, church, government—the values of Afrikaner nationalism and white dominance were reinforced.

South African whites, particularly Afrikaners, believed themselves to be natural overseers of primitive Africans whose shortcomings were not likely to change significantly. White rule, therefore, was seen as a kind of paternalistic relationship in which more intelligent and civilized peoples asserted their domination over inferior peoples in the interests of both (Mulder, 1972). In this view, notes Leonard Thompson, "It was the duty of the Whites, constituting the civilized, Christian race, to use their control of the state to prevent racial friction and racial bastardization by ensuring that the races would be separated from one another" (1985:190).

This essentially racist ideology underwent alteration during the apartheid regime. Although the crude belief in biological superiority had been by no means unvoiced, the white resistance to changes in the system of racial inequality was based mainly on the notion of cultural inferiority and ethnic incompatibility (Adam, 1971a, 1971b; Thompson, 1985; Welsh, 1975). The cultural gap between the races was so vast, it was claimed, that blacks were simply incapable of operating the institutions of a modern industrial society. As Lawrence Schlemmer explained,

> The perception, and to a very great extent the reality, of a conflict of interests is not simply between classes and ethnic or racial interests, but between a first world and a third world social order. In the collective fears of whites, the third world within the country is associated with a lack of sophistication, disorder and lower standards. The perverted vision of a future under majority rule most frequently conjured up among rank-and-file whites is one of irregularities of administration, loiterers in parks, dirt in the street, a lack of public discipline and power-hungry politicians with a fateful sway over slavishly adoring masses. The fact that none of these images is confined to the third world, and that all are gross oversimplifications, does not alter the power of the perception (1988:42).

Hence, the continued domination of the society by the white numerical minority was, in this view, essential.

The view that it was necessary to maintain white rule, based on the understood cultural superiority of whites, was thoroughly incorporated into all institutions, especially the school and the mass media. Through the Bantu Education Act of 1953, complete control of black education was given to the central government, thereby assuring that the history and social science taught would reflect the Nationalist ideology. Blacks, it was thought, would be trained to accept their subordinate position in the society. Prime Minister Hendrik Verwoerd articulated this view:

> The school must equip the Bantu to meet the demands which the economic life of South Africa will impose on him.... There is no place for him in the European community above the level of certain forms of labour. Within his own community, however, all doors are open.... Until now he has been subject to a school system which drew him away from his own community and misled him by showing him the green pastures of European society in which he is not allowed to graze.... What is the use of teaching a Bantu child mathematics when it cannot use it in practice? (quoted in Harrison, 1981:191)

The separation of white and black was deemed necessary to protect whites from cultural pollution by an obviously inferior people. A schoolbook, for example, described the differences between white and black: "The White stands on a much higher plane of civilisation and is more developed. Whites must so live, learn and work that we shall not sink to the cultural level of the non-White" (quoted in Harrison, 1981:204).

In addition to the rationale based on economic and political efficiency, an element of religious mission infused the determination of the Afrikaners, particularly the zealously nationalistic element among them, to maintain white supremacy. In this, the Afrikaners diverged from the English-speaking whites, whose motivations to support the system of ethnic inequality rested mostly on the desire to protect their economic interests. As already noted, the Afrikaners have historically invoked a biblical analogue to portray their place in South Africa, and their self-image of a chosen, superior people is strongly expressed in various aspects of Afrikaner culture and social life. The Dutch Reformed Church has served as an integral part of Afrikaner nationalism. For many, the church is the focal point of Afrikaner culture and the religious expression of its nationalism. Unlike most English-language churches, it maintained a close link with the Nationalist state and abided apartheid.

To better comprehend the determination of Afrikaners to preserve their power and, in so doing, their ethnic culture, one needs to keep in mind the unique status of the Afrikaners on the African continent. Unlike traditional European colonial groups in Africa, Afrikaners have no cultural or political ties to a European homeland. Generations of Afrikaners have known nothing but South Africa, and there is no thought of returning to a society of origin, for there no longer is one. The Afrikaners are, in all ways, a group *sui generis.* The prevailing belief has been that the Afrikaner way of life would be submerged if the numerically dominant nonwhite populace were afforded equal access to social and political institutions. Black political and cultural dominance, therefore, would represent to them the dissolution of their ethnic group.

Although these beliefs were the ideological bases of the apartheid system for a large element of the Afrikaner populace, the practical foundations of the system lay mostly in the resolve of the dominant whites, most Afrikaners and English alike, to sustain their supremacy. Whites well understood that a truly democratic political system would inevitably produce an African-dominated government, given the numerical imbalance between the two groups. This, in turn, would spell the end of white economic and social privilege.

Whatever their psychological and ideological commitment to white supremacy may have been, for most whites, social pressures ordinarily compelled conformity. As we saw in Chapter 3, prejudice and discrimination are most effective and self-sustaining where such thought and behavior are socially expected and rewarded. The economic benefits that flowed to whites from black exploitation were, as we have seen, enormous. Even those who may have been bothered by feelings of moral contradiction would therefore have been reluctant to call for the total dismantling of a system from which they greatly profited. Within the South African context, prejudicial attitudes were therefore easily converted into degrading and humiliating forms of behavior toward nonwhites, enacted through a whole etiquette of intergroup relations, similar to that of the pre-1960s American South. Thus, even after the abolition of the formal apartheid system, the informal system of white supremacy remains evident.

Even nonresidents of South Africa, used to more fluid social relations, might easily be caught up in the everyday workings of a system of extreme social inequality and separation that, at first, may have been repugnant. One American sociologist studying in South Africa in the 1970s reflected with astonishment on his own behavior: "One falls in with the culture and those with whom one must cooperate in assumptions and behavior. Even in my short stay I could observe the process occurring in myself: by the time I left I was beginning to unconsciously accept conditions that I found shocking when I first confronted them" (Mechanic, 1978:137).

Not only is it difficult to resist patterns of thought and behavior that reward conformity so generously, but it is especially hard to question a system of exploitation that, so far as it is recognized at all, is perceived from a distance. Because for whites the standard of living has been so comparatively high and the political order essentially democratic, there can arise among them only a limited understanding of black economic and, until recently, political deprivation. Adding to the incomprehension has been the essentially separate societies in which whites and blacks have lived. For the most part, whites have seen blacks only as workers and servants, not as neighbors and schoolmates. Few whites speak any of the various African languages and few ever venture into black townships. Put simply, blacks and whites in South Africa have lived in different worlds.

Three decades ago, one observer of South Africa declared that "the best way of describing South African society is to say that it consists of groups of peoples thrown together by history, all hating each other, but not enough to want to end their relationship. South Africa is, in other words, the same as everywhere else—only more so" (Feit, 1967:403). Even though the differences among South Africa's groups remain distinct and strong, the relationship among them has changed in a fundamental way. Indeed, if only ten years ago one were to have suggested the scope and depth of change in the ethnic system that has taken place in South Africa since 1989, that person might have been seen as hopelessly naive. The unimaginable, however, has occurred. Today, the government is led by blacks and the most egregious policies of racial separation have been repealed. Official apartheid has ended and the society is moving, however tortuously, toward ethnic democracy. What propelled this amazing change in a system so seemingly uncompromising in its commitment to ethnic inequality?

The radical transformation of South Africa that is now occurring was not a sudden event but had roots that reached back at least a decade when it became evident to the white power structure that changes in the apartheid system would have to be made if the society were to avoid a bloody confrontation. Former Prime Minister P. W. Botha publicly declared in 1978 that South Africa must "adapt or die," and reforms in petty apartheid and in the labor system instituted in the 1980s were manifestations of that view. As a desperate effort to preserve white rule, a new constitution was adopted in 1993, creating a tripartite parliamentary body, representing whites, Coloreds, and Asians. This new governing system, however, assured the continued dominance of whites and, most important, made no provision at all for the representation of Africans. The reformulated constitution was interpreted as essentially an attempt to co-opt the Colored and Indian groups, and most blacks (including Coloreds and Indians) recognized it as a sham.

The exclusion of Africans from the constitution was one of the precipitating factors in the civil unrest that began in 1984 (Adam and Moodley, 1986; Dugard, 1992; Mallaby, 1992; van zyl Slabbert, 1987). Widespread violent protests took place in black townships, the target of which was the system of black local authority. Black officials in the townships were seen as collaborators in the oppressive system and were driven from their communities, forced to resign, or in some cases even murdered. In addition to the violence in the townships, nationwide marches, rent strikes, consumer boycotts, student agitation, and labor protests became commonplace (Meredith, 1987; Murray, 1987). In attempting to control the protests, the government responded with force, using the South African Army as well as the police, who were given broad discretionary powers. Hundreds of blacks were killed in government actions to reassert control in the townships, and in 1985 a state of emergency was declared, providing for arrest and detention without trial and searches without warrant.

South Africa's panoply of internal contradictions brought the society to a watershed beginning in the late 1980s. An oppressive and largely dysfunctional sociopolitical system enforced by coercion no longer seemed viable, and major reforms were now proposed. Much of the impetus behind the reforms was provided by the newly elected President, F. W. de Klerk, who acknowledged the failure of apartheid. As Mallaby describes him, "de Klerk had finally abandoned whites' hopes of extending their privileges beyond the colonial age" (1992:78). In October 1992, de Klerk basically declared an official end to the apartheid system. Referring to his National Party, he stated that "For too long we clung to a dream of separate nation states when it was already clear that it could not succeed. For that we are sorry. That is why we are working for a new dispensation" (South African Institute of Race Relations, 1993:25).

Pressures for change arose primarily from international opprobrium of the apartheid regime, the requirements of economic interdependence, and political movements within the black population.

International Pressures Throughout the Nationalist ascendancy during which apartheid was at its peak, most nations viewed South Africa as a kind of pariah state. Nonetheless, South Africa remained strategically important as a source of vital minerals as well as a major link in the worldwide production and distribution networks of multinational corporations, most of them based in Britain or the United States. Thus, although the apartheid system was condemned, economic relations between South Africa and the Western industrial nations generally were unaffected.

The often brutal suppression of the black protests starting in 1984 led to a stepped-up international campaign that included calls on multinational corporations to divest their interests in South Africa and to force Western nations to impose trade sanctions. Although these measures seemed to only minimally affect the ability of the South African economy to function, they contributed to economic uncertainty and, combined with the civil unrest of the late 1980s, impelled the South African business community to call for greater and more rapid social reforms (Blumenfeld, 1987; Harari and Beaty, 1989).

Economic Forces More significant changes in the system of racial inequality were impelled by the demands of economic interdependence. The South African economy required an increasingly skilled work force that could not be supplied by the white population alone. As noted earlier, this resulted in modifications in the job reservation system, permitting more blacks to fill higher positions. In effect, the system of job reservation, restricting for decades black labor mobility, collapsed (Adam and Moodley, 1986).

Labor needs also compelled relaxation of prohibitions against black trade unions and the right to strike, creating a growing and increasingly significant black labor movement (Thompson and Prior, 1982). Whereas 292,000 Africans were members of registered unions in 1980, by 1991, African union membership was over one million (South African Institute of Race Relations, 1992). Strikes by black workers arose frequently starting in the mid-1980s in certain South African industries, most notably mining.

Internal Political Forces After the Nationalist regime came to power in 1948, opposition black political organizations and movements were declared illegal, forcing them to operate underground or in neighboring African countries. The most important of these were the African National Congress (ANC), the Pan-Africanist Congress (PAC), and the Black Consciousness Movement. In 1990, the de Klerk government restored the legality of these political organizations, providing a new element in the chemistry of change in South Africa.

The ANC is the oldest and most significant of the black political organizations, having been founded in 1912. From the time of its banishment in 1960 until its reinstatement in 1990, it had assumed the status of "government in exile" for most of the African population. Its official leader, Nelson Mandela, was released from prison in 1990 after serving a sentence of twenty-eight years for antiregime political activity. Mandela's release was seen as a threshold. From that point forward, the end of white South African rule and the emergence of a black majority government were only a matter of time. Shortly thereafter, the South African regime entered into talks with the ANC and other black groups to negotiate the entrance of blacks into the political system and to dismantle the apartheid system.[5] The historic elections held in 1994, in which all South Africans, black and white, voted for the first time, were won by the ANC, bringing Mandela to power as the country's president. South Africa was now poised to create a nonracial state.

BUILDING A NEW SOCIETY

Apartheid is gone, but still unsettled are the policies that will replace those abolished and the shape of the society that ultimately will be produced. Further-

[5] An important stimulus to change was a referendum held in 1992 in which white voters presented de Klerk with a mandate to forge ahead with negotiations with black leaders to ultimately construct a new multiracial constitution.

more, although the initial stages of transformation have been relatively peaceful, there is no certainty that violent ethnic conflict may not again erupt. The key issues facing South Africa today, therefore, concern the course the society will follow in its movement toward greater ethnic equalitarianism. In this movement, South Africa is being driven by forces of both change and stability. Here we want to examine those forces and explore the possibilities for South Africa's future ethnic system. It must be reiterated that in recent years, change of great dimensions in South Africa has come swiftly and often unpredictably. Thus, although certain issues and processes seem evident, whatever one may say about the future of ethnic relations can only be conjectured. With that caveat in mind, let us look at how South Africa may move into the twenty-first century.

Today, the ANC's leadership of the government is unchallenged. Moreover, Mandela himself remains the most widely supported and admired figure in South Africa, among whites as well as blacks. But monumental problems face both Mandela and the ANC. First, they are forced to work around a government bureaucracy and police force that are heavily occupied by those associated with the old apartheid regime. There are not enough qualified blacks to displace these personnel, who are essential to running the government. Thus, the ANC-led government does not completely control its destiny.

Perhaps even more important than the continued dependence on whites within the government is the white control of key elements of the economy. Again, the ruling black regime desperately needs the capital and managerial and technical skills that are now possessed primarily by the white community. Thus, it cannot afford to alienate whites, who might threaten to leave and take with them their businesses and economic resources. The new government has put forth a reconstruction and development plan that outlines its goals in transforming South Africa and bringing about a more equitable distribution of jobs, health care, housing, and land. Here the ruling government faces the delicate task of raising public revenues—which must come primarily from white-controlled businesses and the wealthier white community—to finance ambitious public projects designed to lift the economic and social conditions of blacks.

Despite its overwhelming support among the black population, serious political rifts exist within the ANC and, especially, with rival parties. The major political division lies between Mandela's ANC and Chief Mangosuthu G. Buthelezi's Inkatha Freedom party, a political movement of mainly Zulus. Buthelezi has challenged the tactics and objectives of Mandela and the ANC for years, rallying his followers in opposition. Before the election of 1994, Inkatha and the ANC had engaged in a bitter and bloody conflict in many black townships, resulting in the deaths of 6,000 in only four years. These hostilities continue to occur in the province of KwaZulu/Natal, where most Zulus live. This, as well as other less profound divisions, demonstrates the serious differences within the black population regarding future scenarios for South Africa and the strategies that are believed to be most effective in bringing about a transformation of the society.[6]

Future Prospects

As we have now seen, the legal and structural roots of the apartheid system have been removed, and significant policy changes have muted the most blatant forms of racial discrimination. Most importantly, blacks have been brought into the political process. Awarding full and equitable voting rights to the black population was the major prerequisite to any peaceful solution of South Africa's ethnic conflict. The great fear of whites was precisely what was enacted: a one-man, one-vote, majority-rule system, which, as they realized, spelled the end of white dominance. Astonishingly, this change was brought about peacefully. Thus, in light of the generations of racial hostility and the determination of the white regime to retain power, few could have predicted that the resistance to a black majority political system would evaporate so relatively swiftly and nonviolently.

A major step toward the formation of a postapartheid government was taken in early 1993, when ANC and government negotiators agreed to create a coalition government, made up of blacks and whites, that would rule for five years. During that time, a democratically elected assembly, acting as an interim parliament, would write a new constitution (Keller, 1993a).[7]

Whether a continuation of the South African transformation can proceed without violent resistance, however, remains uncertain. Resistance may come from blacks as well as whites. Whites are divided between moderates, like de Klerk and the current National Party, who pragmatically recognize the inevitability of fundamental change in all dimensions of society, and conservative hard-liners who remain unprepared to accept the transfer of political power from whites to blacks. Some among the latter envision the eventual formation of an autonomous, white territory of their own. Also part of the white right wing is a neo-Nazi element that, in the past, threatened to violently resist any fundamental power shift, threats that proved idle following the 1994 election. Although the various right-wing factions represent only a minority of the white population, their possible future appeal, particularly among those who may find themselves economically displaced in the new South Africa, may present a threat to the continuation of a peaceful transition (Adam and Moodley, 1993).

---

[6]Also of concern is the question of who will succeed Mandela. He is aging and his charisma may make it difficult to find a suitable successor equally capable of mollifying disparate groups.

[7]The different political scenarios for postapartheid South Africa are discussed by Adam and Moodley (1993) and Horowitz (1991).

Resistance may also come from disgruntled blacks who see change occurring too slowly or who do not want to enter into any kind of power-sharing arrangement with the National Party. Paradoxically, changes effected in the economic and political status of the black population may inevitably provoke more, not less, political unrest. Now that blacks are the politically dominant ethnic group in South Africa, expectations are very high that change in their economic status will follow their newly won political status. If expectations exceed the ability of the new government to meet them, however, blacks may become frustrated and place even greater pressure on leaders to produce. As many have observed, ironically, it is only after conditions have begun to improve that people may be drawn into political action (Hoffer, 1951; Runciman, 1966; Turner and Killian, 1972). Thus, the abolition of official apartheid and the increasing breakdown of white privilege may create growing stresses and strains on the new government.

The new South African regime, then, must walk a precarious line in trying, on the one hand, to allay white fears of loss of economic power and, on the other, to meet the rising economic and social expectations of blacks now that political democracy has been won. Failure to accomplish either may throw the society back into renewed and prolonged violence. Failure to seriously reduce socioeconomic inequality may contribute not only to politically inspired hostilities but also to an already dramatically high crime rate (Brewer 1994).[8]

In addition to the overwhelming problems of trying to create a system that will provide for a stable government and thriving economy, South Africa is faced with ethnic diversity that was camouflaged for decades under the oppression of apartheid. To think of South Africa as a society neatly divided into black and white is, as we earlier noted, a glaring oversimplification. Nothing could better illustrate this than the fact that today South Africa recognizes eleven official languages.

Creating Institutional Change Securing political equality cannot be seen as an end in itself; rather, it is only a first step toward the elimination of ethnically based social inequalities in jobs, education, health, and other life chances. De facto segregation continues to characterize most institutions of South Africa, sustaining the enormous discrepancies between whites and blacks. Removing the legal and political foundations of apartheid, therefore, cannot guarantee rapid changes in other institutions — particularly the economy — that have real bearing on the distribution of power and wealth. As one observer has put it, "South Africa today is a place where the white minority of 5 million owns the economy, the 32 million-strong black majority runs the politics, and both races have a living memory of three centuries of oppression" (P. Taylor, 1995:16).

[8]A soaring crime rate is also contributing to, among other changes, an increase in the emigration of whites, whose wealth and skills are badly needed (Daley, 1995).

An issue of greatest significance, therefore, concerns whether more racially equitable economic relations will emerge following the end of a racially discriminatory political system. Indeed, it may be argued that as long as white South Africans retain control of the bulk of the society's economic resources, the wide differentials between whites and blacks, even after fundamental political changes have been effected, will remain firmly in place. The emergence of genuine racial and ethnic equality, then, will remain a long-term objective.

The kind of multiethnic society into which South Africa will evolve can only be speculated. Few observers, even the most informed, were able to predict the scope and speed of the changes that have occurred in less than a decade. To forecast the specifics of South Africa's future, therefore, would surely be presumptuous. What is clear, however, is that South Africa is in the midst of basic change in its ethnic structures and relations. As that change unfolds, much of the world's attention will be focused on this society in the remaining years of this century and probably beyond.

Despite its lengthy history of extreme ethnic division and conflict, South Africa presents itself today as, ironically, a potential model for other multiethnic societies. If the most oppressive system of ethnic relations in the contemporary world can be fundamentally altered in a mostly peaceful fashion, and if those changes can be sustained, there is strong reason to expect that comparable change can be achieved in other societies with severe ethnic divisions.

SUMMARY

Among contemporary multiethnic societies, South Africa has been the most extreme case of inequalitarian pluralism. The society is a racial dichotomy, black and white, but there are more specific group subdivisions. The whites, roughly 15 percent of the total population, comprise two major subgroups, Afrikaners and English. Blacks comprise three subgroups: Coloreds, a racially mixed group; Asians, mostly Indians; and Africans of several different tribal origins. Africans are by far the largest racial category, making up over 75 percent of the population.

Ethnic stratification in South Africa has traditionally been a castelike system in which each ethnic group is institutionally separated by a customary and, until recently, legal code. The system of ethnic separation and inequality, called *apartheid,* was designed to maintain white supremacy in all areas of social, economic, and political life. Africans were excluded from the political system and were protected by few civil rights. Coloreds and Asians were only slightly better off. On economic measures, the inequality between whites and blacks remains extreme though the gap is beginning to narrow.

Starting in the late 1980s, the apartheid system began to be dismantled as black protest and economic uncertainty created strong pressures for reform on the South African regime. The national election of 1994 was the first in which blacks participated as equals and a black-dominated government, led by Nelson Mandela and his ANC, came to power for the first time. This spelled the end of the apartheid system. Internal divisions within the black majority, however, as well as challenges from dissident whites, continue to create political unease.

Though the official system of racial inequality has been abolished and blacks now dominate the political system, the major components of white economic domination remain in place. The key issues of South Africa's immediate future, therefore, concern the manner in which the tremendous economic gap separating whites and blacks will be reduced. The end product of the radical changes that have occurred during the past decade, therefore, are still uncertain.

Suggested Readings

Beinart, William. 1994. *Twentieth-Century South Africa.* New York: Oxford University Press. A succinct history of modern South Africa through the end of apartheid and an analysis of its attempts to create a new national identity.

de Villiers, Marq. 1987. *White Tribe Dreaming: Apartheid's Bitter Roots as Witnessed by Eight Generations of an Afrikaner Family.* New York: Penguin. The author describes the development of ethnic identity among Afrikaners and their current perspective on South Africa by tracing his own family's history.

Fredrickson, George M. 1981. *White Supremacy: A Comparative Study in American and South African History.* New York: Oxford University Press. Examines the historical development of South African race relations and American race relations, comparing their ideologies, policies, and actions.

Horowitz, Donald L. 1991. *A Democratic South Africa?: Constitutional Engineering in a Divided Society.* Berkeley: University of California Press. Discusses various possible future political systems for South Africa. Argues that democracy can be achieved but considers the daunting impediments that must be overcome.

Lelyveld, Joseph. 1985. *Move Your Shadow: South Africa, Black and White.* New York: Penguin. A noted journalist presents a poignant account of life under the apartheid system.

Mallaby, Sebastian. 1992. *After Apartheid: The Future of South Africa.* New York: Times Books. A journalist explores the social and political forces that brought an end to apartheid. In predicting the ultimate outcome of South Africa's transition, the author draws on the experience of other African nations following the end of white rule.

South African Institute of Race Relations. (published annually). *Race Relations Survey.* Johannesburg: South African Institute of Race Relations. An invaluable source of social and economic data pertinent to all aspects of South African ethnic relations.

Thompson, Leonard. 1990. *A History of South Africa.* New Haven, Conn.: Yale University Press. A comprehensive history of this society, tracing the development of race relations through various periods, written by one of the outstanding historians of South Africa.

van den Berghe, Pierre L. 1967. *South Africa: A Study in Conflict.* Berkeley: University of California Press. Although dated, remains one of the few thorough sociological analyses of South Africa under apartheid.

CHAPTER 13

BRAZIL

Moving from South Africa to Brazil is analogous to traveling from one planet to another. A more striking contrast in ethnic relations could not be found in the modern world. Instead of an ethnic purgatory, Brazil has often been praised, by Brazilians and outside observers alike, as a racial and ethnic paradise where people of varied physical features and with diverse cultural heritages live amicably together. Whereas the dominant ideology of South Africa, until recently, was ethnic pluralism in a most extreme form, in Brazil the predominant philosophy has been ethnic assimilation. Here the overriding belief is that the more quickly the society's various ethnic and racially defined groups can be merged into one, the better the society will be.

Whether Brazil is in fact an ethnic paradise is, as we will see, by no means universally agreed on either by scholars or by Brazilians themselves. Another, less laudatory, view of Brazil places it well within the bounds of other multiethnic societies characterized by intergroup conflict and division. In any case, it is clear that the direction of ethnic relations in Brazil is toward assimilation, in which ethnic convergence and the breakdown of distinct physical and cultural groups are the society's long-range goals. More significant, Brazil has advanced further in the realization of these objectives than almost any other major multiethnic society. This fact alone distinguishes Brazil not only from societies with inequalitarian pluralistic systems but also from the United States and other societies like it that have traditionally proclaimed an assimilationist ideology but

BRAZIL

have often exhibited contradictory patterns of action and policy. If ethnic relations are not so ideally harmonious as Brazilian ideologists would have the world believe, they are nonetheless far less combative than in most other multiethnic societies. In what ways and why Brazil's ethnic relations are more amicable are issues that deserve our careful attention.

MAJOR FEATURES OF BRAZILIAN SOCIETY

Brazil seems especially appropriate to compare with the United States because the two societies exhibit strongly parallel ethnic developmental patterns. Both were originally colonized by Europeans who overwhelmed the indigenous population; both imported vast numbers of Africans who were the mainstays of an institutionalized slave system that lasted until the late nineteenth century; and both were peopled by immigrants from a variety of European societies. There are, of course, important ethnic differences between the two societies as well. Africans taken to Brazil as slaves far exceeded in number those taken to the United States, and as a result, blacks composed the largest element of the Brazilian population until the nineteenth century. Although whites predominate today in both societies, in Brazil the African cultural influence has been more widespread and significant. More important, Brazil's ethnic and racial mix is far more amalgamated than the United States'.

Although Brazil is as ethnically varied as any society of the modern world, its dynamics of race and ethnic relations revolve primarily around its major racially defined groups. Our discussion, therefore, will concentrate mainly on patterns of black–white relations and only secondarily on interrelations among European ethnic groups and indigenous Indians.

Brazil is the largest nation of Latin America, both in area and in population. Yet it is clearly divergent from other societies of the region in at least two important respects. For one, its colonial heritage derives from Portugal rather than Spain. Its language and other major cultural elements are therefore Portuguese. As we will see, some hold that the Portuguese cultural legacy largely accounts for Brazil's seemingly more pacific patterns of ethnic relations.

Second, the ethnic composition of Brazil is different from that of almost all other Latin American societies. Except for a few Caribbean nations (most notably, Cuba), Brazil is unique in that it comprises a combination of European, Indian, and African peoples. Specifically, it is the European–African ethnic synthesis that sets off Brazil so sharply from other Latin American societies. It is important to consider, however, that although European, African, and Indian racial strains are evident, the Brazilian population today is so thoroughly mixed physically that it is difficult to determine the racial derivation of most people with any precision.

Brazil is a huge and regionally diverse nation. It is almost as large as the continental United States, and its population is almost as great as that of all the remainder of Latin America exclusive of Mexico. Given this vastness, social and economic conditions are, predictably, not the same in all regions of the country. Brazil's major ethnic groups are also regionally concentrated, and patterns of ethnic relations therefore vary accordingly.

In the Northeast, a majority of the population is either black or mulatto (a combination of European and African), and the African cultural heritage is very apparent. This is Brazil's poorest region, an agrarian and largely underdeveloped area that has never fully recovered from the decline of the slave-supported sugar economy of the seventeenth and eighteenth centuries, of which it was the center. The aftereffects of slavery are still felt, and paternalistic black–white relations are more evident than in other regions of Brazil.

The Southeast, dominated by the metropolis of São Paulo, is the most industrialized and urbanized region of Brazil and is the core of its wealth and power. Ethnically it is primarily European, especially Italian, though its industrial preeminence has served as a magnet for out-migrants from other areas, especially the depressed Northeast. This has contributed to an increasing ethnic heterogeneity in recent decades though whites remain an overwhelming majority. The other region of greatest European influence is the extreme South, where German, Polish, Russian, Italian, and Portuguese farming communities are numerous alongside the traditional *estancias,* or cattle ranches.

The Indian influence in Brazil's population is most significant in the extreme North and in the interior, especially the Amazon basin. This is an enormous but sparsely populated region where most of the remaining Indians (around 200,000) live, most in isolated and primitive settings. Indian cultural and physical influences, however, are strongly evident here, and most of the population derives from a combination European–Indian ancestry.

DEVELOPMENT OF THE BRAZILIAN ETHNIC MÉLANGE

The shaping of Brazil's ethnic mélange resembles that of the United States in one important respect: Almost all its peoples have been immigrants, either voluntary or involuntary. The indigenous population at the time of Portuguese settlement in the early sixteenth century, perhaps two and a half million, was reduced to virtual insignificance in less than three centuries (Hemming, 1987). Unlike the U.S. colonists, who made an unsuccessful attempt to enslave Indians, the Portuguese used the indigenous peoples as slave laborers almost from the outset of colonization. The Indians were devastated in a relatively short time, however, because of their susceptibility to European diseases such as smallpox and measles and the intensity of agricultural labor, with which they were unfamiliar.

In the late sixteenth century, sugar cane became Brazil's chief cash crop. As a labor-intensive industry, sugar required large plantations, and as the Indian slave population diminished, the introduction of Africans in their place proved vital. Marvin Harris (1964) explains that with the development of the world market for sugar, the importation of African slaves—a costly enterprise—became economically feasible. Black slaves were many times more valuable than their Indian predecessors, for they seemed better prepared for field labor and were more resistant to European diseases. The African slave population in Brazil grew rapidly, and by the end of the eighteenth century, it numbered around a million and a half. Estimates of the actual number of slaves imported into Brazil between 1550 and the 1850s, when the slave traffic was ended, are varied, ranging from three million to eighteen million (Knight, 1974; Pierson, 1967; Poppino, 1968;

Ramos, 1939; van den Berghe, 1978). Because all records of the slave trade were purposely destroyed following the abolition of slavery in 1888, there are no precise data concerning the number of slaves actually imported into Brazil. That far more were brought to Brazil than to the United States, however, is undisputed.

The Brazilian and American Slave Systems

To better understand the roots of Brazil's system of ethnic relations, it is necessary to look at its long experience with slavery. It is especially important to compare slavery as practiced by the Portuguese in Brazil with the American slave system because many have argued that contemporary differences in black–white relations in the two societies are traceable to differences in the practice of slavery in each. Indeed, the similarities and differences in the two slave systems and the subsequent effects of each on patterns of race relations together make up one of the most widely debated issues of comparative social history.

Brazilian Slavery: A Milder System? Gilberto Freyre, Brazil's most renowned social historian, contended that slavery in Brazil was a milder, more humane form than its North American counterpart. This he attributed to several factors, the most important of which pertain to differences between Anglo-Saxon and Portuguese culture and society. Different practices of slavery, in turn, produced divergent patterns of relations between blacks and whites once slavery was ended in each society.

Freyre held that the Portuguese came to Brazil with a more tolerant attitude toward people of color. Portugal had experienced contact with African peoples in its colonial ventures as early as the fifteenth century and had endured a long period of rule by the Moors even earlier. The Portuguese sense of racial difference, therefore, was not so acute and absolute as that of the Anglo-Saxons, who had had little contact with nonwhites. A greater racial tolerance, according to Freyre's thesis, accounts for the more harmonious, less conflict-ridden nature of race relations in Portuguese colonies and the greater tendency among the Portuguese to intermix with native and slave populations. Freyre considered especially significant the extremely high rate of miscegenation between Portuguese slave masters and African slaves in Brazil. This, he contended, is evidence of a more compassionate and essentially altruistic master–slave relationship (1956, 1963a).

The thesis that the Portuguese slave system in Brazil was basically more humane than its North American counterpart is supported by an American historian, Frank Tannenbaum (1947). Tannenbaum explained that a milder form of slavery was evident in Brazil and in the Spanish colonies of Latin America because of differences between the Portuguese (and Spanish) and the Anglo-Saxons in their religious and legal concepts of slavery. He, like Freyre, maintained that there was a preconceived notion among the Portuguese regarding their treatment of slaves, deriving from principles established by both church and state, that fostered a less harsh slave system. In Brazil, the African did not lack a "soul" but was perceived simply as an unfortunate human being, a slave by accident of fate. In the United States, however, the African was perceived as less than human and thus naturally enslaved. Slavery in Brazil was understood as an economic necessity, hardly a relationship whose rationale derived from religious, or later biological, principles as in the United States. In short, in Brazil the human character of slaves was never denied. In the Portuguese view, slavery was a necessary evil. Thus, it was not defended so firmly on the notions of racial inferiority and superiority.

The Brazilian and U.S. legal codes dealing with slaves also differed significantly. Portuguese (and Spanish) laws pertaining to slavery protected the slaves' human rights, at least technically. Thus, slaves could not be dealt with by their masters with total impunity. Whereas slaves in the United States were chattel, reduced to merely property, in Brazil they were recognized as humans with certain legal rights. They were entitled to own property, marry freely, seek out another master if they were dealt with too harshly, and even buy their freedom, which many apparently did. It is this last-mentioned feature of the slaves' status, their greater opportunities for manumission, that Tannenbaum explained as an especially critical difference. Compared with the United States, manumission in Brazil was both common and expected. Its much higher incidence, Tannenbaum held, is in itself evidence of a milder form of slavery. More important, however, the accomplishment of manumission brought the freed slave automatic entry into Brazilian society with full and equal rights. In the United States, by contrast, even if slaves were freed, they did not acquire full citizenship but continued to suffer the economic, social, and legal handicaps that attached to their racial status.

The ability of the crown or the church to enforce these protective slave laws, of course, was quite another matter. These were, in effect, paper rights, whose effectiveness was not sufficient to materially alter the nature of life for Brazilian slaves. "In Brazil, as everywhere in the colonial world," notes Harris, "law and reality bore an equally small resemblance to each other" (1964:77). Neither Freyre nor Tannenbaum contended that slavery in Brazil was not harsh and inherently cruel. The key difference between the Brazilian and American varieties, however, was that "in the Spanish and Portuguese colonies the cruelties and brutalities were against the law, that they were punishable, and that they were perhaps not so frequent as in the British West Indies and the North American colonies" (Tannenbaum, 1947:93). Most important, the openings for freedom were much greater in the Brazilian system, as was the advancement to full citizenship following manumission. Given the different legal and philosophical status of the slave and the freed black in Brazil as compared with the United States, more benign and assimilationist relations were engendered between whites and blacks following the formal abolition of slavery.

Brazilian Slavery: A Harsher System? Revisionist thinking on the matter of Brazilian slavery has seriously questioned whether it was in fact a distinctly more humane system than those established in the United States and the Caribbean. Some have even concluded that in many ways it was a system more dehumanizing than the Anglo-Saxon variety. That slaves in Brazil were acknowledged as human and that they possessed certain legal and religious rights did not seem to discourage the common application of brutal and inhumane forms of punishment by masters and the maintenance of severely debased living conditions by any standard. Historian Charles Boxer (1962) has described the brutish and often sadistic cruelty with which slave discipline was enforced by the Portuguese planters. To tie a slave to a cart, flog him, rub his wounds with salt, lemon juice, and urine, and then place him in chains for a number of days was not considered excessive. Runaways were dealt with by branding for the first offense, cutting off an ear for the second, and, usually, death for the third.

That such punishments did not prevent slaves from continuing to run away on a large scale bespeaks the desperation to which they were driven. Other features of Brazilian slavery provide further evidence of its unusually harsh nature. Slaves commonly committed suicide, and insurrections among them were far more frequent than in the United States. Indeed, compared with the few rebellious actions by American slaves, Brazilian slave insurrections, many of them well organized, appear to have been very common. In fact, organized settlements of fugitive slaves, called *quilombos,* sprang up in various locales.

Historical evidence, then, seems to have proved beyond much doubt that "humane" slaveowners in Brazil were the exception rather than the rule (Boxer, 1962, 1963; Davis, 1966; Graham, 1970). Recent historians have also disputed Freyre's and Tannenbaum's assumption that more frequent miscegenation and manumission as features of the Brazilian slave system substantiate the greater humanity and tolerance of the Portuguese. As for miscegenation, Skidmore (1993) notes that this was a common occurrence within the American system as well. The early enactment of laws prohibiting interracial marriage in the United States confirms the strong sexual attraction between the two groups, as does the wide range of physical traits of American blacks. Moreover, in Portugal's African colonies like Mozambique, the extent of miscegenation was never as significant as it was in Brazil. Clearly, factors other than the natural attraction of Portuguese men to native and slave women and their greater racial tolerance account for the high rate of interracial sexual encounters in Brazil. A more compelling explanation lies in the simple fact that colonial Brazil lacked sufficient Portuguese women. Unlike North American colonists, who came mostly in family units, Portuguese settlers were mainly unaccompanied males who quite naturally turned to native and slave women to meet their sexual needs. Harris has put the matter succinctly: "In general, when human beings have the power, the opportunity and the need, they will mate with members of the opposite sex regardless of color or the identity of grandfather" (1964:68).

As to the question of more common manumission granted Brazilian slaves, historian Carl Degler (1971) has suggested that this may have been simply a method of relieving slaveowners of the responsibilities and costs of caring for elderly or physically unfit slaves. Harris (1964) also notes that although manumission was more common in Brazil, it was not so much more common as to build a case on this fact alone for a presumably milder form of slavery. Degler has pointed out that the American and Brazilian slave systems may not have differed so fundamentally even on the matter of slaves' legal and moral status. Though legal decisions were not consistent from time to time and place to place, both church and state, at least technically, recognized slaves as humans in the United States as well as in Brazil. In neither society, however, was the law, whether church or state, effective in protecting slaves.

The issue of whether the Brazilian system of slavery was more humane than the American may, in the final analysis, be meaningless. How can such an argument be verified in any case? Indeed, the very notion of a "humanized" or "mild" slave system would be interpreted by most as a grotesque contradiction. Whatever the circumstances, slavery was dehumanizing and was resisted by the slaves.

Abolition and Its Aftereffects Although the actual conditions of slavery may not account in large part for the subsequently more benign relations between blacks and whites in Brazil, the conditions leading to the abolition of slavery and the status of blacks following their freedom may tell us a great deal.

Slavery endured in Brazil even longer than it did in the United States, ending formally in 1888. Its demise had been assured for several decades, however, and occurred in a gradual, evolutionary manner. This piecemeal abolition of slavery resulted in less wrenching social aftereffects than occurred in the United States, where slavery was ended abruptly following a cataclysmic civil war.

With the more common manumission of slaves in Brazil, blacks had already been a part of free Brazilian society for many decades before the formal end of slavery. In 1872, sixteen years before abolition, freed blacks were almost three times as numerous as slaves. By comparison, in the United States at the start of the Civil War, less than half a million out of a total black population of four and a half million were free (Skidmore, 1972). This meant that blacks and whites had experienced a long period of social interaction exclusive of the master–servant relationship. As Tannenbaum explained, in Brazil, "The Negro achieved complete legal equality slowly, through manumission, over centuries, and after he had acquired a moral personality. In the United States he was given his freedom suddenly, and before the white community credited him with moral status" (1947:112). As a result, there was no question of the social standing of free blacks in Brazil and their access to full citizenship. In Brazil, freed slaves were in fact free; in the United States, they were only technically free.

Furthermore, blacks in colonial Brazil had played a variety of occupational roles — even as slaves — and their entrance into the competitive labor market was therefore not so precipitous as it was in the United States (Karasch, 1975; Klein, 1969). Harris (1964) argues, in fact, that the continual shortage of skilled and semiskilled labor in colonial Brazil created the need to fill numerous positions with blacks, which, in turn, contributed more than anything else to the ease and frequency of manumission.

Freed blacks in Brazil, then, were not seen as either a social or an economic threat, and resistance from whites, therefore, did not strongly materialize. Brazil thus avoided the development of a racial caste system designed to permanently handicap blacks, as was the eventual outcome of abolition in the United States.

Although Brazilian society was not so fundamentally dislocated by the abolition of slavery as the American South had been, the consequences faced by newly freed slaves were in many ways similar. Freed slaves either wandered back to the rural work force, seeking out their former masters, or migrated to the cities with few urban labor skills. They were further handicapped as they were forced into competition with more skilled European immigrants, especially in the industrializing Southeast.

The release of former slaves into a competitive labor market for which they were ill prepared produced a debilitating effect on black mobility that has remained a basic feature of the Brazilian social structure to the present day. As Brazilian sociologist Florestan Fernandes explains, blacks emerged from slavery materially and psychologically ravaged and "lacked the means to assert themselves as a separate social group or to integrate rapidly into those social groups that were open to them" (1971:28). Further disabling the freed black was the whites' eager relinquishment of any legacy of the slave system and its consequences. The occupational benefits of an industrializing Brazil with its competitive social order was therefore reaped primarily by European immigrants.

European Immigration

Beginning in the late nineteenth century, the ethnic composition of Brazil was altered fundamentally by the large-scale immigration of Europeans. Under the colonial regime, immigration into Brazil had been severely restricted. As a result, the bulk of the population was descended from three major ethnic lines: the Portuguese colonials, the indigenous Indians, and the African slaves. Following Brazilian independence in 1822, immigration remained sporadic until the mid-nineteenth century. An upward trend turned into a veritable flood tide of immigrants in the century's last two decades.

With the realization that slavery was doomed, coffee planters increasingly turned to Europeans, especially Italians, as a labor force. The bulk of European immigration beginning in the late nineteenth century, therefore, was directed to the state of São Paulo, where the major coffee-growing estates were located. Growers saw Europeans as better and more reliable workers than native Brazilians, especially freed blacks from the declining Northeast. São Paulo also became the primary destination for European immigrants because it represented the nucleus of burgeoning Brazilian industry. Here were to be found the industrial jobs for which these immigrants were better prepared than were Brazilian blacks.

European immigration was encouraged for another reason having nothing to do with the needs of labor. Racially, these immigrants were white and therefore contributed to the "whitening" of the society, an ideal that was now prevalent among the Brazilian political leadership and intelligentsia (Skidmore, 1990). Since abolition, the notion of an increasingly whiter population had been a staple of Brazilian intellectual thought. Abetted by the widespread influence of Social Darwinism at the turn of the century, whites felt that Brazil should seek to dilute as quickly and completely as possible the black element of its racial composition. Put simply, the idea was "the whiter Brazil, the better" (Skidmore, 1972). Immigration policy thus reflected this philosophy.

The mass immigration from Europe gave Brazil's whitening process a giant infusion. So limited had immigration been previously and so substantial had been the importation of African slaves that by the end of the eighteenth century, Africans were a majority of the Brazilian population. European immigration not only increased racial mixing but reversed this ratio.

In the century between 1850 and 1950, five million immigrants came to Brazil. The peak era of immigration, however, occurred between 1870 and 1920, when over three million Europeans entered the society (de Azevedo, 1950; Skidmore, 1974). Almost two million entered in the little more than two decades between 1881 and 1903. The majority were absorbed by the state of São Paulo, whose foreign population rose from 3 percent in 1854 to 25 percent by 1886 (Dzidzienyo and Casal, 1979). More than half the population by 1934 were first- and second-generation immigrants (Mörner, 1985). Italians were by far the largest immigrant group, but Portuguese, Spanish, and Germans were also significant. By 1897, Italians outnumbered Brazilians in the city of São Paulo two to one (Morse, 1958). Today, about one-third of São Paulo state's population is of Italian descent (Freyre, 1963a).[1] Europeans were also attracted to Brazil's extreme southern states, where large numbers of Germans and Poles established cohesive ethnic communities (Luebke, 1987).

In addition to non-Portuguese Europeans, another distinct ethnic element was added to the Brazilian mosaic beginning in 1908, when Japanese immigrants entered the society in large numbers for the first time. By 1941, almost 190,000 Japanese had immigrated into Brazil, with another 50,000 arriving after World War II. Today, the Japanese ethnic community of Brazil numbers more than a million. Most originally settled as agricultural workers in the states of São Paulo and Paraná, and the Japanese still remain heavily concentrated in those areas.

ETHNIC STRATIFICATION

In the United States, ethnic stratification, particularly as it pertains to racial-ethnic groups, is clear-cut: Those classified as white are at the top of the ethnic hierarchy and disproportionately occupy most top positions of power and privilege, while those classified as nonwhite are at the lower end of the hierarchy and collectively possess less power and privilege. (Recall the three-tiered hierarchy outlined in Chapter 11.) Furthermore, in social relations, the lines dividing white and black are rigid, with groups coalescing around separate social institutions except for the economy and polity. Brazil's system of ethnic stratification resembles that of the United States in at least one regard, the rank order of ethnic categories. Whites disproportionately hold the society's power, privilege, and prestige. But several unique features of the Brazilian system distinguish it from that of the United States, as well as other multiethnic societies, making it one of the most far-reaching attempts at racial and ethnic democracy in the modern world.

The Brazilian Multiracial System

Perhaps the key factor differentiating the Brazilian system of ethnic relations from those of other societies in which there are several distinct physical types is the very concept of race and the manner in which people are racially classified.

In the United States, one's racial status is a product of one's descent; that is, individuals, regardless of physical appearance, are assigned to a racial status on the basis of ancestry. Such policies are based on what Harris (1964) calls the rule of "hypo-descent": a person with any known or recognized ancestry of the subordinate group is automatically classified as part of that group rather than the

[1] Until the early 1900s, Brazil and especially Argentina—not the United States—were the major destinations of Italian immigrants (Mörner, 1985).

dominant group. There are no provisions for "intermediate" cases, so that the system of racial classification is essentially a dichotomy—white and black. Thus, in the United States one may exhibit generally white physical features yet still be classified as black. For example, in 1983, a Louisiana judge declared a woman who had been raised as a white legally black on the basis of a distant black ancestor. She and her siblings were the great-great-great-great grandchildren of a black slave and a white planter. The decision was based on a Louisiana law declaring anyone with one-thirty-second "Negro blood" legally black. The woman was listed as three-thirty-seconds black (*Miami Herald,* 1983).

In Brazil, however, this strict rule of descent is not operative. One is not automatically placed into a racial category on the basis of family of origin. Instead, one's total physical appearance—skin color, hair texture, facial features—are the "obvious" determinants of one's racial classification. But in addition to these strictly physical aspects of racial identity, there are in Brazil certain social factors that are used to classify people racially. The most important of these is social class (Harris et al., 1993). Thus, as one experiences class mobility, one's racial identity may, to some extent, also change. Popular Brazilian expressions such as "money whitens" or "a rich Negro is a white man, and a poor white man is a Negro" connote that as people improve their class status, they are perceived as lighter racially. To a degree, then, racial status is subject to redefinition in different social circumstances.

The imprecise nature of racial classification in Brazil means that racial groups are not castelike as in the United States, where "black" describes anyone with traceable African heritage, including those of mixed origins. But this understanding does not hold in Brazil. *Branco* and *prêto*—literally, white and black—merely denote people who are predominantly white or black in appearance, regardless of actual racial origin (Nogueira, 1959; Pierson, 1967). Simply having some traces of black ancestry does not result in the classification of someone as black. Indeed, a majority of those classified as white in Brazil exhibit some evidence of black descent. Many American blacks, then, would be considered white in the Brazilian context.

Racial and Ethnic Categories This perception of race, different from American racial perceptions, has produced a correspondingly different system of racial categorization. Whereas the United States maintains an essentially biracial system in which people are classified as either white or black,[2] Brazil's multiracial system provides for more than two categories. Three major racial groupings

[2]The increasing Asian population in the United States has, of course, forced a third major category into the racial-classification scheme used by government and other institutions. The principal racial division, however, remains black and white.

are recognized in Brazil, but the boundaries between them are neither rigid nor clear-cut. *Branco* (white), *prêto* (black), and *pardo* (mulatto) are the most encompassing terms, but Brazilians employ literally dozens of more precise terms to categorize people of various mixed racial origins, depending on their physical features. Racial terminology is strongly localized, and from region to region—and even community to community—standards of classification vary.

The most minute physical features are often used to sort out people racially. The complexity and exactness of Brazilian racial typologies can baffle the American familiar with only a two-part racial scheme. Consider the descriptive categories used by the people of Vila Recôncavo, a town near Salvador in the state of Bahia, studied by anthropologist Harry Hutchinson:

> The *prêto* or *prêto retinto* (black) has black shiny skin, kinky, woolly hair, thick lips and a flat, broad nose.
>
> The *cabra* (male) and *cabrocha* (female) are generally slightly lighter than the *prêto,* with hair growing somewhat longer, but still kinky and unmanageable, facial features somewhat less Negroid, although often with fairly thick lips and flat nose.
>
> The *cabo verde* is slightly lighter than the *prêto,* but still very dark. The *cabo verde,* however, has long straight hair, and his facial features are apt to be very fine, with thin lips and a narrow straight nose. He is almost a "black white man."
>
> The *escuro,* or simply "dark man," is darker than the usual run of *mestiços,* but the term is generally applied to a person who does not fit into one of the three types mentioned above. The *escuro* is almost a Negro with Caucasoid features.
>
> The *mulato* is a category always divided into two types, the *mulato escuro* and *mulato claro* (dark and light mulattoes). The *mulato* has hair which grows perhaps to shoulder length, but which has a decided curl and even kink. . . . The *mulato*'s facial features vary widely; thick lips with a narrow nose, or vice-versa.
>
> The *sarará* . . . has a very light skin, and hair which is reddish or blondish but kinky or curled . . . His facial features are extremely varied, even more so than the *mulato*'s.
>
> The *moreno* . . . is light-skinned but not white. He has dark hair, which is long and either wavy or curly . . . His features are much more Caucasoid than Negroid (1963:28–30).

Figure 13.1 Racial Classification Patterns in Brazil and the United States

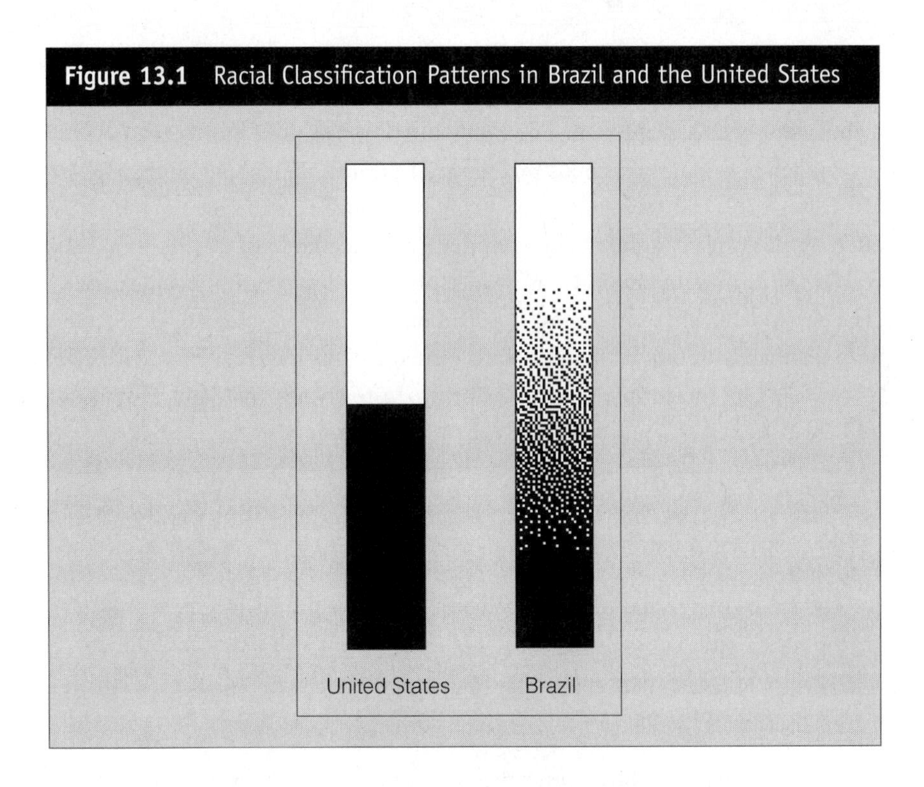

These are only the major categories, explains Hutchinson, each of which is further broken down even more precisely. Even the whites are subclassified on the basis of skin, hair, and facial features.

These categories, however, even the major ones, do not denote discrete racial groupings. Rather, they blend into one another as on a scale (Figure 13.1). As already noted, miscegenation has, from the time of the initial Portuguese settlement, been a constant and widespread practice in Brazil among the society's three major population types—European, African, and Indian. As a result, the racial origins of people are thoroughly interwoven in various combinations. *Black, white,* and *mulatto,* therefore, represent very broad categories, each with a wide range of physical types. Although there is general agreement on the terms *branco* and *prêto,* it is the mulatto, or *pardo,* category, between extreme white and extreme black, that provides the almost endless variety of descriptive combinations throughout Brazil. As is seen in Table 13.1, mulattoes officially composed 39 percent of the population in 1990, but unofficially, the percentage is much higher.[3] Again, it is important to emphasize the indistinct lines between these

---

[3]The *pardo* category consists not only of mulattoes (those of mixed European and African origins) but also those who are mixed European and Indian, usually referred to as *caboclos,* and those of mixed African and Indian origins, *cafusos. Cafusos,* however, are numerically a small element of the Brazilian population.

Table 13.1 Brazilian Population by Color

	1940		1960		1980		1990	
GROUP	*Number*	*Percent*	*Number*	*Percent*	*Number*	*Percent*	*Number*	*Percent*
White	26,171,778	63	42,838,639	61	65,212,759	55	81,407,395	55
Mulatto	8,744,365	21	20,706,431	29	45,779,466	38	57,821,981	39
Black	6,035,869	15	6,116,848	9	7,009,104	6	7,264,317	5
Yellow and other	284,303	1	529,452	1	1,069,536	1	811,181	1

SOURCE: Instituto Brasileiro de Geografia e Estatística, 1981, 1992.

categories. Racial definition is largely subjective and arbitrary. Because, as we will see shortly, it is socially advantageous in Brazil to be considered white, the 1990 population figure of 55 percent white in Table 13.1 is greatly exaggerated in terms of the North American conceptualization of race (Harris et al., 1993). In any case, these census figures indicate a progressive decline of the white percentage, signaling that, even when measured by people's self-identification, Brazil is gradually emerging as a racially mixed society.

Brazilians, then, are very conscious of color distinctions among people; indeed, they are far more conscious than North Americans or even South Africans, perceiving numerous physical types between black and white.[4] These gradations of color, however, make for a fluid system of ethnic stratification. It is the color continuum of Brazil as opposed to the rigid black–white dichotomy of the United States that explains much of the difference in the tone of ethnic relations in the two societies. Degler (1971) contends that the existence of the intermediate — mulatto — category in Brazil, more than anything, accounts for the diminished hostility between blacks and whites. The presence of what he calls "the mulatto escape hatch" makes unfeasible the development of segregated institutions, which have so dominated most of the history of black–white relations in the United States. Where there are so many recognized racial combinations and where people's racial status is so arbitrary and personalized, it is impossible to erect castelike structures based on racial definitions. "The presence of the mulatto," explains Degler, "not only spreads people of color through the society, but it literally blurs and thereby softens the line between black and white" (1971:225).

[4]When asked to describe their color in the 1980 census, Brazilians responded with 136 different labels, and in an earlier national household survey, with almost 200 different labels (Andrews, 1991).

Furthermore, where there is such a varied biological mixture of the population, racial ideologies propounding the notion of superior and inferior races are difficult to sustain.

In a recent article, Skidmore (1993) casts doubt on the assumption that the presumed multiracial classification system of Brazil is today basically different from the U.S. biracial system. He contends that, at the same time that there is an increasing awareness of the racial variability—and thus color gradations—of the U.S. population, there is a growing tendency among Brazilian scholars and others to describe Brazil in bipolar racial terms, that is, simply "white" and "nonwhite."

The Brazilian Ethnic Hierarchy

At first glance, it might seem pointless even to envision an ethnic hierarchy, that is, a ranking structure of ethnic groups, in a society where hard and fast lines between ethnic groups are not recognized. Ethnicity would seemingly play no role in the allocation of social rewards. This, in fact, is "officially" how Brazil has been portrayed. Several aspects of the Brazilian system of ethnic relations, however, belie this ideal picture.

The Class-Ethnic Correlation By looking at the general class positions of Brazil's racial-ethnic categories, one can conclude without hesitation that there is indeed an ethnic rank order in this society. Although the boundaries of the three major groupings are not precise, there is little question that whites, mulattoes, and blacks represent a hierarchy of economic, political, and social standing. Whites are clearly at the top, followed in order by mulattoes and blacks. A stranger in Brazil could quickly confirm through simple observation the validity of an old Brazilian rule of thumb, "The darker the skin, the lower the class" (Wagley, 1971). Studies compiled by social scientists have repeatedly corroborated this visual impression. In his classic study of Salvador, Bahia, in the late 1930s, Donald Pierson (1967) found that Europeans were heavily concentrated at the top of the city's occupational structure even though they were a numerical minority. Though his findings were limited to one region, Pierson maintained that they were generally applicable to Brazilian society. More recent studies have confirmed that his conclusions generally remain valid for all regions of the country (Dzidzienyo and Casal, 1979; Fernandes, 1971; Garcia-Zamor, 1970; Hasenbalg, 1985; Reeve, 1975, 1977; Silva, 1985; Wagley, 1971).

In looking at the class-ethnic relationship, one must be careful to keep in mind the considerably different class structure of Brazil compared with those of the United States and other advanced industrial societies. Brazil, like most Latin American societies, has traditionally maintained a two-class system—an upper

Figure 13.2 Relationship Between Color and Social Status in Brazil

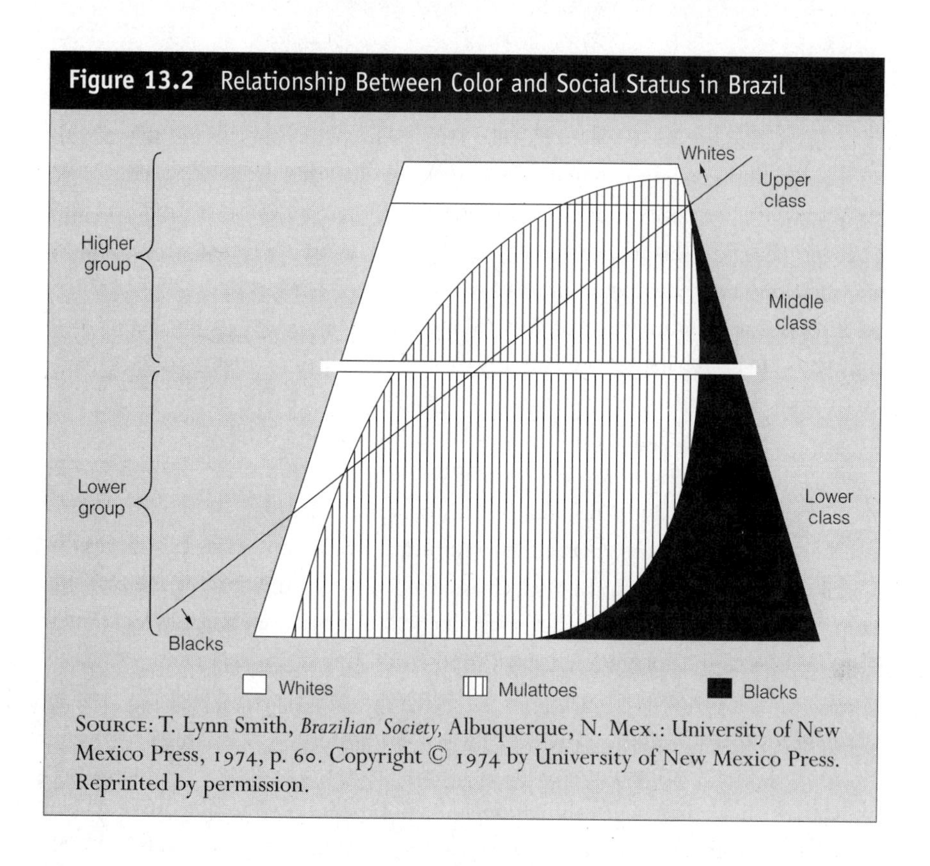

SOURCE: T. Lynn Smith, *Brazilian Society,* Albuquerque, N. Mex.: University of New Mexico Press, 1974, p. 60. Copyright © 1974 by University of New Mexico Press. Reprinted by permission.

class, made up of landowners as well as wealthy merchants and professionals, and a huge lower class, comprising much of the remainder of the population. Moreover, the line between classes has been rigid, making upward mobility difficult. This traditional class system is gradually breaking down with the continued industrialization of the society, and class patterns are beginning to resemble those of advanced industrial societies. As this occurs, an expanding middle class of salaried professionals and white-collar workers is emerging, though it is still small by North American standards.

The ethnic hierarchy in Brazil has customarily paralleled the class system, and this pattern remains largely evident today (Figure 13.2). At the pinnacle of the structure are whites, and "as one moves down the social hierarchy, the number of racially mixed or otherwise nonwhite individuals gradually increases" (Wagley, 1971:121). Although mulattoes and blacks are represented throughout the occupational structure, they are not found in any significant number or anywhere near their proportion of the population in positions of authority or decision making. The elite levels of government, the economy, the military, and education are thoroughly dominated by whites. For example, in 1988, none of the country's 23 state governors were black, and of the 559 members of Congress, only 7 considered themselves black (Simons, 1988). At the opposite end of the hierarchy, whether in rural or urban areas, the bulk of the poor and uneducated are black. Only in sports and entertainment do Brazilian blacks play a prominent role (Schneider, 1996).

Income and education statistics also demonstrate this racial hierarchy. Most simply, whites not only hold a disproportionate share of more prestigious jobs but also receive a disproportionate share of income (Hasenbalg, 1985; Silva, 1985). Nonwhites comprise almost two-thirds of the poorest Brazilians, those earning the minimum wage or less (Margolis, 1988). One study showed that nonwhites in all regions of the country earned about half of what whites earned. And these differentials were shown to be attributed not to differences in schooling or work experience but to racial discrimination (Webster and Dwyer, 1988). Not surprisingly, similar gaps between whites and blacks are evident in the distribution of education and in rates of literacy (Schneider, 1996). Moreover, few nonwhites were part of the significant rise in university enrollment in Brazil during the 1970s and 1980s (Hasenbalg, 1985). Such discrepancies in income and education lead to other social inequalities between racial groups, such as child mortality (Schneider, 1996; Wood and Carvalho, 1988).

The "Whitening" Ideal Another clear-cut indication of an ethnic hierarchy in Brazil is the social preference for white over black. Despite the amalgam of skin colors and other physical features in its population, "white" has always been understood as more desirable than "black." The socialization process is quite effective in conveying the notion that to be white is to be socially favored. The previously quoted adages, implying that one gets "whiter" as one moves up in social class, are manifestations of this racial ideology. Brazil's white partiality has found expression at both individual and societal levels.

At the individual level, the white ideal is most evident in the preference for lighter-skinned marital partners. Such a marriage usually earns the darker-skinned partner a higher social status and in any case produces lighter-skinned children, who are assured of better social opportunities. Degler notes that "when a black or mulatto succeeds in marrying lighter, he speaks of 'purging his blood'—a phrase that in itself is revealing of what the process connotes in the minds of Brazilians" (1971:191–92). The white preference is often imparted to children in blunt fashion. Sociologist Oracy Nogueira explains that "from the start it is impressed upon the white child's mind that negroid characteristics make [the] bearer ugly and undesirable for marriage" (1959:171). Adults frequently tease a white boy that when he grows up he will "marry a Negro." Whereas marrying into a lighter family and having lighter offspring is the most effective form of whitening for Afro-Brazilians, others include upward economic mobility, associating with white friends and acquaintances, and adopting a white middle-class lifestyle (Andrews, 1991).

At the societal level, the "white is best" attitude has been evident in various governmental policies, particularly those pertaining to immigration. As earlier noted, European immigration was encouraged not only to expand the industrial labor force but also, as Roger Bastide puts it, "to submerge the descendants of Africans into a more prolific white population, and, in the last analysis, to change the ethnic composition of the population of the country" (1965:17).

The desire to create a whiter population stems from the response of Brazilian political and intellectual elites in the late nineteenth and early twentieth centuries to racist theories emanating from Europe and America, propounding notions of inferior and superior races. Without having to accept openly the idea of white superiority and black inferiority, they posited that the black (and, presumably, inferior) element of Brazil's population would be submerged through miscegenation and European immigration. These, they assumed, would progressively "whiten" the population, and problems of race, therefore, would be avoided. The whitening process is still seen as desirable and is expressed, as we will see, in the tendency to discount racial inequalities and to squelch the development of racial consciousness. Brazilians, writes historian Thomas Skidmore, "are still implicit believers in a whiter Brazil, even though it may no longer be respectable to say so" (1990:28).

PREJUDICE AND DISCRIMINATION

It would be expected that in a society openly and ardently proclaiming the desirability of amalgamation of its various groups, ethnic prejudice and discrimination would be minimal. Brazilians do customarily assert that racial prejudice and discrimination in Brazil are insignificant. As with ethnic stratification, however, there are glaring inconsistencies between the official accounts of racial democracy and the realities of individual and group relations.

Visitors to Brazil often comment on the seemingly placid and harmonious nature of interethnic relations. People of the entire range of color and of widely varied cultural origins appear to interact at all social levels with little or no difficulty arising from their ethnic differences. An American journalist stationed in Brazil described his surprise at witnessing one such social encounter, a dance at a neighborhood-based club:

> The crowd was interracial, with both blacks and whites in substantial numbers. At first glance, not much different from the crowd at the average Washington Bullets basketball game. But if you looked closer, you saw that blacks and whites were arriving together, dancing together and leaving together in couples and groups, not in separate cliques. This was different.

> And if you went down and made your way through the crowd, you saw that up close it was hard even to tell who was black and who was white. [L]ater . . . I pondered what I had found so striking about the evening. I realized that I had been in the midst of an interracial crowd that didn't really feel like an interracial crowd, at least not like one in the United States (Robinson, 1996:23).

Such impressions extend back even to the period of slavery. Henry Koster, an Englishman traveling in Brazil at the beginning of the nineteenth century, remarked that "it is surprising, though extremely pleasing, to see how little difference is made between a white man, a mulatto and a creole Negro if all are equally poor and if all have been born free" ([1816] 1966:152).

Observant visitors have not been deceived. Relations between ethnic groups in Brazil *are* clearly more benign than those in most of the remainder of the modern world. Overt interethnic conflict is rare, and in no institutional area of the society have measures designed to discriminate against people on the basis of ethnicity ever been common or officially sanctioned. But such manifestations of ethnic harmony do not tell the entire story. Closer analysis reveals that prejudice toward nonwhites is a well-founded aspect of this society's culture. Moreover, these attitudes are translated into stringent rules of social distance, which are often disguised by the customary physical contact between ethnic groups in public places. Particularly at the primary levels of social interaction, the major color groups do not interact to a significant extent, and at the secondary levels, too, interaction is less cordial and acceptable than superficial observations indicate. Patterns of discrimination against nonwhites, though less overt and intensive than in other multiethnic societies, are also evident.

Prejudice

All who have carefully studied Brazilian ethnic relations have attested to the widespread prejudicial beliefs about and attitudes toward blacks and mulattoes. Specifically, ethnic prejudice is manifested in negative stereotyping of nonwhites, in patterns of social distance, and in the general disapproval of intermarriage between those of extremely divergent racial features.

Stereotyping Numerous studies have verified the prevalence of negative stereotypes of Brazilian people of color (Andrews, 1991; Bastide, 1965; Bastide and van den Berghe, 1957; Fernandes, 1971; Harris, 1963; Hutchinson, 1963; Pierson, 1967; Sanders, 1981; Saunders, 1972). Furthermore, similar stereotypes of blacks and mulattoes exist both in regions with great numbers of people of African origin and in areas with few (Bastide, 1965; Cardoso, 1965).

Bastide and van den Berghe (1957) found, for example, that among white middle-class university students in mainly white São Paulo, blacks were commonly viewed as dirty, physically unattractive, superstitious, profligate, immoral, aggressive, lazy, lacking persistence at work, and sexually "perverse." Mulattoes were seen in much the same manner. Almost half of their sample also felt that blacks were intellectually inferior to whites. It is interesting to consider the resemblance of these stereotypes to those historically applied to blacks, Hispanics, and American Indians in the United States.

In Minas Velhas, a small town in the state of Bahia where most of the inhabitants were of mixed European–African origin, Marvin Harris reported generally similar beliefs and attitudes. "The superiority of the white man over the Negro," he wrote, "is considered to be a scientific fact as well as the inconvertible lesson of daily experience" (1963:51). Harris noted the abundance of legends, folk stories, and pseudoscientific notions that supported the negative beliefs about blacks. A school textbook used in Minas Velhas, for example, stated that "of all races the white race is the most intelligent, persevering, and the most enterprising," whereas "the Negro race is much more retarded than the others," a view not disputed by any of the village's six teachers (Harris, 1963:51–52).

Harris explained that although blacks in the town were commonly derided as ignorant and ugly, such views were rarely expressed bitterly or hatefully.

> In Minas Velhas, to the white and to a certain extent to the Negro himself, the Negro is primarily a curious laughable anomaly. He is looked upon as a sport of nature, as a being with certain substandard and grotesque characteristics which make him amusing rather than disgusting. A white man will say: *"Negro desgraçado. Que bicho feio!"* (Miserable *negro*! What an ugly creature!) and smile broadly as though he were speaking of some rare amusing freak (1963:53).

Moreover, the black stereotypes here (as well as in other regions) are often contradictory. Blacks can be seen at once as stupid and cunning, honest and dishonest, naive and shrewd, or lazy yet fit only for hard work. Three white attitudes, however, are constant and basic: (1) the black is inferior to the white; (2) the black does and should play a subservient role to the white; and (3) blacks' physical features are displeasing (Harris, 1963:56). "On the whole," noted Harris, "there is an ideal racial ranking gradient in which whites occupy the favorable extreme, Negroes the unfavorable extreme and mulattoes the various intermediate positions" (1964:60).

Social Distance Studies of social distance also reveal strongly prejudicial attitudes toward nonwhites (Bastide and van den Berghe, 1957; Garcia-Zamor,

1970; Zimmerman, 1963). In their study of white university students in São Paulo, Bastide and van den Berghe (1957) found that almost all accepted the idea of equality of opportunity for everyone, regardless of color. Better than half also accepted the desirability of casual relations between members of different racial groups. But beyond this level, social distance increased markedly. Sixty-two percent opposed intimacy with blacks beyond simple comradeship, and 77 percent opposed miscegenation with blacks (55 percent also opposed miscegenation with mulattoes). The most unwavering attitudes involved intermarriage; 95 percent indicated they would not marry a black, and 87 percent would not marry a light-skinned mulatto.

Endogamy The infrequency of marriages between those of noticeably different color is perhaps the clearest evidence of negative attitudes toward nonwhites. The disapproval of interracial marriage may seem curious in a society that has countenanced miscegenation for its entire social history. Compared with the United States, of course, there is a great deal of intermarriage across color lines, and the racially mixed character of most Brazilians testifies to the persistence of this practice. But the degree of social tolerance for such unions drops sharply when the contrast in color of the partners is great (Cardoso, 1965; Degler, 1971; Fernandes, 1971; Hutchinson, 1963; Reeve, 1977). Most marriages are between persons of similar color, and most intermarriages are between blacks and mulattoes rather than between whites and either blacks or mulattoes. Moreover, most intermarriages occur at the lower end of the class hierarchy. Hutchinson (1963), for example, noted that in the small Bahian community he studied, the aristocratic white upper class was completely endogamous. He went on to explain, however, that "no one in any class likes a marriage in which the skin colours are too far apart. Any marriage of dark with light or white will be referred to as *mosca no leite,* or fly in the milk, and a certain repugnance is felt by all" (1963:40). In marriages of black and white, the union is considered fortunate for the former and a case of lowering oneself for the latter (Nogueira, 1959).

The disapproval of marriage or even romance between persons of extremely divergent colors is revealed by the storm of protest that followed the portrayal on a popular television series of a love affair between a black foreman of a plantation in Haiti and the boss's white daughter. The public outcry was so great that the story was rewritten. A prominent white producer stated that on Brazilian television, "A white man can kiss a black woman, but a black man cannot kiss a white woman" (Vidal, 1978).

Brazilians often cite the tradition of miscegenation as proof of their lack of prejudice toward people of different colors. Yet it has been pointed out that the commonness of this practice may simply reflect the accessibility of black and mulatto women to white men throughout Brazilian history. The mulatto woman is traditionally seen in Brazil as epitomizing female beauty, but this has been interpreted by some as simply a sexual attraction, indicative of an exploitative relationship. Indeed, Bastide (1961) contends that rather than proof of a lack of prejudice, miscegenation is a manifestation of it because it occurred primarily in the context of slavery, not marriage, and certainly not marriage of social equals.

Discrimination

Although studies of Brazilian ethnic relations have rarely failed to note the prevalence of negative stereotypes of dark-skinned persons, they have often pointed out that these attitudes are not ordinarily translated into discriminatory behavior (Freyre, 1963b). Indeed, some hold that the key difference between ethnic relations in the United States and Brazil is that in the latter, prejudicial attitudes do not fundamentally affect people's behavior toward members of particular groups (Harris, 1964). That discrimination against nonwhites in Brazil is essentially unlike ethnic discrimination in the United States is clear enough. The forms of blatant derogation, denial, and physical violence historically so well rooted in American society are simply not present in the same manner or degree in Brazil, and the institutionalized forms of discrimination that characterized the United States before the 1960s are not found at all. Whatever the extent of ethnic discrimination in Brazil, then, it is hardly comparable to discrimination American-style. Its less overt, less socially acceptable, and far less virulent style, however, does not mean that ethnic discrimination is alien to Brazil.

Quite clearly, we have already seen that color plays a fundamental role in the determination of people's life chances. Simple observation as well as careful analysis reveals that Brazil is a society in which those of darker skin are consistently relegated to the lower rungs of the social hierarchy. Although white skin does not guarantee social success, "it always improves one's position over that of a darker person" (Degler, 1971:191). Thus, whites continue to occupy most of the higher-status, better-paying jobs in business and government (Sanders, 1981).

Class or Race? Although the lower place of blacks and mulattoes in the occupational structure of Brazil is undeniable, many observers have customarily attributed this situation not to discrimination but to the effects of class disadvantage (Freyre, 1963a; Pierson, 1967). The argument is essentially this: The lower place of blacks and mulattoes is the product of a heritage of slavery. Whites have enjoyed the historic advantage of being at the top of the social hierarchy and therefore continue to disproportionately occupy the society's better jobs and leadership positions. Unlike the situation in the United States, the disadvantaged place of nonwhites is not attributable to institutionalized and customary forms of racial discrimination. Differences in class, then, generally do reflect differences in color, but this relationship is due to historical accident, not patterns of discrimination. Because there is no castelike system in Brazil, as blacks and mulattoes rise in the class hierarchy, they are not confined to their own communities and institutions but are recognized by and accepted into the dominant (white) society. That is, the individual's class status takes precedence over his or her ethnic status, as opposed to the continued prominence of ethnicity in the United States. In Brazil, a successful black politician or doctor, for example, will not be identified invariably as a *black* politician or *black* doctor but only as a politician or doctor.

There is little question that class factors weigh heavily in the social placement of Brazilians. But the argument that color is simply an expression of class in Brazil has been challenged by many social scientists (Hasenbalg, 1985; Morse, 1953; Russell-Wood, 1968; Silva, 1985; Skidmore, 1974; van den Berghe, 1978). Their studies have shown that although one is not denied opportunities automatically on the basis of racial ancestry, one's color is indisputably significant in the distribution of wealth and power.

In the past, color in Brazil did not seem as overwhelmingly decisive as in the United States. Whereas in the United States, one's racial classification seemed to supersede all other social characteristics, in Brazil, social class seemed more significant, particularly as one approached the lighter pole of the color continuum. That may no longer be the case, however. Through an analysis of census and other data, Andrews (1991) compared the degree of racial inequality in Brazil with that in the United States over several decades and found that in some ways there is a lower degree of inequality between black and white in the United States. Although in past decades racial disparities in life expectancy and occupation were not as great in Brazil, by 1980 those indicators had reversed themselves. Although Brazil showed less residential segregation and proportionately fewer Afro-Brazilians in poverty, these were counterbalanced by greater racial equality in the United States in school enrollment, median years of schooling, and earnings.

Andrews concludes that the United States has generally moved toward a reduction in racial inequality in the last several decades while Brazil has moved in the opposite direction. He attributes this to several factors. First, blacks in Brazil did not move out of the country's most economically backward region (the Northeast) as quickly as African Americans moved from the rural South to the industrial North where economic and educational opportunities favored upward mobility. Second, the benefits of economic growth in Brazil have tended to flow disproportionately to the upper and middle classes, overwhelmingly white; in the United States, a more equitable distribution of wealth generated by the post–World War II expansion created a more favorable climate for black mobility. Finally, an active role on the part of the U.S. federal government in reducing discrimination through civil rights laws and other antidiscrimination measures has had no Brazilian counterpart.[5]

Overt Discrimination In addition to the indisputable relationship between skin tone and social class, blatant and direct forms of ethnic discrimination have frequently been revealed by social scientists and journalists in the past several decades. Blacks may be excluded from hotels and clubs and are commonly harassed by the police. Apartment house elevators in São Paulo, Brazil's largest city, commonly are marked "social," for residents and their guests, or "service," for maids and workmen; blacks, no matter their status, are routinely steered into the latter (Schemo, 1995). Discrimination is particularly evident in the area of employment (Sanders, 1981). In São Paulo, jobs requiring contact with the public, such as receptionists, salespeople, and waiters in better restaurants, are often closed off to Afro-Brazilians. Various subtle techniques are used in screening black job applicants. Employment want ads, for example, will often use the phrase "good appearance" (*boa aparência*), which is understood by all to mean "white." Such restrictions on Afro-Brazilians dealing with the public extend even to higher-ranking jobs, including management and the professions (Andrews, 1991).

There are, of course, regional differences in Brazil regarding the level of discrimination against blacks and mulattoes. In the South (the most heavily white region), discriminatory actions are more obvious, and the social lines between color groups are more rigidly drawn. At times, discrimination can manifest itself in stark fashion. For example, in Rio Grande do Sul, a mostly white state, a mulatto woman won the right to represent her town in the state beauty contest as a preliminary to choosing Miss Brazil. A member of the jury, however, said it would not "look right" for a primarily white state to be represented by a "colored person." The jury, after several more votes, named the runner-up—white—to represent the town (Vidal, 1978). Even in the more socially and physically integrated Northeast, however, patterns of overt discrimination are not uncommon, and certain social and occupational areas are well understood to be off-limits to blacks (Degler, 1971).

Perhaps the most telling evidence of ethnic discrimination in Brazil came forth in 1951 when the Brazilian legislature was prompted to enact an antidiscrimination measure, the Afonso Arinos law. The significance of this legislation lay not so much in its subsequent effect but in the formal recognition for the first time in Brazilian history that discriminatory practices did in fact exist. Ironically, the measure was reportedly prompted by the refusal of a Rio de Janeiro hotel to accommodate a visiting black American entertainer and the clamor that followed. The

[5] Andrews notes, however, that these U.S. trends came to a halt in the 1980s.

ineffectiveness of this measure in alleviating discrimination was generally recognized, and in 1988, the Brazilian Congress passed a stronger antidiscrimination law that provided for increased penalties for discriminatory acts (Andrews, 1991).

STABILITY AND CHANGE

In looking at the forces of stability and change in Brazilian ethnic relations, we might focus on the two most prominent features of this society's ethnic ideology: the commitment to full assimilation of diverse peoples and the denunciation of racial prejudice and discrimination. How fully has Brazil attained these objectives, and will they continue to guide the course of ethnic relations in the future? In no society, of course, should we expect to find consistency between ideology and reality in all aspects of ethnic behavior and attitudes. The United States, as we have seen, has maintained for most of its history an equalitarian ideology even though official policies and individual actions have not always reflected that belief system. Brazil's affirmation of equalitarian and assimilationist principles, however, has been so unfailing that as a multiethnic society it promises to be a positive model for others.

The Goal of Assimilation

In some ways, Brazil seems to epitomize the assimilationist type of multiethnic society.

Amalgamation Without question, the intermixing of diverse peoples—amalgamation—has advanced further in Brazil than in almost any other nation of the modern world. Several factors have contributed to this dilution of racial lines. First, immigration from Europe substantially increased the white element of the population. As we have seen, before the late nineteenth century whites were a numerical minority in most parts of Brazil, and European immigration played a large part in reversing that ratio. Second, a system of sexual exploitation had been in effect since the first Portuguese arrived in the early 1500s, a system whereby white men enjoyed relatively unlimited access to black and Indian women. Finally, as was seen in Table 13.1, both black and white components of the Brazilian population have declined in percentage over the past few decades. Correspondingly, the racially mixed category (mulatto) has grown larger. This may simply be a result of the tendency for people increasingly to identify themselves as neither white nor black, but it may also suggest a steady pattern of increasing intergroup marriage and thus a continuation of racial amalgamation (Sanders, 1982).

Cultural Assimilation The "official" view of assimilation in Brazil is that the society is fusing not only into a single racial group but also into a single Brazilian culture, comprising Portuguese, Indian, and African elements. "The Brazilian ideology of interracial and interethnic relations," wrote Nogueira, "is pro-amalgamation concerning physical traits and pro-assimilation as to cultural characteristics" (1959:172). The assimilationist goals of Brazil have been expressed most articulately by Freyre, who asserted that the society "has not attempted to be or become exclusively European or Christian in its styles" (1963a:156).

This view has largely been verified. African and, to a lesser extent, Indian elements have been blended into the Brazilian culture far more than have African or Native American traits in the United States. African cultural elements are especially evident in religion, food, music, and dance, particularly in those areas of Brazil where the population is heavily of African origin. For example, West African religious rituals have been fused with Catholicism in cults known as *candomblé* and *macumba* in Bahia and Rio de Janeiro.

The fusion of Portuguese with African and Indian cultures, however, should not be overstated. As Wagley has noted, "The only cultural models for the African slave, the Indian, and their mixed descendants were their Portuguese, Catholic, and aristocratic masters" (1971:9). Thus, Brazil has evolved into a society in which African and Indian features are welded into what is fundamentally a Luso (Portuguese)-Brazilian culture.

As for the various European ethnic groups that have populated Brazil since the late nineteenth century, problems of assimilation have seemed less acute than in other multiethnic societies though separatist elements have at times evidenced themselves. For example, the German communities of the southern Brazilian states in the early twentieth century sought to perpetuate their language and to remain unassimilated into what they saw as an "inferior" Luso-Brazilian culture (Luebke, 1983). Even in more recent times they have remained relatively separated from other Brazilian groups, as have the Japanese (Hastings, 1969; Rodrigues, 1967; Suzuki, 1969). Moreover, the first two generations of other European groups have displayed the resistance to assimilation generally characteristic of early generations of voluntary immigrant groups. By the third generation, however, they have become an integral element of the Brazilian economic and political elites and have served as a prime segment of the expanding Brazilian middle class (Horowitz, 1964; Willems, 1960).

Brazil, then, has clearly achieved a level of physical and cultural assimilation surpassing that of most other multiethnic societies. Although there remain in Brazil strong regional patterns of culture and of ethnic composition, giving the society a pluralistic flavor, the commitment to an assimilationist ideology has been constant and has yielded an increasingly integrated society, physically and culturally. Two qualifications, however, bear mention. First, as we saw earlier, the amalgamation of Brazilian peoples has proceeded within the context of a white ideal. This essentially racist notion does not negate Brazil's integrative achievement relative to that of other societies, but it certainly reveals its less-than-democratic stimulus. The contradiction between pride in the development of a "racial democracy" and a social preference for whiteness is transparent. Brazilians officially perceive that they are moving increasingly toward a racially amalgamated society in which color distinctions are gradually eroding. Yet there is no denying that this is an amalgamation in which the darker hues are—desirably—being diluted by the lighter hues.

Second, although cultural assimilation is apparent, *structural* assimilation of blacks at both the primary and secondary levels has lagged far behind. As noted earlier, blacks and even mulattoes have not entered the higher echelons of Brazil's political and economic institutions at anywhere near a level commensurate with their percentage of the population, nor have they begun to interact personally in any significant way with whites. In short, regarding structural assimilation, Brazil is not essentially unlike the United States and other societies with contrasting racial-ethnic groups.

The Denunciation of Prejudice and Discrimination

The most glaring inconsistency between the ideology and the reality of ethnic relations in Brazil involves the irrefutable evidence of racial prejudice and discrimination even though it is of a far milder and less intense form than in the United States. There is no doubt that the belief in equality of opportunity is widely accepted and that public opinion strongly opposes racial discrimination, but it is clear that there is an ethnic hierarchy in which blacks and mulattoes experience differential treatment.

Brazil's nondiscriminatory ideology, however, has been so preeminent that revelations or accusations of discrimination are customarily met with denials or disbelief. There is a well-understood principle that any expression of racial discrimination should be attacked as "un-Brazilian" (Smith, 1963). Even speaking of a "racial problem" is met with disapproval. In the late 1970s, for example, censors banned a program dealing with racial issues from Brazil's largest television network (Vidal, 1978). Official denials of discrimination tend to perpetuate the mystique of racial and ethnic democracy and make any attack on discriminatory practices more difficult. If officially "there is no racial discrimination," public opinion cannot easily be galvanized to confront it.

A further impediment to the elimination of ethnic discrimination in Brazil is the subtlety of its form. Although violations of the antidiscrimination law of 1951 were often prominently reported, most were dismissed by the authorities for lack of evidence. One observer notes that whites were rarely charged under the law, probably because discrimination is so common (Sanders, 1981). Moreover, the subtle and mild forms of prejudice and discrimination may cause blacks themselves to remain relatively unaware of them.[6] Even when they do express awareness of discrimination, blacks lack the economic and political resources to address such issues. Most are part of the lower class, who, like the poor and uneducated everywhere, are ordinarily pawns, not movers, in the society's political processes.[7]

That most blacks in Brazil are part of the lower class makes the issue of discrimination more complex because, as we earlier noted, class and ethnic factors become closely intertwined and difficult to distinguish. Here it is instructive to recall the debate among American sociologists regarding William J. Wilson's thesis of the "declining significance of race" (Chapter 8). As we noted, Brazilian whites—and many blacks—prefer to explain prejudice and discrimination against darker-skinned people as a product of their class, not their racial status. Furthermore, the explanation that blacks are disproportionately occupants of the lower social positions in the society because of class factors is more readily acceptable because this places the blame on "natural" social and economic forces, not those purposely designed to limit the placement of people. Accusations of class prejudice and discrimination are therefore accepted, and accusations of racism are denied. Episodes of discrimination obviously based on race are treated as individual and exceptional cases, not as reflections of a customary pattern (Fontaine, 1981). Also, failure to improve one's social position is seen, as in the popular American ideology, as attributable to the individual's shortcomings, further obscuring the effects of structural disadvantages (Hasenbalg and Huntington, 1982/1983).

Ironically, the difficulties of attacking prejudice and discrimination in Brazil are due not only to the effectiveness of the society's assimilationist and equalitarian ideology, but also to the presence of a color continuum, as opposed to the rigid color lines of the United States. Because racial lines are not inexorably fixed, there is little ground for organizing people on the basis of color. Hence, the development of a strong civil rights movement, aimed at alleviating black deprivation and discrimination, is undermined. How can there emerge a "black movement" when there are varying perceptions and definitions of who is black?

---

[6]Telles found a more moderate level of residential segregation among Brazil's racial groups compared to other countries with large populations of African and European origin. That blacks, mulattoes, and whites live in similar neighborhoods, he concluded, "may strengthen the perception that race has little or no effect on life chances, at least for individuals of the same social class" (1992:195).

[7]It is of note that the limited Brazilian black movement is led primarily by middle-class Afro-Brazilians with comparatively light skin (Burdick, 1995).

Blacks in the United States, by contrast, cannot escape racial prejudice and discrimination no matter what the shade of their skin and no matter what their class standing; all are part of the category "blacks." Creating solidarity among them, therefore, becomes easier despite internal group differences of class and status.[8] Although Brazilian civil rights organizations have emerged from time to time, such as the *Frente Negra Brasileira* (Brazilian Black Front), a movement of the 1930s whose designs were to integrate blacks into mainstream Brazilian society (Fernandes, 1971), they have not displayed the endurance or the effectiveness of American organizations like the NAACP or the Urban League.

What's more, that people may marry lighter and thus produce lighter children further detracts from building enthusiasm for black political activity. Bastide quotes a black woman who, looking at her lighter-skinned children, declares: "They're white already. What's the use of fighting, forming leagues for the defence of the Negro, and all that?" (1961:12). This low level of racial consciousness and militance is also a result, in large measure, of the feeling of political leaders and intellectuals that to heighten racial awareness, as occurred in the United States during the black civil rights movement, would be "racist" and thus a denial of the idea of Brazil as a racial democracy (Skidmore, 1990).

A combination of an effective antiracist ideology, more benign forms of prejudice and discrimination, the complex relationship between color and class, and a multiracial classification scheme has served to make the long-range improvement in the status of Brazilian blacks perhaps even more problematic than raising black status in the United States. In Brazil, as Degler (1971) has pointed out, there is still lacking a widespread acknowledgment that blacks are socially disabled by a heritage of slavery and discrimination. In the United States, this fact has been generally recognized, and compensatory measures, such as affirmative action programs, have been instituted. Comparable measures in Brazil, however, have not yet even been suggested. This, of course, would require the admission of racial prejudice and discrimination, which, Degler notes, "the national myth explicitly denies" (1971:271). A São Paulo city council member, introducing an antidiscrimination measure, stated that "Everybody behaves as if inequality doesn't exist. It's therefore much worse, because unlike in the United States, we've never faced it directly, It's disguised" (Schemo, 1995). Some have suggested, however, that today the myth of racial democracy in Brazil is waning, as the long-standing depressed status of blacks in the society has become more difficult to ignore (Andrews, 1991).

---

[8] As Toplin (1981) has pointed out, however, in the United States, too, lighter-skinned blacks have always fared better than darker-skinned blacks. This pattern began during slave days, when mulattoes were often given house duty and more frequently granted manumission. The continuity of this pattern is evident in the traditional predominance of lighter-skinned blacks as part of the black upper class.

American-Style Ethnic Relations?

In Chapter 4, we saw that in preindustrial or colonial societies, relations between ethnic groups are generally paternalistic. People "know their place" and do not question the ethnic hierarchy. As a result, the level of intergroup hostility may be low, despite gross inequities. As societies gradually shift to industrial capitalist systems, competitive ethnic relations emerge. The demands of economic rationality now require an open labor market, and in such settings, groups use whatever means are at their disposal to protect and improve their economic position. This commonly produces virulent forms of prejudice and discrimination that were previously unnecessary and brings to the fore racist ideologies that serve to rationalize the denial of opportunities to minorities. Many students of Brazilian society have noted the unfolding of competitive ethnic relations, particularly in the industrial São Paulo region, and with it a rising level of intergroup conflict (Bastide, 1965; Cardoso, 1965; Degler, 1971; Fernandes, 1971; van den Berghe, 1978; Wagley, 1963).

Some have even suggested that increasingly competitive ethnic relations in Brazil may create American-style patterns (Bastide, 1965; Fernandes, 1971). This does not mean, however, that Brazil will begin to resemble the United States of the Jim Crow era. A system of institutionalized segregation, for example, is unimaginable. Rather, it means that ethnic conflict will probably intensify and that the ethnic consciousness of both blacks and whites will rise. For example, in the 1970s, a black-consciousness movement arose among young blacks in the *favelas* (slums) and working-class neighborhoods of Rio de Janeiro, much influenced in style and substance by its earlier American counterpart (Fontaine, 1981; Vidal, 1978). What's more, numerous political and cultural organizations developed, stressing black awareness (Sanders, 1981). Although this movement did not succeed in galvanizing poor and working-class blacks, it did force Brazil's major institutions, particularly government, to begin to acknowledge and seriously address issues of racial inequality for the first time (Andrews, 1991).

Some have also noted that ethnic relations in Brazil and the United States seem to be on convergent paths (Degler, 1971; Toplin, 1981). In the United States, the deprived place of the black underclass is increasingly attributed to class factors as the more blatant and legitimized forms of racial discrimination are eradicated. In Brazil, however, the tendency seems to be just the reverse. Racial discrimination is becoming more obvious and acknowledged as a factor in black deprivation, and traditional class explanations have been seriously questioned.

In sum, Brazil presents a complex and often paradoxical case of ethnic relations. In its commitment to racial amalgamation, it has few peers among current societies. Similarly, its embracing of the theme of ethnic equality has been unwavering since the abolition of slavery. Furthermore, the inconsistencies between ideology and reality cannot repudiate the fact that relations between blacks and whites in Brazil *are* more benign than in other societies with diverse racial-ethnic groups. If Brazil is not the racial and ethnic paradise its advocates have proclaimed, it is nonetheless a society in which interethnic relations are less volatile than in most others and in which giant strides have been made toward ethnic assimilation. At the same time, however, Brazil demonstrates the difficulties of mitigating the intergroup conflict that, as we noted in Chapter 4, seems endemic to multiethnic societies.

SUMMARY

Brazil presents perhaps the best case for comparison with the United States. Its ethnic composition and the historical development of its ethnic diversity closely parallel the American experience. As in the United States, the main focus of ethnic relations is the confrontation of blacks and whites. Other ethnic elements are present, however, including many European ethnic groups that entered as voluntary immigrants, and indigenous Indians.

Brazil was an integral part of the African slave trade, and its system of slavery endured even longer than its American counterpart. Debate among historians and social scientists concerns differences in the practice and aftereffects of slavery in the two societies. Some hold that the Portuguese brand of slavery in Brazil was more humane than the Anglo-American version and consequently produced more benign black–white relations following its abolition.

Ethnic stratification in Brazil features relatively fluid boundaries between groups. Racial classification is not a dichotomy but a continuum in which groups blend into one another. Racial categories are based not on ancestry but on physical appearance and social class. Despite blurred lines of division between groups, there is an ethnic hierarchy in which whites maintain a generally advantageous status over blacks and mulattoes, the latter a large intermediate racial grouping.

Although Brazilians proclaim themselves a society free of prejudice and discrimination based on race and ethnicity, these phenomena are present, though in forms far milder than those found in the United States. Blacks are the main objects of this largely subtle ethnic antagonism.

Brazil is most clearly an assimilationist society, whose long-range objective is ethnic convergence. And it has, in fact, seemed to progress further toward both physical and cultural assimilation than most other multiethnic societies of the world. Hence, although the historical conditions of race and ethnicity are similar to those of the United States, the outcome of ethnic relations in Brazil and the ideology on which they are founded have been quite different, particularly as they pertain to the relationship between blacks and whites.

Suggested Readings

Andrews, George Reid. 1991. *Blacks & Whites in São Paulo, Brazil 1888–1988.* Madison: University of Wisconsin Press. Analyzes changes in Brazilian race relations over the past 100 years, looking specifically at individual and institutional forms of prejudice and discrimination and patterns of racial inequality.

Degler, Carl. 1971. *Neither Black nor White: Slavery and Race Relations in Brazil and the United States.* New York: Macmillan. Explains how the emergence of a mulatto, or intermediate, racial category between black and white in Brazil produced a set of race relations markedly different from that in the United States, with its dichotomous system of racial classification.

Fernandes, Florestan. 1971. *The Negro in Brazilian Society.* New York: Atheneum. The major work of one of Brazil's leading sociologists, explaining the development of race relations following the abolition of slavery in 1888. Questions the notion of Brazil as a "racial democracy."

Fontaine, Pierre-Michel (ed.). 1985. *Race, Class, and Power in Brazil.* Los Angeles: Center for Afro-American Studies, UCLA. Essays by social scientists that document the critical role of race in the distribution of jobs, income, education, and other life chances in Brazil.

Freyre, Gilberto. 1963a. *New World in the Tropics: The Culture of Modern Brazil.* New York: Vintage. A brief work by Brazil's most celebrated historian, with a valuable discussion of slavery and the evolution of Brazil's racial-ethnic system. Serves as an introduction to Freyre's lengthier and more detailed works, *The Mansions and the Shanties* (1963) and *The Masters and the Slaves* (New York: Knopf, 1956).

Hellwig, David J. (ed.). 1992. *African-American Reflections on Brazil's Racial Paradise.* Philadelphia: Temple University Press. A unique collection of observations of Brazil by African-American scholars from the early 1900s through the 1980s.

Page, Joseph A. 1995. *The Brazilians.* Reading, Mass.: Addison-Wesley. A lively description of Brazil's culture and of the major racial and ethnic elements of its population.

Smith, T. Lynn. 1974. *Brazilian Society.* Albuquerque, N. Mex.: University of New Mexico Press. A good general introduction to Brazilian society, especially its racial-ethnic system.

ASK AMERICANS TO NAME the capital of Russia, a society very distant from the United States both geographically and culturally, and most will probably answer correctly. Ask them to name the capital of Canada, a society bordering the United States for the entire length of the continent and culturally almost as close, and most will be hard-pressed. One reason Americans pay so little attention to and know so little about Canada is that its sociological features are assumed to be replicas of their own. In the American mind, Canada might easily pass for the fifty-first state.

Undeniably, the economic and cultural hegemony of the United States on the North American continent for two centuries has created great similarities in the two societies. But the fact that Americans can easily feel at home in Canada tends to disguise some fundamental differences. Common cultural preferences, consumer patterns, political alliances, and for two-thirds of Canada, language, do not make Canadian society simply a northern microcosm of the United States. These differences are nowhere more apparent than in the realm of ethnic relations.

Canada is a society whose ethnic structure is today extremely diverse. Moreover, it displays some features of the corporate pluralistic model described in Chapter 4. This state of affairs stems primarily from the presence of two founding groups, British and French, each of whose language and culture have been legally protected since the eighteenth century. The present language division in

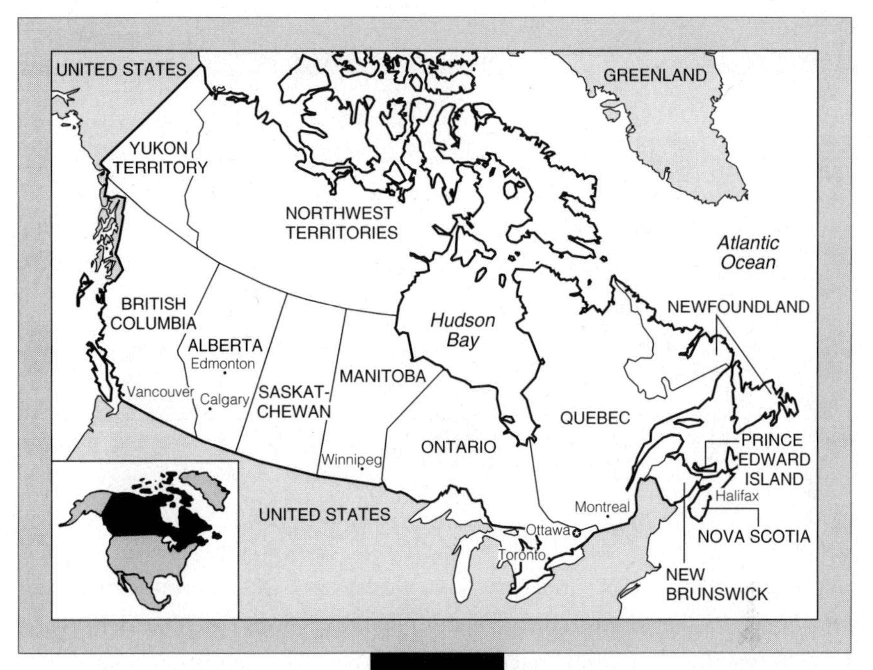

CANADA

the society is shown in Table 14.1. Most of Canada's French-speaking people (Francophones) live in the province of Quebec, creating for them, in addition to their linguistic and cultural distinction, a territorial base. English-speaking Canadians (Anglophones), by contrast, are dispersed throughout Canada and include groups of various national origins. Those of British origin, however, are recognized as the cofounding group and remain the largest segment of English Canada.

In a sense, then, Canada is not one nation but two. As we saw in Chapter 1, when ethnic groups occupy a definable territory—as do French Canadians—they also commonly maintain or aspire to some degree of political autonomy. Hence, they are "nations within nations." The Canadian experiment in national dualism is still being conducted, and the issues separating the English and French fragments have by no means been resolved. Indeed, the historical discord between these two major cultural groups has in recent decades intensified, making the union of Canada more precarious than ever.

The relations between English and French Canadians constitute the first and most important dimension of ethnicity in Canada, cutting across other lines of ethnic diversity. As such, English–French relations will occupy most of our attention in this chapter. Two other dimensions of ethnicity in Canada, however,

Table 14.1 Mother Tongue of the Canadian Population

LANGUAGE	NUMBER	PERCENT
English	16,169,880	59.9
French	6,502,860	24.1
Italian	510,990	1.9
Chinese	498,845	1.8
German	466,240	1.7
Other	2,845,235	10.6
TOTAL	26,994,050	100.0

SOURCE: Statistics Canada, 1995.

are extremely important, even if their manifestations have not been so dramatic. One concerns the various ethnic groups whose origins are neither British nor French, together composing one-third of the Canadian population. Canada is a society of virtually dozens of distinct ethnic groups varying in national origin, culture, and racial features. The important social issues pertaining to these groups involve the nature and extent of their cultural and structural assimilation. Because Canada is, at least officially, a bilingual society, however, these groups have fallen within the realm of either Anglophones or Francophones and are therefore part of the dual-nation issue as well.

The third ethnic dimension involves the relations of Native (Aboriginal) peoples with the rest of Canadian society. Canadian Indians and Inuit (Eskimos) are, along with their American counterparts, the original peoples of the continent. Today they are a relatively tiny proportion of the Canadian population. Their official status, their cultural systems, and the nature of their relations with other groups, however, largely set them apart from the rest of the society. Hence, they must be looked at as a unique component of the Canadian ethnic configuration.

These three dimensions create a complex ethnic picture in which the different elements are interwoven in complicated and, to American eyes, often bewildering combinations (Figure 14.1). Moreover, this ethnic diversity, particularly in recent decades, has yielded public policies more pluralistic in approach than those of the United States. Whereas the United States has been committed to an assimilationist approach, at least for those groups of European origin, the Canadian philosophy, as we will see, has traditionally been more tolerant of the continued expression of cultural differences among diverse groups.

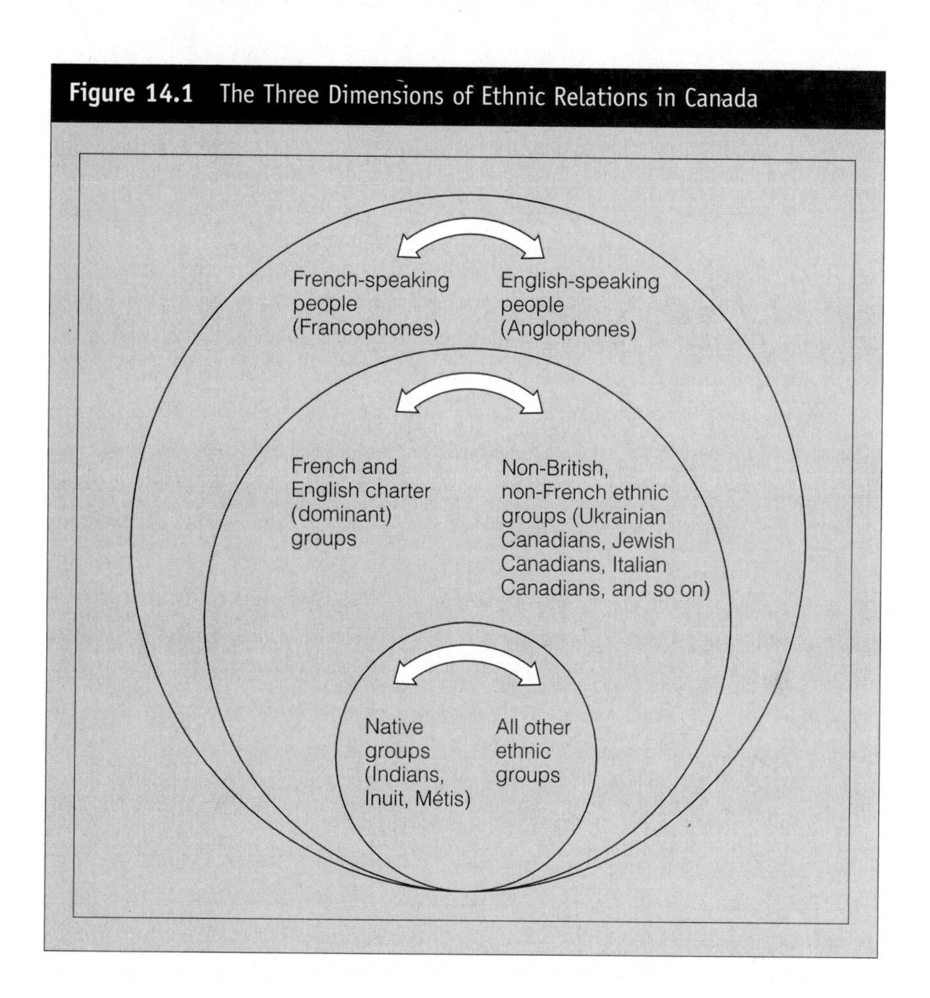

FORMATION OF THE CANADIAN ETHNIC MOSAIC

We can begin to make sense of Canada's ethnic mosaic by looking first at the historical process by which ethnic diversity in the society has taken shape.

The Evolution of Two Nations: The English–French Schism

Most Americans have in the last few years become at least vaguely aware of ethnic stirrings in Canada, specifically, the conflict in Quebec concerning French and English speakers. The roots of this conflict, however, reach back over 200 years. Although other ethnic groups would subsequently contribute to Canada's population, the confrontation of English and French groups consumed the affairs of state from the outset and continues to play the preeminent role in internal relations. French-speaking Canadians are about 24 percent of the Canadian population, and as we noted, they live for the most part in a single province, Quebec. The historical and contemporary relations between French and English Canadians form the major focus of ethnic conflict in Canada.

The English Conquest Both Britain and France established colonies in North America beginning in the early seventeenth century and vied for continental dominance for almost a century and a half. The victory in 1759 of the British forces led by General Wolfe over the French on the Plains of Abraham marked the end of French colonialism in North America, but it did not eliminate the presence of a French cultural group. New France (Quebec) was now made politically part of the British colonial empire, but its French inhabitants were granted the right to certain cultural privileges, including the retention of French civil law, the use of the French language, and the practice of Catholicism. The price of this cultural autonomy, however, was that French Canada was now placed in the position of a permanent minority within an English milieu. From the moment of British ascendancy in Quebec, French Canadians were consumed with avoiding assimilation into the English-speaking North America that surrounded them. And it is this objective of cultural survival that remains at the heart of today's English–French schism.

Following the American Revolutionary War, thousands of English-speaking emigrants from the thirteen American colonies who had remained loyal to Britain entered Canada. Many of these United Empire Loyalists, as they were called, settled in Quebec. This gave the province for the first time a substantial English-speaking population, which quickly came into conflict with the French majority over economic and political issues. In an attempt to separate the two groups, Britain established an Upper and a Lower Canada, the former composed mostly of English-speaking people, the latter of French-speaking. These would later become the provinces of Ontario and Quebec. Hence, the national duality was firmly established, with two peoples, differing essentially in language and culture, juxtaposed geographically and socially.

Although the French in Quebec were not threatened in a numerical sense by the influx of English-speaking people, this period of English–French relations marked the establishment of English dominance of the Quebec economy. With the exit of the French commercial elite following the British conquest, business and financial activities became the domain of the English, and the French Canadians remained mostly on the land as subsistence farmers. This cultural division of labor became a fixed arrangement that lasted into the modern era, making the French a numerical majority but an economic minority in their own province.

The disparities in power, wealth, and social position between English and French in Lower Canada led to the emergence of a French nationalist political element that sought the colony's independence. An open rebellion ensued in 1837, after which Britain consolidated Upper and Lower Canada into a single union. The underlying objective of this move was the eventual assimilation of the French into the English colonial society. To bring about this end, discriminatory measures were enacted assuring Anglo dominance.

In addition, with the failure of the rebellion, the conservative forces of French Canada were now ascendant. The clergy, in particular, reinforced its influence in the life of the French-speaking people and extended the doctrine that the French-Canadian culture could be sustained only through loyalty to the Catholic faith and the French language. This traditional nationalism, combining a rural-oriented way of life with a staunch Catholicism, held sway among the French-Canadian masses until the midtwentieth century (Juteau Lee, 1979). The clergy, then, assumed the dominant institutional role in Quebec, promoting détente with the English colonialists as a means of assuring *la survivance,* the survival of the French culture.[1]

Confederation In 1867, the Canadian provinces were linked in a federal system. The effect of confederation was to further isolate Quebec from the remainder of Canada and to heighten the minority status of French Canadians. Although the French and English languages were both protected by law, only in Quebec was the principle of bilingualism instituted, and there it was strictly one-sided; Francophones were forced to use English in dealings with the Anglo Quebecers, but the reverse was not the case. Outside Quebec, Francophones found themselves defenseless against the Anglophone majority, which chose to ignore the need for French schools and other institutions. The powerlessness of French Canadians was driven home in numerous instances in which minority interests or wishes were swept aside by the dominance of English Canada.

For French Canadians perhaps the most humiliating aspect of the relationship with English Canada was the minority status to which they were relegated in their own province. English control of Quebec's commerce and finance produced an English-speaking business elite that assured that higher-status positions would remain the reserve of Anglophones.

The Catholic church's control of education in the province also handicapped the upward mobility of the French-speaking masses. Expounding the philosophy of French-Canadian cultural survival through retreat to church and land, the clergy emphasized the humanities, classics, and religion in schools and colleges, not the commercial and technical skills that were appropriate to an industrial system. Hence, few French Canadians were prepared for skilled positions in business

---

[1] A classic description of this traditional Quebec culture is found in Miner (1939). More recent essays are contained in Rioux and Martin (1978).

and science. As French Canada increasingly industrialized, Francophones therefore occupied the least-skilled and lowest-ranking jobs (Porter, 1965).

Emergence of a New French Nationalism Industrialization beginning in the early decades of the twentieth century set in motion social processes that would gradually spell the demise of the pastoral, clerical Quebec culture. Urbanization accompanied the growth of industry. Whereas Quebec had been 60 percent rural in 1900, by 1931 it was 63 percent urban (Legendre, 1980). The focus of this rural–urban movement was Montreal, which tripled in size from 1901 to 1921. Industrialization and urbanization conflicted with the traditional Catholic nationalism, which continued to stress agrarian and ecclesiastical values. The customary culture of Quebec society was clearly no longer compatible with the forces of modernization.

The industrialization of Quebec, however, was not sponsored by French-Canadian capital but by outsiders, specifically, English-Canadian, American, and British capitalists. Quebecers thus found themselves in a colonial-like situation. The skilled and more desirable jobs naturally went to the Anglophones, and the French speakers, though a numerical majority, were given the unskilled, menial positions. By the 1960s, French Canadians were the most poorly paid workers in Quebec, below even the newly arrived immigrants from Europe, and their standard of living was below that of the average Canadian (Royal Commission on Bilingualism and Biculturalism, 1969a).

The English of Quebec, as members of an insular group with their own institutional structure—schools, businesses, churches, and neighborhoods—held an attitude toward the Francophones of arrogance and even disdain. They made little effort to learn French, for it was unnecessary. On the contrary, because so many jobs were controlled by the Anglophone business class, all the pressure was on the Francophones to learn English. In his classic account of a French-Canadian community of the 1930s, Everett C. Hughes noted that not only was there no need for the English residents of the town to speak French, but to do so in any case would have upset the subordinate–superordinate relationship:

> In fact, the English do not have to learn French to keep their position in industry. The housewife does not have to learn French to keep her housemaid. If they were to speak French in these relationships—except in a joking or patronizing spirit, as is occasionally done—they would be in some measure reversing roles. For they would then be making the greater effort, which generally falls to the subordinate; and they would speak French badly, whereas the subordinate generally speaks English pretty well (1943:83).

Living in cohesive areas and in control of the most important commercial and financial institutions, Quebec's English community could, by and large, ignore the French around them. It was a situation described by novelist Hugh MacLennan (1945) as "two solitudes," French and English living side by side but essentially in different social worlds.

The Quiet Revolution The decade of the 1960s in Quebec is often referred to as the "Quiet Revolution," for the transformation of French-Canadian society that had begun in the early part of the century now culminated in the emergence of a powerful nationalist ideology espoused by Quebec leaders determined to make Francophones *maîtres chez nous* (masters in our own house). Though revolutionary only in a figurative sense, the changes accomplished during this period basically redefined both the role of government in Quebec and the identity and goals of French-Canadian society (Posgate, 1978).

The key instrument for change was the state. In the past, the provincial government had been viewed as a force in sustaining the status quo, but it now became the principal vehicle of change. Responsibility for health, welfare, and, especially, education now became the concern of the Quebec state, whereas in the past these had been largely church-sponsored institutions. In the economic realm, the state pursued the objective of improving the position of Francophones and eventually ending Anglophone control of the important segments of the economy (McRoberts, 1988). In short, with the development of a dynamic state led by a new middle class, technologically skilled and ideologically committed to a strong and self-sufficient society, Quebec by 1970 was well on the way toward a social and political transformation.

The Rise of the Parti Québécois The new nationalism spawned strong sentiments favoring the separation of Quebec from the rest of Canada and the formation of an independent French-speaking nation. In the late 1960s, elements of the Francophone leadership that had emerged from the Quiet Revolution began to promote the idea that there was no longer any middle ground between partnership within the Canadian confederation and full sovereignty for Quebec; the latter was the only realistic option if Quebecers were to fully control their destiny. To consolidate and advance the achievements of the Quiet Revolution, Quebec, they felt, had to fully sever its ties with Canada (Posgate, 1978). Led by René Lévesque, they left the ruling Liberal party (which, during the regime of Jean Lesage, had initiated the radical changes of the Quiet Revolution) and formed the Parti Québécois, declaring as its objective the political liberation of the Francophone community. Recognizing the need to retain an economic link with the rest of Canada, however, Lévesque created the notion of sovereignty-association, which proposed that a politically independent Quebec would retain close economic ties to Canada. Nothing short of the removal of Quebec from its historical place in the Canadian union, however, was the avowed aim of the Parti Québécois. In its view, the split with English Canada was absolute and irreversible. "First and foremost I am a *Québécois,*" declared Lévesque, "and second—with a rather growing doubt—a Canadian" (Saywell, 1977:4).

By the early 1970s, the growing support among Quebecers for the new, radically nationalist Parti Québécois was evident and alarming to English Canada. Even more alarming were the activities of a small extremist separatist group, the *Front de Libération du Québec* (FLQ), whose violent tactics and revolutionary ideology provoked a crisis in 1970 prompting the federal prime minister, Pierre Trudeau, to briefly invoke the War Measures Act, giving the police sweeping powers to arrest and detain suspected persons.

This was a wrenching experience for Quebec and all of Canada and in part forced the separatists to moderate their proposals for change. Lévesque now stressed the democratic processes by which Quebec's independence was to be attained and emphasized the "economic association" aspect of the envisioned sovereignty-association relationship with Canada. The Parti Québécois declared that if it were elected to power, it would put the issue of sovereignty-association before the people of Quebec as a referendum, to accept or reject.

In 1976, only eight years after its formation, the Parti Québécois won election as the government of Quebec. Its stunning victory sent shock waves throughout English Canada. For the first time, the contemporary Quebec nationalist movement now drew the serious attention of even the United States. What had been feared since 1970 had occurred—the election of a provincial government ideologically committed to the separation of Quebec from Canada, led by the charismatic Lévesque. It now seemed that the division of Canada into two separate nations, in a real rather than a symbolic sense, was possible.[2]

Sovereignty-Association: The Recurrent Issue In 1980, the referendum on sovereignty-association that Lévesque had promised was held. Lévesque downplayed the independence aspects of the measure and emphasized that approval of sovereignty-association meant only that Quebec would enter into "a new partnership" with Canada. Moreover, the referendum was worded so that approval was only a first step toward political independence. It would give the Quebec government merely a mandate to negotiate a new arrangement with Canada, approval of which in any case would be left to the people through another referendum.

---

[2]Much of the support for the Parti Québécois was based not on its advocacy of separatism but on its promise to deliver "good government" and on public disaffection with the relationship between Quebec and the federal government (McRoberts, 1988).

By a margin of three to two, Quebecers rejected the idea of sovereignty-association. Not unexpectedly, virtually all Anglophones voted no, but a slight majority of Francophones did as well. For the moment, then, the intensity of the Quebec nationalist movement had been quelled. In 1985, the Parti Québécois was defeated and replaced by a government led by Robert Bourassa, the same Bourassa whom, nine years earlier, it had defeated.

A decline in Quebec nationalism was induced as well by the economic changes that had occurred in the province during the previous decade. Business participation by Francophones was encouraged by the government, and managerial positions were increasingly made available to them. With the decline of English economic dominance, a growing number of Quebecers took control of much of the Quebec economy, forming a kind of Francophone economic elite, replacing the traditional Anglophone economic leadership.

The fundamental issues of the relationship between Quebec and the remainder of Canada, however, had not been resolved. In the late 1980s, these issues surfaced again resoundingly. The catalyst was an effort to induce Quebec to sign the new Canadian constitution, adopted in 1982. To this end, an accord was reached in 1987 between Quebec and the federal government, giving the province broader independent powers. Most important, under the agreement, Quebec was to be given special recognition among Canadian provinces as a "distinct society." Some in English Canada, acknowledging Quebec's cultural uniqueness, believed that such exceptional political recognition was correct, whereas others held that Quebec should be accorded no special status. This pact, called the Meech Lake Accord, failed to be ratified by all the Canadian provinces in 1990 and thus was not adopted. This was widely interpreted in Quebec as a rebuff by English Canada, again igniting strong support for the nationalist cause of separation from Canada.

Two years later, another attempt was made at bringing Quebec into the Canadian constitution. Much like the Meech Lake Accord, the 1992 proposal would have given Quebec special status in Canada as a distinct society and certain numerical guarantees of political and judicial representation in the federal government. In a national referendum, Canadians rejected the proposal by 60 to 40 percent. This seemed to put to rest temporarily the issue of Quebec sovereignty, but the forces in Quebec working for independence were given renewed strength and promised to put the issue before Quebecers again.

The opportunity for a revisit to the issue came in 1994 with the election once again of the Parti Québécois. The party leaders had made the sovereignty issue, as before, the core of their electoral program. As in the past, their message was familiar: Quebec would never be afforded the kind of unique place in the Canadian confederation that would allow it to protect and promote its language and culture; thus, political independence was the only logical alternative. To temper the uncertainties of separation, however, Quebecers were promised that, after separation, the province would be able to continue to interact closely with the rest of Canada in economic matters such as trade and use of a common currency. Canadian federal leaders warned Quebecers against making such an assumption. The Parti Québécois promised to present the issue to Quebecers soon after the election and, in October 1995, fifteen years after the 1980 proposal was defeated, a referendum on sovereignty was once again held in the province.

As in 1980, the measure was defeated, but this time by the barest of margins, 50.6 percent to 49.4 percent. A closer analysis of the results revealed that Francophones in the province had supported the separatist proposal by 60 percent to 40 percent while Anglophones and Allophones (immigrants speaking neither French nor English) had roundly voted against the measure by over 90 percent. The closeness of the vote and the inability of the federal government to mollify the concerns of those supporting separatism seemed to augur another referendum on Quebec sovereignty in the near future. Francophone Quebecers continue to want recognition in the Canadian constitution as a distinct society and, accordingly, want veto power over any future constitutional changes. The rest of Canada, however, remains strongly opposed to granting Quebec unique privileges. The view of the majority outside of Quebec is that Canada is a federation of ten equal provinces (*Maclean's,* 1996).

Thus, despite the defeat of two efforts within fifteen years to bring sovereignty to Quebec, the issue remains unresolved. The intransigence of an element of Quebec society to accept nothing short of political independence, as well as the low-level animosity that seems so often to define the relations between Quebec and English Canada, promise to push the issue forward again in the future. The question of whether Quebec will separate from Canada is, after two and a half decades of irresolution, still uncertain. As one observer has put it, the issue of Quebec separation may simply remain "a recurrent component of the Canadian agenda" (Banting, 1992:161).

The Language Issue Although Quebec remains, perhaps tenuously, part of Canada, the policies prompted by the ongoing separatist movement have created certain basic changes in the political and social status of French Canadians within Quebec and within the society as a whole. In one way or another, all of these policies revolve around the issue of language. In Quebec itself, the most far-reaching measures prompted by the movement have involved the primary use of French in all spheres of public life, including, most important, business and education. Because language continues to evoke such strong emotions on both sides and so thoroughly encompasses all facets of division and hostility between English-speaking Canadians and Quebecers, let us briefly review the issue as it has evolved in the past twenty-five years.

Language is the very foundation of any people's culture. Hence, it was long felt that assimilation into the dominant English-Canadian society would be inevitable for French Canadians if measures were not taken to assure the preservation of the French language. Moreover, language for the French Canadians, as for many ethnic collectivities in multiethnic societies, is the key symbolic marker setting them off from other groups. Quite simply, that they speak French is what defines most essentially the uniqueness of the French Canadians. As Lévesque put it, "Everything else depends on this one essential element and follows from it or leads us infallibly back to it" (1968:14). It would not be overstating the case to assert that the question of language was at the core of all historical divisions between English and French Canada from the time of the British conquest. It is no less so today.

In the 1970s, aggressive pro–French language policies were enacted by successive Quebec governments. These policies must be seen in the context of *la survivance,* the survival of the French-Canadian culture. French Canadians in the 1970s were faced with an overarching demographic problem: How could they remain a numerical majority in their own province? This problem was created by two population trends—a lowered birthrate and the influx of non-French-speaking immigrants. Quebec had historically maintained the highest birthrate in Canada. Thus, even though few immigrants from France were attracted to Quebec after the colonial period, the ratio of French to English in Canada held relatively constant for two centuries. This was derisively referred to as the "revenge of the cradle," the implication being that the complete English conquest was foiled by the high French birthrate. Quebec's extraordinarily high birthrate, however, dropped dramatically in the 1960s, becoming the lowest in all of Canada. In addition to this precipitous fall, the arrival of large numbers of European immigrants, most adopting English as their new language, threatened to further dilute the Francophone population.

In the late 1960s, the federal government under Trudeau began to promote the notion of a truly bilingual Canada. Responding to the recommendations of the Royal Commission on Bilingualism and Biculturalism, which had been charged with studying how a more equitable balance could be established between the two founding peoples, the government adopted and avidly supported a policy of official bilingualism. But this effort was seen as essentially meaningless to Quebecers, who in many cases were already bilingual—though involuntarily so. Even today, it is only in Quebec where bilingualism is characteristic of a large percentage (over one-third) of the population. Although 16 percent of the total Canadian population is able to converse in both English and French, over half of these bilinguals live in Quebec (Albert, 1989). Moreover, it was the French language that served as the very basis of a distinct French-Canadian society and, as Dale Thomson notes, "the equality of French and English was incompatible with that goal" (1995:76).

The first move in the direction of French-language preeminence in Quebec came in 1974, when the Liberal provincial government of Robert Bourassa instituted a new language law, declaring French the official language of the province and making access to any but French-language schools difficult for those whose native tongue was not English. This created strong opposition from Montreal's English business community as well as the city's large number of European immigrants, who sought to maintain their right to choose the language of instruction for their children. Neither was the measure acceptable to the separatists, who felt that it did not go far enough in establishing the primacy of the French language. In 1977, therefore, Lévesque's government enacted a far more sweeping and radical language law, Bill 101, which mandated that French would be the prime, if not sole, language used not only in official matters but also in commerce and industry. Limitations on the rights of parents to educate their children in English were also made far more stringent.

The aim of the new language law was to alter the structure of opportunity in favor of Francophones, who, it was felt, had been kept in a subordinate position by the Anglophone business elite. Legendre (1980) explains that the use of the language issue as the spearhead for Francophone control of the economy reflected a basic difference in philosophy between the Parti Québécois leadership of the 1970s and the leadership of the Quiet Revolution in the 1960s. Whereas the latter saw the past failure of Francophones to improve their economic status as the result of an outdated and ill-focused educational system as well as the continued adherence to traditional values, the new leaders placed the blame more fundamentally on the structure of Quebec's economy, which fostered institutional discrimination against Francophones. By requiring the use of French in business and industry, it was felt, the Anglophone domination of economic institutions might be attacked directly. No longer could managerial positions be denied Quebecers on the basis of language.

The response of the English-dominated business world of Montreal was indignation. Rather than acquiescing to the stringent new language regulations, many large corporations chose to move their offices to Toronto and other Canadian cities or at least to limit their Montreal operations. This proved economically costly to the province.

Even more controversial was the effect of the language law on Quebec's educational system. Restrictions on instruction in English prompted an outmigration of Anglophones. But the law's primary target, many believed, was the large European immigrant population in Montreal, whose mother tongue was neither French nor English (Arnopoulos and Clift, 1980; Levine, 1990). Most of these groups had ordinarily chosen English-language schools for their children, believing that in the North American context, fluency in English would afford them greater occupational and status opportunities. Under the requirements of the new law, however, almost all were now required to send their children to French schools. Because most immigrants as well as the English community in Quebec resided in Montreal, the conflict centered primarily in that city.

Montreal continues to be the focal point of the language issue today (Thomson, 1995). However, the context of the issue has been fundamentally changed. Despite periodic controversies, clearly, the primacy of the French language in Quebec is no longer disputed.

French-Canadian Ethnic Identity French Canadians are not simply one more ethnic group within the Canadian mosaic. The French in Canada are a people with a linguistic and cultural autonomy formally recognized from the very founding of the society. Moreover, unlike other Canadian ethnic groups, French Canadians maintain a territorial base. For Canada's French-speaking people who live outside the province of Quebec, the tendencies toward assimilation into the dominant English-speaking society are very evident (Joy, 1972). More important, the contemporary nationalist movement in Quebec has jettisoned these French-speaking non-Quebecers. It is Quebec alone that has become, in the eyes of the separatists, the only meaningful base for a sovereign French-speaking society in North America.

Manifestations of ethnic identity among Quebecers also illustrate their uniqueness within the Canadian ethnic system. Unlike English Canadians, French Canadians, specifically, those in Quebec, are faced with a dual national identity. Whereas those outside Quebec think of themselves simply as "Canadians," French Canadians are both *Canadiens* and *Québécois* (Quebecers).[3] In recent years, the latter identity has, for most, seemed to take precedence. This dual identity, however, demonstrates well the idea of a unique French-Canadian nation in which people's allegiance and consciousness of kind do not focus necessarily on the same national unit perceived by other Canadians (Brunet, 1969). One must keep in mind, of course, that there are different degrees of nationalism among Quebecers. Some are vehemently and radically nationalistic and are unrelenting advocates of separatism, whereas others see themselves more as part of a unified Canada.

There is great irony in the desire today of Quebecers for full independence from the Canadian federation. Most important, there is no longer any question of the dominance of the French language in Quebec. Moreover, never before has Quebec achieved such great economic development, and at no other time have Quebecers played such a strong role in the federal government. The crux of the

---

[3]The inhabitants of the original French colony in Canada called themselves *Canadiens.* Only later was the term *Canadian* adopted by the English. Hence, *Canadien* connotes, in the French-Canadian view, the idea of the original or true Canadians (see Brunet, 1969).

problem, however, is a view of Canada, and thus Quebec's place within it, that is very different from the view of English-speaking Canadians. The latter conceive of Canada as a nation of individual citizens who are equal before the law — regardless of their ancestors' language or ethnic origin — and who live in a federation of provinces with equal constitutional status. Quebecers, on the other hand, view Canada as a nation of collectivities defined primarily by language, specifically, English and French (Bercuson, 1995). Because they are heavily outnumbered by English-speaking Canadians, most Quebecers feel that they are entitled to special status and not subject to the same institutions that are dominated by Anglophones. This leads to the view on the part of many that only independence can assure the survival of the French language and culture.

The consequences of the dual nationhood of Canada are not simply problems of identity and cultural maintenance. As we have already seen, there is a historical relationship in Canada between language and culture, on the one hand, and class, status, and power, on the other. Indeed, it is precisely this relationship that in the past made French Canadians a minority group. We will look later at these patterns of ethnic stratification.

Canada's Other Ethnic Groups: The Third Force

Although the division between English and French Canadians overshadows all other ethnic issues, we must consider the second dimension of ethnicity in Canada — those groups that are neither French nor British in origin. Canada is one of the most heterogeneous multiethnic societies in the world, and the noncharter groups stand out increasingly in the Canadian ethnic picture. As we already noted, they affect the English–French schism on the issue of language, but perhaps more important, they constitute one-third of the total Canadian population and thus play an enormous role of their own in all aspects of ethnic relations and policy. These groups represent an extremely diverse spectrum, including Germans, Ukrainians, Italians, Jews, Poles, Greeks, Portuguese, Chinese, Indians, Pakistanis, and West Indians (Table 14.2). Americans, too, are among these groups, but they are hardly distinguishable except in a formal sense.[4]

In a very real sense, the size of these groups as well as their cultural variety have basically altered Canada's ethnic composition. To speak any longer of "English Canada" is a misnomer. Those who see themselves as "British" in origin continue to decline and those whose ethnic origins are neither British nor French continue to comprise a greater and greater share of Canada's population. Today,

[4]Americans make up less than 6 percent of the foreign-born population (Statistics Canada, 1992).

Table 14.2 Canadian Population by Ethnic Origin

ETHNIC GROUP	SINGLE ORIGINS	MULTIPLE ORIGINS[a]
British (incl. Irish)	5,611,050	6,436,870
French	6,146,600	2,242,580
Western European	1,355,485	2,396,200
Dutch	358,180	603,415
German	911,560	1,882,220
Other	85,750	188,205
Northern European	213,605	591,950
Eastern European	946,810	1,230,600
Hungarian	100,725	112,975
Polish	272,805	467,905
Ukrainian	406,645	647,650
Other	166,625	272,480
Southern European	1,379,030	602,875
Greek	151,150	40,330
Italian	750,055	397,720
Portuguese	246,890	45,295
Other	230,935	153,990
Other European	251,140	130,265
Jewish	245,840	123,725
Other	5,300	8,275
Arab	144,050	50,830
West Asian	81,660	16,315
South Asian	420,290	68,080
East Indian	324,845	54,435
Other	95,445	30,270
East and South East Asian	961,225	105,840
Chinese	586,645	66,000
Filipino	157,250	17,725
Indo-Chinese	116,535	14,355
Other	100,805	25,925
African	26,430	13,180
Pacific Islands	7,210	4,605
Latin, Central and South American	85,535	34,445
Caribbean	94,395	72,225
Black	224,620	127,045
Aboriginal	470,615	532,060
North American Indian	365,375	418,605
Other	105,240	156,665
Other origins	780,040	312,090

SOURCE: Statistics Canada, 1993a.
[a]Respondents with multiple origins were counted as part of each ethnic category of their particular combination of origins. For example, a respondent giving the ethnic origin combination of French and Italian was counted in both French and Italian multiple origin totals.

as one senior government official has put it, "Canada outside of Quebec is an immigrant society" (Rodal, 1991:159).

The shaping of Canada's ethnic diversity can be seen historically as a process encompassing two major periods of immigration: the influx of peoples during the late nineteenth and early twentieth centuries, and the post–World War II stream, which continues today. Each period differed in the origins of the major immigrant groups and in the groups' settlement patterns in Canada.

The Period of Western Settlement Although non-British and non-French groups had been present in Canada before the late nineteenth century, they were not large enough to make a significant impact on the ethnic composition of the society. In 1871, of a total population of 3.7 million, the largest of these groups were the Germans, who numbered 200,000; others were considerably smaller (Hawkins, 1972). Throughout the nineteenth century, the British—English, Scots, Welsh, and Irish—dominated immigration and reinforced the Anglo-Canadian preeminence (exclusive of Quebec, of course), both numerically and culturally.

In the latter years of the century, spurred by the promotion of western settlement, large numbers of European immigrants were attracted to the prairie provinces of Saskatchewan, Manitoba, and Alberta. Germans and, to a lesser extent, Dutch were the major immigrant groups until the early 1900s, when people from central and eastern Europe, especially Ukrainians, entered in large numbers. Americans, eager to acquire lands that were no longer available after the closing of the U.S. frontier, were also a significant immigrant group during this time (Troper, 1972).

Although immigrants did settle in Canada's cities, the objectives of nineteenth-century immigration policy were mainly to attract people who would develop the agricultural potential of the vast western domain. This did not mean an open-door policy, however. British peoples, or at least those from northwestern Europe, were deemed most desirable. Ukrainians and others from eastern and central Europe, though technically not among these, were recruited nonetheless because their agricultural background was considered more critical than their national origin. What the West needed above all were farmers; therefore, even nonconforming religious sects like Hutterites, Mennonites, and Doukhobors were promised parcels of land to settle, often as entire communities. Expressly rejected, however, were non-Europeans.

Post–World War II Immigration After World War II, the next great phase of immigration began, changing considerably the Canadian ethnic composition. Both their societies of origin as well as their Canadian destinations distinguished the new immigrant groups from those of previous eras.

Table 14.3 The Top Ten Countries of Origin of Canadian Immigrants

1961–1965	1971–1975	1993
United Kingdom	United Kingdom	Hong Kong
Italy	United States	India
United States	Portugal	Philippines
West Germany	Hong Kong	Taiwan
Greece	Jamaica	China
Portugal	India	Sri Lanka
France	Philippines	Vietnam
Netherlands	Italy	United States
Poland	Greece	United Kingdom
Hong Kong	Trinidad and Tobago	Poland

SOURCE: Employment and Immigration Canada, 1988; Citizenship and Immigration Canada, 1994.

Whereas British, northwestern European, and, selectively, central and eastern European groups had been the major immigrants of the past, the new arrivals were mainly from southern and eastern European countries, in addition to the continued large-scale entry of the British. The largest among the non-British groups were the Italians, almost a half million of whom had come by 1971 (Iacovetta, 1992).

The contemporary period of immigration also introduced an entirely new dimension to Canada's ethnic mosaic because it included for the first time non-white peoples in significant numbers. Although not officially stated, Canada's immigration policy before 1962 had been, in effect, "white only" (Hawkins, 1989; Richmond, 1976). Discriminatory measures favoring northwestern Europeans, particularly those from the British Isles, were now dropped, profoundly affecting the makeup of immigration. Whereas blacks and Asians before 1967 had constituted fewer than 4 percent of immigrants, by 1973 they were approximately one-third, and by 1981, one-half. The development of a broader immigration into Canada in the post–World War II era is seen in Table 14.3. Whereas European countries were the major source of immigrants in 1961, by 1971 Asian and Caribbean countries were among the leading sources, a trend that was even more pronounced in the 1980s and 1990s.

Asian immigrants in the contemporary period have come from almost every country of the region, with especially large numbers of Chinese (from Hong Kong, Taiwan, and the mainland), Filipinos, and Vietnamese, as well as large numbers of Indians, Sri Lankans, Lebanese, and Iranians. Blacks have come mostly from the West Indies, especially Jamaica, Trinidad and Tobago, and Haiti, and have settled almost entirely in Toronto and Montreal. Before this period, most of Canada's tiny black population was of American origin. Most Canadian blacks during the colonial period had come with the American loyalists during the Revolutionary War or during the War of 1812, settling mainly in Nova Scotia and New Brunswick (Clairmont and Magill, 1974; Winks, 1971). Many American slaves also sought refuge in Canada before the Civil War via the "underground railroad."

In addition to their diversity, the immigrants of the post–World War II era differed from those of previous eras in that their destinations were almost entirely urban, specifically, Canada's three major metropolises: Toronto, Montreal, and Vancouver. More than a third of all the new immigrants entered Toronto alone, making that city the core of Canada's new ethnic character. What had been a staid, conservative, and relatively provincial British-dominated city was transformed into a cultural and linguistic panorama not unlike American cities of the East and Midwest at the turn of the century.

As we will see, Canada's groups that are neither British nor French in origin have established themselves as an important political and social element in their own right, a so-called "Third Force." Their growing numbers and diversity more than anything else have contributed to the current philosophy and government policy of multiculturalism, which we will discuss later.

Aboriginal Peoples

The third major element in the Canadian ethnic picture is the Native, or Aboriginal, groups, who together make up about 3 percent of the total population. Among these peoples are Native Indians (the so-called "First Nations"), Inuit, and Métis, the last of mixed racial origins. As can be seen in Table 14.4, about one million Canadians claimed full or partial Native origins in the 1991 census.

Culture and physical features no longer define Native Indians in Canada; rather, "Indianness" is legally defined (Frideres, 1983). Canadian Indians are officially designated as being "status" or "nonstatus." Status Indians, of whom there are about 365,000, have been classified as Indians under the Canadian Indian Act and are the direct responsibility of the federal government. About three-quarters live on reserves (the counterpart of reservations in the United States) established by the government. Because nonstatus Indians are not officially recognized, their number can only be approximated. Estimates range between 100,000 and one million (Gottesman, 1985).

The Métis are marginal to both white and Native Indian societies. During the eighteenth and early nineteenth centuries, fur traders, mostly French, often lived with or married Indian women. The offspring of these mixed unions de-

Table 14.4 Canadian Aboriginal Population

	SINGLE ORIGINS	MULTIPLE ORIGINS[a]
Total	470,615	532,060
Inuit	30,090	19,165
Métis	75,150	137,500
North American Indian	365,375	418,605

SOURCE: Statistics Canada, 1993a.
[a]Respondents with multiple origins were counted as part of each ethnic category of their particular combination of origins.

veloped a unique culture, partly European and partly Indian. (*Métis* is a French word similar in meaning to the Spanish *mestizo,* "mixed.") In western Canada, where the intermixture occurred in a geographically and socially isolated environment, the Métis grew in number and developed a distinct ethnic identity, even declaring themselves a nation. Following an abortive rebellion in 1885, however, they became a highly stigmatized group. The Métis were recognized officially as a distinct people until 1940, when the Canadian government reversed this position. Today, numbering about 125,000, they are reemerging as an ethnic group. But because they are neither whites nor status Indians, they tend to suffer the same discrimination as status Indians while enjoying no special rights to land or other government benefits (Sealey and Lussier, 1975).

Canada's Inuit have remained relatively isolated geographically in the far North and historically have not been subject to colonization to the extent that Native Indians have. Thus, although they do qualify for special political status, technically they are not Indians.

Natives and Whites: Changing Relations Except for being markedly less violent, the historical relations between Canada's Native peoples and white settlers have been, tragically, not essentially unlike those that evolved in the United States. The technological superiority of the whites and their desire for land spawned policies and practices that led to deculturation, dependency, and impoverishment among the Aboriginal populace. The history of Native–white relations in Canada can be conveniently divided into pre- and postconfederation eras.

Accommodation characterized the white attitude toward Native peoples from the time of initial contact until 1867. The indigenous peoples were vital to the fur trade, and as enterprises like the Hudson's Bay Company expanded their operations, Indians and Inuit were engaged in trapping. As they were drawn into this trade, they became increasingly dependent on the same companies for their subsistence, and, in the process, their traditional cultures were basically overturned. During this period, however, there was little direct intent by whites to impose their cultural ways on the Native peoples. Their prime concern was profit maximization. As Breton, Reitz, and Valentine have put it, "The traders and whalers wanted productive trappers and hunters, not North American versions of themselves" (1980:73). White domination, then, was real though not necessarily deliberate.

With confederation and the western expansion, this general attitude of accommodation changed to one of domination. The federal government entered into treaties with the Native peoples for the acquisition of occupied lands, and Indians were increasingly relegated to reserves, isolated from the mainstream society. The government subsequently assumed the role of patron, regulating various aspects of Indian life. Moreover, the reserves were increasingly incapable of supporting hunting and fishing economies, forcing Indians into an even greater dependence on the state.

Concurrent with government jurisdiction of reserve lands were efforts by missionary schools to assimilate Natives into either the English or French culture, resulting in the denigration of indigenous cultures. Thus, while they were being more and more isolated from white society, they were expected to assimilate into the white culture (Breton, Reitz, and Valentine, 1980). Indeed, until the 1970s, the goal of assimilation, though largely unsuccessful, had been the foundation of government policies toward Native peoples since the nineteenth century (Ponting and Gibbins, 1980).

In recent years, the Native peoples of Canada have developed a resurgent group consciousness and pride and have become more politically mobilized in movements that seem to parallel those among Indians in the United States, with whom they share generally common origins and cultures. Viewing themselves as a colonized minority in Canada, they have sought the renegotiation of land claims, and, in recent years, policies have been instituted that have led to greater Native control over their reserves (Asch, 1984; Frideres, 1990). In a sense, the political movement among Aboriginal peoples in Canada has displayed many basic similarities to the Quebec nationalist movement. Each has sought greater freedom from the control of the federal government, and each has espoused an assertive and sometimes militant nationalism.

Several events occurred in the early 1990s that significantly altered the relations between Native Canadians and non-Natives. All involved land claims. In 1989, in the small Quebec town of Oka, about twenty miles west of Montreal, the town council announced plans to expand their golf course to eighteen holes. To do that, however, they needed part of a disputed land area that included a Mohawk cemetery. In March 1990, Mohawks resisted by occupying the area, and when Quebec provincial police moved on the occupying group, one policeman lay dead. The siege then spread as Mohawks blocked the Mercier Bridge, part of a freeway into Montreal used as a major commuter artery. Before a settlement was reached, the occupation by Mohawks lasted several months and involved the eventual deployment of 4,000 Canadian troops and hundreds of Quebec provincial police. The confrontation provoked strong protest across Canada not only among Indians but also among many prominent Canadians who condemned the government's actions in the affair and its failure to negotiate with the Mohawks (Wright, 1992). This was an agonizing affair for Indian–white relations in Canada and sparked a public awareness of the land claims and political status of Native Indians. It also contributed strongly to the growing sense of solidarity among Indians not only in Canada but across North America. A series of confrontations between Native Indians and government authorities occurred again in the summer of 1995 across Canada, over issues ranging from fishing rights to land claims.

An important event in 1993 moved some of those claims to actualization. The federal government signed an agreement with Inuit leaders that provided for the creation of a new territory called Nunavut ("our land" in the Inuit language), to be carved out of the Northwest Territories. Under the terms of the agreement, in 1999 the Inuit are to have effective control of 135,000 square miles (one-fifth of Canada's land mass) and also receive the equivalent of about $1 billion over a fourteen-year period. Although the pact had no bearing on the movement for self-government among Native Indians of southern Canada, it nonetheless marked an important step in giving back control of land to indigenous Canadians.

ETHNIC STRATIFICATION

Because Canada is one of the most diverse multiethnic societies of the modern world, it would be foolish to expect it not to exhibit a system of ethnic stratification. One looking superficially at Canadian society, however, might be hard put to perceive such a system. For example, the current prime minister, Jean Chrétien, is a Quebecer, and many other Quebecers serve in the highest political positions of the federal government. What's more, nonwhite peoples in Canada are only a small, though growing, percentage of the population, and the presence of racial ghettos, so common to the United States, is not part of the Canadian urban scene. But to see these cases as proof of the absence of dominant–minority relations is misguided, for there is indeed an ethnic hierarchy in Canada. In its basic outline, it is not unlike the ethnic stratification system of the United States: an Anglo-Canadian group dominant, other white ethnic groups in intermediate positions, and nonwhites—with a few exceptions—at the bottom.

The Vertical Mosaic

In a study that has become a classic of modern Canadian sociology, John Porter (1965) described the system of ethnic stratification in Canada as a "vertical mosaic." As he explained it, a clear relationship was evident between ethnicity and various measures of social class. The general pattern that emerged from his analysis was a three-part structure made up of British and French charter groups, the former at the top of the income and occupational hierarchies, followed by later-arriving European ethnic groups and, finally, Native peoples. The notion of a vertical mosaic suggested that Canadian society was not only ethnically differentiated but was ethnically stratified as well. As his work proceeded, Porter explained, "The hierarchical relationship between Canada's many cultural groups became a recurring theme in class and power" (1965:xiii).

The Royal Commission on Bilingualism and Biculturalism in 1969 reaffirmed Porter's basic findings regarding the class hierarchy and revealed the concentration of particular ethnic groups in certain occupational areas. British, Jewish, and Asian groups were overrepresented in the managerial, professional, and technical occupations. Ukrainian, Scandinavian, Dutch, German, and Russian groups were overrepresented among farmers (a reflection of these groups' significant presence in the agricultural provinces of Manitoba, Saskatchewan, and Alberta). And Italians, French, Polish, and Hungarians were noticeably overrepresented among blue-collar workers (Porter, 1985; Royal Commission on Bilingualism and Biculturalism, 1969b).

The English–French Difference Perhaps the most glaring difference in class position, historically, was between the two charter groups, English and French. Whereas the English were traditionally at the top of the hierarchy, the French were no higher than most noncharter groups and actually lower than some.

Surprisingly, French Canadians traditionally occupied a lower collective economic position in Quebec, where they had always been a numerical majority, than in Canada as a whole. Porter found that in Quebec "by and large the British run the industrial life" (1965:92). As occupational status rose, he noted, so did the proportion of English personnel. The tendency for French Canadians to fill the lower-level working-class positions in the Quebec economy had been vividly demonstrated over two decades earlier by Everett Hughes in his study of a Quebec town. In the town's major industries, Hughes found, the English held "all positions of great authority and perform[ed] all functions requiring advanced technical training" (1943:46). The French, by contrast, predominated in the lower occupational ranks and eventually disappeared in number as the degree of authority rose. "French Canadians as a group," noted Hughes, "do not enjoy that full confidence of industrial directors and executives that would admit them easily to the inner and higher circles of the fraternity—and fraternity it is—of men who run industry" (1943:53). Hughes concluded that the same situation prevailed throughout the province generally.

The English–French variance shown by Hughes and Porter was later corroborated by the Royal Commission on Bilingualism and Biculturalism. It was shown, for example, that in Quebec, over 30 percent of men of British origin were part of the professional and managerial occupational categories, compared with less than 15 percent of men of French origin.[5] This disparity was also apparent in wages. Canadians of French origin were shown to earn incomes that were on the average 20 percent lower than those of English Canadians. In Quebec, their incomes were 35 percent lower (Royal Commission on Bilingualism and Biculturalism, 1969a).

To see the historically lower place of the French Canadians in Quebec's economy and that of Canada as a whole as a product simply of discrimination by a powerful Anglo elite, however, is a gross oversimplification. The factors that created and sustained the lower occupational levels and incomes of French Canadians are complex and subject to much debate. Were they the result of the traditional clerical emphasis on nontechnical and nonadministrative skills that permeated French-Canadian education until the 1960s? Were they the result of Francophone geographic concentration, making it difficult for French Canadians to move into English-speaking areas and to compete with English speakers? Were they the product of levels of achievement motivation among the English and other ethnic groups that were different from those of the French Canadians? Or were they the product of a basic power differential that did not permit the French to acquire the necessary capital to challenge English control of the economy? Elements of each of these explanations may be involved (Blishen, 1970; Fenwick, 1982; Levine, 1990; Porter, 1985; Richer and Laporte, 1979; Royal Commission on Bilingualism and Biculturalism, 1969a).

In the last two decades, these patterns have changed dramatically. The income discrepancies between Anglophones and Francophones have narrowed significantly, and occupational differences have declined. From an extensive study of mobility in Canadian society, Pineo and Porter (1985) concluded that there is little meaningful difference in the degree of equality of opportunity enjoyed by Francophones and Anglophones throughout Canada. Within Quebec specifically, however, Anglophones, they found, continue to be overrepresented in higher-ranked occupations. But even this situation appears to be changing as more Francophones assume professional, technical, and managerial positions in all sectors

[5]Even as late as the early 1980s, only 26 of 165 enterprises in Quebec with an annual production in excess of $10 million were French-Canadian owned (McRoberts, 1988).

of Quebec's economy (Levine, 1990). Particularly striking changes have occurred at the elite level. As French has become the primary language of business and commerce, a new Francophone economic elite has emerged, making Anglophone dominance largely a phenomenon of the past. "There is now a genuine Francophone capitalist class headquartered in Montreal," notes Marc Levine, "strongly influencing not only the Quebec economy, but also emerging as powerful actors in the entire North American economic system" (1990:194). Most significant, the state-run hydroelectric power industry, Quebec's most important, is thoroughly Francophone at all managerial levels.

Ethnicity and Class in Canada Today Although the historically consistent pattern of English-Canadian dominance of the higher occupational ranks and the most important decision-making positions of Canada's major institutions remains apparent, it is not the dominance of the past. Class differences within particular ethnic groups are probably as great as differences between them (Darroch, 1979). The growing need for a technologically proficient labor force has broken down the ability of the dominant Anglo-Canadian group to monitor the occupational structure as it traditionally had been able to do. Even though collective differences among the various ethnic groups remain evident, ethnic inequality in Canada is declining (Brym, 1989; Isajiw et al., 1993; Lautard and Guppy, 1990; Pineo and Porter, 1985). "The relationship between ethnic origin and class position is in flux," concludes a recent study, "and no ethnic group unequivocally dominates the Canadian class structure" (Nakhaie, 1995:187). Some, in fact, maintain that intergroup differences among white ethnic groups are now virtually nonexistent (Breton, 1989).

The large influx of non-British immigrants since World War II has provided further pressures to afford upward mobility and a greater share of power to minority Canadians. Some, like Jews and Asians, have experienced relatively rapid and substantial mobility into high occupational positions and incomes. Others, like southern and eastern European groups, have not exhibited such significant upward movement though their overall occupational distribution has improved and become more like the national average (Breton, Reitz, and Valentine, 1980; Brym, 1989; Nakhaie, 1995; Pineo and Porter, 1985). Those immigrants who have come to Canada in recent decades have fared better than earlier non-British immigrants because the Canadian economy has improved considerably and because the new immigrants were more carefully selected on the basis of the society's occupational needs (Statistics Canada, 1988).

Even at the elite level, the dominance of Anglo-Canadians appears to be eroding with the advent of a more ethnically diverse institutional leadership. In the past, non-British groups had not been overevident in the power elites of major

Canadian institutions (Clement, 1975; Kelner, 1970; Lautard and Guppy, 1990; Newman, 1975, 1979; Presthus, 1973). New evidence indicates that, increasingly, elites in all major institutions—political, economic, educational, labor—are ethnically varied (Ogmundson and Fatels, 1994; Ogmundson and McLaughlin, 1992).[6]

Aboriginal groups merit a special note. They continue to occupy the lowest rung on all dimensions of social class and consequently have the worst life chances of all Canadians. They exhibit the highest unemployment rates in the country, have the lowest levels of income and education, and disproportionately occupy low-level jobs. Unemployment among Aboriginals, for example, is twice as high as for the total Canadian labor force (Barsh, 1994). The standard of living on Indian reserves is characterized by poor and overcrowded housing, extremely low incomes, and low participation in the labor force (Siggner, 1986). For Indians who migrate to the cities, a social pattern has developed that is similar to that of American urban Indians: high unemployment, high criminality, and high alcoholism (Dosman, 1972; Frideres, 1993; Stanbury, 1975; Statistics Canada, 1993b).

The social consequences of these conditions are predictable. Life expectancy rates are considerably lower and infant mortality rates, though declining, remain higher among Canadian Indians than among the general Canadian populace (Barsh, 1994; Norris, 1990). Standards of housing, education, nutrition, and general health remain below the national averages (Barsh, 1994; Blue, 1985; Siggner, 1986; Statistics Canada, 1993b). Moreover, Aboriginal people make up a highly disproportionate number of inmates in Canadian prisons. Although only 3.6 percent of Canada's population, they account for about 12 percent of male and 17 percent of female convicts in federal penitentiaries. In provincial prisons, the figures are even higher. In Saskatchewan, where Aboriginals account for less than 11 percent of the population, they make up 72 percent of prisoners (Moon, 1995).

Despite these chilling facts, in both income and education, Indians have clearly improved their status relative to the total Canadian population. In 1980, the average annual income of Indians was 60 percent of that of the general population; this was up, however, from 1971 when it was only 33 percent. In 1971, only 3 percent of Indians had attained some postsecondary education; by 1981, almost 19 percent had (Siggner, 1986). In sum, the life chances of Native peoples in Canada have improved, but they continue to represent a markedly deprived segment of the society and, in many ways, are divorced from mainstream institutions.

---

[6]The class and ethnic origins of Canadian elites, however, remain a contested issue among social scientists. See Ogmundson (1990) and Clement (1990).

When comparing their own country with their American neighbor in the conduct and character of ethnic relations, Canadians usually see themselves in a favorable light. The prevalent view is that their treatment of and attitudes toward minority ethnic groups have been and remain more tolerant and equitable. This popular Canadian perspective is to some degree valid. Though incidents of racial and ethnic violence are scattered throughout Canadian history, they have never reached the magnitude, frequency, or intensity of ethnic violence in the United States. Moreover, as we will see in the following section, an ethnic ideology more pluralistic in content has traditionally been proclaimed in which the society's various cultural groups have not been forced into a monolithic "Canadian" mold. That there have been, in essence, two Canadian nations, English and French, no doubt played an important role in establishing and sustaining this seemingly greater tolerance of ethnic diversity.

Despite its more subdued character, however, ethnic conflict in Canada has been revealed in patterns that in some ways are parallel to those of the United States. Like most other Western societies, Canada does not lack a racist tradition, though it has been more muted in expression and less malignant in consequence. Historically, Canadian racism has been evidenced in its immigration policies and in its treatment, both official and unofficial, of nonwhites. Indeed, prejudice and discrimination have traditionally been aimed primarily at the highly visible racial-ethnic groups and only secondarily at white ethnic minorities. Thus, it is the Indians, Métis, Inuit, Chinese, Japanese, East Indians, and blacks—groups referred to in Canada as "visible minorities"—who have borne the brunt of negative attitudes as well as individual and institutional forms of discrimination.

Immigration and Racism

Historian Howard Palmer has written that "the more one scratches the surface of the period up to 1920, the more difficult it becomes to differentiate between the immigration histories of Canada and the United States" (1976:499). Official government policies regarding who and how many would be admitted and what place they would take in the occupational structure, as well as nativist sentiments and attitudes toward immigration in general, were all basically similar.

One important aspect of the early history of immigration in both Canada and the United States was the decidedly racist character of the selection process. Like the dominant Anglo group in the United States, "English Canadians," notes Evelyn Kallen, "have, from the beginning, exercised control of federal immigration policies responsible for determining which ethnic groups would be allowed into

Canada, where they would settle, what jobs they could assume, and what ranking and social position would be accorded them within the existing system of ethnic stratification" (1995:131). Those deemed most suitable for Canada, therefore, were, not surprisingly, most like the English Canadians, culturally and physically.

After 1870, when large-scale immigration to Canada began, the preference for British, or at least northwestern European, immigrants was an outspoken national policy. The pretext for this selectivity was the matter of assimilation. Those from Britain and culturally similar societies, it was asserted, were more easily absorbed into the mainstream Canadian society. Others were not culturally or, as some believed, biologically fit.

That prejudice and discrimination against southern and eastern European groups are no longer significant should not imply that they have no heritage in Canada. In fact, Italians, Jews, and other such groups, especially those who immigrated into Canadian cities at the turn of the century, met with a brand of nativism very similar to the U.S. variety (Avery, 1979; Harney and Troper, 1977). But it was nonwhites, specifically, Asians, who were least welcome and against whom the most blatantly racist policies and actions were directed.

The Chinese were the major target group. They had been permitted entry into Canada as workers in the construction of the Canadian Pacific Railway. Once this project was completed, however, they became a surplus labor force, seen by white workers and organized labor as an economic threat. During the late nineteenth and early twentieth centuries, discriminatory policies were enacted to discourage their entry, and in 1923 an exclusionary act totally barred their immigration (Elliott, 1979). Other restrictive measures severely limited entry into Canada of those deemed "unassimilable," understood to mean nonwhites. In 1911, for example, blacks seeking to emigrate from the United States were rejected because they were presumably unable to adapt to Canada's harsh winters (Palmer, 1976).

Discriminatory immigration policies restricting Asians and other nonwhites were not basically changed until the 1960s. Indeed, as recently as 1947, Prime Minister Mackenzie King stated bluntly Canada's intention to encourage population growth through selective immigration. Large-scale Asian immigration, he stated, "would change the fundamental composition of the Canadian population"; therefore, no changes in immigration regulations were to be made (Corbett, 1957:36).

Seeking to enhance its international image and to more realistically meet its need for human resources, Canada abolished its racially discriminatory immigration policy in legislative acts of 1962 and 1967. The selection of immigrants was no longer to be based on nationality or race but on a system of points that objectively evaluated each immigrant's potential economic and social contribution to Canadian society (Hawkins, 1989). As in the United States after it discarded restrictive immigration quotas in 1965, the ethnic origins of immigrants entering Canada changed radically: most were now non-European.

Ethnic Attitudes

As we have seen, Canada in the last three decades has become a society far more ethnically diverse than at any time in its history. The most visible newcomers are large numbers of nonwhites. This has prompted attitudes and actions that, though mild by American standards, indicate some degree of racism. Such patterns, of course, are not of recent vintage. In addition to the virulent anti-Asian feelings of an earlier period, whites' generally abusive interrelations with Native groups as well as episodes of antisemitism have resembled American historical patterns (Henry and Tator, 1985). Studies of social distance, however, indicate that Canadian racial attitudes remain somewhat less intense than those of Americans.

In their national attitudinal study, Berry, Kalin, and Taylor (1977) found that respondents in general reacted very favorably to English and French Canadians but less favorably to non-British and non-French groups. Specifically, northwestern Europeans were judged most favorably, central and southern European groups next, and nonwhite groups least favorably, except for Japanese. This social distance scale has been demonstrated by other studies as well (Mackie, 1980; Pineo, 1987). It is interesting to observe its essential similarity to the ethnic scale that, as was noted in Chapter 3, has been consistently reaffirmed in the United States, though the differences among groups are declining in both societies (Reitz and Breton, 1994).

Despite this clear acknowledgment among Canadians of an ethnic hierarchy, Berry and his colleagues found no evidence of extreme ethnic prejudice. Though the rank order of groups was very apparent, the differences among them were not exceedingly great. Race (that is, physical differences among groups) was found to be an important dimension of group perception among Canadians, but the researchers concluded that "Canadians reject explicit racism" (1977:206). Bibby's surveys (1995) indicate increasing ethnic tolerance on several measures, including intermarriage and perceptions of visible minorities. Acceptance of marriage between whites and blacks, for example, rose from 57 percent in 1975 to 81 percent in 1995.

Despite the seemingly lower level of racism among Canadians, at least by U.S. standards, some suggest that Canadian prejudice, especially toward nonwhites, may simply be more subtle and covert and may reveal itself when carefully probed (Frideres, 1976; Kallen, 1995). In a survey of Toronto residents, for example, Henry (1978) found that half of 617 respondents expressed racist attitudes to some degree, and 16 percent were extremely racist.

Discriminatory Actions

Although the overt suppression of minority peoples is generally lacking in Canadian history, there are nonetheless some parallels between Canada and the United States in the treatment of ethnic minorities, particularly nonwhites. In Canada, these racial-ethnic groups have suffered various forms of discrimination, including, at one time or another, restrictions in voting, employment, land ownership, housing, and public accommodations (Davis and Krauter, 1971; Kallen, 1995).

We have already described anti-Asian immigration measures. In addition to this official discrimination, violence against Chinese and Japanese workers in British Columbia was common in the late nineteenth and early twentieth centuries. The most blatant discriminatory action against Asians, however, occurred during World War II. As in the United States, those of Japanese origin were forcibly removed from their homes and businesses and placed in internment camps for the duration of the war (Adachi, 1976).

Serious racist actions have also been directed at people of East Indian origin. Along with the Chinese and Japanese, East Indians were barred from entry into Canada earlier in the century, and those remaining were subjected to discrimination in almost all areas of social life. East Indians in the past were seen as intrinsically dirty, sinister, immoral, prone to overcrowding, and generally inferior to whites. Buchignani suggests that East Indians in Canada before World War II were "an almost ideal type subordinate racial caste" (1980:129). They were denied entry into various occupational fields and could not vote or hold citizenship. Today there are over 300,000 East Indians in Canada, but informal discrimination against them still surfaces at times, and well-worn negative stereotypes persist (Berry, Kalin, and Taylor, 1977; Henry, 1978; Ramcharan, 1982).[7]

Although at no time have blacks made up a significant element of Canada's ethnic population (today they are about 1.5 percent), their plight has in many ways conformed to that of American blacks. Negative actions and attitudes toward blacks in Canada have never equaled in scope and intensity those in the United States, but this may be simply the result of the limited size of the black populace in Canada. Immigration restrictions and discrimination in schools, housing, and public accommodations have been common features of the Canadian black experience (Walker, 1980). Today, however, although blacks rank at or near the bottom of the Canadian ethnic hierarchy (Berry, Kalin, and Taylor, 1977; Pineo, 1987), the degree to which their low class and status positions are the result of discrimination is not easily verified because milder and more subtle

---

[7]A number of physical attacks against East Indians occurred in the 1970s in some of Canada's largest cities (Ferrante, 1977; Hill, 1977; Pitman, 1977).

forms of discrimination have replaced official and obtrusive ones (Bolaria and Li, 1988; Hill, 1977; Scott, 1975; Thompson and Weinfeld, 1995).

The 1990s have produced a number of incidents that may augur more serious problems for Canada's visible minorities. These have been limited primarily to Toronto, Montreal, and Vancouver, where, because of immigration, nonwhites constitute an increasing presence. In Montreal, with a growing black population made up primarily of immigrants from the Caribbean, especially Haiti and Jamaica, charges of police brutality against blacks have surfaced several times, and other racially inspired incidents have arisen with some frequency (Farnsworth, 1995). Police actions against blacks also were protested in Toronto, and in Vancouver, where most immigrants have been Asian, a backlash against newcomers from Hong Kong has become evident (Farnsworth, 1991; Katz, 1992; McGraw, 1990). The frequency of hate crimes, directed mostly at visible minorities and Jews, has also increased in recent years (Corelli, 1995).

Such episodes prompt serious consternation among those who have viewed ethnic discrimination as uncharacteristic of Canadians. Over two decades ago, Hughes and Kallen explained that "Canadians, consciously or unconsciously, attempt to boost their feelings of self- and national identity by contraposing their 'peaceful and just' society to that of their 'violent and racist' neighbour to the south" (1974:213). A 1994 survey indicates the continuing strength of that attitude (Dwyer, 1994). Whether there are fundamental differences between Canadian and American beliefs and actions toward nonwhite peoples, however, is an arguable matter. Reitz and Breton (1994) suggest that, when black–white relations are put aside, the differences between Canada and the United States regarding racial and ethnic tolerance are not great. They cite the cases of the historical treatment of Native peoples as well as the treatment of Japanese Canadians during World War II, which they see as fairly similar in the two societies.[8]

STABILITY AND CHANGE

As a multiethnic society, Canada faces even greater problems in seeking societal unification among its various ethnic groups than does the United States. But the attempts to solve these problems do not necessarily resemble those of the United States because Canada is more clearly pluralistic in philosophy. The ethnic end product envisioned by Canadians is not a duplicate of that envisioned by Americans. Whereas the United States can be categorized as more assimilationist than

[8]It is of note that a national poll conducted in 1992 indicated a public perception of growing racial intolerance in Canada (Bozinoff and MacIntosh, 1992b).

pluralistic in orientation (more so, of course, for white ethnic groups than for nonwhites), Canada displays many elements of a corporate pluralistic society.

Melting Pot Versus Mosaic

It has often been noted that one of the key factors differentiating the Canadian and American ethnic systems is how the two societies regard relations among diverse groups and the eventual absorption of these groups into the larger society. Canada has commonly been observed as a society in which ethnic group differences are tolerated more so than in the United States. The popular phrases of comparison are *melting pot,* supposedly characteristic of the United States, and *mosaic,* supposedly characteristic of Canada.

The Melting Pot In the United States, as we saw in Chapter 5, the idea of the melting pot—the fusing of many immigrant groups into an American hybrid culture—became popular beginning in the early part of this century. Never, however, was it translated into public policy to any serious extent. In reality, the expectation that immigrants would conform to the dominant Anglo-Protestant culture was always assumed. Even in recent years, the assumption that new groups will quickly learn English and generally adopt dominant norms and values has continued to guide social thinking despite increased pluralistic rhetoric.

The Mosaic The Canadian ideology, in contrast, has historically favored a more pluralistic outcome of the massing of various ethnic groups. There have been and remain a greater awareness and tolerance of ethnic separateness. A simplified view is "unity in diversity." Canada, in this ideal view, is a mosaic, the various pieces of which fit together within a common political and economic framework. What has accounted for this mosaic ideology, and how valid is it in light of the reality of Canadian ethnic relations?

To begin with, the dual national character of Canada has made ideas of ethnic assimilation problematic. Given the historical fact of two founding groups, neither the melting pot nor the Anglo-conformity model could have the same meaning in the Canadian context as in the American. The question is, how can ethnic groups assimilate into the dominant culture and society when there are two dominant groups? The idea of "Canadianizing" people becomes an empty notion when there is no uniform "Canadian way of life" to serve as a societal reference point. As Hiller writes, "If one group had dominated, there would have been more accord about the specific nature of the dominant culture; but since the British and French were in conflict themselves, the society had a greater built-in tolerance for the perpetuation of ethnic identities" (1976:107–8).

Although it is also true that one would be hard put to clearly define an "American way of life," there is nonetheless no essential ambiguity with regard to language and other major elements of American culture. In Canada, however, the presence of two founding peoples with distinctly different cultural systems has allowed ethnic groups that entered the society after the British and French to parlay this basic schism into significant freedom to retain their culture and group structure.

Some have suggested that one explanation for Canada's greater tolerance of ethnic pluralism lies in the fact that Canada has historically retained a more aristocratic and particularistic social system, in contrast to the equalitarian and universalistic system of the United States (Clark, 1950; Lipset, 1968, 1990). A traditionally greater emphasis on hierarchy and status has restrained pressures to melt down group differences. The necessity for ethnic groups in the United States to "become like others" has therefore not found a strong counterpart in Canada.

Finally, the impact of immigrant groups on Canadian society has in a way been even more profound than their impact in the United States, making assimilation extremely challenging in any case. Though not as numerous in an absolute sense, immigrants in Canada have been a considerably higher percentage of the population than immigrants in the United States. In the early years of this century, the foreign-born in some Canadian provinces outnumbered the native-born by two to one (McKenna, 1969). Since World War II as well, the relative impact of sizable immigration on Canada has been greater than on the United States. Today, whereas the foreign-born in the United States are about 9 percent of the total population, in Canada they are 16 percent.

In comparing the ethnic ideologies of Canada and the United States, it is important to recognize that, just as the American melting pot has been more myth than reality, so, too, the Canadian ethnic mosaic is not a true reflection of public attitude and policy toward ethnic differences. For both societies, the reality of ethnic relations lies somewhere between these two ideals. As Porter points out, "In practice neither [ideal] has been practicable and neither has been particularly valued by the respective societies despite the rhetoric in prose and poetry that has been devoted to it" (1979:144). Just as Anglo-conformity was the dominant public policy with respect to ethnic groups in the United States after its emergence as a heterogeneous society in the late nineteenth century, so, too, in Canada the first period of heavy immigration at the turn of the century produced an ethnic policy that encouraged assimilation into the dominant Anglo-Canadian group. This basic policy did not change for the next three decades (Burnet, 1976, 1981; Palmer, 1976). Although immigrants entered a society in which assimilation could not be enforced as strongly as it was in the United States, there was never any question that Anglo-conformity was, except in Quebec, the guiding force of government policy toward the newcomers.

In a recent comparison of ethnicity in the two societies, Reitz and Breton (1994) suggest that the differences between Canada and the United States regarding assimilation, at least as it pertains to European-origin groups, are today not so great as has been commonly assumed. "The fact is," they note, "in *both* Canada and the United States, the public discourse on immigration reflects *both* a tolerance for diversity and a bias toward assimilation" (1994:10). Moreover, their analysis suggests that there is no evidence to support the notion that ethnic minorities in Canada retain their ethnic identity and culture longer or more strongly than American ethnic minorities. In Canada, as in the United States, ethnicity declines in significance with each passing generation, taking on a largely symbolic function.

Multiculturalism in the Modern Era

The post–World War II changes in Canadian society, especially the influx of hundreds of thousands of immigrants from various nations, created the setting for a more robust ethnic mosaic than had existed at any previous time. In the 1960s, Canada experienced a flurry of ethnic activity and an upsurge in ethnic consciousness, much as did the United States and many other Western societies.

The chief catalyst of these tendencies was the Quiet Revolution in Quebec and the subsequent rise of Quebec nationalism. The appointment of the Royal Commission on Bilingualism and Biculturalism to study ways of reconciling the age-old but newly freshened rift between the two Canadas, English and French, reflected the acknowledgment of a society faced with a powerful and potentially destructive ethnic conflict. The commission supported ways of strengthening the bilingual and bicultural character of the society and of equalizing the place of English and French cultures.

But the non-British and non-French groups, the other major element in the Canadian ethnic amalgam, were not content with the commission's recommendation that Canada was and should remain a bicultural society. The special cultural and language privileges of the two founding groups were, they believed, being furthered in disregard of the historical and contemporary role of other Canadian ethnic groups. Also being ignored was that these groups now constituted almost one-third of the total population. In their view, they had been relegated to the status of second-class citizens. Hence, this so-called "Third Force" demanded not a *bi*cultural but a *multi*cultural Canada, in which the retention of a unique ethnic culture would be recognized and supported not only for the French but also for all other ethnic groups.

The Royal Commission subsequently added to its report a volume dealing with Canada's non-British and non-French groups (Royal Commission on Bilingualism and Biculturalism, 1969b). In it, these groups' contribution to the Canadian mosaic was clearly acknowledged. The commission recommended that ethnic groups need not surrender their cultures to either Anglo or French Canada but, on the contrary, should be encouraged, if they wished, to maintain them strongly (Burnet, 1976). In response, the federal government of Prime Minister Trudeau announced in 1971 the policy of "multiculturalism within a bilingual framework." This appeared to be a compromise measure to satisfy the French concerns about language retention and the other ethnic groups' concerns about their cultural status. One important aspect of the policy was the awarding of grants to ethnic groups to help them preserve their cultures. An even stronger commitment to the policy was made in 1988 when Canada became the world's first country to enact a national multiculturalism law (Fleras and Elliott, 1992). Multiculturalism, then, was firmly established as a public doctrine and set of public policies that recognize, protect, and encourage the retention of ethnic cultures within the larger Canadian society (Elliott and Fleras, 1990; Hawkins, 1989).

Multiculturalism, however, has not been met with overwhelming enthusiasm. Many have seen it as merely part of a political strategy designed to gain the support of ethnic minorities. There also is concern that it is moving Canada away from its traditional open, individualistic political system toward one in which, as one observer has described it, "the cake has to be sliced very carefully among powerful and competing groups" (Hawkins, 1989:216). Moreover, the notion of multiculturalism within a bilingual framework has seemed to satisfy neither the non-English, non-French ethnic minorities nor the Quebec nationalists. The former argue that without protection and support of language rights (that is, multilingualism), multiculturalism is rendered meaningless because group culture cannot be sustained without the ethnic language (as the Quebecers have so effectively argued). The French in Quebec, on the other hand, argue that multiculturalism at the national level reduces them from charter-group status to simply one more ethnic group in Canada (Rocher, 1976).[9]

There is general agreement that multiculturalism as a government policy has been at least symbolically beneficial in that Canada now officially celebrates its ethnic diversity and encourages the expression of unique ethnic cultures. Minority ethnic groups have established themselves as part of the Canadian identity and are now recognized in the school curriculum and in national patriotic events. "Tolerance towards diversity and the acceptance of pluralism," notes political commentator Richard Gwyn, "have become the defining characteristic of the country and its citizens" (1995:203). But in the final analysis, multicultural-

---

[9]In recent years, however, Quebec, in light of its growing minority ethnic population, has adopted its own policy of multiculturalism, recognizing the unique needs of these ethnic subcultures while trying to integrate them into Quebec institutions (Levine, 1990).

ism touches only the most superficial aspects of ethnic relations in Canada and does not speak to the more significant issues of ethnic inequality (the vertical mosaic) and of prejudice and discrimination toward the visible minorities.

Moreover, whether succeeding generations of minority ethnic groups in Canada can (or will want to) resist the forces of assimilation into the dominant Anglo group seems doubtful. Some have pointed out that immigrants and their children are usually determined to take advantage of the opportunities of their new society rather than preserve the same kind of life they left in their societies of origin. For them, as Freda Hawkins has claimed, "the preservation of cultural heritage is a lesser concern and . . . the whole concept of multiculturalism can be confusing" (1989:217). Here the experience of white ethnic groups in the United States may be instructive. Although at times tortuous, a high level of assimilation, both cultural and structural, has been the fate of most groups by the third generation. This seems to be the pattern followed as well by Canadian groups (Reitz and Breton, 1994). Moreover, assimilation expectations seem solidly part of the view of most Canadians. In 1975, 85 percent of Canadians agreed with the statement that "immigrants to Canada have an obligation to learn Canadian ways." Twenty years later the percentage agreeing was 88. Furthermore, this view was no less evident among those who had come to Canada since the 1960s, that is, immigrants themselves (Bibby, 1995).

Finally, not all see the aims of multiculturalism as socially beneficial. Some view it as divisive in effect and stultifying as well for minority ethnic groups (Porter, 1975, 1979). Sustaining and enhancing ethnic pluralism, they maintain, can only hinder the movement of these groups into mainstream institutions, thus perpetuating the system of ethnic stratification. Moreover, as writer Neil Bissoondath has pointed out, rather than instilling cultural understanding among ethnic groups, the policy of multiculturalism may actually have the opposite effect by reinforcing and perpetuating stereotypes. "Multiculturalism," he writes, "with all its festivals and celebrations, has done—and can do—nothing to foster a factual and clear-minded vision of our neighbours. Depending on stereotype, ensuring that ethnic groups will preserve their distinctiveness in a gentle and insidious form of cultural apartheid, multiculturalism has done little more than lead an already divided country down the path to further social divisiveness" (Bissoondath, 1994:89–90).

It is instructive to consider public opinion on this issue. In comparing mosaic and melting pot models, in 1985, 56 percent of Canadians said they preferred the mosaic, only 28 percent the melting pot. Ten years later, the preference for the mosaic had declined to 44 percent while the melting pot had risen to 40 percent (Bibby, 1995). The Canadian commitment to multiculturalism, then, is hardly firm and unequivocal.

The Future of Canadian Ethnic Relations

Canada is a North American experiment in ethnic pluralism. But the experiment is still very much in the testing stage, and its outcome is uncertain. In considering the future course of ethnic relations in Canada, we might return to the theme of the three ethnic dimensions denoted at this chapter's outset.

The overriding ethnic issue of Canadian society remains, of course, the French–English schism. This is the most encompassing dimension of ethnicity in Canada and continues to present the most vexing problems of intergroup relations. Whether Quebec will retain its place in the Canadian union or eventually go the way of independence is a question that in the 1990s remains unresolved. More than any other, this ethnic conflict will dominate internal politics in Canada in the coming years and will, in a real sense, define the future of Canada as a nation-state. Whether a politically and economically centralized society can sustain what are in essence two nations is a question that has been addressed in other corporate pluralistic societies of territorially based ethnic groups divided along the lines of language and culture. Judging from those cases, a continuing separatist movement of some scope can be expected even if conciliatory arrangements are established.

The roots of English–French discord in Canada are, as we have seen, historically deep and abiding. But this is a conflict that, compared with the ethnic strife in the former Yugoslavia or even black–white relations in the United States, is subdued. Whatever the nature of its resolution, it is not likely to degenerate into warfare, as in Yugoslavia, or to erupt periodically in violence, as in the United States. The democratic context of Canada makes ethnic conflict amenable to political solution.

The second major issue concerns the place of the non-British, non-French groups in the Canadian ethnic order. Will multiculturalism become more than an appeasement to these groups and eventually translate into a real alteration of the ethnic hierarchy—affording them greater, and eventually proportional, power, privilege, and prestige? As we have seen, this appears to be unmistakably the trend. Whether these groups will be able, or will choose, to preserve their ethnic cultures in a more corporate pluralistic society, however, is questionable. In any case, it is now clear that these groups have fundamentally altered the ethnic composition and flavor of Canada. As journalist Peter Newman has written, the members of Canada's two founding groups are no longer in the ascendancy, which means that "a new and radically different country has been created" (1995:34).

Another ethnic issue that will absorb the attention of Canadians in the next decade is the racial and ethnic characteristics of its new immigrants: mostly non-European and nonwhite. This, as we have seen, is a critical ethnic issue in the United States as well. Although it is clear that Canada, with a very low birthrate, must continue to attract new immigrants in order to maintain its population, there is a growing restiveness over the shift in immigration patterns. Already the change in the racial and ethnic makeup of new immigrants has called into question public support for the official policy of multiculturalism. A 1989 national poll indicated that though 63 percent of Canadians supported the policy, 61 percent also felt that immigrants should change their distinct culture in order to "blend with the larger society." This response was even higher than the 51 percent of American respondents who, in a parallel poll, felt the same about new immigrants to the United States (*Maclean's,* 1989). A later national survey (Edwards and Hughes, 1995) revealed that nearly half of the Canadian public felt that Canada should accept fewer immigrants.

Finally, the place of Aboriginal peoples has yet to be clearly determined. Whether they will participate with other Canadians as equals in all aspects of citizenship or whether they will be granted greater autonomy is still an unfolding issue.

Canada will hold the attention of students of ethnic relations in the coming decades, for it may provide an answer to the question of how much diversity can be accommodated by a society that remains a centralized nation-state. Factors irrespective of ethnicity have hindered the development of a definitive Canadian national identity. Great regional differences as well as the constant specter of the United States, with its enormous economic and cultural influences, continue to divide Canadian society and lead to divergent visions of the country's future. But above all, it is issues of ethnic diversity that today Canada must address in assuring its national survival.

SUMMARY

Canada is geographically and culturally close to the United States, but its ethnic structure is decidedly different in several respects. In its ethnic makeup, Canada is a dual-nation society, French and English, within which additional ethnic groups have taken their place. There are three dimensions of ethnicity in Canada: the French–English bifurcation; other ethnic groups, neither French nor English in origin; and Aboriginal peoples.

The overriding issue of ethnic relations is the relentless conflict between French and English Canadians. The division between the two derives from the French colonial defeat by the British in the eighteenth century. Conflict has revolved around the efforts of French Canadians to retain their language and culture within the context of a surrounding English majority. The rupture intensified in the 1970s with the emergence of a separatist movement in Quebec, where most French Canadians live, and is today wider than at any time in recent years.

Although non-British, non-French ethnic groups have always been a part of the Canadian society, a particularly large and varied influx of European, Asian, and West Indian immigrants entered beginning in the late 1950s. Today, these groups make up one-third of the Canadian population. Aboriginal peoples — Indians, Inuit, and Métis — are a numerically tiny element of the society.

Ethnic stratification in Canada was, in the past, a vertical mosaic in which those of British origin occupied disproportionately the higher occupations and the most important decision-making positions of major institutions. Today, French Canadians and, increasingly, those of neither British nor French origin have narrowed those discrepancies considerably and are beginning to play a more prominent role in Canadian economic and political affairs.

Certain patterns of prejudice and discrimination in Canada, although milder in form and substance than those expressed in the United States, are evident in past immigration policies and in the treatment of nonwhite peoples.

Canada has maintained a greater tolerance toward ethnic diversity than has the United States and can be placed among those societies we have called corporate pluralistic. Rather than a melting pot, Canada historically has advanced a mosaic ethnic ideal, according to which ethnic minorities are not expected to adopt in full the cultural ways of the dominant group. Whereas in past eras this philosophy was mostly disregarded in favor of Anglo-conformity, today, with Canada's growing ethnic diversity, the mosaic has become the society's official policy, though questions regarding its full acceptability by the Canadian public remain evident.

Suggested Readings

Bissoondath, Neil. 1994. *Selling Illusions: The Cult of Multiculturalism in Canada.* Toronto: Penguin. The author, himself a first-generation immigrant, argues that the Canadian policy of multiculturalism is fundamentally flawed, leading only to greater societal divisiveness and the perpetuation of ethnic stereotypes.

Fleras, Augie, and Jean Leonard Elliott. 1992. *Multiculturalism in Canada: The Challenge of Diversity.* Toronto: Nelson Canada. Explains the various dimensions of multiculturalism, how it has been implemented in policy, and how Canadians have responded to it.

Gwyn, Richard. 1995. *Nationalism Without Walls: The Unbearable Lightness of Being Canadian.* Toronto: McClelland & Stewart. One of its most noted political commentators addresses Canada's ability to forge a national identity from its increasingly diverse population and its future as a multiethnic society.

Hiller, Harry H. 1991. *Canadian Society: A Macro Analysis.* 2d ed. Scarborough, Ont.: Prentice-Hall of Canada. Presents a good survey of the numerous divisions that characterize Canadian society, especially those that revolve around language and ethnic identity.

Levine, Marc V. 1990. *The Reconquest of Montreal: Language Policy and Social Change in a Bilingual City.* Philadelphia: Temple University Press. A very thorough examination of the critical role of language in sustaining Quebec nationalism and the French–English schism.

Li, Peter S. (ed.). 1990. *Race and Ethnic Relations in Canada.* Toronto: Oxford University Press. A collection of essays by social scientists that deal with the full range of ethnic issues in Canada, including language, immigration, multiculturalism, political economy, and state policies.

McRoberts, Kenneth. 1988. *Quebec: Social Change and Political Crisis,* 3d ed. Toronto: McClelland and Stewart. A comprehensive analysis of the economic and political transformation of Quebec society, addressing the overriding issue of political sovereignty.

Reitz, Jeffrey G., and Raymond Breton. 1994. *The Illusion of Difference: Realities of Ethnicity in Canada and the United States.* Toronto: C. D. Howe Institute. On the basis of survey data, the authors suggest that the ethnic systems of the two societies, on such issues as ethnic tolerance, acceptance of assimilation, and the economic incorporation of immigrants, are not so different as social scientists have assumed.

Resnick, Philip. 1994. *Thinking English Canada.* Toronto: Stoddart. Argues that English Canada (that is, Canada exclusive of Quebecers and Aboriginal peoples) constitutes a coherent sociological nation, the distinguishable features of which are described.

Rioux, Marcel, and Yves Martin (eds.). 1978. *French-Canadian Society.* Toronto: Macmillan of Canada. A collection of sociological studies that trace the historical development of Quebec culture, economy, and social organization.

AS IS THE CASE with other ethnic conflicts that erupt in periodic violence, Americans are faintly aware of the participants in and the basis of the ethnic strife in Northern Ireland, which have been described so often by the news media. Media narratives of ethnic relations in Northern Ireland, however, give a simplistic and somewhat distorted view, for they are sensational and largely superficial. Indeed, little information about this society ever reaches the American public (or, for that matter, the rest of the world) other than accounts of the latest rash of shootings or bombings. Such accounts create the impression of a society in perpetual turmoil. In reality, although strife is certainly closer to the surface than it is in most other multiethnic societies, it is not ceaseless and does not affect all people in the same manner or degree. Of greater importance than body counts or gruesome descriptions of the physical and psychological damage wrought by the recurrent terrorism of both sides are the historical antecedents and the current structural conditions that underlie the ethnic chasm between Protestants and Catholics. These are the constant features of life in Northern Ireland that we need to look at to gain a more thorough understanding of the ethnic relations of this small but highly polarized society.

Although the protagonists in Northern Ireland's tragic conflict are Protestants and Catholics, we must understand at the outset that the struggle is not a religious one per se. Rather, it is an ethnic conflict in which religious identities mark off the boundaries of the two major ethnic groups. Although religion certainly

NORTHERN IRELAND

impinges on group relations, matters of doctrine and dogma are not the basic issues. Regarding the role of religion in the conflict, Harold Jackson and Anne McHardy write that it is simply "the handiest identifying mark available to the two sides" (1984:6). Or, as John McGarry and Brendan O'Leary have explained, religion in Northern Ireland "is best seen as an ethno-national marker rather than as an important independent motivator of violent conflict" (1995b:852).

Political scientist Richard Rose (1971) has referred to Northern Ireland as a "biconfessional society" because almost all people identify themselves as either Protestant or Catholic. Those claiming to be atheists, it is joked, must say whether they are Protestant atheists or Catholic atheists. These identities, however, do not mean the same as they do in the United States or other Western nations. In those societies, *Protestant* and *Catholic* denote denominational affiliations that no longer carry with them significant social consequences. In Northern Ireland, these are ethnic categories that determine in a fundamental manner one's place in the society and the nature of one's relations with others. Religious affiliation is, quite simply, the most important social characteristic, taking precedence over one's social class and even one's sex (Harris, 1972). In some measure, it is not unlike the significance of racial identity for persons in the United States or South Africa. The relationship of religious affiliation to almost every other aspect of social life—occupation, education, politics, residence, leisure activity—makes Northern Ireland a case of ethnic pluralism in the extreme. Not only do separate institutional structures exist for each group, but each also maintains basically different perspectives on how and by whom the society should be governed.

The essence of the ethnic conflict in Northern Ireland is this: In 1920, six counties of the Irish province of Ulster, which today make up Northern Ireland, were adamant in retaining their tie to Britain following Ireland's successful movement for independence. Protestants outnumbered Catholics in this area, unlike the remainder of Ireland, and were resolved to remain apart from a sovereign, Catholic-dominated society. As a result, Ireland was partitioned into two states, the Irish Free State (later the Republic of Ireland), with a Catholic majority, and Northern Ireland, with a Protestant majority. Unfortunately, this left a substantial Catholic minority in the six counties.[1] The dispute since that time has concerned this artificially created boundary. Most Protestants seek to maintain the boundary as it is drawn and to remain linked with Britain, whereas most Catholics do not recognize the permanence of the dividing line and visualize an eventually united Ireland.

Today, Protestants are a numerical majority within Northern Ireland by about 10 percent, but in a united Ireland they would be outnumbered by about three and a half to one. (Ninety-five percent of the Republic's three million people are Catholic.) Hence, the root of the Protestants' resistance to anything that threatens the current political status of Northern Ireland is the fear that they will be absorbed into a unified Ireland and thus become the minority within a Catholic-majority state. The nub of the matter, then, is that Protestants wish to remain the dominant group in Ulster. (In this chapter, *Ulster* and *Northern Ireland* are used interchangeably.)

Although these conditions are the immediate sources of the contemporary conflict, the Protestant–Catholic division has much deeper roots reaching back many centuries. Ethnic polarization well antedates the partition of 1920. Evidence of this long history of sectarian struggle is visually abundant on the streets and buildings of Belfast and Londonderry. Defiant anti-Catholic slogans scrawled on the walls and fences of Protestant areas are rejoined by anti-

---

[1]The province of Ulster originally comprised nine counties, but with partition it was reduced to six to assure a more comfortable Protestant majority.

Protestant shibboleths in Catholic areas. Perhaps the most telling messages, however, are those with a historical content. Visitors to Northern Ireland may be baffled by the sight of graffiti exhorting people to "Remember 1690" or "Remember 1916." What conceivable relevance could events of over 300 or even 80 years ago have on current societal issues? As puzzling as it may seem to those unfamiliar with Northern Ireland, events of past ages in this society may just as well have occurred yesterday. They are constantly resurrected as a means of keeping the struggle alive.

HISTORICAL DEVELOPMENT OF ETHNIC RELATIONS

The conflicting ethnic identities and loyalties of Protestants and Catholics in Northern Ireland emerge out of what both groups see as very different and opposing heritages: British for Protestants, Irish for Catholics. Each, therefore, maintains a distinctly different view of history. Let us look briefly at the major events of Ulster's past that have created these contrasting perspectives.

The Plantation of Ulster

The beginnings of conflict between Protestants and Catholics in Northern Ireland are traceable to the early seventeenth century, when England embarked on the colonization of Ireland. The most comprehensive colonizing effort was made in Ulster, where Irish lands were confiscated and Scottish Presbyterians and English Anglicans settled on them. The native Catholics were either driven out of Ulster or subdued into a totally subordinate position.

The period of English and Scottish colonization, known as the Plantation of Ulster, marked a group confrontation that clearly illustrates the process by which ethnic stratification is established: ethnocentrism on the part of each group, competition between them for scarce resources, and, finally, the establishment of dominance by one through superior power.

In the English view, the native Irish were uncivilized people against whom the most brutal tactics could be employed. Indeed, Fredrickson (1981) suggests that the experience of the English in Ireland foreshadowed their colonizing practices and policies toward indigenous peoples in North America. If the Natives were uncooperative in ceding their lands, the "savage" image could be invoked, and they could then be either exterminated or confined to reservations. Unlike the whites' experience with the Indians in North America, however, the Scottish and English settlers could not so easily dispose of the Catholics. For one, there were simply too few settlers to impose total control. Furthermore, Catholics resisted, and in 1641 they revolted in a massacre of Protestants. This was avenged with a countermassacre a few years later by the armies of the English dictator Oliver Cromwell, and most of the remaining land held by Catholics was confiscated. By 1703, Catholics possessed less than 14 percent of all land in Ireland, and in Ulster, land was almost entirely held by Protestants (Rose, 1971).

The colonization and suppression of the Catholics in Ulster produced antipathies and armed hostilities that continue, so evidently, to the present day. For Protestants, a siege mentality was created, not unlike that of the Afrikaners in South Africa, that remains the core of Protestant militance. Historian Liam de Paor vividly describes the trepidation of the English and Scottish settlers, who harbored a constant dread that the indigenous Catholics would someday wreak their revenge:

> Fear, contempt and hatred animated the opinion held by the planters of the ragged native Irish who worked on the land, or passed homeless on the roads, fear, and a deep-seated uneasiness compounded of cant, guilt, and a determination to hold what they were building where before there had been nothing. Like white farmers in Kenya watching their Kikuyu workers and thinking of the midnight advent of the Mau Mau, the English and Scottish planters watched and waited (1970:23).

At the end of the seventeenth century, the Protestant ascendancy in Ulster was firmly secured as the result of several political and military events. Most important of these was the defeat of James II, the last Catholic English monarch, by William of Orange at the Battle of the Boyne in 1690. The festivities in Protestant communities commemorating this battle each July 12 are replete with parades and bonfires; they are also blatantly anti-Catholic in tone.

What followed William's victory at the Boyne was an even more sustained dominant–subordinate relationship between Protestants and Catholics. Penal laws were enacted, further suppressing the Catholic populace. The Catholic clergy was also restricted, as was the Irish educational system. Although the Penal Laws were applied most severely against the Catholics, many were also employed by the dominant English against the Scottish Presbyterians. This prompted the mass emigration of thousands of Ulster Presbyterians to the United States. These were the so-called Scotch-Irish, who made up such a major element of the colonial American population in the eighteenth century.

At the close of the eighteenth century, one of the few nonsectarian efforts by the Irish to secure independence from British rule emerged in an alliance of radical dissenting Presbyterians and Catholics. The movement was short-lived, however, and led subsequently to the abolition of the Irish parliament. Thereafter, Ireland was ruled directly from London.

Home Rule and Irish Rebellion

Late in the nineteenth century, sentiment grew in England for the establishment of home rule for Ireland. This was strongly resisted by the Protestants of Ulster, who saw themselves at the brink of absorption into a Catholic majority. The slogan "Home rule is Rome rule" expressed this fear, and an effort was begun to defeat any movement toward Irish autonomy. This campaign against home rule in the 1880s produced a set of institutions that united Protestant Ulster and that remain the political foundations of the Protestant community. Chief among these was the Orange Order, an organization combining political, military, religious, and fraternal elements, whose raison d'être was to defend the Protestant ascendancy. Originally formed as an anti-Catholic vigilante group, it emerged in the 1880s with a newfound respectability as the nucleus of the anti–home rule movement, uniting disparate unionist factions (Darby, 1976). The Orange Order would remain one of the strongest forces in Northern Ireland's politics and social life. Even today it is the largest Protestant fraternal organization and still wields much social and political influence.

Closely allied to the Orange Order and developed as a means of parliamentary resistance to home rule was the unionist movement. With the victory of a Conservative government in Britain in 1886, home rule, having been favored by the Liberals but not by the Conservatives (with whom the unionists of Northern Ireland had aligned), was shelved until the advent of a new Liberal government in 1906. Once again, militant Protestant resistance arose. This time, Ulster Protestants were prepared to resist with arms, and they formed a paramilitary group, the Ulster Volunteer Force. Moreover, almost all adult Protestants in Ulster (nearly half a million) signed a covenant declaring their refusal to recognize any home rule measure passed by the British Parliament.

A home rule bill was enacted in 1914, but with the onset of World War I, British attention was deflected from the Irish issue. In 1916, however, impatient Irish nationalists sparked an armed rebellion against British rule in what has become known as the Easter Rising. Like the Battle of the Boyne for Protestants, this event has assumed great symbolic meaning for Catholics. The rebellion eventually led to the Government of Ireland Act of 1920, providing for the partition of Ireland into North and South, with the latter granted full sovereignty from Britain a year later.

The partition was a compromise measure that fully satisfied neither the North nor the South. The Ulster Protestants wanted no form of autonomy for Ireland, preferring retention of the British union, and the South desired a fully unified Irish nation, including Ulster. Northern Ireland, then, was essentially created by the British to avoid bloodshed between Protestants and Catholics. As one historian has written, "No one in Ireland, of any political persuasion, wanted or welcomed it. Its creation was an expedient imposed on the country by a hard-pressed British cabinet, which sought only to hold two groups of Irishmen from each other's throats and to give them an opportunity of living peaceably apart, since they could not live peaceably together" (Beckett, 1972:14).

A recalcitrant nationalist element led by the Irish Republican Army (IRA) provoked a bitter civil war over the issue of partition, refusing to accept its legitimacy. Thus, the new Irish Free State as well as the politically questionable entity of Northern Ireland were founded amid the violence and bloodshed its creators had hoped to avoid. Not until 1925 did all parties—Britain, Ireland, and Northern Ireland—finally accept the arrangement as permanent. Nonetheless, Irish Catholics, in both the North and South, never fully relinquished their belief that Ireland would eventually be unified.

The Civil Rights Movement

From the late 1920s until 1969, Northern Ireland experienced four decades of relatively unperturbed ethnic relations. Although anti-Catholic violence flared in the depression years of the 1930s, World War II brought economic prosperity to Ulster with its important shipbuilding and aircraft industries. After the war, new industries were attracted, many of them international firms whose hiring practices did not conform to the common anti-Catholic discriminatory patterns of the past. The opening of new occupational fields to Catholics served as a hopeful sign of increasingly equitable economic status for the minority as well as the possibility of more harmonious ethnic relations in general.

The increased economic prosperity of the 1950s and 1960s was paralleled by a substantial government emphasis on education that benefited greatly the Catholic minority. Many now went on to a university education, seeking upward mobility in jobs commensurate with their educational qualifications (McAllister, 1977). A political leadership emerged from this new Catholic middle class determined to arrest the inequities long suffered by Catholics in various institutional areas. Taking its cue from blacks in the United States, a civil rights movement now petitioned for equal rights for Catholics. Peaceful protest tactics similar to those employed by American blacks were adopted, including marches and sit-ins. Even the American civil rights movement's anthem, "We Shall Overcome," was sung by protesters. Specifically, the movement sought to end discriminatory practices in voting procedures, housing distribution, and employment, those areas in which Catholics had been most severely and continually aggrieved.

What made this movement different from previous political efforts of the Catholic minority was that its ultimate objective was not the end of partition and the eventual unification of Ulster with the Republic. The movement's focus was on issues pertinent to Catholics in Northern Ireland, that is, treatment as equal citizens under British rule. Implicit in this focus was, for the first time, a recognition of the legitimacy of the Northern Irish state and a willingness to participate in its operation. This marked a significant departure from past Catholic political attitudes. Previously, most had felt that participation was meaningless within governmental institutions, specifically, Ulster's parliament, which were thoroughly dominated by Protestants who refused to yield an inch to Catholic interests. Moreover, Catholics had traditionally been reluctant to participate in an institution whose essential legitimacy they did not recognize. Now it appeared that they were prepared to think and act in terms of Northern Ireland alone rather than a unified Ireland.

The prospects for a rapprochement between the Protestant and Catholic communities in the 1960s were given further stimulus by the views and actions of Northern Ireland's prime minister, Terence O'Neill. Although not deviating basically from the standard loyalist policies and objectives of the Protestant-dominated Unionist party, O'Neill seemed intent on creating an economically sound Ulster that would usher in an era of cooperative relations between Protestants and Catholics. As Rose has described him, in the context of Northern Ireland, O'Neill "was an innovator, even a revolutionary, for he sought nothing less than a fully legitimate regime in which Catholics would support the Constitution as well as comply with its basic laws" (1971:97). In 1965, O'Neill pulled off a startling political act of reconciliation by meeting publicly with the prime minister of the Irish Republic. Given the suspicion and animus of past Unionist governments toward the Irish state in the South, this was an extraordinary symbolic act. It was an event that "had all the overtones and anxieties, in Irish terms, of a summit meeting between Russian and American leaders" (Rose, 1971:99).

Tragically, both the civil rights movement and O'Neill's reformist government fell victim to Protestant paranoia and extremism — the politics of the old. That the civil rights movement had been founded primarily on issues of minority rights in Northern Ireland, not on the issue of unification with the South, was ignored; Protestant extremists, led by the demagogic preacher Ian Paisley, interpreted the movement as a Republican effort and mobilized a militant Protestant countermovement. Furthermore, Paisley and other Protestants of the extreme right had even earlier accused O'Neill of appeasement and sought his political defeat. With the outbreak of Protestant violence against civil rights marchers, the highly volatile situation rapidly deteriorated, and O'Neill was forced to resign in 1969.[2] What had promised to be the onset of an era of peace and conciliation between Protestants and Catholics degenerated into the most stressful period of Northern Ireland's history.

---

[2]For a view of the Catholic civil rights movement and responses to it, expressed by one of its key participants, see Devlin (1977).

The Current Troubles

People of Northern Ireland refer to the contemporary phase of the conflict, beginning in 1969, simply as "the troubles." By August of that year, the situation had become extremely tense, and civil rights marchers were attacked by Protestant extremists. Rioting began in Londonderry, in turn producing a brutal police response against Catholics. The riots spread to Belfast, where Protestant mobs invaded the Catholic Lower Falls area; seven people were killed, and over 3,000 were left homeless as entire neighborhoods were set afire (Darby, 1976). These events prompted the British government to send troops into both Belfast and Londonderry to protect Catholics from what seemed to be turning into a bloody pogrom. At this time, Catholics welcomed the British Army because there was no longer any assurance that the Royal Ulster Constabulary (Northern Ireland's police) could or would protect them. Indeed, in some cases the RUC appeared to be in collusion with Protestant attackers (Coogan, 1980).

With the formation of the Provisional IRA in 1970, the major protagonists in the current troubles were in place. Although the IRA had been sporadically active in attacks on the Unionist government since the 1920s, it had gained little support among the Catholic community, and by the 1960s it was seen as something of an anachronism (Holland, 1982). With the emergence of renewed violence, the group split into two factions, the Official IRA, committed to a primarily nonviolent political course, and the Provisional IRA, which favored the use of violent tactics to effect unification with the South. Eventually the Provisionals (or, as they are commonly known, "Provos") gained ascendancy and became the major paramilitary group among the Catholics.

The rebirth of the IRA was given considerable impetus by the actions of the British Army. Even though resistance to reforms and much provocation of the violence was coming from the Protestant side, in June 1970 the Catholic Falls Road area of Belfast was sealed off by the British Army, which proceeded to ransack homes in search of arms. The Army's foray into the Catholic area was unnecessarily severe and was afterward referred to as the "Rape of the Falls." Several people were killed in the operation, many were injured, and the search produced very few arms (Coogan, 1980). After that incident, the British Army was viewed by Catholics no longer as protectors but as oppressors. The Catholic ghettos of Belfast and Londonderry increasingly turned to the IRA for protection. Although the IRA's initial purpose was essentially to defend Catholic areas, it later began an offensive campaign against the British forces in Ulster, often involving particularly callous and shocking actions, that has continued to the present.

The rise of the IRA was countered by Protestants with vigilantes of their own, and from that point on these paramilitary groups on both sides were the chief forces in a turbulent and violent political environment that often seemed to defy rational explanation.

Two other events hardened Catholic attitudes toward the British Army. One was the 1971 reactivation of the policy of internment, carried over from the time of the Irish rebellion in the earlier part of the century. It provided that anyone suspected of terrorist activities could be arrested without charge, jailed, and held without bond or legal aid. Although the reinstitution of internment was ostensibly intended to control the violence emanating from Protestant as well as Catholic groups, it was applied in a less than evenhanded manner. Further adding to the fallen image of the British Army among Catholics was the shooting of thirteen unarmed demonstrators in Londonderry in 1972, an event now annually recalled by Catholics as "Bloody Sunday."

With the intensification of hostilities, the British government in 1972 took full control of Northern Ireland, suspending Ulster's parliament and establishing direct rule from London. This proved no antidote to the level of violence, however. An attempt was made a year later to establish a new Northern Ireland Assembly, which would have provided a form of power sharing among the major political parties, but a general strike led by intransigent Protestant workers brought a return to direct British rule.

Until 1994, little had changed in the basic situation. The British Army, though reduced in size, remained in Northern Ireland. The paramilitary groups on both sides continued to engage in a kind of urban guerrilla warfare, intent on protecting their communities and interests and promoting the causes of either unification, on the one hand, or the status quo, on the other. From time to time, sensationalistic events, such as the hunger strike of imprisoned IRA men in 1981, provoked a round of demonstrations and counterdemonstrations, violence and counterviolence. Meanwhile, a political middle course, aimed at reason and reconciliation, was stymied. A cease-fire was effected in 1994, which represented the first real break in the cycle of violence and held out promise that the troubles might end. We will return to the current politics of the Northern Ireland conflict in a later section.

ETHNIC STRATIFICATION

Having briefly traced the history of intergroup relations in Northern Ireland, we now turn to a more detailed sociological look at this ethnically polarized society. In several ways, Northern Ireland illustrates almost classically many of the notions we discussed in Chapters 2, 3, and 4. To begin with, there is a system of ethnic stratification among Protestants and Catholics in which the former maintain their power not only through superior numbers but also through dominance of the society's vital institutions. Second, the extreme social division of the two groups is supported by a high level of prejudice and discrimination and the implementation of an ideology by the dominant group that rationalizes and justifies its actions toward the minority. Finally, Northern Ireland is a prototypical pluralistic society in which each group maintains a relatively cohesive and institutionally complete community.

The Protestants

Northern Ireland's population of about 1.5 million is made up of approximately 51 percent Protestants and 43 percent Catholics (Table 15.1). The Catholic community is growing more rapidly and its proportion of the total population has increased considerably in the past several decades (Darby, 1995; McGarry and O'Leary, 1995a). Nonetheless, Protestants remain both a numerical as well as a sociological majority.

Despite their generally undivided opposition to any measure that threatens the political status quo, the Protestant population should not be seen as a thoroughly unified community. There are not only class differences among them but also significant sectarian differences. The two major sects, however, are Presbyterians, most of whom derive from the Scottish settlers of the Plantation era, and those belonging to the Church of Ireland, the Irish counterpart of the Anglican church. These two Protestant subgroups constitute a certain degree of cleavage within the dominant group though in no way as acute as between Afrikaner and English South Africans.

The Protestants of Northern Ireland are often portrayed as illiberal, anti-Catholic bigots, eagerly following demagogic leaders. The majority, however, are not political or religious extremists (Heskin, 1980). Their religious practices and beliefs, though, are attended to quite seriously. The intentness with which they approach religion is reflected in the fact that church attendance is higher among Northern Ireland's Protestants than among any other Protestant groups in Europe. Almost two-thirds attend church at least monthly, a phenomenal figure by comparison with other Western societies (Barritt and Carter, 1972; Moxon-Browne, 1983; Rose, 1971). Even in the United States, where church attendance remains higher than in European societies, only 40 percent of Protestants say they regularly attend church (Princeton Religion Research Center, 1990).

Most important, religion for Protestants (and, as we will see, for Catholics) is the key organizing feature of social life. Little is unaffected by religion, including politics. Although the ethnic conflict in Northern Ireland is hardly reducible to religious matters, certain Protestant elements do see their cause as the defense of their religion against an all-powerful Catholic church (Darby,

Table 15.1 Religious Distribution of the Population of Northern Ireland

RELIGIOUS	1926		1951		1961		1971		1981		1991	
PROFESSION	*Number*	%	*Number*	%	*Number*	%	*Number*	%	*Number*	%	*Number*	%
Roman Catholic	420,428	33.5	471,460	34.4	497,547	34.9	477,919	31.4[a]	414,532	28.0[a]	605,639	38.4
Presbyterian	393,374	31.3	410,215	29.9	413,113	29.0	405,719	26.7	339,818	23.0	336,891	21.4
Church of Ireland	338,724	27.0	353,245	25.8	344,800	24.2	334,318	22.0	281,472	19.0	279,280	17.7
Methodist	49,554	3.9	66,639	4.9	71,865	5.0	71,235	4.7	58,731	4.0	59,517	3.8
Others	52,177	4.1	63,497	4.6	69,299	4.9	87,938	5.8	113,244	7.5	122,448	7.8
None	—	—	—	—	—	—	—	—	—	—	59,234	3.8
Unstated	2,304	0.2	5,865	0.4	28,418	2.0	142,511	9.4	274,162	18.5	114,827	7.3
Total	1,256,561	100.0	1,370,921	100.0	1,425,042	100.0	1,519,640	100.0	1,481,959	100.0	1,577,836	100.0

SOURCES: Doherty, 1993; Northern Ireland, Department of Health and Social Services, 1984; *Ulster Year Book,* 1963–1965, 1983.
[a]The Catholic population in both 1971 and 1981 is seriously understated by these figures because of the large percentage of the general population who chose not to respond to the voluntary census question on religious affiliation (9.4 percent in 1971 and 18.5 percent in 1981). The actual Catholic population was estimated to be about 37 percent in 1971 and possibly higher in 1981 (see Compton, 1982). The problem of no response was not so serious in 1991, when only 7.3 percent failed to declare their religious preference (see Doherty, 1993). Most of the latter, however, are assumed to be Catholic, thus raising the actual Catholic percentage.

1976). This view is epitomized by Paisley, whose anti-Catholic rhetoric is replete with admonitions that the aim of the Catholic church is to engulf Protestantism. For most Protestants, however, the major fear of unification with the Republic of Ireland centers on the social and political influence of the Catholic church in that nation.

Although Protestants look to Britain for their historical identity, there is among them a curious ambiguity regarding their relationship with the British. There is, on the one hand, a resistance to certain aspects of English life as well as the prevalence of strongly negative English stereotypes; on the other hand, however, Protestants maintain an unusually strong British political identity—some would say an overidentity (Arthur, 1980; Harris, 1972).

The Catholics

As both the numerical and sociological minority, Catholics in Northern Ireland are a more homogeneous community than are Protestants. If religion is the key organizing ingredient of social life for Protestants, it is even more so for Catholics. Almost all Catholic children, for example, attend parochial schools. Whereas church attendance for Protestants is uncommonly high, it is even higher for Catholics, 90 percent of whom attend at least once a week (Barritt and Carter, 1972; Moxon-Browne, 1983; Rose, 1971). This figure is considerably higher than that for other strongly Catholic nations.

The most telling difference between Protestants and Catholics, aside from matters of faith, involves their national identities and aspirations. Catholics identify themselves overwhelmingly as "Irish," whereas Protestants more commonly see themselves as "British" or "Ulster" (Boyle and Hadden, 1994; Moxon-Browne, 1983; Pollak, 1993). Many Catholics continue to recognize the Republic of Ireland as the rightful government of all of Ireland and thus do not view local Unionist or British governments as entirely legitimate. Hence, they retain the hope that unification will someday be realized. The past tendency for Catholics to reject the legitimacy of the Ulster government is a key factor in explaining Protestant militancy. Catholics are often seen as "disloyal," and this perception served in the past as one rationale for discrimination in employment and government services. As we will see, however, the Catholics are far more divided internally in their political views and aspirations than Protestants have assumed.[3]

[3]Boyle and Hadden (1994) suggest that there is really a third community, made up of those who do not identify with either Protestant or Catholic communities, or who reject the hard boundaries between the two. The percentage of those who fall into this category may vary, depending on different social and political issues.

Ethnic Boundaries

In the United States, although there are no "official" black or white ethnic groups, ethnic or racial status is clearly defined in a customary and rigid fashion; rarely are there questions about who is white and who is black. In Northern Ireland, however, there are no apparent physical distinctions between the two ethnic groups, and with the exception of religion, even cultural differences are not readily evident. How, then, do interacting persons perceive ethnicity in a society in which it is the most critical social identity? Social interaction with members of the out-group calls for a very different set of behavioral responses than interaction with one's coreligionists. Some technique is therefore necessary by which religious affiliation can be determined in the course of social interaction.

With obvious physical marks such as skin color not available, cultural cues and symbols, to which the people of Northern Ireland are finely attuned, serve effectively in their place. Names, for example, often suffice as indications of ethnic identity. Certain surnames indicate an English or Scottish (and thus Protestant) or Irish (and thus Catholic) ancestry, and even Christian names are commonly associated with one side or the other. Other cues are accent, dress, demeanor, and, some would even insist, facial features (Fraser, 1973). Protestants sometimes assert that they can recognize Catholics because of their "generally dark Celtic features," and Protestants may be described as having "an Anglo-Saxon or Scottish 'look'" (Sheane, 1977:67).

The ethnic symbolism is so strong that even certain words or references can alert people to one's religious affiliation. For instance, Londonderry, Northern Ireland's second largest city, has great historic meaning for both groups. Catholics refer to it as Derry, its pre-Plantation name, whereas Protestants prefer Londonderry. Thus, maneuvering a conversation to include the city's name becomes a ploy to ascertain religion (Darby, 1976). Because residence and education are so rigidly segregated, however, the surest indicators of religious affiliation are one's address or school attended. Knowing either of these social facts is ordinarily sufficient to identify a person ethnically.

Burton (1978) refers to the process by which Protestants and Catholics distinguish each other in their day-to-day interactions as "telling." "Telling," he explains, "is a necessary social skill for triggering the cognitive store of sectarianism. It is a trigger which brings into play the typifications and stereotypes about the nature of the 'other side'" (1978:37). Although each cue is subject to various interpretations and is often unclear and imprecise, what remains constant is "the desire and necessity to tell." Within the rural village she studied, anthropologist Rosemary Harris (1972) also noted the need to determine the ethnicity of strangers in order to facilitate interaction. Whereas outsiders might explain as bigotry the tendency for people, on meeting others for the first time, to ask about religious affiliation, Harris points out that "in fact this question is not in itself an expression of religious prejudice but an obvious response to the necessities of social existence in this area" (1972:148).

These ethnic cues by which people attempt to identify others and thus avoid uncomfortable interactions (or perhaps even dangerous situations) reinforce the effects of physical segregation in restricting the knowledge each group has of the other. Interpreting these cues, notes Burton, "constitutes mental bricks and mortar. It is the conceptual and cognitive ghetto of Northern Irish ideological social relations" (1978:66). Even more than Canada, Northern Ireland demonstrates dramatically that visible physical differences or even vast cultural differences are not necessary to create and sustain fierce ethnic hostility. Groups need only perceive their differences as unbridgeable.

Class Divisions

Class differences between Protestants and Catholics are noticeable, though, in light of the intensity of social conflict, they are not as severe as might be expected. Discrimination in employment was one of the major grievances of Catholics in Northern Ireland and was an important factor in energizing the Catholic civil rights movement in the late 1960s. Traditionally, higher-status jobs were reserved for Protestants, and the work force was highly segregated. Today, those discriminatory patterns have been rectified to a great extent.

As is seen in Table 15.2, Protestants are more strongly represented in the higher occupations, and Catholics in the lower, but the disparity is not great. The percentages for each occupational area are fairly close to Protestant and Catholic proportions of the total population. The differences are in no way comparable to those between blacks and whites in South Africa or even between whites and racial-ethnic minorities in the United States. There is an entire range of classes among both groups, including large numbers of Protestant poor. The most serious discrepancy is in the rate of unemployment, which is markedly higher among Catholics (Darby, 1995; Pollak, 1993; Smith and Chambers, 1991).

Recent studies indicate a continuation of the trend toward greater ethnic equity in the Northern Ireland class structure (Boyle and Hadden, 1994; Cormack and Osborne, 1994; Osborne and Cormack, 1991). Catholics are moving into higher managerial and supervisory positions in significant numbers, and the Catholic middle class in general is expanding. Moreover, Catholics have achieved educational parity with Protestants. Also apparent, however, is the comparatively little change between Protestants and Catholics at the working-class level and especially among the unemployed. Catholics are twice as likely to be unemployed as Protestants (Darby, 1995). This sharply higher unemployment rate

Table 15.2 Composition of the Northern Ireland Work Force[a]

OCCUPATIONS	PROTESTANT	CATHOLIC
Managers and Administrators	65.6%	34.4%
Professional Occupations	60.7	39.3
Associate Professional and Technical Occupations	58.0	42.0
Clerical and Secretarial Occupations	63.3	36.7
Craft and Skilled Manual Occupations	64.5	35.5
Personal and Protective Service Occupations	69.0	31.0
Sales Occupations	64.1	35.9
Plant and Machine Operatives	58.6	41.4
Other Occupations	59.6	40.4

SOURCE: Fair Employment Commission for Northern Ireland, 1995.
[a]Percentages are based on those whose religion was determined. This was over 90 percent for almost all occupational categories.

has become a focus of grievance for Catholics, who see it as the result of continued discrimination. We will consider this issue in a later section.

Political Power

A key to understanding the minority status of Catholics in Northern Ireland lies in the political sphere. As previously noted, the correlation between political preference and religious affiliation is extremely high. This is similar to other highly stratified multiethnic societies where various forms of inequality stem from great differences in political power.

Essentially there have been two diametrically opposed sides to the political conflict in Northern Ireland, each identified with either the Protestant or Catholic communities. One is the unionist movement, whose basic position, since its emergence in the late nineteenth century, has been preservation of the union between Northern Ireland and Great Britain (McGarry and O'Leary, 1995a). The Unionist party (in recent years there have been, in fact, two unionist political parties) is made up primarily of Protestants who, in keeping with their intention to keep Northern Ireland British, have staunchly opposed any efforts at developing closer ties with the Republic of Ireland. Any such efforts have been interpreted as the first steps toward eventual unification. This position has remained fundamentally unchanged over the years. Protestants, whether they are politically active or not, are seen as supporters of unionism. The nationalist movement, on the other hand, is made up primarily of Catholics whose objective is eventual unification of Northern Ireland with the Republic of Ireland. The Nationalist party and, in recent years, its successor, the Social Democratic and Labour party, have been identified with Catholic issues and interests, especially the question of unification.[4]

Because Protestants were always a numerical majority, the power of the unionists was unchallengeable. In a sense, before 1972, when Britain took full control of Northern Ireland's central government, Northern Ireland had been a one-party state. This dominance was exacerbated by the tendency of Catholics not to participate in parliamentary politics because they found themselves a permanent minority.

As a result of the political dominance of the unionist Protestants, Catholics had been, in effect, excluded from the political process at the national level and seriously underrepresented at the local level (Aunger, 1981; Rose, 1971). Before direct British rule in 1972, Catholics had not been appointed to any cabinet positions since the founding of the Northern Irish state in 1921. Senior civil service and police appointments had also been almost exclusively Protestant. Unionist control of local councils had a similar discriminatory effect. In Omagh, for example, with over 60 percent Catholics, almost all public jobs and offices were held by Protestants (Campaign for Social Justice in Northern Ireland, 1972).

The wide discrepancy in political power between Protestants and Catholics resulted in severe inequalities in other areas of social life, especially jobs and housing. When Britain assumed control of the Northern Ireland government in 1972, however, Protestant domination was ended. Now, Britain, not unionist Protestants, controlled the central government. In addition, Britain imposed political reforms at the local level, which resulted in substantially increased Catholic representation on local councils. Catholic employment in civil service jobs subsequently increased substantially, and Catholics today are represented in such occupations proportionate to their share of the general population (Cooper, 1991; Cormack and Osborne, 1994; Fair Employment Commission for Northern Ireland, 1995). Changes also occurred in the distribution of housing. Political control of public housing was an especially significant power held by Protestants because a high percentage of people in Northern Ireland rent their dwellings from public agencies. A centralized housing agency replaced local agencies in 1972, resulting in a more equitable distribution of public housing (Darby, 1995).

---

[4]Protestants are commonly referred to as "unionists" or "loyalists," in accordance with their desire to maintain Northern Ireland's union with Great Britain. Catholics are described as "nationalists" or "republicans," reflecting their favor of an eventual unification with the Republic of Ireland.

As we have now seen, in most multiethnic societies the dominance of one ethnic group over others is enforced through discriminatory practices and is rationalized through a set of negative beliefs regarding minority groups. In some societies, this system is officially sanctioned and rigidly administered; moreover, negative beliefs about the subordinate groups are encompassed in an elaborate racist ideology. This characterized South Africa under apartheid. In other societies, like the United States, the system of ethnic dominance depends on custom rather than law, and the ideology on which such practices are based is subject to strong dispute. In still other societies, like Brazil, ethnic dominance may be denied entirely though its manifestations are subtly evident. Where, in this range of ethnic patterns, does Northern Ireland stand? Prejudice and discrimination in this society are perhaps most similar to patterns of black–white relations in the United States. They are neither officially legitimized nor enforced through the application of coercive power; but neither are they difficult to detect.

Protestant and Catholic Antagonism

In explaining ethnic prejudice and discrimination in Northern Ireland, two views have been put forth. One is that Protestants harbor deep suspicions of Catholicism. The Catholic church is seen as wielding great influence on its parishioners, and, because it is the central institution for the Catholic population, it becomes the focus of Protestant hostility (Bruce, 1986). "For many Protestants," notes Heskin, "it seems that Catholic Churches are mysterious and forbidding places and Catholicism is a powerful, sinister and potentially engulfing creed" (1980:27). For extremists, he explains, Catholicism symbolizes evil, much like some Americans have perceived Communism. A more prevalent view, however, rejects the notion that anti-Catholic antagonism derives from Protestants' disdain of Catholicism per se. Rather, it is seen as stemming mainly from the nationalist sentiments of Catholics: If Ulster were to be united with the heavily Catholic Irish Republic, the cultural identity and political power of Ulster Protestants would disappear and their economic position decline (Jackson and McHardy, 1984; McGarry and O'Leary, 1995a).

Because most Catholics continue to hope for the eventual unification of Northern Ireland with the Republic, they are commonly viewed suspiciously by Protestants as disloyal citizens whose reluctance to accept the regime fully justifies their minority position. In 1933, Sir Basil Brooke, while minister of agriculture (he was later to become Lord Brookeborough and prime minister), called on Protestant employers not to hire Catholics because "99 percent are disloyal"

(Wallace, 1971). Thus, the dynamics of prejudice and discrimination by Protestants against Catholics are fueled in part by the perception of Catholics as seditious persons intent on destroying the present political arrangement and thus turning the Ulster Protestant majority into an Irish minority. Protestant fears also stem from the belief that Catholics may eventually become the numerical majority in Ulster, a belief that has a solid basis in demographic trends. Since the 1960s, the Catholic population has been steadily growing as a proportion of the total population, due in large measure to their higher birthrate (Doherty, 1993; McGarry and O'Leary, 1995a).

For Catholics, anti-Protestant animosity stems primarily from a heritage of discriminatory treatment rather than from any fears of Protestantism per se. The objections to and suspicions of Protestants focus on their political power and the society's social and economic issues, which tend to be shaped along religious lines (Barritt and Carter, 1972; Bell, 1976; Heskin, 1980; Pollak, 1993).

It should be remembered, of course, that in any multiethnic society not all members of one group exhibit equal levels of negative beliefs about members of the other. In each ethnic community, there is always a range of thought among individuals, extending from those who are highly prejudiced to those who genuinely desire a mixed society. So segregated are the two ethnic groups in Northern Ireland, however, that little opportunity arises for even moderate thought to take hold among the majority of the population.

Stereotypes

Many of the stereotypes commonly attributed to Catholics are part of what seems to be an almost standard litany employed by dominant groups in Western multiethnic societies—lazy, dirty, shiftless, oversexed, ignorant, quarrelsome, and so on (Bell, 1976; Fraser, 1973). In the rural village of Ballybeg, Harris (1972) found that Protestants routinely regarded Catholics as poor and as intent on cheating the welfare system. She quotes one of the villagers as believing that "Catholics have their own people in the Civil Service in Belfast and they tell them how to get money they don't deserve." Catholic doctors, the villager thought, "were always ready to say a Catholic was sick even when he was healthy, if only it would get him government money" (1972:174).

Fraser writes that of all negative characteristics, the size of Catholic families "seems to provoke Protestant hostility more than anything else" (1973:94). This stems primarily from the fear of being displaced as the numerical majority but also from the resentment of larger Social Security benefits paid to Catholic parents. Fraser quotes one young Protestant boy of eleven, already well imbued with anti-Catholic conviction:

> The Catholics have big families and they won't work. Daddy says we're poor because the Government takes his money to give to the Catholics and all their children and their priests to keep them in luxury because they don't work. In the South they wouldn't get any money and would have to work for it. Here they keep us all poor and don't appreciate what they get (1973:96).

In a study conducted among both Protestants and Catholics, O'Donnell found that out-group stereotypes were not entirely negative. Members of both groups, he reported, saw each other basically as "decent, fine, ordinary people" (1977:165). As in other multiethnic societies, negative stereotypes are ordinarily applied to out-groups as a whole rather than to specific individuals. One's out-group neighbor, therefore, may be seen as "different from the others." Negative group images, however, are clearly expressed. Protestants, on the one hand, see Catholics as being Irish nationalists who are brainwashed by priests, as having too many children, and as being superstitious and bitter. Catholics, on the other hand, view Protestants as holders of power who are unwilling to relinquish any of it, whatever the cost. The images each side has of the other, then, are manifestations of the ideological basis of the conflict—Protestants' fear of unification and the Catholic church and Catholics' fear of Protestant political power.

Discrimination

Because the two groups are, as we shall see, so segregated in their own communities, it is difficult to ascertain the actual degree of discrimination applied by Protestants against Catholics, or vice versa. In a social environment in which each group maintains its own complete communities, it is natural for institutions to become ethnically homogeneous. Schools, businesses, recreational organizations, and so on all cater to a clientele made up primarily of members of one group or the other. The question, then, is how much of the separation of the two groups and the resultant social differences between them is the result of voluntary forces and how much the result of forces imposed on them.

There is also the question of institutional discrimination. How much of the deprived condition of the minority is the result of direct actions by the dominant group, and how much is the product of the workings of institutions that, through historical evolution and unchanged manner of operation, discriminate against the minority routinely?

Although these questions are certainly pertinent in the case of Northern Ireland, it is generally agreed that ethnic discrimination has been direct as well as institutional, particularly in employment and politics, and in both areas Catholics have been the chief victims.

Employment It is generally recognized that, in the past, discrimination in the work force on the basis of religion was widespread (Barritt and Carter, 1972; Bell, 1976; Boyle and Hadden, 1994; Hoare, 1982). In 1989, a fair employment act, designed to monitor the employment practices of companies with 250 or more workers, was instituted in Northern Ireland. A similar measure in 1976 seemed to have minimal effect on reducing ethnic discrimination in employment, but the new act provided for a kind of affirmative action as a means of promoting ethnic equality (Edwards, 1995; Smith and Chambers, 1991). No longer could ethnic discrimination in the work force be viewed as acceptable or unavoidable (Boyle and Hadden, 1994). Such efforts at overcoming discrimination appear to have had a positive impact, as Catholics have increasingly moved into higher-status jobs and into occupations once closed to them. During 1990–1994, Catholics increased their percentage in all job categories, and the largest increases occurred in more prestigious occupations such as professionals and managers (Edwards, 1995).

Unemployment in Northern Ireland, however, remains, as noted earlier, much higher among Catholics, and it is here that questions of discrimination continue to arise. Although Catholics see chronically higher Catholic unemployment rates as a result of continued Protestant discrimination, structural factors unquestionably contribute as well. Residential segregation and differences in educational attainment, for example, play some part in explaining different rates of unemployment between the two groups. Regional location is another important structural factor. Northern Ireland as a whole is divided economically along east-west lines. In the east, it is dominated by Belfast, where most industry and, therefore, jobs are found. The west has remained largely agricultural and, hence, much less economically advanced. Protestants are a majority of the population in the more prosperous eastern sector, and Catholics are a majority in the western sector. Given this demographic imbalance, Catholics are automatically subject to less favorable occupational opportunities and higher rates of unemployment.

Although past anti-Catholic discrimination is unquestioned, many Catholics in Northern Ireland may now be in much the same situation occupationally as the black underclass in the United States—the victims no longer of a system of blatant ethnic discrimination but of an *institutional* system of discrimination that has left them with inadequate skills to compete successfully in the modern labor market. In addition, Catholics may hesitate to take jobs in historically Protestant companies because of fear of violence or are reluctant to leave their ethnic communities and venture into Protestant areas.[5]

---

[5]Hewitt (1981) asserts that the degree of anti-Catholic discrimination before the civil rights movement and the subsequent troubles has been exaggerated. Moreover, he points out that though discrimination by Protestants has been most widely reported, there is comparable anti-Protestant discrimination in politics and employment where Catholics are in an advantageous position.

Perceived Discrimination Whatever its extent and whatever its source, Catholics *perceive* discrimination by Protestants. Thus, it may matter little that the socioeconomic conditions of Protestants are only marginally better than those of Catholics or that the most flagrant devices used to deny Catholics job opportunities have been increasingly checked. Because one's religious affiliation is not easily camouflaged, Catholics will simply not be convinced that "failure to get a job or a home is not governed by that fact" (Jackson and McHardy, 1984:4).

Rose in 1971 confirmed the significant difference between Protestants and Catholics in their perceptions of discrimination in Northern Ireland. Whereas nearly three-quarters of Protestants saw no discrimination on the basis of religion, three-quarters of Catholics did. Twenty years later, another survey reaffirmed the perception of Catholics as the victims of discrimination in employment. Surprisingly, a significant percentage of Protestants, too, believed that discrimination in the labor market on the basis of religion was prevalent. However, Protestants seemed to view the situation as one in which members of each ethnic community "look after their own," whereas Catholics saw it as societal discrimination against only their group (Osborne, 1991).

Sociologist John Darby suggests that in recent years there have been signs of moderating attitudes of each community toward the other. As political, educational, economic, and social reforms have been implemented, "there has been a much greater acceptance that they are irreversible, and also that they are fair" (1995:25). Protestants, in other words, have accepted, perhaps reluctantly, the legitimacy of minority rights.

NORTHERN IRELAND AS A PLURALISTIC SOCIETY

In Chapter 4, multiethnic societies were seen as falling generally into three types: colonial, corporate pluralistic, and assimilationist. In some multiethnic societies certain characteristics of two or perhaps all three of these types may be apparent. Northern Ireland is such a case, displaying features of both colonial and corporate pluralistic systems. Not at all ambiguous, however, is the virtual absence of assimilationist features. This is clearly a society in which the centrifugal forces of pluralism—some imposed, some voluntary—predominate in the relations between the two ethnic groups.

Cultural Pluralism

Although most cultural differences such as language or style of life seem minor or nonexistent, the religious division is so critical that it often creates the perception of difference among members of the two groups where none or little is present. Political attitudes and intergroup relations are colored by the predominant belief that Protestants and Catholics are the historical product of diverse cultures that have coexisted uneasily since the Plantation of Ulster in the seventeenth century (Buchanan, 1982). De Paor, for example, writes that the confrontation of the Scottish and English settlers with the native Irish in that period represented a difference not simply in religion but also "in language, law, custom, economy, thought, and art" (1970:24). This understanding of a fundamental cultural gap between the two groups is sustained by all the society's major institutions. Protestants and Catholics see each other, quite simply, as two sharply different peoples.

Some, however, have suggested that the cultural differences between the two groups are not as historically clear as has commonly been believed. Buchanan points out that because Scotland and Ireland had been in close contact for more than a thousand years before the colonization of the seventeenth century, the folk culture the Scottish settlers found when they arrived was essentially "very similar to that of their homeland" (1982:67). Moreover, he asserts that what differences did exist between the native Irish and the Scottish and English colonists were resolved through a process of acculturation within a century. The religious breach, however, proved intractable and thus served to sustain group division. Buchanan also suggests that the growth of Irish nationalism in the nineteenth century accompanied by the liberation of Catholic education provided Catholics with a new sense of pride in Irish culture and history. The place of the Protestants in Irish history, however, was only as colonialists and usurpers, and, as a result, Protestants turned to Britain for their historical identity—hence the development of the two contrasting historical views.

Given the rigorous ethnic segregation of schooling in Northern Ireland, these views are promulgated from generation to generation. Noticeably different histories are taught Protestant and Catholic students, one emphasizing the Irish connection, the other the British. Events, personalities, and political institutions are not interpreted in the same manner, and thus the discordant societal views of the two groups are perpetuated.

In addition to sectarian education, the cultural chasm separating Protestants and Catholics is deepened by uniquely ethnic activities in other social areas. Team sports played by the two communities, for example, are oriented to British games, for Protestants, and Irish games, for Catholics. Rugby or cricket, games of English origin, are rarely played in Catholic secondary schools, and Irish games like Gaelic football or hurling are not played in Protestant schools (Darby, 1976).

Cultural differences are reinforced by various symbols displayed prominently by each group. It is common, for example, for flags to be flown from houses, either the Union Jack, for Protestants, or the Irish Tricolour, for Catholics. Protestants will respond with pride to a playing of "God Save the Queen," whereas

Catholics may ignore it and favor the Irish national anthem. Such symbols provide cultural boundary markers that reinforce in-group feelings and sharpen the perception of out-group differences.

The result of all this is an inability to understand the other side. In Ballybeg, Harris discovered that, although members of the two ethnic communities might in some contexts have close and friendly contacts, for the most part they "managed to remain in almost complete ignorance concerning each other's beliefs" (1972:146). In situations where they interacted, Catholics and Protestants went to great lengths to avoid discussing any controversial topics, especially involving religion and politics. "The obvious aim," Harris notes, "was to prevent social relations being ruffled by disagreements over subjects on which it was accepted that there could be no consensus" (1972:146). Thus, no effort was made to bridge their differences. Although group differences may be tolerated in situations where Protestants and Catholics meet, rarely are they disregarded (Barritt and Carter, 1972).

Structural Pluralism

The establishment of separate institutional structures for each ethnic group is a key feature of highly pluralistic societies, and in this regard Northern Ireland is almost a pure case. Protestants and Catholics are ethnic communities with a high degree of identification and institutional completeness. Segregation is the general rule in most areas of social life. Each group maintains its own neighborhoods, schools, businesses, voluntary associations, and even newspapers. Much social and leisure activity occurs within the churches of both Protestants and Catholics, which further restrains intergroup contact. Social interaction across ethnic lines, therefore, is confined mostly to secondary, instrumental contexts like work and, even there, is limited. Although social separation is not dictated by law, as it was in South Africa under apartheid or the pre-1960s American South, it is enforced by a well-founded system of custom.

Residential Segregation Perhaps the most evident form of separation of Protestants and Catholics, having the most far-reaching social effects, is residential segregation. Segregated living areas can be traced to the Plantation era, when English and Scottish colonizers displaced the native Irish Catholics. Scots and English lived within the walls of the town or around the market, and the Irish were driven into outlying areas or beyond the town gates. In the rural areas, a similar pattern developed, with the Irish excluded from the more productive valley lands and forced into the mountains (Barritt and Carter, 1972).

Today, the most acute patterns of residential segregation are found in the larger towns, especially in the major city of Belfast. By North American standards,

Belfast is the only truly large urban area in Northern Ireland, with a population of roughly 500,000. Londonderry, the second largest city, has only 80,000. The industrial expansion of Belfast in the nineteenth century produced a huge and rapid influx of people displaced from the land throughout Ireland (Jones, 1960). A large percentage of these urban migrants were Catholics and represented a threat to Protestant workers. Intolerance and conflict between the two working-class groups gave rise to sectarian colonies within the city, a pattern duplicated in other Ulster towns. "Catholic and Protestant workers lived in the same general urban areas," writes Darby, "but were divided into religious ghettos, compact, mutually exclusive and leaning on each other" (1976:28; see also Boal, 1982).

Recurrent religious conflict during the nineteenth and early twentieth centuries tended to reinforce ethnic housing areas. Families of "the other side" were driven out during each conflagration, purifying, in a sense, these ethnic enclaves. Darby and Morris (1974) found that this pattern had continued with the onset of the current troubles, with Catholics and Protestants moving into more homogeneous areas. Between August 1969 and February 1973, some 8,000 families in the greater Belfast area alone were forced to leave their homes because of direct intimidation by the other side combined with general fear and a feeling of vulnerability. In actuality, the number displaced may have been much greater because the recorded figure included only those cases that had come to the attention of relief agencies. This represented between 5 and 10 percent of the entire Belfast population (Darby, 1976). The exchange of houses between Protestants and Catholics was commonplace in towns like Londonderry where violence was especially intense. In some cases, families exchanged their houses on different sides of the same street, so clearly were the ethnic boundaries drawn (Darby and Morris, 1974).

Ethnic residential segregation, always common in the past, has become more pronounced in recent times. About one-half of Northern Ireland's 1.5 million people live in areas more than 90 percent Protestant and 95 percent Catholic (Pollak, 1993). Some working-class areas of Belfast are divided literally by a twenty-foot-high reinforced wall (ironically referred to as a "peace line"). Although there has been some residential mixing, particularly within the middle class, this is limited in scope (Cormack and Osborne, 1994).

Segregated residential patterns are created not only by fear and intimidation but also in part by the natural inclination to live among one's own. In the city of Belfast and in other large towns, where urban villages have evolved, each serves as the focus of city life for its residents. A kind of territoriality has thus developed, in which people feel more comfortable in home areas where their churches and schools are located. Added to this natural territoriality, however, is the ethnic factor, which further strengthens the homogeneity of the neighborhood. Segregation, then, once established, "shows a strong tendency towards inertia" (Darby, 1976:37).

The dynamics of Northern Ireland's residential patterns seem quite similar to those of blacks and whites in American cities, where deliberate efforts by whites were most instrumental in creating black ghettos and subsequently preventing their dispersal (Marger, 1989). Poole (1982) shows, for example, that residential segregation is highest in Northern Ireland towns where the numerical minority (usually Catholics) is a large percentage of the population. In such situations, the minority is ordinarily perceived as threatening the ethnic integrity of the community or neighborhood. Again, this is not unlike the black–white patterns of American cities. What differs in the case of Northern Ireland, however, is the intensity of intercommunal violence that has traditionally accompanied ethnic residential segregation (Darby, 1976). Also different is the fact that, as we previously mentioned, residential segregation in Northern Ireland is impelled by both Protestants' and Catholics' strong tendency to seek residence among fellow ethnics and to patronize ethnic institutions. This tendency is abetted by government housing policy, which has been forced to consider ethnic divisions in the building and distribution of public housing units. In the American case, blacks' desire to reside primarily among fellow ethnics is not what propels residential segregation. Recall the studies cited in Chapter 8 indicating that, given the choice, blacks prefer an integrated neighborhood. Racial segregation in the United States, then, is less voluntaristic than it is in Northern Ireland.

The high level of residential segregation in Northern Ireland has far-reaching effects on all other aspects of group interrelations, just as it does in the United States. Segregated living conditions contribute greatly to the continued polarization in all other institutional areas. Social contacts are minimized, and the cycle of misunderstanding, suspicion, and at times violence between the two groups is preserved. As one Derry resident put it, "Most Catholics and Protestants don't meet each other. We Catholics project the worst of ourselves on to Protestants and vice-versa" (Pollak, 1993:42).

Education Ethnic segregation in education is even more complete than in housing. Almost all Catholic children attend Catholic schools, and Protestants attend state (de facto Protestant) schools. Little more than 1 percent of students are schooled in an integrated setting (Boyle and Hadden, 1994; Cormack and Osborne, 1994; Murray, 1983, 1986). In the case of Catholics, however, the origin and continuation of this pattern are mostly voluntary because the church has traditionally been adamant regarding the education of Catholic children in parochial schools (Darby, 1976; Wallace, 1970). Only at the university level and in special schools for the handicapped do students of the two groups mix to any significant degree. Even teachers' training colleges are essentially denominational.

The result is that most Ulster students "are being taught alongside their co-religionists by teachers of their own religion and have minimal educational contact with children from the other main religious group" (Darby, 1976:137–38). As we earlier noted, Protestant and Catholic schools encourage different cultural traditions, do not teach the same political history, and even sponsor different extracurricular activities such as sports and games.[6] Although the attitudinal effects of this educational environment are difficult to measure, it is reasonable to conclude that group prejudices are heightened. Some see educational segregation as the key to explaining the perpetuation of ethnic conflict and believe that only through school integration can community barriers begin to break down (Fraser, 1973; Irwin, 1994). Others, however, believe that there is no precedent for integrated education in the society and that even if it were implemented, it would not guarantee greater intergroup understanding and tolerance (Rose, 1971).

Endogamy At the most primary level of interaction, the social segregation of the two groups is enforced by the norm of endogamy. In his important survey of the late 1960s, Rose (1971) reported that only 4 percent married across religious lines. Later studies have indicated a rising but still low rate of Protestant–Catholic intermarriage (Gallagher and Dunn, 1991; McGarry and O'Leary, 1995a). Interfaith marriages are discouraged in much the same way as interracial marriages are inhibited in the United States. Such marital unions create a degree of social ostracism and hardship that few are prepared to endure (Barritt and Carter, 1972; Donnan and McFarlane, 1983). Harris (1972) noted that in Ballybeg, in cases of intermarriage, kinship ties were not established between members of the two families. An intermarrying couple, she explained, was usually married in the wife's church, and, as a result, tensions were created between the husband and his kinfolk, "especially between them and his wife, that put all normal kin-based contacts between them out of the question" (1972:143). In the 1970s in neighborhoods of Belfast where interethnic violence flared, mixed-religion families were usually among those forced to flee (Darby and Morris, 1974). In any case, the separation of the two groups in most institutional spheres inhibits the development of intimate relations between Protestants and Catholics ultimately leading to marriage.

Given the relatively tiny geographical and demographic dimensions of Northern Ireland, the extent to which ethnic segregation has been established and enforced is extraordinary. A Protestant student at one of the few integrated

---

[6]Murray (1983) claims, however, that the differences between Protestant and Catholic schools may be overstated. Although there is no question that each school system reflects a different culture, he suggests that there are more similarities of practice within them than differences between them. See also Belfast Workers Research Unit, 1980.

schools in Belfast confesses that "I'd never met a Catholic before I sat beside one in class" (Irwin, 1994:106). Interaction between Protestants and Catholics is more common in smaller communities, of course, where daily contact is a routine occurrence. In Belfast, Derry, and other larger towns, however, the Northern Ireland case demonstrates clearly how in-group members can seal themselves off in homogeneous neighborhoods and, with little effort, avoid out-group members. It is a situation not unfamiliar to most Americans, who may reside in urban areas whose communities and neighborhoods are sharply divided along racial lines.

Although the ethnic division in Northern Ireland supersedes all others, class divisions should not be dismissed as insignificant. In fact, this is a society highly stratified by class, and the sociocultural differences between classes *within* each ethnic group may be as great as the differences between ethnic groups themselves (Darby, 1976). The intensity of the ethnic conflict has been highest among the poor and the working classes of both groups, and most of the violence has occurred within those communities. It is among poor and working-class communities that the paramilitaries, on both sides, have successfully recruited. The middle classes, relatively untouched by the violence, can, in a sense, opt out of the conflict (Jackson and McHardy, 1984; Pollak, 1993). Moreover, residential integration and more frequent intergroup contact in general are evident among middle-class Protestants and Catholics (Boyle and Hadden, 1994).

STABILITY AND CHANGE

Is Northern Ireland destined for a future of unresolvable ethnic conflict? Can the divide between Protestants and Catholics ever be bridged so that ethnicity no longer plays the paramount role in intergroup relations? The conflict in this society is perhaps the most confounding among the four at which we have looked.

The absence to date of effective efforts to mobilize the bulk of the population in ending the hostilities stems largely from the fact that Northern Ireland is a political entity in limbo, neither totally sovereign nor totally part of Britain. Thus, in a real sense, the Northern Irish state lacks the ability to impose social order through legitimate power. It is this lack of consensus regarding political legitimacy that makes the situation "a problem to every solution" (Barritt, 1982:5). Most Protestants and Catholics are in basic disagreement about what the Northern Ireland state is or from where political legitimacy derives. As a result, the British Army, the Provisional IRA, and the Protestant paramilitary groups have been, in recent decades, the chief political forces in the society.

As Boal and Douglas (1982) point out, however, although the situation appears utterly without solution, it is more accurate to say that any possible solution is either unpleasant or extremely difficult to attain—or both. It is chimerical to think that significant changes in attitudes and behavioral patterns can be brought about in the foreseeable future. It is obvious that the two groups maintain an extraordinarily high degree of institutional and identificational separateness that, rather than breaking down, has actually solidified in recent years. Hence, so far as any solution can be achieved, it must lie in some form of political settlement that acknowledges interests on both sides.

One important fact contributing to the frustration of a solution to the conflict is that neither of the two conflicting groups wants political autonomy; on the contrary, most Protestants want to cement their tie to Britain, and most Catholics presumably want unification with the Republic of Ireland. This makes the situation considerably different from the Quebec movement for sovereignty as well as from contemporary ethnic movements in Europe. It also, of course, introduces two parties to the conflict that are not part of the indigenous population: Britain and the Republic of Ireland. The fate of Northern Ireland, then, is by no means in its own hands.

Britain is committed to maintaining its link with Northern Ireland as long as the majority of the people of Ulster prefer it, and, given the Protestant numerical majority, this is automatically assured, at least for the immediate future.[7] The removal of the British from Northern Ireland, therefore, although probably inevitable, is likely to be a protracted affair. As for the Irish in the South, enthusiasm for unification is tempered by a realization of the considerable political, economic, and social costs involved.

The desires of the Catholic population of Ulster regarding political unification with the Republic are also more complex than they appear. It is by no means certain that a majority of Catholics do in fact harbor an eagerness to join the South. Surveys have indicated that perhaps little more than half would actually vote themselves into the Republic if given the opportunity (Darby, 1995; Davis and Sinnott, 1979). For one, the Republic is less advanced economically than is Northern Ireland; therefore, unification might create significantly lowered material standards (Compton, 1991). Furthermore, the British tie presently provides considerable social benefits and economic advantages that would be relinquished with unification (Heskin, 1980; Pollak, 1993).

It has often been observed that Ulster's Catholics are most concerned with receiving equal economic and political opportunities and would accept the Northern Ireland state if such conditions were met. But feeling that Protestant

---

[7]As noted earlier, the Catholic population is growing much faster than the Protestant and some forecast a Catholic numerical majority sometime in the twenty-first century. Should that occur, Catholics would be in a position to vote to sever the British link.

intransigence precludes their receiving fair treatment, many are driven to see unification with the Republic as their only hope. There is, of course, a range of opinion among Catholics on this most critical issue (Moxon-Browne, 1983; Rose, McAllister, and Mair, 1978). Yet Protestants continue to perceive the Catholic populace generally as intent on unifying Ireland and respond to them on the basis of that perception.

The level and form of support given the IRA by the majority of the Catholic populace is also questionable. Although some hold that Protestant violence galvanizes popular support among Catholics for the IRA, others contend that this support is not widespread and in any case fluctuates with changing events (Moxon-Browne, 1983). In addition, it is unlikely that violent IRA tactics have ever been supported by more than a small minority. Again, however, these nuances of opinion may go unrecognized by Protestants, who tend to associate Catholics generally with activities of the IRA.[8]

What, then, are the alternative political scenarios that might lead to a resolution of the conflict, at least in some measure? Numerous schemes have been proposed, ranging from, at one extreme, a fully independent Northern Ireland to, at the other extreme, unification of the North with the Republic of Ireland. In between, proposals have included full integration of Northern Ireland with Great Britain; a federal system of some kind, encompassing the Republic and the North; some form of joint rule of Northern Ireland by Britain and the Republic; repartition of Northern Ireland, in which Protestants and Catholics would be placed into ethnically cohesive areas; and a return to simple majority rule within the context of a devolved Northern Ireland government, similar to the situation before Britain assumed direct rule. None of these formulas is given much chance of realization at present.[9]

Essentially, a resolution to the conflict lies along one of two lines: a power-sharing arrangement, guaranteeing political participation and protection to both communities, or the creation of institutions that will more uniformly separate them.

Power Sharing Regarding the first option, the various schemes that have been suggested are of two types: those involving internal power sharing in a consociational-type arrangement, or externally shared authority in which both

---

[8]At the height of the violence in the mid-1970s, a poll showed that 97 percent of Protestants and 96 percent of Catholics expressed disapproval of the use of violence to achieve political ends (*Belfast Telegraph,* 1974).

[9]These have been discussed most fully by Barritt (1982), Boyle and Hadden (1994), Palley (1981), Pollak (1993), and Rose (1976).

British and Irish governments would maintain joint authority over Northern Ireland on behalf of Protestant and Catholic communities (Boyle and Hadden, 1994).

Before 1972, when Britain assumed direct rule of Northern Ireland, a semi-autonomous parliament, the Stormont, acted as Ulster's major ruling body. Because political preference was almost perfectly correlated with ethnicity, a permanent Unionist—and thus Protestant—majority was assured. Protestants, therefore, maintained a continuous monopoly of political power. To return to such an arrangement would again condemn Catholics to a permanent minority political status. A power-sharing system in which both Protestants and Catholics will participate has been seen, therefore, as a necessary ingredient in the establishment of any future viable Ulster government.

As we noted in Chapter 4, a form of political accommodation, consociationalism, has been worked out in certain multiethnic societies in which power is shared proportionately between majority and minority groups (Lijphart, 1977). Since 1972, several initiatives along this line have been made by Britain. All have been rejected mainly because adamant Protestants have not accepted Catholic participation (Aunger, 1981; Lijphart, 1975).

Despite the inherent difficulties in achieving any form of power sharing, numerous proposals continue to be offered. McGarry and O'Leary (1995a) have suggested that, given the increasing demographic parity between Catholics and Protestants, practical politics will likely move Northern Ireland closer to some kind of power-sharing arrangement.

Greater Separation

Rather than moving toward a more cooperative political structure, it is possible that Northern Ireland may move in the direction of greater separation of Protestant and Catholic communities. Living peacefully by minimizing contact may, as Boyle and Hadden (1994) point out, be preferable to a state of perpetual conflict. Some have even suggested that, without some form of power sharing, only repartition might bring a degree of political order to Ulster. That is, the boundaries of Northern Ireland might be redrawn so as to include a majority of its Catholics in the Republic, and the Protestants would be confined to a smaller and more cohesive area (Boal and Douglas, 1982; Lijphart, 1975; Schmitt, 1994).

This scheme has precedents in ethnically heterogeneous societies. India, for example, was partitioned in 1947 into two states, India and Pakistan, one with a Hindu majority, the other with a Muslim. Indeed, partition is the very manner in which Northern Ireland was created in the first place. Some believe that the original partition of Ireland in 1920 might have worked had the areas designated been more ethnically homogeneous (Boal and Douglas, 1982; Compton, 1982;

Lijphart, 1975). Any such repartition, of course, would create enormous logistical problems involving large exchanges of population and, judging from other historical cases (including Northern Ireland), would probably not be peaceful. Obviously, then, this is a measure that might be considered only if all other efforts had failed.

An Enduring Conflict?

Barring a radical change in the near future, it is likely that a condition of political "dissensus" will continue to characterize Northern Ireland, in which the extremist elements of both sides maintain the upper hand, requiring the continuation of direct British rule. Some observers of Northern Ireland, like Richard Rose, pessimistically see no long-term solution, "if a solution is defined as a form of government that is consensual, legitimate, and stable" (1991:133). If that assumption is made, writes John Darby (1991), it might be more sensible to focus not on a "solution," but on addressing the grievances of both groups through concrete and realizable internal reforms. Furthermore, maintaining the status quo, as Aunger (1981) has noted, is not the preferred solution, but neither is it the worst. For Catholics, direct British rule precludes, at least temporarily, political dominance by Protestants, and for Protestants, it precludes unification with the Republic.

What's more, the magnitude of ethnic violence should not be overdrawn. As Sheane (1977) has pointed out, given the availability of modern weapons, the degree of violence today may actually be lower than it was in 1920. Although over 3,000 people have been killed by the warring factions since 1969, homicides during that period in many large American cities have exceeded that number. Moreover, compared to other violent ethnic conflicts in recent times—in Bosnia, Rwanda, Sri Lanka—Northern Ireland's shrinks in significance.

In his study of three small communities, Darby (1986) showed that social mechanisms are devised locally to control the violence between Protestants and Catholics, making life tolerable for both sides. It is the existence of such restraints, Darby maintains, that accounts for the enduring nature of the conflict, preventing its resolution through either some form of settlement or a full-scale internal war. As a recent citizens' inquiry on the Northern Ireland conflict concluded, despite its intractability and devastating effect on civil society, "the situation is not bad enough for people to demand that their elected representatives make the compromises necessary for a settlement" (Pollak, 1993:10). Both Protestants and Catholics have, in a sense, learned to live with the conflict. Media reporting on Northern Ireland tends to create the impression of a society in a state of virtual anarchy. This is not the case. For most people, life continues to revolve around the common activities and concerns of work, family, and leisure. As in any society, activist politics remain confined mostly to a relatively small number; for the majority of people, on both sides, the conflict has had little immediate impact on their daily lives.

In addition, although the discordant aspects of ethnicity in Northern Ireland have necessarily engaged the attention of journalists and social scientists alike, one should not overlook those situations in which Protestants and Catholics interact harmoniously. Indeed, most interaction between the two groups, limited though it is, is not hostile. In addition, nonsectarian relations are not unusual among professional people, in the arts, and within the university environment (Barritt and Carter, 1972; Darby, 1976; Wallace, 1971). Public opinion polls also indicate, among both Protestants and Catholics, increasingly conciliatory attitudes and an inclination to accept and even encourage the development of integrated institutions, like schools and neighborhoods (Gallagher and Dunn, 1991; Moxon-Browne, 1983).[10]

The depth of ethnic division in Northern Ireland, however, is undeniably greater than it is in most other multiethnic societies. Hence, it is difficult to envision a settlement in which all parties can be fully satisfied. It is unlikely, for example, that any solution other than full unification will ever satisfy militant Catholic elements like the IRA. Protestant extremists are equally resolute in rejecting any form of relationship with the Republic.

The intractable nature of the ethnic division is well illustrated by the consistent breakdown of past efforts at reconciling Catholic and Protestant communities. In 1985, for example, a historic agreement was signed by the British and Irish governments, giving the Republic of Ireland a consultative role in certain policy matters pertaining to Northern Ireland. This represented the first joint effort by Britain and the Republic of Ireland toward ending the conflict. It was hoped that permitting the Republic a role in the province's affairs would isolate the IRA and its supporters and compel unionists to finally come to terms with the Catholic minority. Its effect, however, was to further inflame the fears of Protestants, many of whom felt that this marked the beginning of British abandonment of Northern Ireland and the first step toward eventual unification with the Republic. Even Catholics viewed the agreement as a disappointment in practice (O'Malley, 1991; Whitney, 1988). Again in early 1991, talks aimed at creating some formula for power sharing broke down amid an eruption of heavy violence by both the IRA and Protestant paramilitaries (Clarity, 1992).

---

[10]It is, of course, possible that such views are merely reflective of acceptance in principle, with no corresponding proclivity to convert such opinions into practice (Davis, 1994; McGarry and O'Leary, 1995a).

Recent Efforts at Political Conciliation

In 1994, the most serious efforts in twenty-five years at bringing an end to Northern Ireland's ethnic conflict were begun. The IRA declared a cease-fire in its terrorist campaign, which was followed shortly afterward by a similar declaration on the part of the Protestant paramilitary groups. Meanwhile, efforts were being advanced to bring together representatives of the Irish and British governments and all the political parties in Northern Ireland to begin to craft a settlement that would bring a permanent peace. All observers agreed that these developments seemed to represent the most promising possibility of ending the conflict since the beginning of its modern phase in 1969. Additional impetus for the peace process was provided by the United States, particularly the personal efforts of President Clinton.

For more than a year and a half, the cease-fire upheld by the IRA as well as the Protestant paramilitaries brought Northern Ireland an extended relief from ethnic violence. In early 1996, however, the IRA resumed its terrorist attacks by exploding a massive bomb in a London commercial district. It was not certain if this marked the onset of a new phase of violence. Efforts were made, nonetheless, to continue the peace process. Whether another halt to the violence might be effected and whether a permanent settlement of some kind might emerge from those efforts remained unanswered questions in late 1996. However, after two and a half decades of ceaseless, often fierce, conflict, the taste of peace seemed to make both Protestant and Catholic communities in Northern Ireland strongly determined to prevent a return to the violence. A definitive settlement, in any case, would require many years.

But if a political resolution is ultimately worked out, this does not mean that ethnic divisions will be quickly erased. So deeply rooted are the major differences—real or putative—between this society's two ethnic communities that one cannot realistically see, even in the long run, their full integration. "Communal tensions, territorial segregation, and endogamous social and sexual relations," notes Brendan O'Leary, "will remain dominant features of social and political life in Northern Ireland, with or without a definitive constitutional settlement" (1995:713). The most viable objective, therefore, may be the establishment of a social order in which maximum material and psychological security and optimum political power are provided people of both groups, divided though they remain.

In Chapter 4, we noted that race and ethnic relations always involve a synthesis of conflict and competition, on the one hand, and cooperation and accommodation, on the other. The tragedy of Northern Ireland is the severe imbalance in this fusion, sustained by a tendency to create the present mainly in terms of the past. Only by transcending its history, therefore, can this society achieve a higher degree of ethnic harmony among the majority of its people.

SUMMARY

Northern Ireland is one of the most polarized societies of the modern world, one where ethnic conflict often seems to defy explanation. Antagonism between dominant Protestants and minority Catholics, although especially intense since the late 1960s, is rooted in events and processes that extend back over three centuries.

The crux of the contemporary conflict stems from the partition of Ireland in 1920 following the achievement of Irish independence from Britain. Within the province of Ulster, Protestants were numerically dominant and resisted inclusion in the new Republic. Hence, Britain created a semiautonomous Northern Ireland state, leaving Catholics in this newly formed political entity a numerical minority. Today, most Protestants of Northern Ireland desire to retain the link with Britain, and most Catholics seek a unification with the Republic.

Northern Ireland is a society in which ethnicity affects almost all aspects of social life. Protestants and Catholics are highly segregated in education, residence, work, and leisure activities. Much of this segregation is rooted in past and present practices of discrimination, but much is also voluntary. The two groups perceive each other as products of fundamentally different cultural heritages and do not recognize a basis for political consensus. In the past, Protestants, given their numerical superiority, maintained political supremacy and commonly used that power to favor Protestant economic and social interests.

A mutually satisfactory solution to the ethnic conflict in Northern Ireland seems remote given the depth of social division between the two groups and the uncertain political status of Ulster. Most Protestants recognize Ulster as part of Britain, whereas most Catholics recognize it as part of the Republic of Ireland. As a result, a state recognized as legitimate by both groups, with power to enforce social order, cannot be created. For the past twenty-five years, this has left the political arena to extremist paramilitary groups, both Catholic and Protestant, and the British forces. New hopes were stirred in 1995 as the major parties to the conflict seemed to move closer to negotiating a permanent settlement and ending the violence.

Of the four non-American societies at which we have looked, Northern Ireland provides the most distinctive case and the one most difficult to compare with the United States. Little in the American ethnic experience can compare with either the historical basis or the political context of Ulster's ethnic strife. Also, this is a society in which ethnic diversity is quite limited. Nonetheless, certain features of intergroup relations in the United States, particularly those between blacks and whites, parallel those between Protestants and Catholics in Northern Ireland. These include the maintenance of high levels of residential segregation, little intermarriage, and low levels of primary structural assimilation in general.

Suggested Readings

Boyle, Kevin, and Tom Hadden. 1994. *Northern Ireland: The Choice.* New York: Penguin. A concise description of Ulster's two communities with an exploration of possible solutions to the troubles.

Burton, Frank. 1978. *The Politics of Legitimacy: Struggles in a Belfast Community.* London: Routledge & Kegan Paul. An ethnographic study of a Belfast Catholic working-class neighborhood, which analyzes the ideological basis of the ethnic conflict, particularly as expressed by the IRA.

Darby, John. 1995. *Northern Ireland: Managing Difference.* London: Minority Rights Group International. An up-to-date, comprehensive description and analysis of Northern Irish society and the economic, political, and social bases of the troubles.

Fraser, Morris. 1973. *Children in Conflict: Growing Up in Northern Ireland.* New York: Basic Books. A child psychiatrist explores the effect of the ethnic conflict in Northern Ireland on children.

McGarry, John, and Brendan O'Leary. 1995. *Explaining Northern Ireland: Broken Images.* Oxford: Blackwell. Considers various explanations—external and internal—of the roots and intractability of the Northern Ireland conflict.

O'Brien, Conor Cruise. 1995. *Ancestral Voices: Religion and Nationalism in Ireland.* Chicago: University of Chicago Press. Describes the linkage of religion and Irish nationalism. Draws a bleak prognosis of the latest peace process, asserting that Irish nationalists will settle for nothing less than unification of the two Irelands.

O'Malley, Padraig. 1983. *The Uncivil Wars: Ireland Today.* Boston: Houghton Mifflin. A very complete analysis of the contemporary politics of Northern Ireland, explaining the various factions and major protagonists of the troubles. O'Malley's *Northern Ireland: Questions of Nuance* (Belfast: Blackstaff Press, 1990) is a brief update.

Smith, David J., and Gerald Chambers. 1991. *Inequality in Northern Ireland.* Oxford: Clarendon Press. Traces the inequality of Ulster's Protestants and Catholics in various areas of social and economic life and the effect of policies designed to combat discrimination.

Stewart, A. T. Q. 1977. *The Narrow Ground: Patterns of Ulster History.* Belfast: Pretani Press. Contributes to an understanding of the current ethnic conflict in Northern Ireland by examining its deep historical roots.

CHAPTER

16

GLOBAL ISSUES OF ETHNIC CONFLICT AND CHANGE

SINCE THE END of World War II, societies in all regions of the world have become more ethnically diverse. Those, like the United States and Canada, with a history of absorbing many varied ethnic groups, emerged as even more heterogeneous. Many others that had been relatively homogeneous, as in western Europe, now evolved into truly multiethnic societies. This tendency has continued unabated.

In addition to increasing diversity, societies that have had a multiethnic structure and tradition for many decades have experienced a rise in the level of ethnic nationalism among their constituent parts. Whereas civic nationalism had promised to diminish the significance of race and ethnicity in diverse societies, ethnic nationalism has reemerged with great intensity, sometimes at a ferocious level.

In this chapter, we look at these two most evident trends of ethnicity in the contemporary world: the proliferation of ethnic diversity and the resurgence of ethnic nationalism. Both, as we will see, have contributed to a marked rise in ethnic conflict.

THE GLOBAL EXPANSION OF ETHNIC DIVERSITY

The expansion of ethnic diversity throughout the world in recent decades is attributable primarily to two developments: shifting patterns of immigration and inter- and intrasocietal political conflicts.

Shifting Patterns of Immigration

With the movement toward industrialization throughout the West, during the 100 years between the midnineteenth and midtwentieth centuries, global migration occurred mostly within Europe and between Europe and North and South America. Beginning in the 1950s, however, that traditional pattern was replaced by an international migration system encompassing virtually all world regions. Three major immigration streams evolved: (1) from Latin America and the Caribbean countries primarily to the United States and Canada, and secondarily to western Europe; (2) from south and east Asian countries primarily to the United States, Canada, and Australia, and secondarily to western Europe; and (3) from North Africa, the Middle East, and southern Europe primarily to northwestern Europe, and secondarily to the United States and Canada. The traditional east-to-west pattern of immigration, then, was replaced by a primarily south-to-north system involving the movement of people from the developing societies, or Third World, primarily to the developed, industrialized societies (Figure 16.1). As part of this global system, few countries today are not either sources or recipients of immigrants.

Not only did the direction of world migration change starting in the 1950s, but its scope expanded greatly and its pace accelerated. Whereas international migration had been sharply reduced during the 1930s as a result of a worldwide economic depression and then by the outbreak of World War II, immigration now began anew with great impetus. Nothing comparable had occurred since the great migrations to North America in the nineteenth and early twentieth centuries. By the early 1990s, estimates of people living in a country other than their country of origin were as high as 100 million, almost 2 percent of the world's population (Castles and Miller, 1993; UNFPA, 1993).

Several factors account for the reemergence of large-scale immigration after an interim of several decades. First, immigration accelerated as a consequence of the widening gap between rich and poor nations. As with previous movements, the primary driving force of the new international migration was the push–pull factors of labor markets. Workers from less developed countries migrated to those countries where jobs were more plentiful and lucrative. As we noted in Chapters 11 and 14, this, more than anything else, accounts for the unprecedented levels of immigration to the United States and Canada during the past three decades. As we will see, it also accounts for the immigration to western Europe of southern Europeans, North Africans, and people from the Middle East. The discrepancy between rich and poor nations became evident throughout the 1950s and 1960s when many former colonies in Africa, Asia, and the Caribbean gained their independence. The economies of these newly independent countries were generally incapable of supporting rapidly growing populations,

Figure 16.1 Major Contemporary Immigration Flows

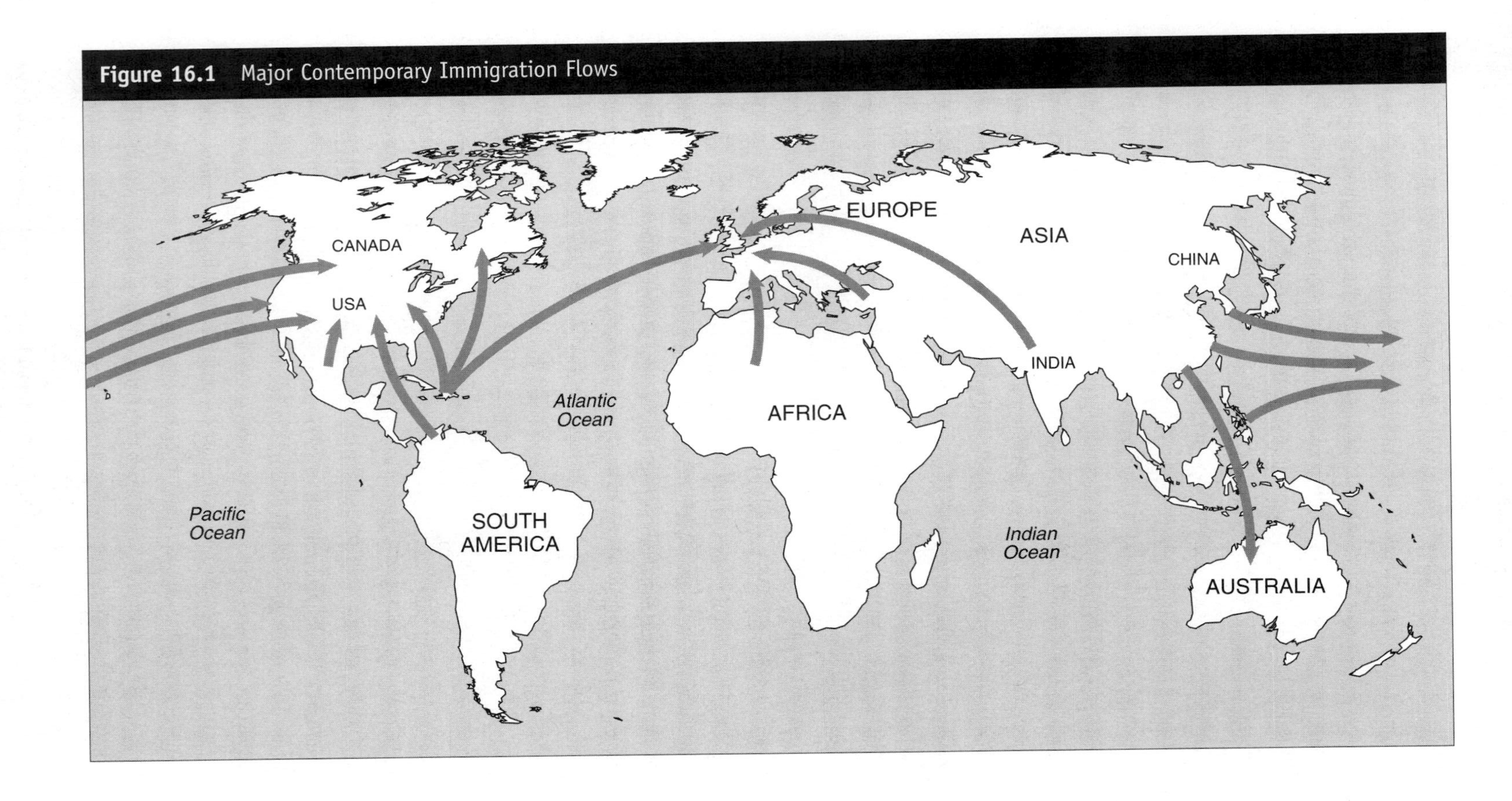

thereby creating a migration push. Most of the world's population growth has occurred since the end of World War II, and the overwhelming majority of that growth has taken place in the less developed countries. Moreover, gains in economic development made by many of these countries in the 1960s and 1970s slowed or reversed in the 1980s due to economic recession, growing debt, and internal conflicts. These conditions induced further pressures to migrate to the wealthier countries.

Political Conflict

Wars and internal political strife in many societies created additional push factors that further stimulated global migration. From the late 1950s to the present, civil wars, internal uprisings, revolutions, and political violence in various forms created a worldwide flow of refugees unprecedented in modern history. Virtually every part of the developing world, as well as a number of countries in the developed world, were affected. Uprisings in the 1950s and 1960s against Soviet occupation in eastern Europe, revolution in Cuba in 1959, civil wars in the 1970s and 1980s throughout Central America, political violence in parts of the Caribbean and the South American continent, the war in Vietnam, civil wars in many parts of Africa, ongoing wars in the Middle East, the Persian Gulf War in 1992, and the collapse of the Soviet Empire in the late 1980s all contributed to the swell of refugees alongside the more conventional economic migrants.

Immigration and Ethnic Change in Western Europe

Arguably, nowhere in the contemporary world has the impact of immigration and the creation of greater ethnic diversity been more pronounced than in Europe. Today the European continent also demonstrates vividly the durability of ethnicity in the modern world and its often malignant potential to create and sustain societal conflict.

Contemporary Immigration to Europe After World War II, workers from southern European countries, such as Italy, Portugal, Greece, Spain, Turkey, and Yugoslavia, or from newly independent former colonies such as Algeria or Indonesia migrated to Germany, France, Switzerland, the Netherlands, Belgium, and the Scandinavian countries as these nations experienced their postwar industrial expansion. A similar immigration occurred in Britain with the entry of thousands of people from the newly independent countries of the British Commonwealth—India, Pakistan, Jamaica, and others (Castles, 1992; Castles, Booth, and Wallace, 1984; Castles and Kosack, 1973; Castles and Miller, 1993; Garson, 1992; Paine, 1974). Labor was in short supply, and immigrants were therefore recruited to work in the economic reconstruction of these countries. This population movement accelerated in the 1960s, leaving previously homogeneous western European societies with substantial ethnic communities made up of immigrants of vastly different cultures and physical characteristics.

Although by the late 1970s immigration had been severely curtailed by restrictive measures, many of those who had come as temporary workers stayed on, establishing themselves and their families as permanent residents. Today approximately fifteen million foreign workers and their families live in western Europe. Germany alone has over six million immigrants and France about four million, making up between 5 and 10 percent of the population. In each case, a majority are Muslim (Gedmin, 1994; Peach and Glebe, 1995; Walsh, 1992). In Switzerland, the foreign-born constitute 15 percent of the population and nearly a quarter of the labor force. In Britain, ethnic minorities, most of them nonwhite, are almost 5 percent of the total population. Italy, traditionally Europe's major immigrant-sending society, is now an attractive destination for immigrants (Campani, 1993; Sciortino, 1992; Stille, 1992). Even in Sweden, long considered a culturally undiluted corner of Europe, one person in eight was a first- or second-generation immigrant by the late 1980s.

The Societal Impact of Immigration to Europe When they began to arrive in the 1960s, immigrant workers were seen by the native populations as temporary residents who would return to their home countries once their labor was no longer needed. This was made especially poignant in Germany (then West Germany), where immigrants were labeled *Gastarbeiter,* literally "guest workers." Although the workers were a vital component of the economic system, little effort was made to integrate them socially or culturally into the host society. Today, the workers and their families are marginal people, and their social conditions are reminiscent of those experienced by successive waves of European immigrants and southern black migrants to northern U.S. industrial cities in the late nineteenth and early twentieth centuries. Their housing is the oldest and most dilapidated; their jobs are the lowest paying and least prestigious; and, except for work, they do not participate in the society's mainstream institutions. Moreover, in Germany, where the largest number of foreign workers has been employed, the eventual assimilation of the immigrant population is not seen as an appropriate societal goal. Very few of Germany's three million Turks, for example, have been afforded citizenship, even those who are second- or even third-generation German residents.

Perhaps most seriously, immigrants in western European countries have encountered high levels of prejudice and discrimination (Campani, 1993; Castles, Booth, and Wallace, 1984; Gurr, 1993; Power, 1979; Skellington, 1996). As previously homogeneous nations, western European countries lacked the historical experience of the United States, Canada, and other traditional immigrant societies in absorbing large numbers of ethnic newcomers. The result has been, at best, a begrudging tolerance of immigrants and, at worst, outbreaks of ferocious hostility. In Germany, physical attacks against immigrants by the early 1990s had become commonplace; in France, North Africans, who compose fully half the immigrant population, are the primary targets of ethnic violence. Even in countries with fewer immigrants and solid traditions of tolerance, like Italy and Sweden, racially motivated incidents have surfaced.

It is clear that regardless of the effort to limit immigration, sizable ethnic populations will remain a permanent part of western European societies. The children and grandchildren of the original immigrants are European born or reared and for them a society of origin, or "mother country," is only an abstraction. Moreover, although few will acknowledge it, the economies of Germany, France, and other countries in the region could not function without foreign workers. With extremely low birthrates, these societies can fill their future labor needs only through substantial immigration. Western Europe is a central link in the emergent global economy in which the flow of capital and goods and, increasingly, workers is unhindered by national boundaries. As in North America, the tendency toward greater ethnic and racial diversity, therefore, is accelerating and appears irreversible. These countries are now permanently multiethnic.

As a result of this new heterogeneity, western European ethnic patterns have begun to converge with those of the United States (Lambert, 1981). In some ways, the impact of increasing ethnic diversity on western European societies in recent years has been even more profound than on the United States, and its social, political, and economic consequences more wrenching. Questions concerning the social and economic integration of large and distinct ethnic minorities are now in the forefront of political discourse. Few parties at any point on the political spectrum have not openly advocated, at minimum, a halt to further immigration; extreme right-wing parties, most notably in Germany and France, have called for immigrants to be repatriated to their countries of origin. In sum, it is quite reasonable to expect that for many years increasing ethnic diversity will continue to be the most serious internal issue facing most western European countries.

THE RESURGENCE OF ETHNIC NATIONALISM

Most people in North America have only a vague understanding of political and social affairs that occur outside the boundaries of their own society. But even those who kept only casually informed of world events in the 1980s and 1990s could not help but become aware of the increasing level of ethnic nationalism, accompanied by political conflict, that erupted in various parts of the world. Some situations, as in Canada, were dealt with through peaceful political means. Others, as in Rwanda, Sri Lanka, and Russia, were violent.

As the world watched the dramatic end of the Cold War in the late 1980s, expectations were high that a new era of global tranquility was on the horizon. With the political map of the world changing frequently and rapidly, however, it became obvious that a new force had been unleashed, one that augured not a more peaceful globe but one enmeshed in ethnic conflict. Ethnic nationalism seemed to replace civic nationalism as a stimulant of identity and political action. Indeed, at the close of the twentieth century, ethnic nationalism appears to be the most ubiquitous, intractable, and devastating global force.

Forms of Ethnic Nationalism

Nationalism, as a form of political and social expression, emerged in the eighteenth century as the first modern nation-states appeared in Europe and the United States (Kohn, 1944). Loyalties to family, kin, tribe, or community were extended so as to encompass entire peoples, marked off by common language, religion, and other cultural elements. Benedict Anderson (1991) has referred to these emergent nations as "imagined communities." In the twentieth century, and especially since the end of World War II, nationalism has spread throughout the world among formerly colonized peoples, intent on achieving political self-determination. It has also advanced among well-established nations where diverse ethnic groups have asserted their identity and have sought greater social and political autonomy.

In its modern-world expressions, nationalism has been of two types: civic and ethnic (Geertz, 1994). *Civic* nationalism, Michael Ignatieff notes, "maintains that the nation should be composed of all those—regardless of race, color, creed, gender, language, or ethnicity—who subscribe to the nation's political creed" (1993:6). Here, people possess equal political and social rights and choose to be members of a nation, along with others, regardless of ethnicity, who share broadly similar beliefs and values. American patterns of nationalism, as well as those of most other multiethnic societies, have historically been expressed along these lines. *Ethnic* nationalism, by contrast, holds that people's allegiance is to an ethnic group or nationality into which they have been born or assigned, not to a larger political entity encompassing many different ethnic groups or nationalities. Here, "an individual's deepest attachments are inherited, not chosen" (Ignatieff, 1993:7–8). In the past several decades, every continent on the globe has witnessed cases of intense ethnic conflict in one or more countries, based on ethnic nationalism.

One form of ethnic nationalism involves political actions based on claims of the superiority of a dominant, or majority, ethnic group over others and its portrayal of rival groups as potentially dangerous. In such cases, the dominant group seeks to entrench its power by further subordinating, expelling, or perhaps even exterminating rival ethnic groups. The two most unmistakable examples of this type of ethnic nationalism in recent years have occurred in Rwanda and the former Yugoslavia, the latter which we will describe in greater detail in the next section. In Rwanda, majority Hutus in 1994 slaughtered at least half a million rival Tutsis within just a few weeks, an action described as a genocide — a systematic, orchestrated effort to exterminate an entire people (African Rights, 1994; Destexhe, 1995).[1] Journalist David Rieff recounts the killing as it began and gained momentum:

> Over the radio, the call kept going out for the extermination of every Tutsi man, woman and child in Rwanda, and of the Hutus who opposed this final solution. In the villages, people were hunted down. They died by the tens of thousands in their homes, in their fields and in the churches . . . in which they had sought refuge. And no matter how many died, the radios kept blaring out the calls for all good Hutus to kill the *imyenzi,* the "cockroaches," who were polluting the Rwandan nation and preventing it from living in peace (1996:31).[2]

Another form of ethnic nationalism, more common in the contemporary world, involves the assertion of ethnic identity on the part of minority ethnic groups and their challenges to majority-group dominance. Ethnic groups may view themselves as culturally and politically oppressed and seek either greater powers or independence from the nation-state dominated by a rival group. Political and cultural movements by Basques in Spain, Bretons in France, Flemings in Belgium, Tamils in Sri Lanka, Ibo in Nigeria, Kurds in Iraq and Iran, Chechens in Russia, Scots and Welsh in Britain, and, as we have seen, Quebecers in Canada all illustrate this form of ethnic nationalism.[3]

In the past decade, ethnic nationalism has been particularly virulent and explosive in eastern Europe. A new and ominous factor was introduced into European patterns of immigration and ethnic heterogeneity with the breakup of

---

[1] No one will ever know the number actually killed. Some estimate as many as a million.

[2] Ethnic conflict between Hutus and Tutsis has also been ongoing for many years in neighboring Burundi.

[3] Even the rise of cultural pluralism in the United States can be seen, in a broad sense, as part of this global trend.

the Soviet Union and the fragmentation of eastern European states in the late 1980s and early 1990s. Almost all countries of eastern Europe — those that had been part of the Soviet sphere of influence following World War II — have lengthy histories of ethnic diversity, each made up of a congeries of culturally distinct populations occupying rough territorial concentrations. Those ethnic divisions, however, had been held in check by Soviet dominance. As these countries asserted their national identities and secured political autonomy beginning in the late 1980s, that social glue dissolved.

The Breakup of Yugoslavia

Nowhere in eastern Europe was ethnic nationalism and ensuing conflict more intense and tragic than in Yugoslavia as it experienced political breakup. Indeed, so stark and violent was the conflict that it held the attention of world media for several years. Thus, few could avoid becoming aware of the sheer brutality exhibited by all sides.

The Attempt at Corporate Pluralism Yugoslavia represents a failed attempt of a multiethnic society to construct some form of corporate pluralism. As we saw in Chapter 4, corporate pluralism is a situation in which a number of ethnic groups coexist in a loosely bounded system. These culturally, and perhaps linguistically, distinct groups are linked together in a federation in which each is recognized officially by a central government that attempts to distribute national political power and economic resources proportionally among them. Each, however, enjoys some degree of political autonomy.

In such cases, each ethnic group also usually occupies a distinct territory or geographical area, where it becomes the dominant group. However, these ethnically concentrated populations usually spill over or are dispersed to other areas of the country, where a majority of the population is composed of members of another ethnic group. As a result, minority ethnic groups develop in those territories. When corporate pluralism breaks down, the federation begins to crumble and each ethnic territory seeks full political independence. As ethnic rivalries intensify, each ethnic group attempts to maximize its territory and to embrace group members who are living as ethnic minorities in other areas. Previously subdued historic rivalries are regenerated and intergroup hostilities heighten, perhaps leading to violent confrontations. The contemporary Yugoslav situation illustrates this process vividly.

Prior to 1991, Yugoslavia, a country of about 24 million, had been comprised of six republics and two provinces, each different in social and political makeup and history, and each dominated by a particular ethnic group. Constituting the six republics were Serbs, Croats, Muslim Slavs, Slovenes, Macedonians, and

Montenegrins. All of these groups (comprising about 80 percent of the total population) were ethnically south Slavs, but they differed, sometimes radically, in language, religion, and other cultural aspects. Other, smaller ethnic groups were Albanians, less than 8 percent of the total population, and Hungarians, about 2 percent. The remaining 10 percent was made up of numerous tiny ethnic populations.

The ethnic differences among the former Yugoslavia's peoples cannot be overstated. Religion in particular has been an ancient divide. In a sense, the region represents an area in which there is a confluence of three civilizations: in the west, one that derives from the Roman Empire, in the east, from the Byzantine Empire, and in the south, from the Turkish Ottoman Empire. Three religions—Roman Catholicism, Eastern Orthodox, and Islam—therefore divide the populace. Each major ethnic group is strongly identified with one of these: Croats are Roman Catholic, Serbs are Eastern Orthodox, and almost half of the Bosnian population is Muslim.

For almost thirty-five years following World War II, this ethnically disparate and politically divided country had been held together delicately by Josip Broz Tito (usually referred to simply as Tito), a World War II hero who led the victorious Partisan guerrilla army against the Nazis. After the war, Tito emerged as a unifying authoritarian figure, acting as prime minister and, later, president. Tito, though an avowed communist, having shaped the Yugoslav economic and political systems along communist lines, moved the country toward nonalignment and away from the Soviet hegemony in eastern Europe.

Tito's strong rule and ability to bring the various ethnic factions together had seemed to create a multiethnic society that fulfilled the major features of corporate pluralism. Institutions were carefully crafted to preserve a sense of ethnic balance and fairness—no easy task in a state in which ethnic diversity characterized almost every facet of social and political life. Equal political and social rights were constitutionally guaranteed for all ethnic groups, and measures were aimed at reducing ethnic intolerance and discrimination. A federal system was created in which the six republics and two provinces enjoyed much autonomy, though they shared national power and authority. The federal and republican constitutions provided for proportional representation of the nations and nationalities in all governing bodies and institutions, in a kind of consociational arrangement. Each group was entitled to use its language in its schools and in its political and commercial life. Even at the national level, linguistic diversity was accommodated by the federal government, which sanctioned three official languages: Serbo-Croatian, Slovenian, and Macedonian. Two minority languages, Albanian and Hungarian, were also given state recognition. Finally, despite the society's sharp tripartite religious division, religious freedom was protected. In addition to Roman Catholics, Orthodox, and Muslims, small numbers of Jews and members of other minor religions were also part of the Yugoslav social landscape.

Table 16.1 Ethnic Makeup of the Major Republics of the Former Yugoslavia, Before 1991[a]

REPUBLIC	MAJOR CITY	ETHNO-RELIGIOUS MAKEUP
SERBIA	Belgrade	Primarily Serbs (Eastern Orthodox), but significant numbers of Albanians, Croats, Muslims, and Hungarians
CROATIA	Zagreb	Croats (Roman Catholic) the majority, but a large Serb minority
BOSNIA and HERZEGOVINA	Sarajevo	The most ethnically diverse region—about 44 percent Muslim, 31 percent Serb, and 17 percent Croatian
SLOVENIA	Ljubljana	Relatively homogeneous with Slovenes (Roman Catholic) a solid majority

[a]The populations of Montenegro and Macedonia, two of the former Yugoslavia's six republics, were mostly unaffected by the ethnic warfare of the 1990s.

None of these measures, however, created more than superficial ethnic harmony and, with Tito's death in 1980, the rancor that had been simmering beneath the surface gradually intensified, erupting into full-blown warfare in 1991.

The Dissolution of the Yugoslav Federation With the end of the Tito regime, Yugoslavia dissolved into an extremely loose federation in which the various republics increasingly dominated the central government. Most people identified themselves with their specific ethnic group, not as "Yugoslavs." Ethnic rivalries previously held in check now reemerged, stirred by political leaders in each republic. These rivalries were exacerbated by widening economic differences among regions and by the breakup of East European communism in the late 1980s. The ethno-religious makeup of the Yugoslav federation prior to its dissolution in 1991 can be seen in Table 16.1.

The major ethnic split was between the republics of Croatia and Slovenia, on the one hand, and Serbia, on the other. The former are both Roman Catholic and strongly oriented culturally toward western Europe; Serbia, in contrast, is Orthodox Christian in religion and more culturally attuned to eastern Europe. With the collapse of the Yugoslav Communist party in 1990, Croatia and Slovenia each moved toward independence and seceded from the federation a year later. This left Yugoslavia composed mainly of Serbia. The smaller and less economically viable republics took sides or became pawns in the ensuing confrontations that developed as the three major ethnic states sought to establish boundaries that each interpreted as corresponding to their historical ethnic dominance.

The Slovenes were a more westernized, industrialized, and prosperous population, occupying a relatively homogeneous area. As a result, resistance to their exit from the federation and declaration of independence was short-lived. Most important, few members of the other ethnic populations of Yugoslavia resided in the Slovenians' traditional territory.

The Serbs and Croats, then, remained as the major ethnic rivals. The rivalry between them that erupted in the early 1990s was in some ways an extension of animosities that had run deep for many decades. The Serbs and Croats, along with the Slovenes, had been linked together in a constitutional monarchy after World War I, but, lacking a tradition of political compromise, the three were unable to achieve a national unity and, by the start of World War II, the country was severely divided ethnically. As a result, fighting occurred not only with the German invaders but also among the various ethnic factions themselves, notably between Serbs and Croats, who took opposing sides during the Nazi occupation. More than 1.5 million Yugoslavs died during the war, most as a result of interethnic fighting.

During the war, Hitler established a puppet fascist regime in Croatia that was run by the Ustasha party. The Ustashas were vehemently anti-Serb and antisemitic. Their essential objective was to purge Croatia of Serbs and, of course, Jews as well, though the latter represented a relatively small population. As a result, the Serbs in Croatia suffered many indignities, including the forced wearing of colored armbands to identify themselves. More significantly, hundreds of thousands of Serbs were executed by the Ustasha regime. Many Croat nationalists and clergy were active in that regime and were participants in the oppression. At the infamous Jasenovac concentration camp, the Ustashas murdered 85,000 people, most of them Serbs (Bennett, 1995). Following the war, these savage killings were bitterly remembered by the Serbs and, in their minds, all Croats were associated with them, whether or not they had been Ustashas.

Similar actions took place during World War II in Serbia, but against resident Croatians. Croatians today point to Serbian guerrilla bands, called Chetniks, who fought against the Nazis (and thus the Ustashas) but who, in the process, oppressed and slaughtered many Croatians. As a result of the atrocities on both sides, Serbs and Croats found themselves in a perpetual state of mutual distrust (Schöpflin, 1993). Indeed, so bloody had been the conflict inside Yugoslavia during World War II and so venomous were the emotions it created on all sides that, as journalist Misha Glenny accurately forecast, "were it ever to revive, it was always likely to be merciless" (1993:171).

The Contemporary Conflict Some argue that Yugoslavia's attempt at corporate pluralism was always precarious and, without Tito, was doomed. Others, however, maintain that it did not fall apart of its own volition but rather was destroyed by a few people acting in their political interests, especially Serb President Slobodan Milosevic (Bennett, 1995; Rieff, 1995).

With the election of the party of Franjo Tudjman in Croatia in 1990, a new wave of Croatian nationalism emerged. Serbia, led by Milosevic, experienced its own resurgence of nationalism, one of the key components of which was a rekindled hatred of the Croats and a remembrance of the World War II horrors. Through a concentrated propaganda campaign in the media, the Serbian government instilled fear and suspicion among Serbs of their Croatian and Bosnian Muslim neighbors. Serbian television, for example, frequently showed films of the atrocities committed by the Croatian Ustashas during World War II. Serbian authorities also censored the press and imposed a blockade on outside information, allowing the government to manipulate facts and communicate misinformation as the conflict unfolded. Croatian and Bosnian governments offered their own forms of propaganda (Human Rights Watch, 1995).

At the outbreak of the conflict, the Serbs were not only the largest of Yugoslavia's ethnic groups, but the most dispersed, living in all areas. Serbia alleged that Serbs living outside the Serbian nation, that is, in other parts of Yugoslavia, were victims of oppression, especially in Croatia and Bosnia-Herzegovina. Serb leaders and the media engendered fear among Croatian and Bosnian Serbs that, if the territories in which they were living were to become independent states, they would be divested of their rights and property (Bennett, 1995). The conflict, then, essentially became one in which these minority populations were "liberated" by Serbs. The first of the wars erupted in the summer of 1991 in a Serb-populated region of Croatia. There, Serbs refused to accept living in an independent Croatia and formed their own rebel state. Not until four years later did the Croatian army smash the resistance and drive out most Serbs who had been living there.

Although they lived in all of Yugoslavia's republics, Muslims were heavily concentrated in Bosnia-Herzegovina (often referred to simply as Bosnia). There they were about 44 percent of the population. Sarajevo, the republic's capital and largest city, was not only heavily Muslim in numbers, but also very much dominated by Islamic culture. Bosnia's Muslims, however, should not be seen as comparable to Muslims in other parts of the world. The Bosnian Muslims are secular, having only a distant link with other, more traditional, Islamic populations. Most are Slavs (that is, Serbs or Croats) who were converted to Islam during the five centuries of Ottoman rule in Bosnia (Glenny, 1993). Most, moreover, are Sunni Muslims, not in sympathy with fundamentalist groups in the Middle East and North Africa.[4] As with the Serbs and Croats, Bosnian Muslims speak a variant of Serbo-Croatian.

---

[4]On the social history of Bosnia's Muslim population, see Duran (1995).

Ethnic Cleansing in Bosnia The second of the wars occurred in Bosnia. When Bosnia declared its independence from Yugoslavia in 1992, Bosnian Serbs refused to accept this arrangement, preferring instead to remain part of Yugoslavia (now essentially only Serbia). Thus began the efforts of the Bosnian Serbs, aided by Serbia itself, to drive Muslims from the territory. The campaign waged by the Serbs was referred to as "ethnic cleansing." This meant killing Bosnian Muslims and Croatians, or driving them from their homes, in order to prepare their areas for a future "Greater Serbia." Bosnian Serbs commonly held the belief that Bosnian Muslims were intent on creating a fundamentalist Muslim state in Bosnia and thus rationalized their actions as simply self-defense. In addition to the Serb attack, the Croats also launched an attack on western Bosnia, driving out Serbs and Bosnians.

One of the ironies of the Yugoslav conflict is that Bosnia, the area affected most severely, had for decades been an example of ethnic harmony, a society of "pluralism and tolerance" (Rieff, 1995). The three major groups—Serbs, Croats, and Muslims—had coexisted peacefully and had created a relatively integrated society. In Sarajevo, the major Bosnian city, Croatian Roman Catholic and Serbian Orthodox churches stood side by side with Muslim mosques, ethnically integrated neighborhoods were commonplace, and intermarriage among the three populations was not at all unusual.

The sudden destruction of the relative harmony in which Yugoslavia's ethnic groups had been living, particularly in Bosnia, is a reminder of how ethnic conflicts that are commonly seen as ancient and unending may, in fact, be the product of relatively recent events and may arise or fade with changing political, economic, and social events or with the emergence of new leaders. As political scientists Susanne and Lloyd Rudolph have put it, "The question . . . is not why old conflicts are flaring up anew, but rather why traditionally harmonious mosaics have been shattered" (1993:25). In the case of Yugoslavia, writes Ignatieff, "Nationalist politicians on both sides took the narcissism of minor difference and turned it into a monstrous fable according to which their own side appeared as blameless victims, the other side as genocidal killers" (1993:22). Contrary to popular belief, the notion that the Serb–Croat rivalry has ancient historical roots is simply not the case. In fact, until this century, the two peoples had had virtually no contact. Past wars were with the multinational empires of the Hapsburgs and Ottomans, not with each other (Bennett, 1995; Rieff, 1995).[5]

---

[5]In Rwanda as well, the cultural differences between Hutus and Tutsis were, historically, virtually nonexistent. Ethnic categorization was essentially a creation of the colonial powers that ruled Rwanda. After independence, this artificial division was exploited by rival political factions (Destexhe, 1995).

Neighbors and family members, who had been comfortably integrated just weeks before, voiced puzzlement at how quickly their social harmony had dissipated and had been reduced to vicious and bloody hatred. "Growing up before the war here, my boys played with all the neighborhood kids," said a Serbian woman living in Bosnia. "We never knew who was Georgy and who was Muhammed. And I baked cakes for them all. Now neighbors we've known for years don't greet us anymore" (cited in Sudetic, 1994). So intense and bitter were the feelings that now surfaced among those of the different ethnic groups that people were driven to exchange their homes with each other in order to create more ethnically homogeneous areas. Later, as Serbs took control of parts of Bosnia, Croats and Muslims were driven from their homes, which were given to Serb refugees who had fled similar incursions in other areas by Croats or Muslims. Often, as Serbs or Croats advanced on each other, houses of those fleeing were summarily destroyed. Ignatieff describes what he saw as he drove through ravaged villages:

> Toward Novska, you pass Serb house after Serb house, neatly dynamited, beside undisturbed Croat houses and gardens. When you turn toward Lipik, it is the turn of all the Croat houses to be dynamited or firebombed, next to their untouched Serb neighbors. Mile upon mile, the deadly logic of ethnic cleansing unfolds (1993:35).

Much more dreadful actions were to occur, however, as the fighting intensified. Indeed, quickly the hostilities between Serbs, Croats, and Muslims degenerated into the most ferocious conflict in recent European history. Each side accused the other of atrocities and in many cases these were confirmed by the United Nations and other nonparticipating observers. The most horrific actions, however, were committed by Serbs in Bosnia as part of their campaign of ethnic cleansing. Mass executions, rape camps, savage mutilations and torture, and wanton destruction of homes, churches, and mosques filled news reports from Bosnia as the war there intensified starting in 1993.[6] Consider this account (corroborated by both sides in the war), given by a Serbian guard who later deserted, of a Serb concentration camp:

> Men were loaded into the back of a truck, taken up to the edge of the ravine, about five miles away, and then shot as they got out of the vehicle.

---

[6]In 1994, the United Nations documented 187 mass graves, each containing the bodies of between 3,000 and 5,000 Muslims murdered by Serbs. Also confirmed were 962 prison camps in which more than a half million people had been held and where rapes and mass killings had occurred (Pomfret, 1995b). As of early 1996, other mass graves in Bosnia were being uncovered.

> Groups of young soldiers were brought in to perform the executions. The bodies fell into the ravine and bulldozers were later used to cover them over (Cohen, 1994).

Just months before, many of the camp guards and victims had been neighbors.

The horrors of the Yugoslav war were eerily reminiscent of the genocide committed by the Nazis fifty years earlier. The massacre in the city of Srebrenica was one of the more thoroughly documented of incidents that seemed so monstrous as to border on the surreal. Srebrenica had served as a United Nations refuge where more than 40,000 people had taken shelter from the war. After it was overrun by the Bosnian Serb army, the aftermath was described as the worst war crime in Europe since World War II. Although no one can be sure how many were slaughtered, American intelligence analysts estimate between 5,000 and 8,000. A few survivors as well as Serbian civilians living in surrounding villages recounted the mass killings. Hundreds were herded into trucks, taken to a field, and shot; others were jammed into a warehouse where they were killed with automatic rifles and shoulder-held grenade launchers. Mass graves of the victims were uncovered later (Engelberg and Weiner, 1995). This incident more than any other prompted the military and diplomatic efforts of the United States and NATO (North Atlantic Treaty Organization) to stop the war. But it was only one of many such massacres of entire villages.

Muslims were victimized as well by Croats seeking to secure what they claimed was their territory. As with the Serbs, a policy of ethnic cleansing produced gruesome incidents of terror and brutality. One such affair, documented by United Nations investigators, occurred in late 1993. A so-called "death platoon" of a few dozen Croatian paramilitaries entered the village of Stupni Do shouting "Let's kill the Muslims" and "Where are all the pretty girls for us to rape? Bring them out." A journalist describes what ensued:

> With avenging fury, the Croatian death squad crushed the skulls of Muslim children, slit the throats of women and machine-gunned whole families at close range. The next morning Croatian reinforcements finished up the job, torching all 52 houses and dynamiting the community's one small mosque. In less than 48 hours Stupni Do simply ceased to exist (Nordland, 1993:48–49).

Although Bosnia suffered the most appalling conditions and was the site of the most intense fighting, other regions were affected, to a lesser degree, in much the same way. Following the Croatian invasion of the Krajina, an area of Croatia populated mostly by Serbs, 100,000 Serbs fled what they feared was the

Croatian version of ethnic cleansing. Like the Bosnian Muslims, the Krajina Serbs expressed the same disbelief that former Croat friends and neighbors with whom they had lived together and even intermarried for decades could so viciously turn on them without provocation. "I had a very good friend, a Catholic Croat," said a refugee Serb. "I would have given blood for him. But three years ago, he said: 'I like you a lot. But please don't come and see me anymore. My neighbors object to a Serb visiting.'" Another put it: "I'm sorry I didn't burn down my house. I'm sorry I left it to the Croats to burn" (quoted in Perlez, 1995a, 1995b).

Toward Resolution of the Conflict In late 1995, after three and a half years of war, a fragile peace was established, in large part through the efforts of the United States, that divided Bosnia into semiautonomous regions along ethnic lines. Under the terms of the agreement, Bosnia was split into two parts—a Muslim–Croat federation made up of 51 percent of the territory, and a Bosnian Serb republic, with 49 percent. NATO forces, including 20,000 Americans, were deployed to enforce the treaty. Although this temporarily stopped the killing, few seemed to feel that anything short of a more complete partition of Bosnia would secure a long-term resolution to the conflict (Mearsheimer and Van Evera, 1995).

The legacy of the war was that the ethnic mosaic of Yugoslavia's population had been radically—probably irreversibly—altered and interethnic relations reduced to mutual fear and antipathy. In Bosnia, the most severely affected region, at least 200,000 people had been killed or had disappeared and almost half of its prewar population of 4.2 million had been displaced. This, it should be remembered, had been Yugoslavia's most ethnically diverse region. Wholesale population transfers occurred in parts of Croatia and Serbia as well. Most Croats living in Serbia were forced to leave as were Serbs living in Croatia. Refugees fleeing regions as they repeatedly changed hands reached epic proportions, unseen in Europe since World War II. As many as three million were driven from their homes.

For those who endured the war, a residue of distrust, resentment, and terror on all sides makes the reemergence of interethnic tolerance and accommodation improbable for many decades in the future. Sarajevo, previously the showpiece of ethnic harmony, symbolized this legacy of ethnic hatred. There, Serbs refused to accept Muslim governance of the city, as the peace agreement provided, and thousands left their homes in protest, some burning them in a final act of defiance. Some even exhumed the remains of dead family members to carry with them, fearing that their graves would be desecrated (Hedges, 1996).

Dominance by one group over another, noted Philip Mason, "is as old as the Pharaohs and springs from passions that are common to all men" (1970:337). The question, then, is not whether dominant–subordinate relations will continue to be an integral feature of diverse human societies but whether and to what degree those relations will be based on ethnicity. Given that multiethnicity is an advancing condition of most societies today on all continents, the ethnic basis of inequality appears unwavering. That, in turn, makes it likely that continued ethnic conflict will be a prevailing condition of our time.

Early on, we suggested that conflict and change are fundamental themes of ethnic relations. We also noted, however, that societies are always admixtures of not only conflict and change but also cooperation and stability. Thus, these two tendencies are always variable. Stephen Steinberg has written that, "If there is an iron law of ethnicity, it is that when ethnic groups are found in a hierarchy of power, wealth, and status, then conflict is inescapable" (1989:170). That cooperation and stability will characterize ethnic relations for long periods is most likely, therefore, in multiethnic societies where groups coexist with relatively minor differences in power, wealth, and prestige among them. In the contemporary world, however, such cases are uncommon. Instead, salient economic and political inequalities characterize most multiethnic societies. Moreover, conflict is induced when groups see their position as the product of discrimination (Gurr, 1993). This, too, is a prevalent condition in most multiethnic societies.

Where there are sharp inequalities in power and wealth among ethnic collectivities, dominant groups may induce stability through various techniques of prejudice and discrimination (discussed in Chapter 3) and the effective transmission of ideology. Eventually, however, a minority group or groups challenge that domination, and conflict then prevails. The United States and each of the other societies we have examined have experienced in recent times a challenge, in some degree, to their prevailing ethnic orders.

It is important to consider, however, that inequalities that induce conflict need not be severe or even objectively salient. Only the *perception* of inequality among ethnic groups is necessary. As we saw in Chapter 14, for example, the socioeconomic differences between French Canadians in Quebec and other Canadians are no longer evident. Despite this shrinking gap over the past several decades, however, the intensity of Quebec nationalism has not been reduced but has actually risen, creating the most serious threat to Canadian unity. Also, the grievances of minority ethnic groups may become more impassioned as they actually experience progressive change in their socioeconomic position. As conditions improve, the expectation of and desire for further improvements grow. If those expected changes do not materialize as rapidly as anticipated, minorities may focus on unmet goals rather than on their improved condition relative to that of an earlier period.[7]

In addition to economic and political grievances, ethnic conflict is generated by sustained cultural differences. Most basically, where ethnic groups speaking different languages or abiding by divergent norms and values must confront each other in various facets of social life—housing, schools, government, media—conflict is a given. Only its severity is variable. Repeatedly, we discovered such conditions in each of the societies at which we looked. Here, too, however, the reality of cultural difference among groups is less critical than the *perception* of difference. As we saw in Northern Ireland, for example, two groups whose cultural variance has faded over many generations (and who even speak a common language) nonetheless have clung to historical perceptions of difference that continue to shape their views and attitudes toward each other. As with economic and political inequality, sociocultural differences, or their perception, are fundamental in contemporary multiethnic societies.

Where there are physical distinctions among groups, of course, an additional basis of division is in place. But racial differences among groups are not an essential ingredient in the emergence of ethnic divisions and conflict. In Canada, Northern Ireland, Rwanda, and the former Yugoslavia, we saw that ethnic boundaries between racially similar groups can be shaped in a number of ways and, once established, adamantly sustained, often for many generations.

Finally, Heribert Adam has pointed out that conflict in multiethnic societies is also induced by differences in how groups are seen and evaluated. "Strife arises," he notes, "when groups are differentially valued in terms of their intrinsic self-worth" (1992:20). Although a group's prestige or esteem is usually tied to its political and economic power, growing equality in these does not in all cases guarantee similar growth in social esteem. As we saw in the United States, African Americans, for example, despite real collective political and economic mobility in recent decades, in many cases continue to see themselves held in lower social esteem.

Not only within but also between societies, certain trends portend continued ethnic conflict. Economies of global scale and world-encompassing communications networks make national boundaries increasingly permeable, thus creating ever-greater ethnic diversity within most of the world's nation-states. Indeed, as we have noted, few societies today cannot, to some degree, be called multiethnic. These

---

[7]Susan Olzak (1992) has postulated that, as inequalities among ethnic groups decline, greater conflict may be generated through intensified competition that ensues among the groups. As ethnic stratification breaks down, more powerful groups seek to prevent weaker, minority groups from getting their share of the society's resources.

tendencies might lead to greater understanding among diverse peoples—and thus to a breakdown of ethnic barriers—as they come to recognize their problems and motivations as commonly human (Shibutani and Kwan, 1965). But these trends might just as easily serve to rigidify ethnocentric and nationalistic views by demonstrating more dramatically the economic and cultural differences among them. "In deeply divided territories," note McGarry and O'Leary, "increased exposure to the 'other' may make group members more aware of what their group has in common and what separates them from the others. Exposure may cement group solidarity rather than diffuse it" (1995b:855). The persistence of ethnic conflict in almost all parts of the world seems to confirm the latter outcome.

Given the pervasiveness of ethnic diversity and inequality, it is perhaps most realistic to think in terms of the management of ethnic tensions and hostilities rather than their elimination. While some ethnic conflicts are resolved with finality, observes political scientist Milton Esman, "the great majority rise, subside, and recur" (1994:23). It is important to reiterate that conflict is a fundamental condition of human societies and is inherent in all human affairs. It should not be confused with violence. Most social conflict is low-intensity and controlled through the society's political process and legal system. Conflict may, of course, involve the use of violence, but this is ordinarily episodic and does not characterize most intergroup relations. Thus, although ethnic conflict is an essential, ongoing feature of all multiethnic societies, it is usually controlled and limited. Cases in which extreme violence occurs (Rwanda or Yugoslavia, for example) are exceptional, not typical. As Darby has put it, "The use of violence marks the failure of normal and functional means of conflict resolution—political exchange, negotiation, compromise" (1995:29). The most achievable objective of multiethnic societies, then, is not to eliminate ethnic conflict entirely but to contain it as much as possible. "The real issue," writes Darby, "is not the existence of conflict, but how it is regulated" (1995:29).

The ability of diverse ethnic groups in pluralistic societies to live together in relative harmony while maximizing equality in the distribution of political and economic resources among them is dependent in large measure on the presence of a central state with sufficient power to contain intergroup conflict to a tolerable degree, but fair enough to win the allegiance of all groups. As Ignatieff has observed, "What keeps ethnic and racial tension within bounds in the world's successful modern multi-ethnic societies is a state strong enough to make its authority respected" (1993:243). Recent developments in South Africa, as we have seen, illustrate that efforts in this regard may in some cases yield significant changes. Yugoslavia, on the other hand, demonstrates the disaster that may result from the collapse of a strong state holding diverse groups together.

Whatever successes may be achieved in the pursuit of ethnic harmony will probably be overshadowed by discord. It is obvious that, with ethnic diversity the prevailing global trend, nations everywhere will, for the foreseeable future, struggle with issues of political, social, and economic inequalities among racially and culturally diverse groups. To some degree, then, ethnic conflict will be a recurring theme. The spread of pluralistic ideas and movements among ethnic groups within multiethnic societies is an outgrowth of the nationalistic fervor that has swept both the developed and developing worlds, and of international population movements that are occurring on a historically unprecedented scale. It is possible, of course, that with the advent of a global economy and the emergence of global communications networks ethnic nationalism may prove to be only a passing phenomenon, destined to recede as people everywhere are increasingly linked by common needs and cosmopolitan cultural systems. This outcome, however, cannot yet be envisioned. The crucial fact, notes Anthony Smith, "is that interethnic conflict has become more intense and endemic in the twentieth century than at any time in history" (1981:10). Today there is little indication of the development of a countercurrent. Rather than diminishing, ethnicity as a global phenomenon persists as an exceptionally powerful and tenacious force motivating human behavior. And it will follow us into the twenty-first century.

SUMMARY

Conflict and change are inherent features of multiethnic societies. Where there are significant economic, political, and social inequalities among groups, challenges to the dominant social order arise periodically. Also, cultural differences among groups may continue to impel ethnic conflict. The conditions of the contemporary world—increasing diversity and ethnic nationalism—make the continuation and even intensification of ethnic conflict very likely.

Shifting patterns of global immigration have emerged in the past forty years. Today, the movement of people is from the developing societies to the developed, industrialized societies. Changing labor markets and political strife in many societies are the major factors that have stimulated global immigration. This has molded societies into increasingly diverse ethnic configurations.

Western European societies have undergone great social changes in the past few decades as a result of immigration and the subsequent creation of ethnic heterogeneity. As societies where there had been relative homogeneity, the introduction of large numbers of culturally and racially diverse people has created ethnic minority communities that have not been assimilated into the mainstream. These ethnic minorities have been the targets of severe prejudice and discrimination.

The breakup of the Soviet Union and the fragmentation of east European countries starting in the late 1980s added a volatile element to the European ethnic mix. Ethnic nationalism swept over groups that had lived relatively harmoniously under the domination of communist governments, resulting in violent conflict in many states. Nowhere was the conflict more virulent or bloody than in Yugoslavia, where Serbs, Croats, and Muslims engaged in the most ferocious warfare in Europe since World War II. Yugoslavia seems to be a case of a failed attempt at the formation of a corporate pluralistic system. Once the glue of a strong central government was removed, underlying ethnic hostilities reemerged with overwhelming force.

Ethnic conflict, while most apparent in Europe, is today a global phenomenon, with no world region immune. The spread of ethnic nationalism and the increasing scope of racial and cultural heterogeneity are conditions that make for the continuation of ethnic conflict in the foreseeable future.

Suggested Readings

Anderson, Benedict. 1991. *Imagined Communities: Reflections on the Origin and Spread of Nationalism.* Rev. ed. London: Verso. Examines the evolution of nationalism and the factors that create a preparedness of people to defend their society or perceived community against others.

Bennett, Christopher. 1995. *Yugoslavia's Bloody Collapse: Causes, Course and Consequences.* New York: New York University Press. Deftly sorts out the complex political and social issues that led to the breakup of Yugoslavia and offers insight into the cultural similarities and differences of the groups involved.

Castles, Stephen, and Mark J. Miller. 1993. *The Age of Migration: International Population Movements in the Modern World.* New York: Guilford. Describes today's global immigration routes and analyzes the economic, political, and social impact of large-scale immigration on receiving societies.

Destexhe, Alain. 1995. *Rwanda and Genocide in the Twentieth Century.* New York: New York University Press. A succinct analysis of the slaughter of Tutsis by Hutus in 1994 in Rwanda and the historic relations between these groups that preceded it. Discusses the Rwandan case in the context of other genocides of the twentieth century.

Ignatieff, Michael. 1993. *Blood and Belonging: Journeys into the New Nationalism.* Toronto: Penguin. The author travels to six locations of contemporary ethnic conflict, relating his observations and his conversations with those involved.

Rieff, David. 1995. *Slaughterhouse: Bosnia and the Failure of the West.* New York: Touchstone. A journalist describes his odyssey through war-plagued Croatia, Serbia, and Bosnia, probing the thoughts of citizens as he tries to understand the ethnic conflict among them.

Smith, Anthony D. 1981. *The Ethnic Revival.* Cambridge: Cambridge University Press. An analysis of ethnic nationalism and its global resurgence in the twentieth century.

REFERENCES

Aboud, Frances E. 1987. "The Development of Ethnic Self-Identification and Attitudes." Pp. 32–55 in Jean S. Phinney and Mary Jane Rotheram (eds.), *Children's Ethnic Socialization: Pluralism and Development.* Newbury Park: Sage.

Abrams, Charles. 1966. "The Housing Problem and the Negro." Pp. 512–524 in Talcott Parsons and Kenneth B. Clark (eds.), *The Negro American.* Boston: Houghton Mifflin.

Abramson, Harold. 1980. "Assimilation and Pluralism." Pp. 150–160 in Stephen Thernstrom (ed.), *Harvard Encyclopedia of American Ethnic Groups.* Cambridge, Mass.: Harvard University Press.

Acuña, Rodolfo. 1972. *Occupied America: The Chicano's Struggle Toward Liberation.* San Francisco: Caufield.

Adachi, Ken. 1976. *The Enemy That Never Was.* Toronto: McClelland & Stewart.

Adam, Heribert. 1971a. "The South African Power-Elite: A Survey of Ideological Commitment." Pp. 73–102 in Heribert Adam (ed.), *South Africa: Social Perspectives.* London: Oxford University Press.

———. 1971b. *Modernizing Racial Domination: South Africa's Political Dynamics.* Berkeley: University of California Press.

———. 1975. "Internal Constellations and Potentials for Change." Pp. 303–326 in Leonard Thompson and Jeffrey Butler (eds.), *Change in Contemporary South Africa.* Berkeley: University of California Press.

———. 1978. "The Political Sociology of South Africa: A Pragmatic Race Oligarchy." Pp. 13–54 in Ian Robertson and Phillip Whitten (eds.), *Race and Politics in South Africa.* New Brunswick, N.J.: Transaction.

———. 1992. "The Politics of Identity: Nationalism, Patriotism, and Multiculturalism." *Journal of Ethno-Development* 1(2):10–22.

———. 1995. "The Politics of Ethnic Identity: Comparing South Africa." *Ethnic and Racial Studies* 18:457–475.

Adam, Heribert, and Kogila Moodley. 1986. *South Africa Without Apartheid: Dismantling*

Racial Domination. Berkeley: University of California Press.

———. 1993. *The Opening of the Apartheid Mind: Options for the New South Africa.* Berkeley: University of California Press.

Adorno, T. W., et al. 1950. *The Authoritarian Personality.* New York: Harper & Row.

African Rights. 1994. *Rwanda: Death, Despair and Defiance.* London: African Rights.

Agbayani-Siewert, Pauline, and Linda Revilla. 1995. "Filipino Americans." Pp. 134–168 in Pyong Gap Min (ed.), *Asian Americans: Contemporary Trends and Issues.* Thousand Oaks, Calif.: Sage.

Alba, Richard D. 1981. "The Twilight of Ethnicity Among American Catholics of European Ancestry." *Annals of the American Academy of Political and Social Science* 454:86–97.

———. 1985. *Italian Americans: Into the Twilight of Ethnicity.* Englewood Cliffs, N.J.: Prentice-Hall.

———. 1990. *Ethnic Identity: The Transformation of White America.* New Haven, Conn.: Yale University Press.

———. 1992. "Ethnicity." Pp. 575–584 in Edgar F. and Mary L. Borgatta (eds.), *Encyclopedia of Sociology.* Vol. 2. New York: Macmillan.

———. 1994. "Identity and Ethnicity Among Italians and Other Americans of European Ancestry." Pp. 21–44 in Lydio Tomasi, Piero Gastaldo, and Thomas Row (eds.), *The Columbus People: Perspectives in Italian Immigration to the Americas and Australia.* Staten Island, N.Y.: Center for Migration Studies.

———. 1995. "Assimilation's Quiet Tide." *The Public Interest* (Spring):3–18.

Alba, Richard D., and Reid M. Golden. 1986. "Patterns of Ethnic Marriage in the United States." *Social Forces* 65:202–223.

Alba, Richard D., and Gwen Moore. 1982. "Ethnicity in the American Elite." *American Sociological Review* 47:373–383.

Albert, Luc. 1989. "Language in Canada." *Canadian Social Trends* (Spring):9–12.

Albini, Joseph L. 1971. *The American Mafia: Genesis of a Legend.* New York: Appleton-Century-Crofts.

Allen, Irving Lewis. 1975. "WASP—From Sociological Concept to Epithet." *Ethnicity* 2:153–162.

Allen, Robert L. 1970. *Black Awakening in Capitalist America.* Garden City, N.Y.: Doubleday.

Allport, Gordon W. 1958. *The Nature of Prejudice.* Garden City, N.Y.: Doubleday.

Alvarez, Rodolfo. 1973. "The Psycho-Historical and Socioeconomic Development of the Chicano Community in the United States." *Social Science Quarterly* 53:920–942.

American Enterprise. 1994. "Immigration." (January/February):97–100.

American Jewish Committee. 1987. *American Jewish Year Book 1988.* New York: American Jewish Committee.

———. 1989. *American Jewish Year Book 1989.* New York: American Jewish Committee.

———. 1994. *American Jewish Year Book 1994.* New York: American Jewish Committee.

Anderson, Benedict. 1991. *Imagined Communities: Reflections on the Origin and Spread of Nationalism.* Rev. ed. London: Verso.

Anderson, Charles H. 1970. *White Protestant Americans: From National Origins to Religious Group.* Englewood Cliffs, N.J.: Prentice-Hall.

Andrews, George Reid. 1991. *Blacks & Whites in São Paulo, Brazil 1888–1988.* Madison, Wis.: University of Wisconsin Press.

———. 1992. "Racial Inequality in Brazil and the United States: A Statistical Comparison." *Journal of Social History* 26:229–263.

Anti-Defamation League. 1993. *Highlights from an Anti-Defamation League Survey on Racial Attitudes in America.* New York: Anti-Defamation League of B'nai B'rith.

Antonovsky, Aaron. 1960. "The Social Meaning of Discrimination." *Phylon* 21:81–95.

Aoyagi, Kiyotaka, and Ronald P. Dore. 1964. "The Buraku Minority in Urban Japan." Pp. 95–107 in *Transactions of the Fifth World Congress of Sociology.* Vol. 3. Louvain: International Sociological Association.

Arias, M. Beatriz. 1986. "The Context of Education for Hispanic Students: An Overview." *American Journal of Education* 95:26–57.

Arnopoulos, Sheila McLeod, and Dominique Clift. 1980. *The English Fact in Quebec.* Montreal: McGill-Queen's University Press.

Arthur, Paul. 1980. *Government and Politics of Northern Ireland.* Burnt Mill, England: Longman House.

Asch, Michael. 1984. *Home and Native Land: Aboriginal Rights and the Canadian Constitution.* Toronto: Methuen.

Aunger, Edmund A. 1981. *In Search of Political Stability: A Comparative Study of New Brunswick and Northern Ireland.* Montreal: McGill-Queen's University Press.

Avery, Donald. 1979. *Dangerous Foreigners: European Immigrant Workers and Labour Radicalism in Canada 1896–1932.* Toronto: McClelland & Stewart.

Bahr, Howard M., Bruce A. Chadwick, and Robert C. Day (eds.). 1972. *Native Americans Today: Sociological Perspectives.* New York: Harper & Row.

Bahr, Howard M., Bruce A. Chadwick, and Joseph H. Stauss. 1979. *American Ethnicity.* Lexington, Mass.: Heath.

Baltzell, E. Digby. 1958. *Philadelphia Gentlemen: The Making of a National Upper Class.* New York: Free Press.

———. 1964. *The Protestant Establishment: Aristocracy and Caste in America.* New York: Vintage.

———. 1994. *Judgment and Sensibility: Religion and Stratification.* New Brunswick, N.J.: Transaction.

Banfield, Edward C. 1968. *The Unheavenly City.* Boston: Little, Brown.

———. 1990 (1974). *The Unheavenly City Revisited.* Prospect Heights, Ill.: Waveland Press.

Banting, Keith G. 1992. "If Quebec Separates: Restructuring Northern North America." Pp. 159–178 in R. Kent Weaver (ed.), *The Collapse of Canada?* Washington, D.C.: Brookings Institution.

Banton, Michael. 1967. *Race Relations.* London: Tavistock.

———. 1970. "The Concept of Racism." Pp. 17–34 in Sami Zubaida (ed.), *Race and Racialism.* London: Tavistock.

———. 1983. *Racial and Ethnic Competition.* Cambridge, England: Cambridge University Press.

Baron, Dennis. 1990. *The English-Only Question: An Official Language for Americans?* New Haven: Yale University Press.

Baron, Harold. 1969. "The Web of Urban Racism." Pp. 134–176 in Louis L. Knowles and Kenneth Prewitt (eds.), *Institutional Racism in America.* Englewood Cliffs, N.J.: Prentice-Hall.

Barone, Michael. 1985. "Italian Americans and Politics." Pp. 378–384 in Lydio F. Tomasi (ed.), *Italian Americans: New Perspectives in Italian Immigration and Ethnicity.* Staten Island, N.Y.: Center for Migration Studies.

Barrera, Mario. 1979. *Race and Class in the Southwest: A Theory of Racial Inequality.* South Bend, Ind.: University of Notre Dame Press.

Barringer, Herbert R., Robert W. Gardner, and Michael J. Levin. 1993. *Asians and Pacific Islanders in the United States.* New York: Russell Sage Foundation.

Barritt, Denis. 1982. *Northern Ireland: A Problem to Every Solution.* London: Quaker Peace and Service.

Barritt, Denis P., and Charles F. Carter. 1972. *The Northern Ireland Problem: A Study in Group Relations.* 2d ed. London: Oxford University Press.

Barsh, Russel Lawrence. 1994. "Canada's Aboriginal Peoples: Social Integration or Disintegration?" *Canadian Journal of Native Studies* 14:1–46.

Barta, Russell. 1979. "The Representation of Poles, Italians, Latins and Blacks in the Executive Suites of Chicago's Largest Corporations." Pp. 418–430 in U.S. Commission on Civil Rights, *Civil Rights Issues of Euro-Ethnic Americans in the United States: Opportunities and Challenges.* Washington, D.C.: U.S. Government Printing Office.

Barth, Ernest A. T., and Donald L. Noel. 1972. "Conceptual Frameworks for the Analysis of Race Relations." *Social Forces* 50:333–348.

Barth, Fredrik. 1969. *Ethnic Groups and Boundaries.* Boston: Little, Brown.

Bastide, Roger. 1961. "Dusky Venus, Black Apollo." *Race* 3(November):10–18.

———. 1965. "The Development of Race Relations in Brazil." Pp. 9–29 in Guy Hunter (ed.), *Industrialization and Race Relations.* London: Oxford University Press.

Bastide, Roger, and Pierre van den Berghe. 1957. "Stereotypes, Norms and Interracial Behavior in São Paulo, Brazil." *American Sociological Review* 22:689–694.

Bean, Frank D., Barry Edmonston, and Jeffrey S. Passel (eds.). 1990. *Undocumented Migration to the United States: IRCA and the Experience of the 1980s.* Washington, D.C.: Urban Institute Press.

Bean, Frank D., and Marta Tienda. 1987. *The Hispanic Population of the United States.* New York: Russell Sage Foundation.

Beck, Roy. 1994. "The Ordeal of Immigration in Wausau." *Atlantic Monthly* 273(April):84–97.

———. 1996. *The Case Against Immigration: The Moral, Economic, Social, and Environmental Reasons for Reducing U.S. Immigration Back to Traditional Levels.* New York: Norton.

Beckett, J. C. 1972. "Northern Ireland." Pp. 11–24 in Institute for the Study of Conflict, *The Ulster Debate.* London: Bodley Head.

Beinart, William. 1994. *Twentieth-Century South Africa.* New York: Oxford University Press.

Belfast Telegraph. 1974. "Results of BBC Poll." (April 19):1.

Belfast Workers Research Unit. 1980. *The Churches in Northern Ireland.* Bulletin No. 8. Belfast: Belfast Workers Research Unit.

Bell, Daniel. 1962. *The End of Ideology.* Rev. ed. New York: Free Press.

———. 1975. "Ethnicity and Social Change." Pp. 141–174 in Nathan Glazer and Daniel P. Moynihan (eds.), *Ethnicity: Theory and Experience.* Cambridge, Mass.: Harvard University Press.

Bell, Geoffrey. 1976. *The Protestants of Ulster.* London: Pluto Press.

Belluck, Pam. 1995a. "Healthy Korean Economy Draws Immigrants Home." *New York Times* (August 22):A1, B4.

———. 1995b. "High School for Immigrants Is Proposed." *New York Times* (March 20):A13.

Benedict, Ruth. 1959 (1940). *Race: Science and Politics.* New York: Viking.

Bennett, Christopher. 1995. *Yugoslavia's Bloody Collapse: Causes, Course and Consequences.* New York: New York University Press.

Bennett, John. 1967. *Hutterite Brethren: The Agricultural Economy and Social Organization of a Communal People.* Stanford, Calif.: Stanford University Press.

Bercuson, David J. 1995. "Why Quebec and Canada Must Part." *Current History* 94:123–126.

Berger, Joseph. 1992. "South African Students Learn to Learn Together." *New York Times* (May 12):A4.

Berger, Peter. 1963. *Invitation to Sociology: A Humanistic Perspective.* Garden City, N.Y.: Doubleday.

Bergesen, Albert. 1980. "Official Violence During the Watts, Newark and Detroit Race Riots of the 1960s." Pp. 138–174 in Pat Lauderdale (ed.), *A Political Analysis of Deviance.* Minneapolis: University of Minnesota Press.

Berkhofer, Robert F., Jr. 1979. *The White Man's Indian.* New York: Vintage.

Bernard, Jesse. 1951. "The Conceptualization of Intergroup Relations with Special Reference to Conflict." *Social Forces* 19:243–251.

Bernstein, Richard. 1988. "Asian Newcomers Hurt by Precursors' Success." *New York Times* (July 10):12.

Berreman, Gerald D. 1966. "Structure and Function of Caste Systems." Pp. 277–307 in George De Vos and Hiroshi Wagatsuma (eds.), *Japan's Invisible Race: Caste in Culture and Personality.* Berkeley and Los Angeles: University of California Press.

———. 1972. "Race, Caste, and Other Invidious Distinctions in Social Stratification." *Race* 13:385–414.

Berry, Brewton, and Henry L. Tischler. 1978. *Race and Ethnic Relations.* 4th ed. Boston: Houghton Mifflin.

Berry, Brian, and John Kasarda. 1977. *Contemporary Urban Sociology.* New York: Macmillan.

Berry, John W., Rudolf Kalin, and D. M. Taylor. 1977. *Multiculturalism and Ethnic Attitudes in Canada.* Ottawa: Minister of Supply and Services.

Betancur, John J., Teresa Cordova, and Maria de los Angeles Torres. 1993. "Economic Restructuring and the Process of Incorporation of Latinos into the Chicago Economy." Pp. 109–132 in Rebecca Morales and Frank Bonilla (eds.), *Latinos in a Changing U.S. Economy.* Newbury Park, Calif.: Sage.

Béteille, André. 1969. *Castes Old and New: Essays in Social Structure and Social Stratification.* Bombay: Asian Publishing Home.

Bibby, Reginald W. 1995. *The Bibby Report: Social Trends Canadian Style.* Toronto: Stoddart.

Biemiller, Lawrence. 1986. "Asian Students Fear Top Colleges Use Quota Systems." *Chronicle of Higher Education* 23(November 19):1, 34–35, 37.

Billingsley, Andrew. 1968. *Black Families in White America.* Englewood Cliffs, N.J.: Prentice-Hall.

Bissoondath, Neil. 1994. *Selling Illusions: The Cult of Multiculturalism in Canada.* Toronto: Penguin.

Black Enterprise. 1990. "B. E. Survey Report: A View of the Past, A Look to the Future." 21(August):77–94.

———. 1995a. "B. E. Industrial/Service 100." 25(June):99–108.

———. 1995b. "B. E. Financial 25." 25(June):163.

Blackwell, James E. 1976. "The Power Basis of Ethnic Conflict in American Society." Pp. 179–196 in Lewis A. Coser and Otto N. Larsen (eds.), *The Use of Controversy in Sociology.* New York: Free Press.

Blalock, Hubert M., Jr. 1967. *Toward a Theory of Minority-Group Relations.* New York: Wiley.

Blau, Joseph L., and Salo W. Baron (eds.). 1963. *The Jews of the United States 1790–1840: A Documentary History.* New York: Columbia University Press.

Blau, Peter M. 1977. "A Macrosociological Theory of Social Structure." *American Journal of Sociology* 83:26–54.

Blauner, Robert. 1969. "Internal Colonialism and Ghetto Revolt." *Social Problems* 16:393–408.

———. 1972. *Racial Oppression in America.* New York: Harper & Row.

———. 1992. "Talking Past Each Other: Black and White Languages of Race." *American Prospect* 10(Summer):55–64.

Blishen, Bernard R. 1970. "Social Class and Opportunity in Canada." *Canadian Review of Sociology and Anthropology* 7:110–127.

Bloomgarden, Laurence. 1957. "Who Should Be Our Doctors?" *Commentary* (January):506–515.

Blue, Art. 1985. "Native People, Social Conditions." Pp. 1223–1224 in *The Canadian Encyclopedia.* Vol. II. Edmonton: Hurtig.

Blumenfeld, Jesmond. 1987. "Economy Under Siege." Pp. 17–33 in Jesmond Blumenfeld (ed.), *South Africa in Crisis.* London: Croom Helm.

Blumer, Herbert. 1958. "Race Prejudice as a Sense of Group Position." *Pacific Sociological Review* 1:3–6.

———. 1965. "Industrialization and Race Relations." Pp. 200–253 in Guy Hunter (ed.), *Industrialization and Race Relations.* London: Oxford University Press.

Boal, Frederick W. 1982. "Segregating and Mixing: Space and Residence in Belfast." Pp. 249–280 in Frederick W. Boal and J. Neville H. Douglas (eds.), *Integration and Division: Geographical Perspectives on the Northern Ireland Problem.* London: Academic Press.

Boal, Frederick W., and J. Neville H. Douglas. 1982. "Overview." Pp. 333–359 in Frederick W. Boal and J. Neville H. Douglas (eds.), *Integration and Division: Geographical Perspectives on the Northern Ireland Problem.* London: Academic Press.

Bogardus, Emory S. 1925a. "Social Distance and Its Origins." *Journal of Applied Sociology* 9:216–226.

———. 1925b. "Measuring Social Distances." *Journal of Applied Sociology* 9:299–308.

———. 1930. "A Race Relations Cycle." *American Journal of Sociology* 35:612–617.

———. 1959. *Social Distance.* Yellow Springs, Ohio: Antioch Press.

———. 1968. "Comparing Racial Distance in Ethiopia, South Africa, and the United States." *Sociology and Social Research* 52:149–156.

Bolaria, B. Singh, and Peter S. Li. 1988. *Racial Oppression in Canada.* 2d ed. Toronto: Garamond Press.

Bonacich, Edna. 1972. "A Theory of Ethnic Antagonism: The Split Labor Market." *American Sociological Review* 37:547–559.

———. 1973. "A Theory of Middleman Minorities." *American Sociological Review* 38:583–594.

———. 1976. "Advanced Capitalism and Black–White Race Relations in the United States: A Split Labor Market Interpretation." *American Sociological Review* 41:34–51.

———. 1979. "The Political Implications of a Split Labour Market Analysis of South African Race Relations." Pp. 106–116 in Pierre van den Berghe (ed.), *The Liberal Dilemma in South Africa.* New York: St. Martin's.

Bonacich, Edna, and John Modell. 1980. *The Economic Basis of Ethnic Solidarity: Small Business in the Japanese American Community.* Berkeley: University of California Press.

Bonilla, Frank, and Ricardo Campos. 1981. "A Wealth of Poor: Puerto Ricans in the New Economic Order." *Annals of the American Academy of Political and Social Science* 110:133–176.

Borjas, George J. 1994. "Tired, Poor on Welfare." Pp. 76–80 in Nicolaus Mills (ed.), *Arguing Immigration.* New York: Touchstone.

Bouvier, Leon F. 1979. "International Migration: Yesterday, Today, and Tomorrow." *Population Bulletin* 32(4):3–40.

———. 1992. *Peaceful Invasions: Immigration and Changing America.* Lanham, Md.: University Press of America.

Bouvier, Leon F., and Anthony J. Agresta. 1985. "The Fastest Growing Minority." *American Demographics* 7(May): 31–46.

———. 1987. "The Future Asian Population of the United States." Pp. 285–301 in James T. Fawcett and Benjamin V. Cariño (eds.), *Pacific Bridges: The New Immigration from Asia and the Pacific Islands.* Staten Island, N.Y.: Center for Migration Studies.

Bouvier, Leon F., and Robert W. Gardner. 1986. "Immigration to the U.S.: The Unfinished Story." *Population Bulletin* 41(November):3–50.

Bowles, Samuel, and Herbert Gintis. 1976. *Schooling in Capitalist America.* New York: Basic Books.

Bowser, Benjamin P., and Raymond G. Hunt (eds.). 1981. *Impacts of Racism on White Americans.* Beverly Hills, Calif.: Sage.

Boxer, Charles R. 1962. *The Golden Age of Brazil.* Berkeley: University of California Press.

———. 1963. *Race Relations in the Portuguese Colonial Empire, 1415–1825.* Oxford: Clarendon Press.

Boyle, Joseph F. 1977. "Educational Attainment, Occupational Achievement and Religion in Northern Ireland." *Economic and Social Review* 8:79–100.

Boyle, Kevin, and Tom Hadden. 1994. *Northern Ireland: The Choice.* New York: Penguin.

Bozinoff, Lorne, and Peter MacIntosh. 1989. "Large Increase in Racial Intolerance Perceived." *Gallup Canada* (March 20).

———. 1992a. "Two Solitudes Apparent Concerning Separation Issue." *Gallup Report* (March 2).

———. 1992b. "Majority Believe Racial Intolerance Has Increased." *Gallup Report* (May 25).

Braverman, Harold, and Louis Kaplan. 1967. "A Study of Religious Discrimination by Social Clubs." Pp. 211–221 in Milton L. Barron (ed.), *Minorities in a Changing World.* New York: Knopf.

Braverman, Marilyn. 1988. "English Yes; English Only, No." *Reconstructionist* 52(June):29–31.

Breton, Raymond. 1964. "Institutional Completeness of Ethnic Communities and the Personal Relations of Immigrants." *American Journal of Sociology* 70:193–205.

———. 1989. "Canadian Ethnicity in the Year 2000." Pp. 149–152 in James Frideres (ed.), *Multiculturalism and Intergroup Relations.* New York: Greenwood.

Breton, Raymond, Jeffrey G. Reitz, and Victor F. Valentine. 1980. *Cultural Boundaries and the Cohesion of Canada.* Montreal: Institute for Research on Public Policy.

Brewer, John D. 1994. "Crime and Control." Pp. 53–70 in John D. Brewer (ed.), *Restructuring South Africa.* New York: St. Martin's.

Briggs, John W. 1978. *An Italian Passage.* New Haven, Conn.: Yale University Press.

Brimelow, Peter. 1995. *Alien Nation.* New York: HarperCollins.

Brink, William, and Louis Harris. 1963. *The Negro Revolution in America.* New York: Simon & Schuster.

———. 1967. *Black and White: A Study of Racial Attitudes Today.* New York: Simon & Schuster.

Brown, Cecil. 1981. "Blues for Blacks in Hollywood." *Mother Jones* (January):20–28, 59.

Brown, Douglas. 1966. *Against the World: A Study of White South African Attitudes.* London: Collins.

Bruce, Steve. 1986. *God Save Ulster!* Oxford: Clarendon Press.

Brunet, Michel. 1969. "Canadians and Canadiens." Pp. 284–293 in Ramsay Cook (ed.), *French-Canadian Nationalism.* Toronto: Macmillan of Canada.

Brym, Robert J. 1989. *From Culture to Power: The Sociology of English Canada.* Toronto: Oxford University Press.

Buchanan, Ronald H. 1982. "The Planter and the Gael: Cultural Dimensions of the Northern Ireland Problem." Pp. 49–74 in Frederick W. Boal and J. Neville H. Douglas (eds.), *Integration and Division: Geographical Perspectives on the Northern Ireland Problem.* London: Academic Press.

Buchanan, William, and Hadley Cantril. 1953. *How Nations See Each Other.* Urbana, Ill.: University of Illinois Press.

Buchignani, Norman L. 1980. "Accommodation, Adaptation, and Policy: Dimensions of the South Asian Experience in Canada." Pp. 121–150 in K. Victor Ujimoto and Gordon Hirabayashi (eds.), *Visible Minorities and Multiculturalism: Asians in Canada.* Toronto: Butterworths.

Burck, Charles G. 1976. "A Group Profile of the Fortune 500 Chief Executive." *Fortune* (May):173–177, 308–312.

Burdick, John. 1995. "Brazil's Black Consciousness Movement." Pp. 174–183 in Kevin Danaher and Michael Shellenberger (eds.), *Fighting for the Soul of Brazil.* New York: Monthly Review Press.

Burgess, M. Elaine. 1983. "Race and Social Change in South Africa: Divergent Perspectives." *Journal of Ethnic Studies* 11:47–71.

Burkey, Richard M. 1978. *Ethnic and Racial Groups: The Dynamics of Dominance.* Menlo Park, Calif.: Cummings.

Burkholz, Herbert. 1980. "The Latinization of Miami." *New York Times Magazine* (September 21):44–47, 84–88, 98–100.

Burma, John. 1954. *Spanish-Speaking Groups in the United States.* Durham, N.C.: Duke University Press.

Burnet, Jean. 1976. "Ethnicity: Canadian Experience and Policy." *Sociological Focus* 9:199–207.

———. 1981. "The Social and Historical Context of Ethnic Relations." Pp. 17–35 in Robert C. Garner and Rudolf Kalin (eds.), *A Canadian Social Psychology of Ethnic Relations.* Toronto: Methuen.

Burns, John F. 1988. "Ottawa Will Pay Compensation to Uprooted Japanese-Canadians." *New York Times* (September 23):A10.

Burton, Frank. 1978. *The Politics of Legitimacy: Struggles in a Belfast Community.* London: Routledge & Kegan Paul.

Business Week. 1992. "The Immigrants." (July 13):114–118.

Business Week/Harris Poll. 1992. "America's Welcome Mat Is Wearing Thin." (July 13):122.

Butterfield, Fox. 1986. "Why Asians Are Going to the Head of the Class." *New York Times* (August 3):12, 18–23.

Button, James W. 1978. *Black Violence.* Princeton, N.J.: Princeton University Press.

Cafferty, Pastora San Juan. 1983. "The Language Question: The Dilemma of Bilingual Education for Hispanics in America." Pp. 101–127 in Lance Liebman (ed.), *Ethnic Relations in*

America. Englewood Cliffs, N.J.: Prentice-Hall.

———. 1985. "Language and Social Assimilation." Pp. 87–111 in Pastora San Juan Cafferty and William C. McCready (eds.), *Hispanics in the United States: A New Social Agenda.* New Brunswick, N.J.: Transaction.

Camarillo, Albert. 1979. *Chicanos in a Changing Society.* Cambridge, Mass.: Harvard University Press.

Campaign for Social Justice in Northern Ireland. 1972. *The Plain Truth.* 2d ed. Castlefields: Campaign for Social Justice in Northern Ireland.

Campani, Giovanna. 1993. "Immigration and Racism in Southern Europe: The Italian Case." *Ethnic and Racial Studies* 16:507–535.

Campbell, Angus. 1971. *White Attitudes Toward Black People.* Ann Arbor, Mich.: Institute for Social Research, University of Michigan.

Caplan, Nathan, Marcella H. Choy, and John K. Whitmore. 1991. *Children of the Boat People: A Study of Educational Success.* Ann Arbor, Mich.: University of Michigan Press.

———. 1992. "Indochinese Refugee Families and Academic Achievement." *Scientific American* 266(February):36–42.

Caplan, Nathan, John K. Whitmore, and Marcella H. Choy. 1989. *The Boat People and Achievement in America.* Ann Arbor, Mich.: University of Michigan Press.

Cardenas, Gilbert. 1976. "Los Desarraigados: Chicanos in the Midwestern Region of the United States." *Aztlan* 7:153–201.

Cardoso, Fernando Henrique. 1965. "Color Prejudice in Brazil." *Présence Africaine* 25:120–128.

Cariño, Benjamin V. 1987. "The Philippines and Southeast Asia: Historical Roots and Contemporary Linkages." Pp. 305–325 in James T. Fawcett and Benjamin V. Cariño (eds.), *Pacific Bridges: The New Immigration from Asia and the Pacific Islands.* Staten Island, N.Y.: Center for Migration Studies.

Carmichael, Stokely, and Charles V. Hamilton. 1967. *Black Power: The Politics of Liberation in America.* New York: Vintage.

Carnoy, Martin, Hugh M. Daley, and Raul Hinajosa Ojeda. 1993. "The Changing Economic Position of Latinos in the U.S. Labor Market Since 1939." Pp. 28–54 in Rebecca Morales and Frank Bonilla (eds.), *Latinos in a Changing U.S. Economy.* Newbury Park, Calif.: Sage.

Carter, Gwendolyn M. 1980. *Which Way Is South Africa Going?* Bloomington, Ind.: Indiana University Press.

Castles, Stephen. 1992. "Migrants and Minorities in Post-Keynesian Capitalism: The German Case." Pp. 36–54 in Malcolm Cross (ed.), *Ethnic Minorities and Industrial Change in Europe and North America.* Cambridge, England: Cambridge University Press.

Castles, Stephen, Heather Booth, and Tina Wallace. 1984. *Here for Good: Western Europe's New Ethnic Minorities.* London: Pluto Press.

Castles, Stephen, and Godula Kosack. 1973. *Immigrant Workers and Class Structure in Western Europe.* London: Oxford University Press.

Castles, Stephen, and Mark J. Miller. 1993. *The Age of Migration: International Population Movements in the Modern World.* New York: Guilford.

Castro, Max J. 1992. "The Politics of Language in Miami." Pp. 109–132 in Guillermo J. Grenier and Alex Stepick III (eds.), *Miami Now!: Immigration, Ethnicity, and Social Change.* Gainesville, Fla.: University Press of Florida.

Cataldo, Everett F., Michael Giles, and Douglas S. Gatlin. 1975. "Metropolitan School Desegregation: Practical Remedy or Impractical Ideal?" *Annals of the American Academy of Political and Social Science* 422:97–104.

Caudill, William, and George De Vos. 1956. "Achievement, Culture and Personality: The Case of the Japanese Americans." *American Anthropologist* 58:1103–1126.

Celis, William, III. 1992. "Hispanic Rate for Dropouts Remains High." *New York Times* (October 14):A1, B8.

Central Office of Information. 1994. *Britain 1995: An Official Handbook.* London: Her Majesty's Stationery Office.

Chanes, Jerome A. 1994. "Interpreting the Data: Antisemitism and Jewish Security in the United States." *Patterns of Prejudice* 28:87–101.

Chaney, Elsa M. 1979. "The World Economy and Contemporary Migration." *International Migration Review* 13:204–212.

Chavez, Linda. 1991. *Out of the Barrio: Toward a New Politics of Hispanic Assimilation.* New York: Basic Books.

Chesley, Roger, and Brenda J. Gilchrist. 1993. "Simmering Distrust." *Detroit Free Press* (February 22):1A, 5A.

Chiswick, Barry R. 1984. "The Labor Market Status of American Jews: Patterns and Determinants." Pp. 131–153 in *American Jewish Year Book 1985*. New York: American Jewish Committee.

Christopulos, Diana. 1980. "The Politics of Colonialism: Puerto Rico from 1898 to 1972." Pp. 120–170 in Adalberto Lopez (ed.), *The Puerto Ricans: Their History, Culture, and Society.* Cambridge, Mass.: Schenkman.

Chronicle of Higher Education. 1992. "College Enrollment by Racial and Ethnic Group." (March 18):A35.

Citizenship and Immigration Canada. 1994. *Immigration Statistics.* Ottawa: Employment and Immigration Canada.

Clairmont, Donald H., and Dennis W. Magill. 1974. *Africville: The Life and Death of a Canadian Black Community.* Toronto: McClelland & Stewart.

Clarity, James F. 1992. "Talks Fail, The Irish and British Report." *New York Times* (November 17):A10.

Clark, Kenneth B. 1965. *Dark Ghetto.* New York: Harper & Row.

———. 1980. "The Role of Race." *New York Times Magazine* (October 5):25–33.

Clark, Robert P. 1980. "Euzkadi: Basque Nationalism in Spain Since the Civil War." Pp. 75–100 in Charles R. Foster (ed.), *Nations Without a State.* New York: Praeger.

Clark, S. D. 1950. "The Canadian Community." Pp. 375–389 in George Brown (ed.), *Canada.* Berkeley: University of California Press.

Clement, Wallace. 1975. *The Canadian Corporate Elite: An Analysis of Economic Power.* Toronto: McClelland & Stewart.

———. 1990. "A Critical Response to 'Perspectives on the Class and Ethnic Origins of Canadian Elites.'" *Canadian Journal of Sociology* 15:179–185.

Cohen, Percy S. 1968. "Ethnic Group Differences in Israel." *Race* 9:301–310.

Cohen, Roger. 1994. "Bosnian Camp Survivors Describe Random Death." *New York Times* (August 2):A1, A6.

Cohen, Steven M. 1983. *American Modernity and Jewish Identity.* New York: Tavistock.

———. 1985. *The 1984 National Survey of American Jews: Political and Social Outlooks.* New York: American Jewish Committee.

———. 1987. *Ties and Tensions: The 1986 Survey of American Jewish Attitudes Toward Israel and Israelis.* New York: American Jewish Committee.

———. 1988. *American Assimilation or Jewish Revival?* Bloomington, Ind.: Indiana University Press.

———. 1995. "Jewish Continuity Over Judaic Content: The Moderately Affiliated American Jew." Pp. 395–416 in Robert M. Selter and Norman J. Cohen (eds.), *The Americanization of the Jews.* New York: New York University Press.

Coleman, James S., et al. 1966. *Equality of Educational Opportunity.* Washington, D.C.: U.S. Government Printing Office.

Commission for Social Justice. 1982. "Godfather's Pizza Chain Agrees to Change Image." *Justice* (September/October):1A.

Commission on Minority Participation in Education and American Life. 1988. *One-Third of a Nation.* Washington, D.C.: American Council on Education.

Commonwealth Group of Eminent Persons. 1986. *Mission to South Africa: The Commonwealth Report.* Harmondsworth, England: Penguin.

Compton, Paul. 1982. "The Demographic Dimension of Integration and Division in Northern Ireland." Pp. 75–104 in Frederick W. Boal and J. Neville H. Douglas (eds.), *Integration and Division: Geographical Perspectives on the Northern Ireland Problem.* London: Academic Press.

———. 1991. "The Conflict in Northern Ireland: Demographic and Economic Considerations." Pp. 16–47 in S. W. R. de A. Samarasinghe and Reed Coughlan (eds.), *Economic Dimensions of Ethnic Conflict.* New York: St. Martin's.

Congressional Quarterly. 1979. "Most Members of Congress Claim Religious Affiliation." *CQ Weekly Report* 37(January 20):80–81.

———. 1989. *CQ Guide to Current American Government* (Spring). Washington, D.C.: Congressional Quarterly Press.

———. 1994. "Freshman Class Boasts Résumés to Back Up 'Outsider' Image." *CQ Weekly Reports* (November 12):9–12.

Connor, Walker. 1972. "Nation-Building or Nation-Destroying?" *World Politics* 24:319–355.

———. 1976. "The Political Significance of Ethnonationalism Within Western Europe." Pp. 110–133 in Abdul Said and Lucy R. Simmons (eds.), *Ethnicity in an International Context.* New Brunswick, N.J.: Transaction.

Coogan, Tim Pat. 1980. *On the Blanket: The H Block Story.* Dublin: Ward River Press.

Cook, Christopher. 1989. "Dearborn Sees a New Generation in Politics." *Detroit Free Press* (August 6):3A, 6A.

Cook, Fred J. 1971. "The Black Mafia Moves into the Numbers Racket." *New York Times Magazine* (April 4):26–27, 107–112.

Cooper, Robert. 1991. "The Role of the Fair Employment Agency." Pp. 199–216 in Robert J. Cormack and Robert D. Osborne (eds.), *Discrimination and Public Policy in Northern Ireland.* Oxford: Clarendon Press.

Corbett, David C. 1957. *Canada's Immigration Policies: A Critique.* Toronto: University of Toronto Press.

Corelli, Rae. 1995. "A Tolerant Nation's Hidden Shame." *Maclean's* (August 14):40–43.

Cormack, Robert, and Robert Osborne. 1991. "Disadvantage and Discrimination in Northern Ireland." Pp. 5–48 in Robert J. Cormack and Robert D. Osborne (eds.), *Discrimination and Public Policy in Northern Ireland.* Oxford: Clarendon Press.

———. 1994. "The Evolution of a Catholic Middle Class." Pp. 65–85 in Adrian Guelke (ed.), *New Perspectives on the Northern Ireland Conflict.* Aldershot, England: Avebury.

Cornell, Stephen. 1988. *The Return of the Native: American Indian Political Resurgence.* New York: Oxford University Press.

———. 1990. "Land, Labour and Group Formation: Blacks and Indians in the United States." *Ethnic and Racial Studies* 13:368–388.

Cortés, Carlos E. 1994. "The Hollywood Curriculum on Italian Americans: Evolution of an Icon of Ethnicity." Pp. 89–108 in Lydio F. Tomasi, Piero Gastaldo, and Thomas Row (eds.), *The Columbus People: Perspectives in Italian Immigration to the Americas and Australia.* New York: Center for Migration Studies.

Cose, Ellis. 1993. *The Rage of a Privileged Class.* New York: HarperPerennial.

Coser, Lewis A. 1956. *The Functions of Social Conflict.* New York: Free Press.

Covello, Leonard. 1967. *The Social Background of the Italo-American School Child.* Leiden: E. J. Brill.

Cox, Oliver C. 1948. *Caste, Class and Race.* New York: Monthly Review Press.

Crawford, James. 1992. *Hold Your Tongue: Bilingualism and the Politics of "English Only."* Reading, Mass.: Addison-Wesley.

Crispino, James A. 1980. *The Assimilation of Ethnic Groups: The Italian Case.* Staten Island, N.Y.: Center for Migration Studies.

Crowe, Keith J. 1974. *A History of the Original Peoples of Northern Canada.* Montreal: Arctic Institute of North America, McGill-Queen's University Press.

Curry, David. 1972. "The Frustration of Being Colored." Pp. 401–417 in N. J. Rhoodie (ed.), *South African Dialogue.* Philadelphia: Westminster.

Dahl, Robert A. 1961. *Who Governs?* New Haven, Conn.: Yale University Press.

Dahrendorf, Ralf. 1959. *Class and Class Conflict in Industrial Society.* Stanford, Calif.: Stanford University Press.

———. 1968. *Essays in the Theory of Society.* Stanford, Calif.: Stanford University Press.

Daley, Suzanne. 1995. "As Crime Soars, South African Whites Leave." *New York Times* (December 12):A1, A6.

D'Amico, Ronald, and Nan L. Maxwell. 1995. "The Continuing Significance of Race in Minority Male Joblessness." *Social Forces* 73:969–991.

Daniels, Roger. 1977. *The Politics of Prejudice: The Anti-Japanese Movement in California and the Struggle for Japanese Exclusion.* 2d ed. Berkeley: University of California Press.

———. 1981. *Concentration Camps: North America.* Malabar, Fla.: Krieger.

Darby, John. 1976. *Conflict in Northern Ireland: The Development of a Polarised Community.* Dublin: Gill & Macmillan.

———. 1986. *Intimidation and the Control of Conflict in Northern Ireland.* Dublin: Gill & Macmillan.

———. 1989. "Northern Ireland: Internal Conflict Analyses." Pp. 166–177 in Yonah Alexander and Alan O'Day (eds.), *Ireland's Terrorist Trauma: Interdisciplinary Perspectives.* New York: St. Martin's.

———. 1991. "Northern Ireland: The Persistence and Limitations of Violence." Pp. 151–159 in Joseph V. Montville (ed.), *Conflict and Peacemaking in Multiethnic Societies.* New York: Lexington Books.

———. 1995. *Northern Ireland: Managing Difference.* London: Minority Rights Group International.

Darby, John, and Geoffrey Morris. 1974. *Intimidation in Housing, February 1974.* Belfast: Northern Ireland Community Relations Commission.

Darroch, A. Gordon. 1979. "Another Look at Ethnicity, Stratification and Social Mobility in Canada." *Canadian Journal of Sociology* 4:1–26.

Davis, David Brion. 1966. *The Problem of Slavery in Western Culture.* Ithaca, N.Y.: Cornell University Press.

Davis, E. E., and R. Sinnott. 1979. *Attitudes in the Republic of Ireland Relevant to the Northern Ireland Problem.* Vol. 1, *Descriptive Analysis and Some Comparisons with Attitudes in Northern Ireland and Great Britain,* Paper No. 97. Dublin: Economic and Social Research Institute.

Davis, F. James. 1978. *Minority–Dominant Relations: A Sociological Analysis.* Arlington Heights, Ill.: AHM.

Davis, Morris, and Joseph F. Krauter. 1971. *The Other Canadians: Profiles of Six Minorities.* Toronto: Methuen.

Davis, Richard. 1994. *Mirror Hate: The Convergent Ideology of Northern Ireland Paramilitaries, 1966–1992.* Aldershot, England: Dartmouth.

Dawidowicz, Lucy S. 1975. *The War Against the Jews, 1933–1945.* New York: Holt, Rinehart & Winston.

Dawsey, Darrell. 1992. "Fatal Shooting Escalates Tension Between Blacks, Arab-Americans." *Detroit News* (September 19):1C.

Dean, John P., and Alex Rosen. 1955. *A Manual of Intergroup Relations.* Chicago: University of Chicago Press.

de Azevedo, Fernando. 1950. *Brazilian Culture.* Trans. William Rex Crawford. New York: Macmillan.

Degler, Carl. 1971. *Neither Black nor White: Slavery and Race Relations in Brazil and the United States.* New York: Macmillan.

de la Garza, Rudolph O. 1977. "Mexican-American Voters: A Responsible Electorate." Pp. 63–76 in Frank L. Baird (ed.), *Mexican Americans: Political Power, Influence or Resource.* Lubbock, Tex.: Texas Technical University Press.

de la Garza, Rodolfo O., Louis DeSipio, F. Chris Garcia, John Garcia, and Angelo Falcon. 1992. *Latino Voices: Mexican, Puerto Rican, and Cuban Perspectives on American Politics.* Boulder, Colo.: Westview Press.

Della Pergola, Sergio, and Allie A. Dubb. 1988. "South African Jewry: A Sociodemographic Profile." Pp. 59–140 in *American Jewish Year Book 1988.* New York: American Jewish Committee.

de los Angeles Torres, Maria. 1988. "From Exiles to Minorities: The Politics of Cuban-Americans." Pp. 81–98 in F. Chris Garcia (ed.), *Latinos and the Political System.* Notre Dame, Ind.: University of Notre Dame Press.

Dent, David J. 1992. "The New Black Suburbs." *New York Times Magazine* (June 14):18–25.
de Paor, Liam. 1970. *Divided Ulster.* Harmondsworth, England: Pelican.
Deroche, Constance, and John Deroche. 1991. "Black and White: Racial Construction in Television Police Dramas." *Canadian Ethnic Studies* 23(3):69–91.
Destexhe, Alain. 1995. *Rwanda and Genocide in the Twentieth Century.* New York: New York University Press.
de St. Jorre, John. 1977. *A House Divided: South Africa's Uncertain Future.* New York: Carnegie Endowment for International Peace.
Detroit Free Press. 1992. "Race: A Shifting Barrier." (October 10):1, 10A, 10B.
Detroit News. 1989. "One World, Two Views." (September 22):10A.
———. 1990. "Pointes' Past Bias Still Keeps Some Minorities Away." (February 28): 1A, 6A.
———. 1991. "Black, White Perceptions Differ." (May 22):8A.
Deutsch, Karl W. 1966. *Nationalism and Social Communication.* Cambridge, Mass.: MIT Press.
Deutsch, Martin. 1969. "Happenings on the Way Back to the Forum: Social Science, IQ, and Race Differences Revisited." *Harvard Educational Review* 39:523–557.
Devlin, Bernadette. 1977. "Playboy Interview." Pp. 97–106 in Charles Carlton (ed.), *Bigotry and Blood: Documents on the Ulster Troubles.* Chicago: Nelson-Hall.
De Vos, George, and Lola Romanucci-Ross (eds.). 1975. *Ethnic Identity: Cultural Continuities and Change.* Palo Alto, Calif.: Mayfield.
De Vos, George, and Hiroshi Wagatsuma. 1966. *Japan's Invisible Race: Caste in Culture and Personality.* Berkeley: University of California Press.
Diaz, Guarione M. (ed.). 1980. *Evaluation and Identification of Policy Issues in the Cuban Community.* Miami, Fla.: Cuban National Planning Council.
Dickie-Clark, H. F. 1972. "The Coloured Minority of Durban." Pp. 25–38 in Noel P. Gist and Anthony G. Dworkin (eds.), *The Blending of Races: Marginality and Identity in World Perspective.* New York: Wiley.
Dinnerstein, Leonard, Roger L. Nichols, and David M. Reimers. 1990. *Natives and Strangers: Blacks, Indians, and Immigrants in America.* 2d ed. New York: Oxford University Press.
Dinnerstein, Leonard, and David M. Reimers. 1988. *Ethnic Americans: A History of Immigration.* 3d ed. New York: Harper & Row.
Dion, Stéphane. 1992. "Explaining Quebec Nationalism." Pp. 77–121 in R. Kent Weaver (ed.), *The Collapse of Canada?* Washington, D.C.: Brookings Institution.
Dobzhansky, Theodosious. 1973. "Differences Are Not Deficits." *Psychology Today* (December):97–101.
Doherty, Paul. 1982. "The Geography of Unemployment." Pp. 225–247 in Frederick W. Boal and J. Neville H. Douglas (eds.), *Integration and Division: Geographical Perspectives on the Northern Ireland Problem.* London: Academic Press.
———. 1993. "Agape to Zoroastrian: Religious Denominations in Northern Ireland 1961 to 1991." *Irish Geography* 26:14–21.
Dollard, John. 1937. *Caste and Class in a Southern Town.* New Haven, Conn.: Yale University Press.
Dollard, John, et al. 1939. *Frustration and Aggression.* New Haven, Conn.: Yale University Press.
Donnan, Hastings, and Graham McFarlane. 1983. "Informal Social Organization." Pp. 110–135 in John Darby (ed.), *Northern Ireland: The Background to the Conflict.* Belfast: Appletree Press.
Dorman, James H. 1980. "Ethnic Groups and 'Ethnicity': Some Theoretical Considerations." *Journal of Ethnic Studies* 7:23–36.
Dosman, Edgar J. 1972. *Indians: The Urban Dilemma.* Toronto: McClelland & Stewart.
Dugard, John. 1980. "Political Options for South Africa and Implications for the West." Pp. 17–30 in Robert I. Rotberg and John Barratt (eds.), *Conflict and Compromise in South Africa.* Lexington, Mass.: Heath.

———. 1992. "The Law of Apartheid." Pp. 3–31 in John Dugard (ed.), *The Last Years of Apartheid: Civil Liberties in South Africa.* Ford Foundation.

Duncan, Otis Dudley, and Beverly Duncan. 1957. *The Negro Population of Chicago.* Chicago: University of Chicago Press.

Dunn, L. C. 1956. "Race and Biology." Pp. 31–67 in Leo Kuper (ed.), *Race, Science and Society.* New York: Columbia University Press.

du Pisanie, J. A., and L. Kritzinger. 1985. "The Federal Option." Pp. 443–478 in D. J. van Vuures et al. (eds.), *South Africa: A Plural Society in Transition.* Durban: Butterworths.

Duran, Khalid. 1995. "Bosnia: The Other Andalusia." Pp. 25–36 in Syed Z. Abedin and Ziauddin Sardar (eds.), *Muslim Minorities in the West.* London: Grey Seal.

Dwyer, Victor. 1994. "In Search of Unity." *Maclean's* (July 1):16–19.

Dye, Thomas R. 1995. *Who's Running America? The Clinton Years.* 6th ed. Englewood Cliffs, N.J.: Prentice-Hall.

Dzidzienyo, Anani. 1987. "Brazil." Pp. 23–42 in Jay A. Sigler (ed.), *International Handbook on Race and Race Relations.* New York: Greenwood Press.

Dzidzienyo, Anani, and Lourdes Casal. 1979. *The Position of Blacks in Brazilian and Cuban Society.* London: Minority Rights Group.

Edsall, Thomas Byrne, with Mary D. Edsall. 1992. *Chain Reaction: The Impact of Race, Rights, and Taxes on American Politics.* New York: Norton.

Edwards, John. 1995. *Affirmative Action in a Sectarian Society: Fair Employment Policy in Northern Ireland.* Aldershot, England: Avebury.

Edwards, R. Gary, and Jon Hughes. 1995. "44% Favour Decreased Immigration." *Gallup Poll* 55(June 15):1–2.

Egan, Timothy. 1990. "Irish Emigrés Wait to Emerge from Shadow of Illegality." *New York Times* (November 25):16.

Ehrlich, Howard J. 1973. *The Social Psychology of Prejudice.* New York: Wiley.

Elkin, Frederick, and Gerald Handel. 1978. *The Child and Society: The Process of Socialization.* New York: Random House.

Elkins, Stanley M. 1976. *Slavery: A Problem in American Institutional and Intellectual Life.* 3d ed. Chicago: University of Chicago Press.

Elliott, Jean Leonard (ed.). 1971. *Native Peoples.* Scarborough, Ont.: Prentice-Hall of Canada.

———. 1979. "Canadian Immigration: A Historical Assessment." Pp. 160–172 in Jean Leonard Elliott (ed.), *Two Nations, Many Cultures.* Scarborough, Ont.: Prentice-Hall of Canada.

——— (ed.). 1983. *Two Nations, Many Cultures.* 2d ed. Scarborough, Ont.: Prentice-Hall of Canada.

Elliott, Jean Leonard, and Augie Fleras. 1990. "Immigration and the Canadian Ethnic Mosaic." Pp. 51–76 in Peter S. Li (ed.), *Race and Ethnic Relations in Canada.* Toronto: Oxford University Press.

Employment and Immigration Canada. 1988. *Profiles of Canadian Immigration.* Ottawa: Minister of Supply and Services.

Engelberg, Stephen, and Tim Weiner. 1995. "Srebrenica: The Days of Slaughter." *New York Times* (October 29):1, 6–7.

Engelhardt, Tom. 1975. "Ambush at Kamikaze Pass." Pp. 522–531 in Norman R. Yetman and C. Hoy Steele (eds.), *Majority and Minority: The Dynamics of Racial and Ethnic Relations.* 2d ed. Boston: Allyn & Bacon.

Eschbach, Karl. 1995. "The Enduring and Vanishing American Indian: American Indian Population Growth and Intermarriage in 1990." *Ethnic and Racial Studies* 18:89–108.

Esman, Milton J. 1994. *Ethnic Politics.* Ithaca, N.Y.: Cornell University Press.

Espiritu, Yen Le. 1992. *Asian American Panethnicity: Bridging Institutions and Identities.* Philadelphia: Temple University Press.

Estrada, Leobardo F. 1976. "A Demographic Comparison of the Mexican Origin Population in the Midwest and Southwest." *Aztlan* 7:203–234.

Eversley, David. 1991. "Demography and Unemployment in Northern Ireland." Pp. 72–92 in Robert J. Cormack and Robert D. Osborne (eds.), *Discrimination and Public Policy in Northern Ireland.* Oxford: Clarendon Press.

Ezorsky, Gertrude. 1991. *Racism and Justice: The Case for Affirmative Action.* Ithaca, N.Y.: Cornell University Press.

Fabricant, Florence. 1993. "Riding Salsa's Coast-to-Coast Wave of Popularity." *New York Times* (June 2):B1, B7.

Fagan, Richard R., Richard A. Brody, and Thomas J. O'Leary. 1968. *Cubans in Exile.* Stanford, Calif.: Stanford University Press.

Fair Employment Commission for Northern Ireland. 1995. *Profile of the Workforce in Northern Ireland: Summary of the 1994 Monitoring Returns, Monitoring Report No. 5.* Belfast: Fair Employment Commission for Northern Ireland.

Farb, Peter. 1968. *Man's Rise to Civilization as Shown by the Indians of North America from Primeval Times to the Coming of the Industrial State.* New York: Dutton.

Farley, Reynolds. 1984. *Black and Whites: Narrowing the Gap?* Cambridge, Mass.: Harvard University Press.

———. 1985. "The Residential Segregation of Blacks from Whites: Trends, Causes, and Consequences." In U.S. Commission on Civil Rights, *Issues in Housing Discrimination: A Consultation/Hearing of the United States Commission on Civil Rights. Vol. 1: Papers Presented.* Washington, D.C.: U.S. Government Printing Office.

———. 1993. "The Common Destiny of Blacks and Whites: Observations about the Social and Economic Status of the Races." Pp. 197–233 in Herbert Hill and James E. Jones, Jr. (eds.), *Race in America: The Struggle for Equality.* Madison, Wis.: University of Wisconsin Press.

Farley, Reynolds, and Walter R. Allen. 1987. *The Color Line and the Quality of Life in America.* New York: Russell Sage Foundation.

Farley, Reynolds, Suzanne Bianchi, and Diane Colasanto. 1979. "Barriers to the Racial Integration of Neighborhoods: The Detroit Case." *Annals of the American Academy of Political and Social Science* 441:97–113.

Farley, Reynolds, and William H. Frey. 1992. *Changes in the Segregation of Whites from Blacks During the 1980s: Small Steps Toward a More Racially Integrated Society.* Ann Arbor, Mich.: Population Studies Center, University of Michigan.

Farley, Reynolds, and Albert Hermalin. 1972. "The 1960s: Decade of Progress for Blacks?" *Demography* 9:353–370.

Farley, Reynolds, Charlotte Steeh, Maria Krysan, Tara Jackson, and Keith Reeves. 1994. "Stereotypes and Segregation: Neighborhoods in the Detroit Area." *American Journal of Sociology* 100:750–780.

Farley, Reynolds, and Alma F. Taeuber. 1974. "Racial Segregation in the Public Schools." *American Journal of Sociology* 79:888–905.

Farley, Reynolds, et al. 1978. "Chocolate City, Vanilla Suburbs: Will the Trend Toward Racially Separate Communities Continue?" *Social Science Research* 7:319–344.

Farmer, James. 1967. "The Controversial Moynihan Report." Pp. 409–413 in Lee Rainwater and William L. Yancey (eds.), *The Moynihan Report and the Politics of Controversy.* Cambridge, Mass.: MIT Press.

Farnsworth, Clyde H. 1991. "Quebec's New Minority Issue: Blacks Charge Bias." *New York Times* (August 16):A7.

———. 1995. "Black in Quebec: Escaping the Streets." *New York Times* (October 13):A4.

Fauman, S. Joseph. 1958. "Occupational Selection Among Detroit Jews." Pp. 119–136 in Marshall Sklare (ed.), *The Jews: Social Patterns of an American Group.* Glencoe, Ill.: Free Press.

Feagin, Joe R., and Clairece Booher Feagin. 1978. *Discrimination American Style: Institutional Racism and Sexism.* Englewood Cliffs, N.J.: Prentice-Hall.

Featherman, David L., and Robert M. Hauser. 1976. "Changes in the Socioeconomic Stratification of the Races, 1962–1973." *American Journal of Sociology* 82:621–651.

Feit, E. 1967. "Community in a Quandary: The South African Jewish Community and 'Apartheid.'" *Race* 8:395–468.

Fenwick, Rudy. 1982. "Ethnic-Culture and Economic Structure: Determinants of French–English Earnings Inequality in Quebec." *Social Forces* 61:1–23.

Fernandes, Florestan. 1971. *The Negro in Brazilian Society.* New York: Atheneum.

Ferrante, Angela. 1977. "Racism? You Can't Argue with the Facts." *Maclean's* (February 7):18–21.

Fineman, Howard. 1995. "Race and Rage." *Newsweek* (April 3):23–33.

Fishman, Josua A. 1987. "What Is Happening to Spanish on the U.S. Mainland?" *Ethnic Affairs* 1(Fall):12–23.

Fitzpatrick, Joseph P. 1980. "Puerto Ricans." Pp. 858–867 in Stephen Thernstrom (ed.), *Harvard Encyclopedia of American Ethnic Groups.* Cambridge, Mass.: Harvard University Press.

———. 1987. *Puerto Rican Americans: The Meaning of Migration to the Mainland.* 2d ed. Englewood Cliffs, N.J.: Prentice-Hall.

———. 1995. "Puerto Rican New Yorkers, 1990." *Migration World* 23(1):16–19.

Fixico, Donald L. 1994. "Mining." Pp. 343–345 in Mary B. Davis (ed.), *Native America in the Twentieth Century.* New York: Garland.

Fleras, Augie, and Jean Leonard Elliott. 1992. *Multiculturalism in Canada: The Challenge of Diversity.* Scarborough, Ontario: Nelson Canada.

Foerster, Robert S. 1919. *The Italian Emigration of Our Times.* Cambridge, Mass.: Harvard University Press.

Fogel, Robert William, and Stanley L. Engerman. 1974. *Time on the Cross.* Boston: Little, Brown.

Fogel, Walter. 1965. *Education and Income of Mexican-Americans in the Southwest.* Advance Report 1, Mexican-American Study Project. Los Angeles: University of California, Los Angeles.

Foner, Nancy. 1987. *New Immigrants in New York.* New York: Columbia University Press.

Fontaine, Pierre-Michel. 1981. "Transnational Relations and Racial Mobilization: Emerging Black Movements in Brazil." Pp. 141–162 in John F. Stack, Jr. (ed.), *Ethnic Identities in a Transnational World.* Westport, Conn.: Greenwood.

Fortune. 1936. "Jews in America." 13(February): 79–85, 128–144.

———. 1995a. "500 Largest U.S. Corporations." 128 (May 15):F1–F22.

———. 1995b. "The 1,000 Ranked Within Industries." 128(May 15):F43–F64.

Frady, Marshall. 1968. *Wallace.* New York: World.

Francis, E. K. 1976. *Interethnic Relations.* New York: Elsevier.

Franklin, John Hope. 1980. *From Slavery to Freedom.* 5th ed. New York: Knopf.

Franklin, Raymond S., and Solomon Resnik. 1973. *The Political Economy of Racism.* New York: Holt, Rinehart & Winston.

Fraser, Morris. 1973. *Children in Conflict: Growing Up in Northern Ireland.* New York: Basic Books.

Fraser, Steven (ed.). 1995. *The Bell Curve Wars: Race, Intelligence, and the Future of America.* New York: Basic Books.

Frazier, E. Franklin. 1939. *The Negro Family in the United States.* Chicago: University of Chicago Press.

———. 1949. *The Negro in the United States.* New York: Macmillan.

Fredrickson, George M. 1971. *The Black Image in the White Mind.* New York: Harper & Row.

———. 1981. *White Supremacy: A Comparative Study in American and South African History.* New York: Oxford University Press.

Freedman, Maurice. 1955. "The Chinese in Southeast Asia." Pp. 388–411 in Andrew W. Lind (ed.), *Race Relations in World Perspective.* Honolulu, Hawaii: University of Hawaii Press.

Freyre, Gilberto. 1956. *The Masters and the Slaves.* New York: Knopf.

———. 1963a. *New World in the Tropics: The Culture of Modern Brazil.* New York: Vintage.

———. 1963b. "Ethnic Democracy: The Brazilian Example." *Americas* 15(December):1–6.

Frideres, James S. 1976. "Racism in Canada: Alive and Well." *Western Canadian Journal of Anthropology* 6:124–145.

———. 1983. *Native People in Canada.* 2d ed. Scarborough, Ont.: Prentice-Hall of Canada.

———. 1990. "Policies on Indian People in

Canada." Pp. 98–119 in Peter S. Li (ed.), *Race and Ethnic Relations in Canada.* Toronto: Oxford University Press.

———. 1993. *Native Peoples in Canada: Contemporary Conflicts.* 4th ed. Toronto: Prentice-Hall.

Fried, Morton H. 1965. "A Four-Letter Word That Hurts." *Saturday Review* (October 2):21–23, 35.

Frisbie, W. Parker, and Lisa Neidert. 1977. "Inequality and the Relative Size of Minority Populations: A Comparative Analysis." *American Journal of Sociology* 82:1007–1030.

Fugita, Stephen S., and David J. O'Brien. 1991. *Japanese American Ethnicity: The Persistence of Community.* Seattle, Wash.: University of Washington Press.

Furnivall, J. S. 1948. *Colonial Policy and Practice.* Cambridge, England: Cambridge University Press.

Gallagher, A. M., and S. Dunn. 1991. "Community Relations in Northern Ireland: Attitudes to Contact and Integration." Pp. 7–22 in Peter Stringer and Gillian Robinson (eds.), *Social Attitudes in Northern Ireland.* Belfast: Blackstaff Press.

Gallo, Patrick J. 1974. *Ethnic Alienation: The Italian-Americans.* Rutherford, N.J.: Fairleigh Dickinson University Press.

Gallup, George, Jr. 1982. *The Gallup Poll: Public Opinion 1981.* Wilmington, Del.: Scholarly Resources.

———. 1988. *The Gallup Poll: Public Opinion 1987.* Wilmington, Del.: Scholarly Resources.

———. 1995. *The Gallup Poll: Public Opinion 1994.* Wilmington, Del.: Scholarly Resources Inc.

Gallup Poll Monthly. 1994. "Religious Preference." (December):43.

Gallup Report. 1989. "Tolerance Indices." Nos. 282/283 (March/April).

Gambino, Richard. 1975. *Blood of My Blood: The Dilemma of the Italian-Americans.* Garden City, N.Y.: Anchor.

———. 1977. *Vendetta.* Garden City, N.Y.: Doubleday.

Gamson, William A. 1975. *The Strategy of Social Protest.* Homewood, Ill.: Dorsey.

Gans, Herbert J. 1958. "The Origin and Growth of a Jewish Community in the Suburbs: A Study of the Jews of Park Forest." Pp. 205–258 in Marshall Sklare (ed.), *The Jews: Social Patterns of an American Group.* New York: Free Press.

———. 1967a. *The Levittowners.* New York: Pantheon.

———. 1967b. "Some Comments on the History of Italian Migration and on the Nature of Historical Research." *International Migration Review* 1:5–9.

———. 1968. *People and Plans: Essays on Urban Problems and Solutions.* New York: Basic Books.

———. 1974. "Foreword." In Neil C. Sandberg, *Ethnic Identity and Assimilation: The Polish-American Community.* New York: Praeger.

———. 1979. "Symbolic Ethnicity: The Future of Ethnic Groups and Cultures in America." *Ethnic and Racial Studies* 2:1–20.

———. 1982. *The Urban Villagers: Group and Class in the Life of Italian Americans.* Updated and expanded ed. New York: Free Press.

Garcia, F. Chris, and Rudolph P. de la Garza. 1977. *The Chicano Political Experience: Three Perspectives.* North Scituate, Mass.: Duxbury.

Garcia, John A., and Carlos H. Arce. 1988. "Political Orientations and Behaviors of Chicanos: Trying to Make Sense Out of Attitudes and Participation." Pp. 125–151 in F. Chris Garcia (ed.), *Latinos in the Political System.* Notre Dame, Ind.: University of Notre Dame Press.

Garcia, Juan Ramon. 1980. *Operation Wetback.* Westport, Conn.: Greenwood.

Garcia-Passalacqua, Juan Manuel. 1994. "The Puerto Ricans: Migrants or Commuters?" Pp. 103–113 in Carlos Antonio Torre, Hugo Rodríguez Vecchini, and William Burgos (eds.), *The Commuter Nation: Perspectives on Puerto Rican Migration.* Río Piedras, Puerto Rico: Editorial de la Universidad de Puerto Rico.

Garcia-Zamor, Jean-Claude. 1970. "Social Mobility of Negroes in Brazil." *Journal of Inter-American Studies* 12:242–254.

Gardner, Robert W., Bryant Robey, and Peter C. Smith. 1985. "Asian Americans: Growth, Change, and Diversity." *Population Bulletin* 40(October):1–43.

Garrett, J. T. 1994. "Health." Pp. 233–237 in Mary B. Davis (ed.), *Native America in the Twentieth Century.* New York: Garland.

Garson, Jean-Pierre. 1992. "Migration and Interdependence: The Migration System Between France and Africa." Pp. 80–93 in Mary M. Kritz, Lin Lean Lim, and Hania Zlotnik (eds.), *International Migration Systems: A Global Approach.* Oxford: Clarendon Press.

Gedmin, Jeffrey. 1994. "Germany's Emerging Multiethnic Society." *American Enterprise* 5(July/August):21–25.

Geertz, Clifford. 1994. "Primordial and Civic Ties." Pp. 29–34 in John Hutchinson and Anthony D. Smith (eds.), *Nationalism.* Oxford: Oxford University Press.

Gerdes, Wylie. 1993. "Chaldean Merchants Decry Slaying, Look for Answers." *Detroit Free Press* (December 2):1B, 2B.

Gershman, Carl. 1980. "A Matter of Class." *New York Times Magazine* (October 5):24, 92–96, 98–99, 102–105, 109.

Gerth, Hans. 1940. "The Nazi Party: Leadership and Composition." *American Journal of Sociology* 45:517–541.

Gerth, Hans, and C. Wright Mills. 1946. *From Max Weber.* New York: Oxford University Press.

Geschwender, James A. 1978. *Racial Stratification in America.* Dubuque, Iowa: W. C. Brown.

Gevisser, Mark. 1994. "Who Is a South African?" *New York Times* (April 26):A19.

Ghayur, M. Arif. 1981. "Muslims in the United States: Settlers and Visitors." *Annals of the American Academy of Political and Social Science* 454:150–163.

Giago, Tim, and Sharon Illoway. 1982. "Dying Too Young." *Perspectives* 15(Fall):29–33.

Gilbert, Dennis A. 1988. *Compendium of American Public Opinion.* New York: Facts on File.

Gilbert, G. M. 1951. "Stereotype Persistence and Change Among College Students." *Journal of Abnormal and Social Psychology* 46:245–254.

Giliomee, Hermann, and Lawrence Schlemmer. 1989. *From Apartheid to Nation-Building.* Cape Town: Oxford University Press.

Gillen, John. 1948. "Race Relations Without Conflict: A Guatemalan Town." *American Journal of Sociology* 53:337–343.

Ginsberg, Eli. 1978. "Jews in the American Economy: The Dynamics of Opportunity." Pp. 109–119 in Gladys Rosen (ed.), *Jewish Life in America: Historical Perspectives.* New York: Institute of Human Relations, American Jewish Committee.

Ginwala, Frene. 1977. *Indian South Africans.* London: Minority Rights Group.

Glass, Ruth. 1964. "Insiders-Outsiders: The Position of Minorities." Pp. 141–155 in *Transactions of the Fifth World Congress of Sociology.* Vol. 3. Louvain, Belgium: International Sociological Association.

Glater, Jonathan D., and Martha M. Hamilton. 1995. "Affirmative Action's Defenders." *Washington Post* (March 27–April 2):20.

Glazer, Nathan. 1957. *American Judaism.* Chicago: University of Chicago Press.

———. 1958. "The American Jews and the Attainment of Middle-Class Rank: Some Trends and Explanations." Pp. 138–146 in Marshall Sklare (ed.), *The Jews: Social Patterns of an American Group.* Glencoe, Ill.: Free Press.

———. 1971. "Blacks and Ethnic Groups: The Difference and the Political Difference It Makes." *Social Problems* 18:444–461.

———. 1975. *Affirmative Discrimination: Ethnic Identity and Public Policy.* New York: Basic Books.

———. 1993. "Is Assimilation Dead?" *Annals of the American Academy of Political and Social Science* 530(November):122–136.

Glazer, Nathan, and Daniel P. Moynihan. 1970. *Beyond the Melting Pot.* 2d ed. Cambridge, Mass.: MIT Press.

———. 1975. "Introduction." Pp. 1–26 in Nathan Glazer and Daniel P. Moynihan (eds.), *Ethnicity: Theory and Experience.* Cambridge, Mass.: Harvard University Press.

Gleason, Philip. 1964. "Immigration and American Catholic Intellectual Life." *Review of Politics* 26:147–173.

Glenn, Norvall D. 1963. "Occupational Benefits to Whites from the Subordination of Negroes." *American Sociological Review* 28:443–448.

———. 1966. "White Gains from Negro Subordination." *Social Problems* 14:159–178.

Glenny, Misha. 1993. *The Fall of Yugoslavia: The Third Balkan War.* New York: Penguin.

Glock, Charles Y., and Rodney Stark. 1966. *Christian Beliefs and Anti-Semitism.* New York: Harper & Row.

Gold, Steven J. 1994. "Soviet Jews in the United States." Pp. 3–57 in *American Jewish Year Book 1994.* New York: American Jewish Committee.

Goldberg, David Theo (ed.). 1994. *Multiculturalism: A Critical Reader.* Cambridge, Mass.: Basil Blackwell.

Goldberg, Jeffrey. 1995. "The Overachievers." *New York* (April 10):42–51.

Goldscheider, Calvin. 1986. *Jewish Continuity and Change: Emerging Patterns in America.* Bloomington, Ind.: Indiana University Press.

———. 1995. "Modernization, Ethnicity and the Post-War Jewish World." Pp. 130–143 in Robert S. Wistrich (ed.), *Terms of Survival: The Jewish World Since 1945.* London: Routledge.

Goldsen, Rose. 1977. *The Show and Tell Machine: How Television Works and Works You Over.* New York: Dell.

Goldstein, Sidney. 1980. "Jews in the United States: Perspectives from Demography." Pp. 3–59 in *American Jewish Year Book 1981.* New York: American Jewish Committee.

———. 1992. "Profile of American Jewry: Insights from the 1990 National Jewish Population Survey." Pp. 77–173 in *American Jewish Yearbook 1992.* New York: American Jewish Committee.

Golub, Jennifer L. 1990. *What Do We Know About Black Anti-Semitism?* New York: American Jewish Committee.

Gomez, Rudolph. 1977. "Mexican Americans in American Bureaucracy." Pp. 11–19 in Frank L. Baird (ed.), *Mexican Americans: Political Power, Influence or Resource.* Lubbock, Tex.: Texas Technical University Press.

Gonzalez, David. 1992. "What's the Problem with 'Hispanic'? Just Ask a 'Latino.'" *New York Times* (November 15):IV-5.

Goodman, Mary Ellen. 1964. *Race Awareness in Young Children.* Rev. ed. New York: Collier.

Gordon, Milton M. 1963. *Class in American Sociology.* New York: McGraw-Hill.

———. 1964. *Assimilation in American Life: The Role of Race, Religion, and National Origins.* New York: Oxford University Press.

———. 1975. "Toward a General Theory of Racial and Ethnic Group Relations." Pp. 84–110 in Nathan Glazer and Daniel P. Moynihan (eds.), *Ethnicity: Theory and Experience.* Cambridge, Mass.: Harvard University Press.

———. 1981. "Models of Pluralism: The New American Dilemma." *Annals of the American Academy of Political and Social Science* 454:178–188.

Gossett, Thomas F. 1963. *Race: The History of an Idea in America.* Dallas, Tex.: Southern Methodist University Press.

Gottesman, Dan. 1985. "Native People, Demography." P. 1212 in *The Canadian Encyclopedia.* Vol. II. Edmonton: Hurtig.

Gould, Stephen Jay. 1983. *The Mismeasure of Man.* New York: Norton.

———. 1984. "Human Equality Is a Contingent Fact of History." *Natural History* (November):26–32.

Graham, Richard. 1970. "Brazilian Slavery Reexamined: A Review Article." *Journal of Social History* 3:431–453.

Grebler, Leo, Joan W. Moore, and Ralph C. Guzman. 1970. *The Mexican-American People: The Nation's Second Largest Minority.* New York: Free Press.

Greeley, Andrew M. 1971. *Why Can't They Be Like Us?* New York: Dutton.

———. 1974. *Ethnicity in the United States: A Preliminary Reconnaissance.* New York: Wiley.

———. 1977. *The American Catholic: A Social Portrait.* New York: Harper & Row.

Greeley, Andrew M., and Paul B. Sheatsley. 1974. "Attitudes Toward Racial Integration." Pp. 241–250 in Lee Rainwater (ed.), *Social Problems and Public Policy.* Chicago: Aldine.

Greenbaum, William. 1974. "America in Search of a New Ideal: An Essay on the Rise of Pluralism." *Harvard Educational Review* 44:411–440.

Grenier, Guillermo J., and Alex Stepick III. 1992. "Introduction." Pp. 1–17 in Guillermo J. Grenier and Alex Stepick III (eds.), *Miami Now!: Immigration, Ethnicity, and Social Change.* Gainesville, Fla.: University Press of Florida.

Griffin, Rodman. 1992. "Illegal Immigration." *CQ Researcher* 2(April 24):361–384.

Grobsmith, Elizabeth S. 1994. "Prisons and Prisoners." Pp. 481–483 in Mary B. Davis (ed.), *Native America in the Twentieth Century.* New York: Garland.

Gross, Jane. 1989. "Diversity Hinders Asians' Influence." *New York Times* (June 25):18.

Grundy, Kenneth W. 1995. "South Africa: Putting Democracy to Work." *Current History* 94(April):172–176.

Gupte, Pranay. 1984. "Germany's Guest Workers." *New York Times Magazine* (August 19):88–91, 100–101.

Gurr, Ted Robert. 1993. *Minorities at Risk: A Global View of Ethnopolitical Conflicts.* Washington, D.C.: United States Institute of Peace Press.

Gutman, Herbert G. 1976. *The Black Family in Slavery and Freedom, 1750–1925.* New York: Pantheon.

Gwyn, Richard. 1995. *Nationalism Without Walls: The Unbearable Lightness of Being Canadian.* Toronto: McClelland & Stewart.

Hacker, Andrew. 1992. *Two Nations.* New York: Scribners.

———. 1994. "Education: Ethnicity and Achievement." Pp. 214–229 in Nicolaus Mills (ed.), *Debating Affirmative Action: Race, Gender, Ethnicity, and the Politics of Inclusion.* New York: Delta.

Hagan, William T. 1961. *American Indians.* Chicago: University of Chicago Press.

———. 1992. "Full Blood, Mixed Blood, Generic, and Ersatz: The Problem of Indian Identity." Pp. 278–288 in Roger L. Nichols (ed.), *The American Indian: Past and Present.* 4th ed. New York: McGraw-Hill.

Halpern, Ben. 1958. "America Is Different." Pp. 23–39 in Marshall Sklare (ed.), *The Jews: Social Patterns of an American Group.* Glencoe, Ill.: Free Press.

Hamada, Tarek. 1990. "Many Arab Immigrants Invest Hope in Detroit." *Detroit News* (December 2):1A, 10A.

Hamilton, Charles V. 1989. "On Parity and Political Empowerment." Pp. 111–120 in *The State of Black America 1989.* New York: National Urban League.

Hamilton, Richard. 1972. *Class and Politics in the United States.* New York: Wiley.

Handlin, Oscar. 1957. *Race and Nationality in American Life.* Boston: Little, Brown.

———. 1962. *The Newcomers: Negroes and Puerto Ricans in a Changing Metropolis.* Garden City, N.Y.: Doubleday.

Hanf, Theodore. 1991. "Reducing Conflict Through Cultural Autonomy: Karl Renner's Contribution." Pp. 33–52 in Uri Ra'anan, Maria Mesner, Keith Armes, and Kate Martin (eds.), *State and Nation in Multi-Ethnic Societies.* Manchester, England: Manchester University Press.

Hannerz, Ulf. 1969. *Soulside: Inquiries into Ghetto Culture and Community.* New York: Columbia University Press.

Harari, Oren, and David Beaty. 1989. *Lessons from South Africa: A New Perspective on Public Policy and Productivity.* New York: Harper & Row.

Harding, John. 1968. "Stereotypes." Pp. 259–261 in David L. Sills (ed.), *International Encyclopedia of the Social Sciences.* Vol. 15. New York: Macmillan.

Harney, Robert F., and Harold Troper. 1977. "Introduction." *Canadian Ethnic Studies* 9:1–5.

Harris, Louis. 1991a. "Three-to-One Majority Backs Affirmative Action If No Strict Quotas." *The Harris Poll* (June 2).

———. 1991b. "The Code Language That Now Pervades the Racial Issue." *The Harris Poll* (September 15).

Harris, Marvin. 1963. "Race Relations in Minas Velhas, a Community in the Mountain Region of Central Brazil." Pp. 47–81 in Charles Wagley (ed.), *Race and Class in Rural Brazil.* 2d ed. New York: Columbia University Press.

———. 1964. *Patterns of Race in the Americas.* New York: Norton.

Harris, Marvin, Josildeth Gomes Consorte, Joseph Lang, and Bryan Byrne. 1993. "Who Are the Whites?: Imposed Census Categories and the Racial Demography of Brazil." *Social Forces* 72:451–462.

Harris, Rosemary. 1972. *Prejudice and Tolerance in Ulster: A Study of Neighbours and Strangers in a Border Community.* Manchester, England: Manchester University Press.

Harris Survey, The. 1985. (February 18).

Harrison, David. 1981. *The White Tribe of Africa: South Africa in Perspective.* London: British Broadcasting Corporation.

Hartley, E. L. 1946. *Problems in Prejudice.* New York: Kings Crown.

Hasenbalg, Carlos A. 1985. "Race and Socioeconomic Inequalities in Brazil." Pp. 25–41 in Pierre-Michel Fontaine (ed.), *Race, Class, and Power in Brazil.* Los Angeles: Center for Afro-American Studies, University of California, Los Angeles.

Hasenbalg, Carlos, and Suellen Huntington. 1982/1983. "Brazilian Racial Democracy: Reality or Myth?" *Humboldt Journal of Social Relations* 10:129–142.

Hastings, Donald. 1969. "Japanese Emigration and Assimilation in Brazil." *International Migration Review* 3:32–53.

Hawkins, Freda. 1972. *Canada and Immigration: Public Policy and Public Concern.* Montreal: McGill-Queen's University Press.

———. 1989. *Critical Years in Immigration: Canada and Australia Compared.* Kingston and Montreal: McGill-Queen's University Press.

Heath, Shirley Brice. 1985. "Language Policies: Patterns of Retention and Maintenance." Pp. 257–282 in Walker Connor (ed.), *Mexican-Americans in Comparative Perspective.* Washington, D.C.: Urban Institute Press.

Hechter, Michael. 1975. *Internal Colonialism: The Celtic Fringe in British National Development, 1536–1966.* Berkeley: University of California Press.

Hedges, Chris. 1996. "Serbs Left in Bosnia See No Rest for the Dead." *New York Times* (January 18):A6.

Heer, David M. 1990. *Undocumented Mexicans in the United States.* New York: Cambridge University Press.

Heilman, Samuel C., and Steven M. Cohen. 1989. *Cosmopolitans & Parochials: Modern Orthodox Jews in America.* Chicago: University of Chicago Press.

Heisler, Martin O. 1991. "Hyphenating Belgium: Changing State and Regime to Cope with Cultural Division." Pp. 177–195 in Joseph V. Montville (ed.), *Conflict and Peacemaking in Multiethnic Societies.* New York: Lexington Books.

Hemming, John. 1987. *Amazon Frontier: The Defeat of the Brazilian Indians.* Cambridge, Mass.: Harvard University Press.

Henry, Frances. 1978. *The Dynamics of Racism in Toronto: Research Report.* Toronto: York University.

Henry, Frances, and Carol Tator. 1985. "Racism in Canada: Social Myths and Strategies for Change." Pp. 321–335 in Rita M. Bienvenue and Jay E. Goldstein (eds.), *Ethnicity and Ethnic Relations in Canada.* 2d ed. Toronto: Butterworths.

Herberg, Will. 1960. *Protestant-Catholic-Jew.* Garden City, N.Y.: Doubleday.

Herrnstein, Richard J., and Charles Murray. 1994. *The Bell Curve: Intelligence and Class Structure in American Life.* New York: Free Press.

Hershberg, Theodore, et al. 1979. "A Tale of Three Cities: Blacks and Immigrants in Philadelphia: 1850–1880, 1930 and 1970." *Annals of the American Academy of Political and Social Science* 441:55–81.

Herskovitz, Melville. 1941. *The Myth of the Negro Past.* New York: Harper & Brothers.

Heskin, Ken. 1980. *Northern Ireland: A Psychological Analysis.* New York: Columbia University Press.

Hewitt, Christopher. 1981. "Catholic Grievances, Catholic Nationalism and Violence in Northern Ireland During the Civil Rights Period: A Reconsideration." *British Journal of Sociology* 32:362–380.

Higham, John. 1963. *Strangers in the Land: Patterns of American Nativism 1860–1925*. 2d ed. New York: Atheneum.

Hilberg, Raul. 1979. *The Destruction of the European Jews.* New York: Harper & Row.

Hill, Daniel G. 1977. *Human Rights in Canada: A Focus on Racism.* Ottawa: Canadian Labour Congress.

Hill, Robert B. 1981. "The Economic Status of Black Americans." Pp. 1–59 in *The State of Black America 1981*. New York: National Urban League.

Hiller, Harry H. 1976. *Canadian Society: A Sociological Analysis.* Scarborough, Ont.: Prentice-Hall of Canada.

Hirschman, Charles. 1975. *Ethnic and Social Stratification in Peninsular Malaysia.* Washington, D.C.: American Sociological Association.

Hispanic Business. 1995. "The Corporate Elite." (January):11–13.

Hoare, Anthony G. 1982. "Problem Region and Regional Problem." Pp. 195–223 in Frederick W. Boal and J. Neville H. Douglas (eds.), *Integration and Division: Geographical Perspectives on the Northern Ireland Problem.* London: Academic Press.

Hochschild, Jennifer L. 1995. *Facing Up to the American Dream: Race, Class, and the Soul of the Nation.* Princeton, N.J.: Princeton University Press.

Hoffer, Eric. 1951. *The True Believer.* New York: Harper & Row.

Hoffman, Constance A., and Martin N. Marger. 1991. "Patterns of Immigrant Enterprise in Six Metropolitan Areas." *Sociology and Social Research* 75(April):144–157.

Holland, Jack. 1982. *Too Long a Sacrifice: Life and Death in Northern Ireland Since 1969.* New York: Penguin.

Hornblower, Margot. 1995. "Putting Tongues in Check." *Time* (October 9):40–50.

Horowitz, Donald L. 1975. "Ethnic Identity." Pp. 111–140 in Nathan Glazer and Daniel P. Moynihan (eds.), *Ethnicity: Theory and Experience.* Cambridge, Mass.: Harvard University Press.

———. 1991. *A Democratic South Africa?: Constitutional Engineering in a Divided Society.* Berkeley: University of California Press.

Horowitz, Irving Louis. 1964. *Revolution in Brazil.* New York: Dutton.

Horton, John. 1995. *The Politics of Diversity: Immigration, Resistance, and Change in Monterey Park, California.* Philadelphia: Temple University Press.

Hostetler, John A. 1993. *Amish Society.* 4th ed. Baltimore, Md.: Johns Hopkins University Press.

Howard, John R. 1970. *Awakening Minorities.* New Brunswick, N.J.: Transaction.

Howe, Irving. 1976. *World of Our Fathers.* New York: Simon & Schuster.

———. 1977. "The Limits of Ethnicity." *New Republic* (June 25):17–19.

———. 1991. "The Value of the Canon." *New Republic* (February 18):40–44, 46–47.

Huber, Joan, and William H. Form. 1973. *Income and Ideology.* New York: Free Press.

Hughes, David R., and Evelyn Kallen. 1974. *The Anatomy of Racism: Canadian Dimensions.* Montreal: Harvest House.

Hughes, Everett C. 1943. *French Canada in Transition.* Chicago: University of Chicago Press.

Hughes, Everett C., and Helen M. Hughes. 1952. *Where Peoples Meet: Racial and Ethnic Frontiers.* Glencoe, Ill.: Free Press.

Hugick, Larry. 1992. "Blacks See Their Lives Worsening." *Gallup Poll Monthly* (April):26–29.

Human Rights Watch. 1995. *Slaughter Among Neighbors: The Political Origins of Communal Violence.* New Haven: Yale University Press.

Hunt, Chester, and Lewis Walker. 1974. *Ethnic Dynamics.* Homewood, Ill.: Dorsey.

Hurvitz, Nathan. 1958. "Sources of Middle-Class Values of American Jews." *Social Forces* 37:117–123.

Hutchinson, Harry. 1963. "Race Relations in a Rural Community of the Bahian Reconcavo." Pp. 16–46 in Charles Wagley (ed.), *Race and Class in Rural Brazil.* 2d ed. New York: Columbia University Press.

Hutchinson, John. 1987. *The Dynamics of Cultural Nationalism.* London: Allen and Unwin.

Iacovetta, Franca. 1992. *Such Hardworking People: Italian Immigrants in Postwar Toronto.* Montreal and Kingston: McGill-Queen's University Press.

Ianni, Francis A. J. 1974. *Black Mafia: Ethnic Succession in Organized Crime.* New York: Simon & Schuster.

Ianni, Octavio. 1965. "Race and Class in Brazil." *Présence Africaine* 25:105–119.

Ignatieff, Michael. 1993. *Blood and Belonging: Journeys into the New Nationalism.* Toronto: Penguin.

Institute for Social Research. 1964. *Discrimination Without Prejudice. A Study of Promotion Practices in Industry.* Ann Arbor, Mich.: Institute for Social Research, University of Michigan.

Institute for the Study of Conflict. 1972. *The Ulster Debate.* London: Bodley Head.

Instituto Brasileiro de Geografia e Estatística. 1981. *IX Recenseamento Geral do Brasil — 1980.* Vol. 1—Tomo 2. Rio de Janeiro: IBGE.

———. 1992. *Anuário Estatístico do Brasil 1992.* Rio de Janeiro: IBGE.

Iorizzo, Luciano J., and Salvatore Mondello. 1980. *The Italian Americans.* Boston: Twayne.

Irwin, Colin. 1994. "The Myths of Segregation." Pp. 104–118 in Adrian Guelke (ed.), *New Perspectives on the Northern Ireland Conflict.* Aldershot, England: Avebury.

Isaacs, Harold R. 1989. *Idols of the Tribe: Group Identity and Political Change.* Cambridge, Mass.: Harvard University Press.

Isaacs, Stephen D. 1974. *Jews and American Politics.* Garden City, N.Y.: Doubleday.

Isajiw, Wsevolod W., Aysan Sev'er, and Leo Driedger. 1993. "Ethnic Identity and Social Mobility: A Test of the 'Drawback Model.'" *Canadian Journal of Sociology* 18:179–198.

Jackson, Harold, and Anne McHardy. 1984. *The Two Irelands.* Rev. ed. London: Minority Rights Group.

Jacoby, Russell, and Naomi Glauberman (eds.). 1995. *The Bell Curve Debate: History, Documents, Opinions.* New York: Times Books.

Jacoby, Susan. 1975. "A Dream Grows in Brooklyn." *New York Times Magazine* (February 23):11, 41–50.

Jaffe, A. J., Ruth M. Cullen, and Thomas D. Boswell. 1980. *The Changing Demography of Spanish Americans.* New York: Academic Press.

Jaret, Charles. 1979. "Recent Patterns of Chicago Jewish Residential Mobility." *Ethnicity* 6:235–248.

Jencks, Christopher, et al. 1972. *Inequality: A Reassessment of the Effect of Family and Schooling in America.* New York: Basic Books.

Jensen, Arthur. 1969. "How Much Can We Boost IQ and Scholastic Achievement?" *Harvard Educational Review* 39:1–123.

———. 1973a. "The Differences Are Real." *Psychology Today* (December):80–86.

———. 1973b. *Educability and Group Differences.* New York: Harper & Row.

Jiobu, Robert M. 1988. *Ethnicity and Assimilation.* Albany, N.Y.: State University of New York Press.

Johnson, Colleen Leahy. 1985. *Growing Up and Growing Old in Italian-American Families.* New Brunswick, N.J.: Rutgers University Press.

Johnson, Julie. 1989. "Asian-Americans Press Fight for Wider Top-College Door." *New York Times* (September 9):1, 7.

Johnson, Tim. 1995. "The Dealer's Edge: Gaming in the Path of Native America." *Native Americas* 12(Summer):16–25.

Joint Center for Political and Economic Studies. 1992. *Black Elected Officials: A National Roster.* Washington, D.C.: Joint Center for Political and Economic Studies.

Jones, Edward W., Jr. 1986. "Black Managers: The Dream Deferred." *Harvard Business Review* (May–June):84–93.

Jones, Emrys. 1960. *A Social Geography of Belfast.* London: Oxford University Press.

Jones, Maldwyn A. 1960. *American Immigration.* Chicago: University of Chicago Press.

Jordan, Winthrop. 1969. *White Over Black.* Baltimore, Md.: Penguin.

———. 1995. "Slavery and the Jews." *Atlantic Monthly* 276(September):109–114.

Joy, Richard J. 1972. *Languages in Conflict.* Toronto: McClelland & Stewart.

Juliani, Richard N. 1994. "Identity and Ethnicity: The Italian Case." Pp. 54–57 in Lidio F. Tomasi, Piero Gastaldo, and Thomas Row (eds.), *The Columbus People: Perspectives in Italian Immigration to the Americas and Australia.* New York: Center for Migration Studies.

Juteau Lee, Danielle. 1979. "The Evolution of Nationalism in Quebec." Pp. 60–73 in Jean Leonard Elliott (ed.), *Two Nations, Many Cultures.* 2d ed. Scarborough, Ont.: Prentice-Hall of Canada.

Kahlenberg, Richard. 1995. "Class, Not Race." *New Republic* (April 3):21–27.

Kallen, Evelyn. 1995. *Ethnicity and Human Rights in Canada.* 2d ed. Toronto: Oxford University Press.

Kantrowitz, Nathan. 1973. *Ethnic and Racial Segregation in the New York Metropolis.* New York: Praeger.

———. 1979. "Racial and Ethnic Residential Segregation in Boston 1830–1970." *Annals of the American Academy of Political and Social Science* 441:41–54.

Karasch, Mary. 1975. "From Porterage to Proprietorship: African Occupations in Rio de Janeiro, 1808–1850." Pp. 369–393 in Stanley L. Engerman and Eugene D. Genovese (eds.), *Race and Slavery in the Western Hemisphere: Quantitative Studies.* Princeton, N.J.: Princeton University Press.

Kasarda, John. 1985. "Urban Change and Minority Opportunities." Pp. 46–61 in Paul Peterson (ed.), *The New Urban Reality.* Washington, D.C.: Brookings Institution.

Kashena, Rita. 1980. "The Role of American Indians in Motion Pictures." Pp. 106–111 in Gretchen M. Bataille and Charles L. P. Silet (eds.), *The Pretend Indians: Images of Native Americans in the Movies.* Ames, Iowa: Iowa State University Press.

Katz, Daniel, and Kenneth Braly. 1933. "Racial Stereotypes of One Hundred College Students." *Journal of Abnormal and Social Psychology* 28:280–290.

Katz, Diane. 1992. "Canada Hears 'Same Voice' As Los Angeles." *Detroit News* (May 6): 1A, 6A.

Keely, Charles B. 1983. "Immigration and the American Future." Pp. 28–65 in Lance Liebman (ed.), *Ethnic Relations in America.* Englewood Cliffs, N.J.: Prentice-Hall.

Keller, Bill. 1992. "Schools Are Slightly More Equal, But Still Failures." *New York Times* (November 7):2.

———. 1993a. "Mandela's Group Accepts 5 Years of Power-Sharing." *New York Times* (February 19):A1, A4.

———. 1993b. "Afrikaner Rightists Vow to Resist Rule by Blacks." *New York Times* (May 3):A12.

———. 1995. "After Apartheid, Change Lags Behind Expectations." *New York Times* (April 27):A1, A6.

Keller, Suzanne. 1953. "The Social Origins and Career Lines of Three Generations of American Business Leaders." Ph.D. dissertation, Columbia University.

Kelner, Merrijoy. 1970. "Ethnic Penetration into Toronto's Elite Structure." *Canadian Review of Sociology and Anthropology* 7:128–137.

Kennedy, Randall. 1994. "Persuasion and Distrust: The Affirmative Action Debate." Pp. 48–67 in Nicolaus Mills (ed.), *Debating Affirmative Action: Race, Gender, Ethnicity, and the Politics of Inclusion.* New York: Delta.

Kennedy, Robert E., Jr. 1973. *The Irish: Emigration, Marriage, and Fertility.* Berkeley: University of California Press.

Kennedy, Ruby Jo Reeves. 1944. "Single or Triple Melting Pot?: Intermarriage Trends in New Haven." *American Journal of Sociology* 49:331–339.

———. 1952. "Single or Triple Melting Pot?: Intermarriage in New Haven, 1870–1950." *American Journal of Sociology* 58:56–59.

Kessner, Thomas. 1977. *The Golden Door: Italian and Jewish Immigrant Mobility in New York City 1880–1915.* New York: Oxford University Press.

Kiester, Edwin, Jr. 1968. *The Case of the Missing Executive.* New York: Institute of Human Relations, American Jewish Committee.

Kikumura, Akemi, and Harry H. L. Kitano. 1973. "Interracial Marriage: A Picture of the Japanese-Americans." *Journal of Social Issues* 29:67–81.

Kilborn, Peter T. 1992. "Sad Distinction for the Sioux: Homeland Is No. 1 in Poverty." *New York Times* (September 20):1, 14.

Killian, Lewis M. 1953. "The Adjustment of Southern White Migrants to Northern Urban Norms." *Social Forces* 32:66–69.

———. 1975. *The Impossible Revolution, Phase II: Black Power and the American Dream.* New York: Random House.

———. 1981. "Black Power and White Reactions: The Revitalization of Race-thinking in the United States." *Annals of the American Academy of Political and Social Science* 454:42–54.

———. 1985. *White Southerners.* Rev. ed. Amherst: University of Massachusetts Press.

Kim, Illsoo. 1981. *New Urban Immigrants: The Korean Community in New York.* Princeton, N.J.: Princeton University Press.

———. 1987. "The Koreans: Small Business in an Urban Frontier." Pp. 219–242 in Nancy Foner (ed.), *New Immigrants in New York.* New York: Columbia University Press.

King, Martin Luther, Jr. 1964. *Why We Can't Wait.* New York: Harper & Row.

Kinsley, Michael. 1991. "Class, Not Race." *New Republic* (August 19-26):4.

Kirscht, John P., and Ronald C. Dillehay. 1967. *Dimensions of Authoritarianism: A Review of Research and Theory.* Lexington, Ky.: University of Kentucky Press.

Kirschten, Dick. 1992. "Building Blocs." *National Journal* 24(September 26):2173–2177.

Kitano, Harry. 1976. *Japanese Americans: The Evolution of a Subculture.* 2d ed. Englewood Cliffs, N.J.: Prentice-Hall.

Kitano, Harry H. L., and Roger Daniels. 1995. *Asian Americans: Emerging Minorities,* 2d. ed. Englewood Cliffs, N.J.: Prentice-Hall.

Klaaste, Aggrey. 1984. "Exiles in Their Native Land." *New York Times Magazine* (June 24):34–35, 51–53, 68–69.

Klausner, Samuel Z. 1988. *Succeeding in Corporate America: The Experience of Jewish M.B.A.'s.* New York: American Jewish Committee.

Kleiman, Dena. 1983. "Less Than 40% of Jews in Survey Observe Sabbath." *New York Times* (February 6):45.

Klein, Herbert S. 1969. "The Colored Freedman in Brazilian Slave Society." *Journal of Social History* 3:30–52.

Klineberg, Otto. 1968. "Prejudice: The Concept." Pp. 439–447 in David L. Sills (ed.), *International Encyclopedia of the Social Sciences.* Vol. 12. New York: Macmillan.

Knight, Franklin. 1974. *The African Dimension in Latin American Societies.* New York: Macmillan.

Knowlton, Clark. 1972. "The New Mexican Land War." Pp. 258–270 in Edward Simmen (ed.), *Pain and Promise: The Chicano Today.* New York: New American Library.

———. 1975. "Neglected Chapters in Mexican-American History." Pp. 19–59 in Gus Tyler (ed.), *Mexican-Americans Tomorrow.* Albuquerque, N. Mex.: University of New Mexico Press.

Kohn, Hans. 1944. *The Idea of Nationalism: A Study in Its Origins and Background.* New York: Macmillan.

Konvitz, Milton B. 1978. "The Quest for Equality and the Jewish Experience." Pp. 28–60 in Gladys Rosen (ed.), *Jewish Life in America: Historical Perspectives.* New York: Institute of Human Relations, American Jewish Committee.

Korman, Abraham K. 1988. *The Outsiders: Jews and Corporate America.* Lexington, Mass.: Lexington Books.

Kosmin, Barry A., et al. 1991. *Highlights of the CJF 1990 National Jewish Population Survey.* New York: Council of Jewish Federations.

Koster, Henry. 1966 (1816). *Travels in Brazil.* Carbondale, Ill.: Southern Illinois University Press.

Kovel, Joel. 1970. *White Racism: A Psychohistory.* New York: Pantheon.

Kramer, Judith R., and Seymour Leventman. 1961. *Children of the Gilded Ghetto.* New Haven, Conn.: Yale University Press.

Kristof, Nicholas D. 1995. "Japanese Outcasts Better Off Than in Past but Still Outcast." *New York Times* (November 30):A1, A8.

Kristol, Irving. 1970. "The Negro Today Is Like the Immigrant Yesterday." Pp. 139–157 in Nathan Glazer (ed.), *Cities in Trouble.* Chicago: Quadrangle.

Kruijt, J. P. 1974. "The Netherlands: The Influence of Denominationalism on Social Life and Organizational Patterns." Pp. 128–136 in Kenneth D. McRae (ed.), *Consociational Democracy: Political Accommodation in Segmented Societies.* Toronto: McClelland & Stewart.

Kuper, Leo. 1965. *An African Bourgeoisie: Race, Class, and Politics in South Africa.* New Haven, Conn.: Yale University Press.

———. 1968. "Segregation." Pp. 144–150 in David L. Sills (ed.), *International Encyclopedia of the Social Sciences.* Vol. 14. New York: Macmillan.

———. 1969. "Ethnic and Racial Pluralism: Some Aspects of Polarization and Depluralization." Pp. 459–487 in Leo Kuper and M. G. Smith, *Pluralism in Africa.* Berkeley: University of California Press.

———. 1981. *Genocide.* New York: Penguin.

Kuper, Leo, and M. G. Smith (eds.). 1969. *Pluralism in Africa.* Berkeley: University of California Press.

Kurokawa, Minako (ed.). 1970. *Minority Responses.* New York: Random House.

Kutner, Bernard, Carol Wilkens, and Penny Rechtman Yarrow. 1952. "Verbal Attitudes and Overt Behavior." *Journal of Abnormal and Social Psychology* 47:649–652.

Ladner, Robert A., et al. 1981. "The Miami Riots of 1980: Antecedent Conditions, Community Responses and Participant Characteristics." Pp. 171–214 in Louis Kriesberg (ed.), *Research in Social Movements, Conflicts and Change.* Vol. 4. Greenwich, Conn.: JAI Press.

Lambert, Bruce. 1988. "U.S. Limits Stays of Foreign Nurses." *New York Times* (April 10):1, 15.

Lambert, Richard D. 1981. "Ethnic/Racial Relations in the United States in Comparative Perspective." *Annals of the American Academy of Political and Social Science* 454:189–205.

Landry, Bart. 1987. *The New Black Middle Class.* Berkeley: University of California Press.

Lane, Robert E. 1962. *Political Ideology.* New York: Free Press.

Langberg, Mark, and Reynolds Farley. 1985. "Residential Segregation of Asian Americans in 1980." *Sociology and Social Research* 70(October):71–75.

Langdon, Steven. 1995. "Proof of What Hope Can Do." *Toronto Star* (July 17):A13.

Langer, Gary. 1989. "Polling on Prejudice: Questionable Questions." *Public Opinion* 12(May/June):18–19, 57.

LaPiere, Richard T. 1934. "Attitudes vs. Actions." *Social Forces* 13:230–237.

LaRuffa, Anthony L. 1982. "Media Portrayals of Italian-Americans." *Ethnic Groups* 4:191–206.

Lautard, Hugh, and Neil Guppy. 1990. "The Vertical Mosaic Revisited: Occupational Differentials Among Canadian Ethnic Groups." Pp. 189–208 in Peter S. Li (ed.), *Race and Ethnic Relations in Canada.* Toronto: Oxford University Press.

Learsi, Rufus. 1954. *The Jews in America: A History.* Cleveland, Ohio: World.

Lee, Everett S. 1966. "A Theory of Migration." *Demography* 6:27–37.

Legendre, Camille. 1980. *French Canada in Crisis: A New Society in the Making?* London: Minority Rights Group.

Legum, Colin. 1980. "South Africa in the Contemporary World." Pp. 281–296 in Robert M. Price and Carl G. Rotberg (eds.), *The Apartheid Regime: Political Power and Racial Domination.* Berkeley: Institute of International Studies, University of California.

Lemann, Nicholas. 1988. "Growing Pains." *Atlantic Monthly* 261(January):57–62.

———. 1991. "The Other Underclass." *Atlantic Monthly* 268(December):96–110.

Lemon, Anthony. 1987. *Apartheid in Transition.* Aldershot, England: Gower.

———. 1994. "Educational Desegregation." Pp. 91–110 in John D. Brewer (ed.), *Restructuring South Africa.* New York: St. Martin's.

Lenski, Gerhard. 1963. *The Religious Factor.* Rev. ed. Garden City, N.Y.: Doubleday.

———. 1966. *Power and Privilege: A Theory of Social Stratification.* New York: McGraw-Hill.

Lenski, Gerhard, and Jean Lenski. 1982. *Human Societies.* 4th ed. New York: McGraw-Hill.

Lever, Henry. 1968. "Ethnic Preferences of White Residents in Johannesburg." *Sociology and Social Research* 52:157–173.

———. 1978. *South African Society.* Johannesburg: Jonathan Ball.

Lévesque, René. 1968. *An Option for Quebec.* Toronto: McClelland & Stewart.

Levin, Jack, and William Levin. 1982. *The Functions of Discrimination and Prejudice.* 2d ed. New York: Harper & Row.

Levine, Gene N., and Colbert Rhodes. 1981. *The Japanese American Community: A Three-Generation Study.* New York: Praeger.

Levine, Marc V. 1990. *The Reconquest of Montreal: Language Policy and Social Change in a Bilingual City.* Philadelphia: Temple University Press.

Levine, Naomi, and Martin Hochbaum (eds.). 1974. *Poor Jews: An American Awakening.* New Brunswick, N.J.: Transaction.

Levitan, Sar A., William B. Johnston, and Robert Taggart. 1975. *Still a Dream: The Changing Status of Blacks Since 1960.* Cambridge, Mass.: Harvard University Press.

Levy, Mark R., and Michael S. Kramer. 1973. *The Ethnic Factor: How America's Minorities Decide Elections.* New York: Simon & Schuster.

Lewin, Julius. 1963. *Politics and Law in South Africa.* London: Merlin.

Lewin, Tamar. 1992. "Study Points to Increase in Tolerance of Ethnicity." *New York Times* (January 8):A10.

Lewis, Hylan. 1967. "The Family: Resources for Change." Pp. 314–343 in Lee Rainwater and William L. Yancey (eds.), *The Moynihan Report and the Politics of Controversy.* Cambridge, Mass.: MIT Press.

Lewis, Oscar. 1961. *The Children of Sanchez.* New York: Random House.

———. 1965. *La Vida.* New York: Random House.

———. 1966. "The Culture of Poverty." *Scientific American* 215(October):19–25.

Lichter, S. Robert, and Linda Lichter. 1982. *Italian-American Characters in Television Entertainment.* West Hempstead, N.Y.: Commission for Social Justice.

Lichter, S. Robert, et al. 1987. "Prime-Time Prejudice: TV's Images of Blacks and Hispanics." *Public Opinion* 10(July/August):13–16.

Lieberson, Stanley. 1961. "A Societal Theory of Race Relations." *American Sociological Review* 26:902–910.

———. 1980. *A Piece of the Pie: Blacks and White Immigrants Since 1880.* Berkeley: University of California Press.

Lieberson, Stanley, and Mary C. Waters. 1988. *From Many Strands: Ethnic and Racial Groups in Contemporary America.* New York: Russell Sage Foundation.

Liebman, Charles S. 1973. "American Jewry: Identity and Affiliation." Pp. 127–152 in David Sidorsky (ed.), *The Future of the Jewish Community in America.* New York: Basic Books.

Light, Ivan, and Edna Bonacich. 1988. *Immigrant Entrepreneurs: Koreans in Los Angeles 1965–1982.* Berkeley: University of California Press.

Lijphart, Arend. 1975. "The Northern Ireland Problem: Cases, Theories and Solutions." *British Journal of Political Science* 5:83–106.

———. 1977. *Democracy in Plural Societies: A Comparative Exploration.* New Haven, Conn.: Yale University Press.

———. 1980. "Federal, Confederal, and Consociational Options for the South African Plural Society." Pp. 51–75 in Robert I. Rotberg and John Barratt (eds.), *Conflict and Compromise in South Africa.* Lexington, Mass.: Heath.

Lippmann, Walter. 1922. *Public Opinion.* New York: Macmillan.

Lipset, Seymour Martin. 1960. *Political Man.* New York: Anchor.

———. 1968. *Revolution and Counterrevolution.* New York: Basic.

———. 1987. "Blacks and Jews: How Much Bias?" *Public Opinion* 10(July/August):4–5, 57–58.
———. 1990. *Continental Divide: The Values and Institutions of the United States and Canada.* New York: Routledge.
Lipset, Seymour Martin, and Earl Raab. 1995. *Jews and the New American Scene.* Cambridge, Mass.: Harvard University Press.
Lipton, Merle. 1987. "Reform: Destruction or Modernization of Apartheid?" Pp. 34–55 in Jesmond Blumenfeld (ed.), *South Africa in Crisis.* London: Croom Helm.
Livingston, John C. 1979. *Fair Game? Inequality and Affirmative Action.* San Francisco: Freeman.
Logan, John R. 1988. "Realities of Black Suburbanization." Pp. 231–241 in Roland L. Warren and Larry Lyon (eds.), *New Perspectives on the American Community.* Chicago: Dorsey.
Long, James E. 1977. "Productivity, Employment Discrimination, and the Relative Economic Status of Spanish Origin Males." *Social Science Quarterly* 58:357–373.
Lopez, Adalberto. 1980. "The Puerto Rican Diaspora: A Survey." Pp. 313–343 in Adalberto Lopez (ed.), *The Puerto Ricans: Their History, Culture, and Society.* Cambridge, Mass.: Schenkman.
Lopreato, Joseph. 1970. *Italian Americans.* New York: Random House.
Luebke, Frederick C. 1983. "A Prelude to Conflict: The German Ethnic Group in Brazilian Society, 1890–1917." *Ethnic and Racial Studies* 6:1–17.
———. 1987. *Germans in Brazil: A Comparative History of Cultural Conflict During World War I.* Baton Rouge, La.: Louisiana State University Press.
Lurie, Nancy Oestreich. 1991. "The American Indian: Historical Background." Pp. 132–146 in Norman R. Yetman (ed.), *Majority and Minority: The Dynamics of Race and Ethnicity in American Life.* 5th ed. Boston: Allyn & Bacon.
Lusane, Clarence. 1989. "Black Political Power in the 1990's." *Black Scholar* 20(January/February):38–42.
Lyman, Stanford. 1968a. "The Race Relations Cycle of Robert Park." *Pacific Sociological Review* 11:16–22.
———. 1968b. "Contrasts in the Community Organizations of Chinese and Japanese in North America." *Canadian Review of Sociology and Anthropology* 5:51–67.
———. 1970. "Strangers in the Cities: The Chinese on the Urban Frontier." Pp. 61–100 in Charles Wollenberg (ed.), *Ethnic Conflict in California History.* Los Angeles: Tinnon-Brown.
———. 1974. *Chinese Americans.* New York: Random House.
Maasdorp, Gavin. 1980. "Forms of Partition." Pp. 107–146 in Robert I. Rotberg and John Barratt (eds.), *Conflict and Compromise in South Africa.* Lexington, Mass.: Heath.
MacDonald, J. Fred. 1983. *Blacks and White TV: Afro-Americans in Television Since 1948.* Chicago: Nelson-Hall.
Macías, Ysidro. 1972. "The Chicano Movement." Pp. 137–143 in Edward Simmen (ed.), *Pain and Promise: The Chicano Today.* New York: New American Library.
Mack, Raymond W. 1963. *Race, Class, and Power.* New York: American.
Mackie, Marlene. 1973. "Arriving at 'Truth' by Definition: The Case of Stereotype Inaccuracy." *Social Problems* 20:431–447.
———. 1980. "Ethnic Stereotypes and Prejudice: Alberta Indians, Hutterites and Ukrainians." Pp. 233–246 in Jay E. Goldstein and Rita M. Bienvenue (eds.), *Ethnicity and Ethnic Relations in Canada.* Toronto: Butterworths.
Maclean's. 1989. "*Maclean's*/Decima Poll: A North-South Dialogue." 102(July 3):48–50.
———. 1996. "*Maclean's*/CBC News Poll." (January 1):32–33.
MacLennan, Hugh. 1945. *Two Solitudes.* New York: Duell, Sloan & Pearce.
Makabe, Tomoko. 1981. "The Theory of the Split Labor Market: A Comparison of the Japanese Experience in Brazil and Canada." *Social Forces* 59:786–809.

Mallaby, Sebastian. 1992. *After Apartheid: The Future of South Africa.* New York: Times Books.
Mangiafico, Luciano. 1988. *Contemporary American Immigrants: Patterns of Filipino, Korean, and Chinese Settlement in the United States.* New York: Praeger.
Mannheim, Karl. 1936. *Ideology and Utopia.* New York: Harcourt, Brace & World.
Manning, Wendy, and William O'Hare. 1988. "The Best Metros for Asian-American Businesses." *American Demographics* 10(August):34–37, 59.
Marden, Charles F., and Gladys Meyer. 1978. *Minorities in American Society.* 5th ed. New York: Van Nostrand.
Marger, Martin. 1974. *The Force of Ethnicity: A Study of Urban Elites.* Journal of University Studies, Ethnic Monograph Series. Detroit, Mich.: Wayne State University.
———. 1979. "A Reexamination of Gordon's Ethclass." *Sociological Focus* 11:21–32.
———. 1987. *Elites and Masses: An Introduction to Political Sociology.* 2d ed. Belmont, Calif.: Wadsworth.
———. 1989. "Factors of Structural Pluralism in Multiethnic Societies: A Comparative Case Study." *International Journal of Group Tensions* 19:52–68.
———. 1993. "The Media as a Power Institution." Pp. 238–249 in Marvin E. Olsen and Martin N. Marger (eds.), *Power in Modern Societies.* Boulder, Colo.: Westview.
Margolis, Mac. 1988. "Brazil's Blacks Look Anew at Issue of Race." *Washington Post* (May 13):A25–A26.
Marks, Jonathan. 1995. *Human Biodiversity: Genes, Race, and History.* New York: Aldine de Gruyter.
Marriott, Michel. 1987. "In Jersey City, Indians Face Rise in Violence and Hatred." *New York Times* (October 12):II, 1.
Marston, Wilfred G., and Thomas L. Van Valey. 1979. "The Role of Residential Segregation in the Assimilation Process." *Annals of the American Academy of Political and Social Science* 441:13–25.
Martinez, Thomas M. 1972. "Advertising and Racism: The Case of the Mexican American." Pp. 94–105 in Edward Simmen (ed.), *Pain and Promise: The Chicano Today.* New York: New American Library.
Martire, Greg, and Ruth Clark. 1982. "Anti-Semitism in America." *Public Opinion* (April/May):56–59.
Mason, Philip. 1970. *Patterns of Dominance.* New York: Oxford University Press.
Massarik, Fred, and Alvin Chenkin. 1973. "United States National Jewish Population Study: A First Report." Pp. 264–306 in *American Jewish Year Book 1973.* New York: American Jewish Committee.
Massey, Douglas S. 1979. "Effects of Socioeconomic Factors on the Residential Segregation of Blacks and Spanish Americans in United States Urbanized Areas." *American Sociological Review* 44:1015–1022.
———. 1981a. "Hispanic Residential Segregation: A Comparison of Mexicans, Cubans, and Puerto Ricans." *Sociology and Social Research* 65:311–322.
———. 1981b. "Dimensions of the New Immigration to the United States and the Prospects for Assimilation." *Annual Review of Sociology* 7:57–85.
Massey, Douglas S., and Brooks Bitterman. 1985. "Explaining the Paradox of Puerto Rican Segregation." *Social Forces* 64:306–331.
Massey, Douglas S., and Nancy A. Denton. 1987. "Trends in the Residential Segregation of Blacks, Hispanics, and Asians: 1970–1980." *American Sociological Review* 52:802–825.
———. 1989a. "Residential Segregation of Mexicans, Puerto Ricans, and Cubans in Selected U.S. Metropolitan Areas." *Sociology and Social Research* 73:73–83.
———. 1989b. "Hypersegregation in U.S. Metropolitan Areas: Black and Hispanic Segregation Along Five Dimensions." *Demography* 26:373–391.
———. 1993. *American Apartheid: Segregation and the Making of the Underclass.* Cambridge, Mass.: Harvard University Press.

Mathabane, Mark. 1994. "'Like the Second Coming.'" *Newsweek* (May 9):38.
McAllister, Ian. 1977. *The Northern Ireland Social Democratic and Labour Party.* London: Macmillan.
McAneny, Leslie, and Lydia Saad. 1994. "America's Public Schools: Still Separate? Still Unequal?" *Gallup Poll Monthly* (May):23–27.
McAuley, James W. 1994. *The Politics of Identity: A Loyalist Community in Belfast.* Aldershot, England: Avebury.
McClatchy, Valentine Stuart. 1978 (1919–1925). *Four Anti-Japanese Pamphlets.* New York: Arno Press.
McCleskey, Clifton, and Bruce Merrill. 1973. "Mexican American Political Behavior in Texas." *Social Science Quarterly* 53:785–798.
McCoy, George. 1994. "Mental Health." Pp. 330–335 in Mary B. Davis (ed.), *Native America in the Twentieth Century.* New York: Garland.
McGarry, John, and Brendan O'Leary (eds.). 1990. *The Future of Northern Ireland.* Oxford: Clarendon.
———. 1995a. *Explaining Northern Ireland: Broken Images.* Oxford: Blackwell.
———. 1995b. "Five Fallacies: Northern Ireland and the Liabilities of Liberalism." *Ethnic and Racial Studies* 18:837–861.
McGraw, Bill. 1990. "Montreal Racial Tirade Hits Hard." *Detroit Free Press* (October 8):3A, 6A
McKee, James B. 1993. *Sociology and the Race Problem: The Failure of a Perspective.* Urbana, Ill.: University of Illinois Press.
McKenna, Marian C. 1969. "The Melting Pot: Comparative Observations in the United States and Canada." *Sociology and Social Research* 53:433–447.
McLarin, Kimberly J. 1995. "To Preserve Afrikaners' Language, Mixed-Race South Africans Join Fray." *New York Times* (June 28):A4.
McLemore, S. Dale. 1973. "The Origins of Mexican-American Subordination in Texas." *Social Science Quarterly* 53:656–679.
McRae, Kenneth D. 1964. "The Structure of Canadian History." Pp. 219–274 in Louis Hartz, *The Founding of New Societies.* New York: Harcourt, Brace & World.
———. 1974. *Consociational Democracy: Political Accommodation in Segmented Societies.* Toronto: McClelland & Stewart.
———. 1983. *Conflict and Compromise in Multilingual Societies: Switzerland.* Waterloo, Ont.: Wilfrid Laurier University Press.
———. 1986. *Conflict and Compromise in Multilingual Societies. Vol. 2: Belgium.* Waterloo, Ont.: Wilfrid Laurier University Press.
———. 1991a. "Canada: Reflections on Two Conflicts." Pp. 197–217 in Joseph V. Montville (ed.), *Conflict and Peacemaking in Multiethnic Societies.* New York: Lexington Books.
———. 1991b. "National Identity in Northern Ireland." Pp. 23–30 in Peter Stringer and Gillian Robinson (eds.), *Social Attitudes in Northern Ireland.* Belfast: Blackstaff Press.
McRoberts, Kenneth. 1988. *Quebec: Social Change and Political Crisis.* 3d ed. Toronto: McClelland and Stewart.
McWilliams, Carey. 1948. *A Mask for Privilege: Anti-Semitism in America.* Boston: Little, Brown.
———. 1968 (1948). *North from Mexico: The Spanish-Speaking People of the United States.* New York: Greenwood.
Mearsheimer, John J., and Stephen Van Evera. 1995. "When Peace Means War." *New Republic* (December 18):16–21.
Mechanic, David. 1978. "Apartheid Medicine." Pp. 127–138 in Ian Robertson and Phillip Whitten (eds.), *Race and Politics in South Africa.* New Brunswick, N.J.: Transaction.
Meléndez, Edwin. 1993. "Understanding Latino Poverty." *Sage Race Relations Index* 18(May):3–42.
Meredith, Martin. 1987. "The Black Opposition." Pp. 77–89 in Jesmond Blumenfeld (ed.), *South Africa in Crisis.* London: Croom Helm.
Merton, Robert K. 1949. "Discrimination and the American Creed." Pp. 99–126 in R. H. MacIver (ed.), *Discrimination and National Welfare.* New York: Harper & Row.

———. 1968. *Social Theory and Social Structure.* New York: Free Press.

Metzger, L. Paul. 1971. "American Sociology and Black Assimilation: Conflicting Perspectives." *American Journal of Sociology* 76:627–647.

Miami Herald. 1983. "Slave's Descendant Ruled Legally Black in Louisiana." (May 19):1A, 12A.

Milbrath, Lester. 1982. *Political Participation.* 2d ed. Washington, D.C.: University Press of America.

Miller, James Nathan. 1984. "Ronald Reagan and the Techniques of Deception." *Atlantic Monthly* (February):62–68.

Miller, Randall M. (ed.). 1978. *Ethnic Images in American Film and Television.* Philadelphia: Balch Institute.

Mills, C. Wright. 1956. *The Power Elite.* New York: Oxford University Press.

Mills, C. Wright, Clarence Senior, and Rose Kohn Goldsen. 1950. *The Puerto Rican Journey.* New York: Harper & Brothers.

Min, Pyong Gap. 1988. *Ethnic Business Enterprise: Korean Small Business in Atlanta.* New York: Center for Migration Studies.

———. 1991. "Cultural and Economic Boundaries of Korean Ethnicity: A Comparative Analysis." Paper presented at the Annual Meeting of the American Sociological Association, Cincinnati, Ohio.

———. 1995. "Korean Americans." Pp. 199–231 in Pyong Gap Min (ed.), *Asian Americans: Contemporary Trends and Issues.* Thousand Oaks, Calif.: Sage.

Miner, Horace. 1939. *St. Denis: A French Canadian Parish.* Chicago: University of Chicago Press.

Minority Rights Group. 1992. *The Chinese of South-East Asia.* London: Minority Rights Group.

Mittelbach, Frank G., and Joan W. Moore. 1968. "Ethnic Endogamy: The Case of the Mexican Americans." *American Journal of Sociology* 74:50–62.

Montagu, Ashley. 1963. *Race, Science, and Humanity.* Princeton, N.J.: Van Nostrand.

———. (ed.). 1964. *The Concept of Race.* New York: Free Press.

———. 1972. *Statement on Race.* 3d ed. New York: Oxford University Press.

———. 1974. *Man's Most Dangerous Myth: The Fallacy of Race.* 5th ed. New York: Oxford University Press.

———. (ed.). 1975. *Race and IQ.* New York: Oxford University Press.

Montero, Darrel. 1980. *Japanese Americans: Changing Patterns of Ethnic Affiliation Over Three Generations.* Boulder, Colo.: Westview.

Moodley, Kogila A. 1975. "South African Indians: The Wavering Minority." Pp. 250–279 in Leonard Thompson and Jeffrey Butler (eds.), *Change in Contemporary South Africa.* Berkeley: University of California Press.

———. 1980. "Structural Inequality and Minority Anxiety: Responses of Middle Groups in South Africa." Pp. 217–235 in Robert M. Price and Carl G. Rotberg (eds.), *The Apartheid Regime: Political Power and Racial Domination.* Berkeley: Institute of International Studies, University of California.

Moon, Peter. 1995. "Natives Find Renewal in Manitoba Prison." *Globe and Mail* (July 20):A1, A4.

Moore, Joan W. 1970. "Colonialism: The Case of the Mexican-Americans." *Social Problems* 17:463–472.

Moore, Joan W., and Frank G. Mittelbach. 1966. *Residential Segregation in the Urban Southwest.* Advance Report 4, Mexican-American Study Project. Los Angeles: University of California, Los Angeles.

Moore, Joan W., with Harry Pachon. 1975. *Mexican Americans.* 2d ed. Englewood Cliffs, N.J.: Prentice-Hall.

———. 1985. *Hispanics in the United States.* Englewood Cliffs, N.J.: Prentice-Hall.

Moore, Michael. 1987. "The Man Who Killed Vincent Chin." *Detroit Magazine, Detroit Free Press* (August 30):12–20.

Moquin, Wayne, and Charles Van Doren (eds.). 1974. *A Documentary History of the Italian Americans.* New York: Praeger.

Morales, Armando. 1972. *Ando Sangrando.* Fair Lawn, N.J.: R. E. Burdick.

Morganthau, Tom. 1995. "What Color Is Black?" *Newsweek* (February 13):63–65.

Morin, Richard. 1995. "Feelings That Aren't Set in Concrete." *Washington Post* (March 13–19):37.

Mörner, Magnus. 1985. *Adventurers and Proletarians: The Story of Migrants in Latin America.* Pittsburgh, Pa.: University of Pittsburgh Press.

Morris, C. Patrick. 1988. "Termination by Accountants: The Reagan Indian Policy." *Policy Studies Journal* 16:731–750.

Morris, Raymond N., and C. Michael Lanphier. 1980. "French–English Relations as a Social Problem: Present Inequalities." Pp. 173–183 in Jay E. Goldstein and Rita M. Bienvenue (eds.), *Ethnicity and Ethnic Relations in Canada.* Toronto: Butterworths.

Morse, Richard M. 1953. "The Negro in São Paulo, Brazil." *Journal of Negro History* 38:290–306.

———. 1958. *From Community to Metropolis: A Biography of São Paulo, Brazil.* Gainesville, Fla.: University of Florida Press.

Moxon-Browne, Edward. 1983. *Nation, Class and Creed in Northern Ireland.* Aldershot, England: Gower.

———. 1991. "National Identity in Northern Ireland." Pp. 23–30 in Peter Stringer and Gillian Robinson (eds.), *Social Attitudes in Northern Ireland.* Belfast: Blackstaff Press.

Moynihan, Daniel P. 1965. *The Negro Family: The Case for National Action.* Washington, D.C.: U.S. Government Printing Office.

Mulder, C. P. 1972. "The Rationale of Separate Development." Pp. 48–63 in N. J. Rhoodie (ed.), *South African Dialogue.* Philadelphia: Westminister.

Muller, Thomas. 1993. *Immigrants and the American City.* New York: New York University Press.

Muller, Thomas, and Thomas J. Espenshade. 1985. *The Fourth Wave: California's Newest Immigrants.* Washington, D.C.: Urban Institute Press.

Murguía, Edward. 1975. *Assimilation, Colonialism and the Mexican American People.* Austin, Tex.: Center for Mexican American Studies, University of Texas.

Murguía, Edward, and W. Parker Frisbie. 1977. "Trends in Mexican American Intermarriage: Recent Findings in Perspective." *Social Science Quarterly* 58:374–389.

Murray, Dominic. 1983. "Schools and Conflict." Pp. 136–150 in John Darby (ed.), *Northern Ireland: The Background to the Conflict.* Belfast: Appletree Press.

———. 1986. "Educational Segregation: 'Rite' or Wrong?" Pp. 244–264 in Patrick Clancy et al. (eds.), *Ireland: A Sociological Profile.* Dublin: Institute of Public Administration.

Murray, Martin. 1987. *South Africa: Time of Agony, Time of Destiny.* London: Verso.

Mydans, Seth. 1992a. "New Unease for Japanese-Americans." *New York Times* (March 4):7.

———. 1992b. "A Target of Rioters, Koreatown Is Bitter, Armed and Determined." *New York Times* (May 3):1, 16.

———. 1995. "In Los Angeles, Quandary for Hispanic Voters." *New York Times* (June 2):A7.

Myers, Jerome K. 1951. "Assimilation in the Political Community." *Sociology and Social Research* 35:175–182.

Myrdal, Gunnar. 1944. *An American Dilemma: The Negro Problem and Modern Democracy.* New York: Harper & Row.

Nagel, Joane. 1995. "American Indian Ethnic Renewal: Politics and the Resurgence of Identity." *American Sociological Review* 60:947–965.

Nakhaie, M. Reza. 1995. "Ownership and Management Position of Canadian Ethnic Groups in 1973 and 1989." *Canadian Journal of Sociology* 20:167–192.

Nash, Manning. 1962. "Race and the Ideology of Race." *Current Anthropology* 3:285–288.

National Advisory Commission on Civil Disorders. 1968. *Report of the National Advisory Commission on Civil Disorders.* New York: Bantam.

National Conference. 1995. *Taking America's Pulse: A Summary Report of the National Conference Survey on Inter-Group Relations.* New York: National Conference.

National Conference of Christians and Jews. 1978. *A Study of Attitudes Toward Racial and Religious Minorities and Toward Women.* Conducted by Louis Harris and Associates, Inc.

National Roster of Hispanic Elected Officials. 1994. National Association of Latino Elected and Appointed Officials.

National Urban League. 1981. *The State of Black America 1981.* New York: National Urban League.

———. 1992. *The State of Black America 1992.* New York: National Urban League.

Nava, Julian. 1975. "Educational Challenges in Elementary and Secondary Schools." Pp. 107–136 in Gus Tyler (ed.), *Mexican-Americans Tomorrow.* Albuquerque, N. Mex.: University of New Mexico Press.

Neil, Andrew. 1982. "America's Latin Beat: A Survey of South Florida." *Economist* (October 16):3, 26.

Nelli, Humbert S. 1970. *The Italians in Chicago, 1880–1930.* New York: Oxford University Press.

Nelson, Candace, and Marta Tienda. 1985. "The Structuring of Hispanic Ethnicity: Historical and Contemporary Perspectives." Pp. 49–74 in Richard D. Alba (ed.), *Ethnicity and Race in the U.S.A.: Toward the Twenty-first Century.* London: Routledge & Kegan Paul.

Newcomer, Mabel. 1955. *The Big Business Executive.* New York: Columbia University Press.

Newman, Peter C. 1975. *The Canadian Establishment.* Vol. 1. Toronto: McClelland & Stewart.

———. 1979. *The Canadian Establishment.* Vol. 2. Toronto: McClelland & Stewart.

———. 1995. "A Country of Many Cultures and Flavors." *Maclean's* (July 24):34.

Newman, William M. 1973. *American Pluralism: A Study of Minority Groups and Social Theory.* New York: Harper & Row.

Newsweek. 1988. "A Conflict of the Have-Nots." 112(December 12):28–29.

———. 1989. "The New Voice of America." (June 12):28–29.

———. 1991. "Classrooms of Babel." (February 11):56–57.

———. 1992a. "Most U.S. Youths in Survey View Race Relations as Poor." (March 7):9.

———. 1992b. "Views on Race Relations." (May 11):12.

———. 1993. "Newsweek Poll." (August 9):19.

———. 1995a. "Newsweek Poll." (February 13):67.

———. 1995b. "Newsweek Poll." (July 10):31.

New York Times. 1982. "Business Aide to Reagan Criticizes Food Stamps." (May 28):A12.

———. 1987. "Black Woman Who Lost Job Is Asked Back by Pharmacist." (August 15):A7.

———. 1989. "Anti-Semitic Acts Put at 5-Year High." (January 29):1, 20.

———. 1992. "Student Killed After Objecting to Racial Slur." (August 19):6.

———. 1994. "Harassment of Jews Rose in '93, Anti-Defamation League Reports." (January 25):A11.

Nichols, Roger L. 1992. "Indians in the Post-Termination Era." Pp. 289–301 in Roger L. Nichols (ed.), *The American Indian: Past and Present.* 4th ed. New York: McGraw-Hill.

Niebuhr, Gustav. 1996. "Anti-Semitic Acts Down After Climb." *New York Times* (February 17):9.

Noble, Kenneth B. 1995. "Attacks Against Asian-Americans Are Rising." *New York Times* (December 13):A14.

Nodín Valdés, Dennis. 1989. "The New Northern Borderlands: An Overview of Midwestern Chicano History." Pp. 1–28 in *Mexicans in the Midwest, Perspectives in Mexican American Studies.* Tucson, Ariz.: Mexican American Studies and Research Center, University of Arizona.

Noel, Donald L. 1968. "A Theory of the Origin of Ethnic Stratification." *Social Problems* 16:157–172.

Nogueira, Oracy. 1959. "Skin Color and Social Class." In Vera Rubin (ed.), *Plantation Systems of the New World.* Pan American Union Social Science Monograph No. 7. Washington, D.C.: Pan American Union.

Nordheimer, Jon. 1987. "Black Cubans: Apart in Two Worlds." *New York Times* (December 2):26.

———. 1996. "Nesting Beyond the Snowbirds." *New York Times* (February 20):C1, C4.

Nordland, Rod. 1993. "'Let's Kill the Muslims!'" *Newsweek* (November 8):48–51.

Norris, Mary Jane. 1990. "The Demography of Aboriginal People in Canada." Pp. 33–59 in Shiva S. Halli, Frank Trovato, and Leo

Driedger (eds.), *Ethnic Demography: Canadian Immigrant, Racial and Cultural Variations.* Ottawa: Carleton University Press.

North, Robert D. 1965. "The Intelligence of the American Negro." Pp. 334–340 in Arnold M. Rose and Caroline B. Rose (eds.), *Minority Problems.* New York: Harper & Row.

Northern Ireland, Department of Health and Social Services. 1984. *The Northern Ireland Census, 1981.* Summary Report. Belfast: Her Majesty's Stationery Office.

Nostrand, Richard L. 1973. "Mexican American and Chicano: Emerging Terms for a People Coming of Age." *Pacific Historical Review* 62:389–406.

Novak, Michael. 1972. *The Rise of the Unmeltable Ethnics.* New York: Macmillan.

O'Donnell, E. E. 1977. *Northern Irish Stereotypes.* Dublin: College of Industrial Relations.

O'Dowd, Liam. 1991. "Social Class." Pp. 39–50 in Peter Stringer and Gillian Robinson (eds.), *Social Attitudes in Northern Ireland.* Belfast: Blackstaff Press.

Ogmundson, R. 1990. "Perspectives on the Class and Ethnic Origins of Canadian Elites: A Methodological Critique of the Porter/Clement/Olsen Tradition." *Canadian Journal of Sociology* 15:165–177.

Ogmundson, R. and L. Fatels. 1994. "Are the Brits in Decline? A Note on Trends in the Ethnic Origins of the Labour and Church Elites." *Canadian Ethnic Studies* 26:108–112.

Ogmundson, R., and J. McLaughlin. 1992. "Trends in the Ethnic Origins of Canadian Elites: The Decline of the BRITS?" *Canadian Review of Sociology and Anthropology* 29:227–242.

O'Hare, William. 1987. "Best Metros for Hispanic Businesses." *American Demographics* 9(November):30–32.

O'Leary, Brendan. 1995. "Introduction: Reflections on a Cold Peace." *Ethnic and Racial Studies* 18:695–714.

Oliver, Melvin L., and Thomas M. Shapiro. 1995. *Black Wealth/White Wealth: A New Perspective on Racial Inequality.* New York: Routledge.

Olsen, Marvin E. 1970. "Power Perspectives on Stratification and Race Relations." Pp. 296–305 in Marvin E. Olsen (ed.), *Power in Societies.* New York: Macmillan.

Olson, James S., and Raymond Wilson. 1984. *Native Americans in the Twentieth Century.* Provo, Utah: Brigham Young University Press.

Olzak, Susan. 1992. *The Dynamics of Ethnic Competition and Conflict.* Stanford, Calif.: Stanford University Press.

O'Malley, Padraig. 1990. *Northern Ireland: Questions of Nuance.* Belfast: Blackstaff Press.

———. 1991. "Northern Ireland: Political Strategies for the Management of Conflict." Pp. 161–175 in Joseph V. Montville (ed.), *Conflict and Peacemaking in Multiethnic Societies.* New York: Lexington.

Omi, Michael, and Howard Winant. 1986. *Racial Formation in the United States: From the 1960s to the 1980s.* New York: Routledge & Kegan Paul.

Onishi, Norimitsu. 1995. "Japanese in America Looking Beyond Past to Shape Future." *New York Times* (December 25):1, 9.

Orfield, Gary. 1985. "Ghettoization and Its Alternatives." Pp. 161–196 in Paul E. Peterson (ed.), *The New Urban Reality.* Washington, D.C.: Brookings Institution.

———. 1988. "Separate Societies: Have the Kerner Warnings Come True?" Pp. 100–122 in Fred R. Harris and Roger W. Wilkins (eds.), *Quiet Riots: Race and Poverty in the United States.* New York: Pantheon.

Osborne, R. D. 1991. "Discrimination and Fair Employment." Pp. 31–38 in Peter Stringer and Gillian Robinson (eds.), *Social Attitudes in Northern Ireland.* Belfast: Blackstaff Press.

Osborne, Robert, and Robert Cormack. 1991. "Religion and the Labour Market: Patterns and Profiles." Pp. 49–71 in Robert J. Cormack and Robert D. Osborne (eds.), *Discrimination and Public Policy in Northern Ireland.* Oxford: Clarendon Press.

Owen, Carolyn A., Howard C. Eisner, and Thomas R. McFaul. 1981. "A Half-Century of Social Distance Research: National Replication of the Bogardus Studies." *Sociology and Social Research* 66:80–99.

Pachon, Harry P. 1988. "Hispanic Underrepresentation in the Federal Bureaucracy: The Missing Link in the Policy Process." Pp. 306–313 in F. Chris Garcia (ed.), *Latinos and the Political System.* Notre Dame, Ind.: University of Notre Dame Press.

Pachon, Harry P., and Joan W. Moore. 1981. "Mexican Americans." *Annals of the American Academy of Political and Social Science* 454:111–124.

Padilla, Elena. 1958. *Up from Puerto Rico.* New York: Columbia University Press.

Padilla, Felix. 1987. *Puerto Rican Chicago.* Notre Dame, Ind.: University of Notre Dame Press.

Padilla, Fernando V., and Carlos B. Ramirez. 1974. "Patterns of Chicano Representation in California, Colorado and Nuevo Mexico." *Aztlan* 5:189–235.

Page, Joseph A. 1995. *The Brazilians.* Reading, Mass.: Addison-Wesley.

Paine, Suzanne. 1974. *Exporting Workers: The Turkish Case.* London: Cambridge University Press.

Palley, Claire. 1981. "Ways Forward: The Constitutional Options." Pp. 183–206 in David Watt (ed.), *The Constitution of Northern Ireland: Problems and Prospects.* London: Heinemann.

Palmer, Howard. 1976. "Mosaic vs. Melting Pot?: Immigration and Ethnicity in Canada and the United States." *International Journal* 31:488–528.

Park, Robert E. 1924. "The Concept of Social Distance." *Journal of Applied Sociology* 8:339–344.

———. 1928. "Human Migration and the Marginal Man." *American Journal of Sociology* 33:881–893.

———. 1950. *Race and Culture.* Glencoe, Ill.: Free Press.

Parkin, Frank. 1971. *Class Inequality and Political Order.* New York: Praeger.

Patterson, Orlando. 1995a. "The Culture of Caution." *New Republic* (November 27):22–26.

———. 1995b. "The Paradox of Integration." *New Republic* (November 6):24–27.

Peach, Ceri, and Günther Glebe. 1995. "Muslim Minorities in Western Europe." *Ethnic and Racial Studies* 18:26–45.

Pear, Robert. 1992. "New Look at the U.S. in 2050: Bigger, Older and Less White." *New York Times* (December 4):A1, A10.

Pearlin, Leonard I. 1954. "Shifting Group Attachments and Attitudes Toward Negroes." *Social Forces* 33:47–50.

Pedraza-Bailey, Silva. 1985. *Political and Economic Migrants in America: Cubans and Mexicans.* Austin, Tex.: University of Texas Press.

Peñalosa, Fernando. 1967. "The Changing Mexican-American in Southern California." *Sociology and Social Research* 51:405–417.

Perez, Lisandro. 1980. "Cubans." Pp. 256–261 in Stephen Thernstrom (ed.), *Harvard Encyclopedia of American Ethnic Groups.* Cambridge, Mass.: Harvard University Press.

———. 1986. "Cubans in the United States." *Annals of the American Academy of Political and Social Science* 487:126–137.

Pérez-Stable, Marifeli, and Miren Uriarte. 1993. "Cubans and the Changing Economy of Miami." Pp. 133–159 in Rebecca Morales and Frank Bonilla (eds.), *Latinos in a Changing U.S. Economy.* Newbury Park, Calif.: Sage.

Perlez, Jane. 1995a. "Serbs Become Latest Victims in Changing Fortunes of War." *New York Times* (August 7):A1, A4.

———. 1995b. "Thousands of Serbian Civilians Are Caught in Soldiers' Gunfire." *New York Times* (August 9):A1, A6.

Petersen, William. 1980. "Concepts of Ethnicity." Pp. 234–242 in Stephen Thernstrom (ed.), *Harvard Encyclopedia of American Ethnic Groups.* Cambridge, Mass.: Harvard University Press.

Petit, Arthur G. 1980. *Images of the Mexican American in Fiction and Film.* College Station, Tex.: Texas A&M Press.

Pettigrew, Thomas F. 1960. "Social Distance Attitudes of South African Students." *Social Forces* 38:246–253.

———. 1964. *A Profile of the Negro American.* Princeton, N.J.: Van Nostrand.

———. 1975. "Black and White Attitudes Toward Race and Housing." Pp. 92–126 in Thomas Pettigrew (ed.), *Racial Discrimination in the United States.* New York: Harper & Row.

———. 1979. "Racial Change and Social Policy." *Annals of the American Academy of Political and Social Science* 441:114–131.

———. 1980. "Prejudice." Pp. 820–829 in Stephen Thernstrom (ed.), *Harvard Encyclopedia of American Ethnic Groups.* Cambridge, Mass.: Harvard University Press.

———. 1981. "Race and Class in the 1980s: An Interactive View." *Daedalus* 110:233–255.

Peukert, Detlev J. K. 1987. *Inside Nazi Germany: Conformity, Opposition and Racism in Everyday Life.* London: Penguin.

Philliber, William W., and Clyde B. McCoy (eds.). 1981. *The Invisible Minority: Urban Appalachians.* Lexington, Ky.: University Press of Kentucky.

Pierson, Donald. 1967 (1942). *Negroes in Brazil: A Study of Race Contact at Bahia.* Carbondale, Ill.: Southern Illinois University Press.

Pineo, Peter C. 1987. "The Social Standing of Ethnic and Racial Groupings." Pp. 256–272 in Leo Driedger (ed.), *Ethnic Canada: Identities and Inequalities.* Toronto: Copp Clark Pitman.

Pineo, Peter C., and John Porter. 1985. "Ethnic Origin and Occupational Attainment." Pp. 357–392 in Monica Boyd et al., *Ascription and Achievement: Studies in Mobility and Status Attainment in Canada.* Ottawa: Carleton University Press.

Pinkney, Alphonso. 1963. "Prejudice Toward Mexican and Negro Americans: A Comparison." *Phylon* 24:353–359.

———. 1984. *The Myth of Black Progress.* New York: Cambridge University Press.

———. 1993. *Black Americans.* 4th ed. Englewood Cliffs, N.J.: Prentice-Hall.

Pitman, Walter. 1977. *Now Is Not Too Late.* Toronto: Council of Metropolitan Toronto, Task Force on Human Relations.

Pitt-Rivers, Julian. 1987. "Race, Color, and Class in Central America and the Andes." Pp. 298–305 in Celia S. Heller (ed.), *Structured Social Inequality: A Reader in Comparative Social Stratification.* 2d ed. New York: Macmillan.

Piven, Frances Fox, and Richard A. Cloward. 1971. *Regulating the Poor: The Functions of Public Welfare.* New York: Random House.

Plamenatz, John. 1970. *Ideology.* New York: Praeger.

Poll, Solomon. 1969. *The Hasidic Community of Williamsburg.* New York: Schocken.

Pollack, Andrew. 1992. "Asian Immigrants New Leaders in Silicon Valley." *New York Times* (January 14):A1, C5.

Pollak, Andy. 1993. *A Citizens' Inquiry: The Opsahl Report on Northern Ireland.* Dublin: Lilliput Press for Initiative '92.

Pomfret, John. 1995a. "Between War and Peace." *Washington Post* (December 25–31):6–7.

———. 1995b. "For the War's Victims, the Search for Justice Begins." *Washington Post* (December 25–31):8–9.

Ponting, J. Rick, and Roger Gibbins. 1980. *Out of Irrelevance: A Socio-Political Introduction to Indian Affairs in Canada.* Toronto: Butterworths.

Poole, Michael A. 1982. "Religious Residential Segregation in Urban Northern Ireland." Pp. 182–308 in Frederick W. Boal and J. Neville H. Douglas (eds.), *Integration and Division: Geographical Perspectives on the Northern Ireland Problem.* London: Academic Press.

Poppino, Rollie E. 1968. *Brazil: The Land and People.* New York: Oxford University Press.

Porter, John. 1965. *The Vertical Mosaic: An Analysis of Social Class and Power in Canada.* Toronto: University of Toronto Press.

———. 1975. "Ethnic Pluralism in Canadian Perspective." Pp. 267–304 in Nathan Glazer and Daniel P. Moynihan (eds.), *Ethnicity: Theory and Experience.* Cambridge, Mass.: Harvard University Press.

———. 1979. *The Measure of Canadian Society.* Toronto: Gage.

———. 1985. "Canada: The Societal Context of Occupational Allocation." Pp. 29–65 in Monica Boyd et al., *Ascription and Achievement: Studies in Mobility and Status Attainment in Canada.* Ottawa: Carleton University Press.

Porter, Judith D. R. 1971. *Black Child, White Child.* Cambridge, Mass.: Harvard University Press.

Portes, Alejandro. 1969. "Dilemmas of a Golden Exile: Integration of Cuban Refugee Families in Milwaukee." *American Sociological Review* 34:505–518.

———. 1979. "Illegal Immigration and the International System: Lessons from Recent Legal Mexican Immigrants to the United States." *Social Problems* 26:425–438.

Portes, Alejandro, and Robert L. Bach. 1985. *Latin Journey: Cuban and Mexican Immigrants in the United States.* Berkeley: University of California Press.

Portes, Alejandro, J. M. Clark, and R. L. Bach, 1977. "The New Wave: A Statistical Profile of Recent Cuban Exiles to the United States." *Estudios Cubanos/Cuban Studies* 7:1–32.

Portes, Alejandro, Juan M. Clark, and Robert D. Manning. 1985. "After Mariel: A Survey of the Resettlement Experiences of 1980 Cuban Refugees in Miami." *Estudios Cubanos/Cuban Studies* 15:37–59.

Portes, Alejandro, and Rafael Mozo. 1985. "The Political Adaptation Process of Cubans and Other Ethnic Minorities in the United States: A Preliminary Analysis." *International Migration Review* 19:35–63.

Portes, Alejandro, and Rubén Rumbaut. 1990. *Immigrant America: A Portrait.* Berkeley: University of California Press.

Portes, Alejandro, and Alex Stepick. 1993. *City on the Edge: The Transformation of Miami.* Berkeley: University of California Press.

Portes, Alejandro, and Cynthia Truelove. 1987. "Making Sense of Diversity: Recent Research on Hispanic Minorities in the United States." *Annual Review of Sociology* 13:359–385.

Posgate, Dale. 1978. "The Quiet Revolution." Pp. 50–57 in Norman Penner et al. (eds.), *Keeping Canada Together.* Toronto: Amethyst.

Poston, Dudley L., and David Alvírez. 1973. "On the Cost of Being a Mexican American Worker." *Social Science Quarterly* 53:697–709.

Poston, Dudley L., and Mei-Yu Yu. 1990. "The Distribution of the Overseas Chinese in the Contemporary World." *International Migration Review* 24(Fall):480–508.

Power, Jonathan. 1979. *Migrant Workers in Western Europe and the United States.* Oxford: Pergamon.

Presthus, Robert. 1973. *Elite Accommodation in Canadian Politics.* Toronto: Macmillan.

Price, John A. 1973. "The Stereotyping of North American Indians in Motion Pictures." *Ethnohistory* 20(Spring):153–171.

Price, Robert M. 1991. *The Apartheid State in Crisis: Political Transformation in South Africa, 1975–1990.* New York: Oxford University Press.

Princeton Religion Research Center. 1990. "Church Attendance Unchanged as We Enter the 1990s." *Emerging Trends* 12(June):4.

Public Opinion. 1982. "Opinion Roundup." (June/July):34.

———. 1987. "Opinion Roundup." (July/August):25–39.

Purcell, Victor. 1980. *The Chinese in Southeast Asia.* 2d ed. Oxford: Oxford University Press.

Quinley, Harold E., and Charles Y. Glock. 1979. *Anti-Semitism in America.* New York: Free Press.

Raab, Earl. 1989. *What Do We Really Know About Anti-Semitism and What Do We Want to Know? Working Papers on Contemporary Anti-Semitism.* New York: American Jewish Committee.

Raab, Earl, and Seymour Martin Lipset. 1971. "The Prejudiced Society." Pp. 31–45 in Gary T. Marx (ed.), *Racial Conflict: Tension and Change in American Society.* Boston: Little, Brown.

Ramcharan, Subhas. 1982. *Racism: Nonwhites in Canada.* Toronto: Butterworths.

Ramirez, Anthony. 1986. "America's Super Minority." *Fortune* 114(November 24):148–161.

Ramirez, J. Martin, and Bobbie Sullivan. 1987. "The Basque Conflict," pp. 120–139 in Jerry Boucher, Dan Landis, and Karen Arnold Clark (eds.), *Ethnic Conflict: International Perspectives.* Newbury Park, Calif.: Sage.

Ramos, Arthur. 1939. *The Negro in Brazil.* Trans. Richard Pattee. Washington, D.C.: Associated Publishers.

Ramsey, Patricia G. 1987. "Young Children's Thinking About Ethnic Differences." Pp. 56–72 in Jean S. Phinney and Mary Jane Rotheram (eds.), *Children's Ethnic Socialization: Pluralism and Development.* Newbury Park: Sage.

Redfield, Robert. 1939. "Culture Contact Without Conflict." *American Anthropologist* 41:514–517.

———. 1958. "Race as a Social Phenomenon." Pp. 66–71 in Edgar T. Thompson and Everett C. Hughes (eds.), *Race: Individual and Collective Behavior.* Glencoe, Ill.: Free Press.

Reed, John. 1982. "Black America in the 1980s." *Population Bulletin* 37(December). Washington, D.C.: Population Reference Bureau.

Reeve, R. Penn. 1975. "Black Economic Mobility in a Brazilian Town. *Plural Societies* 6:45–50.

———. 1977. "Race and Social Mobility in a Brazilian Industrial Town." *Luso Brazilian Review* 14:236–253.

Reich, Michael. 1978. "The Economics of Racism." Pp. 381–388 in Richard C. Edwards, Michael Reich, and Thomas E. Weisskopf (eds.), *The Capitalist System.* 2d ed. Englewood Cliffs, N.J.: Prentice-Hall.

Reid, Ira De A. 1939. *The Negro Immigrant.* New York: Columbia University Press.

Reitz, Jeffrey G., and Raymond Breton. 1994. *The Illusion of Difference: Realities of Ethnicity in Canada and the United States.* Toronto: C. D. Howe Institute.

Republic of South Africa. 1983. *South Africa 1983: Official Yearbook of the Republic of South Africa,* 9th ed. Pretoria: Department of Foreign Affairs and Information.

———. 1988. *Suid-Afrikaanse Statistieke/South African Statistics, 1988.* Pretoria: Central Statistical Service.

Rex, John. 1970. *Race Relations in Sociological Theory.* New York: Schocken.

Richard, Alfred Charles, Jr. 1994. *Contemporary Hollywood's Negative Hispanic Image: An Interpretive Filmography, 1956–1993.* Westport, Conn.: Greenwood Press.

Richer, Stephen, and Pierre E. Laporte. 1979. "Culture, Cognition, and English–French Competition." Pp. 75–85 in Jean Leonard Elliott (ed.), *Two Nations, Many Cultures.* Scarborough, Ont.: Prentice-Hall of Canada.

Richmond, Anthony H. 1969. "Immigration and Pluralism in Canada." *International Migration Review* 4:5–24.

———. 1976. "Immigration, Population, and the Canadian Future." *Sociological Focus* 9:125–136.

Rieff, David. 1991. "The New Face of L.A." *Los Angeles Times Magazine* (September 15):14–20, 46–48.

———. 1993. *The Exile: Cuba in the Heart of Miami.* New York: Touchstone.

———. 1995. *Slaughterhouse: Bosnia and the Failure of the West.* New York: Touchstone.

———. 1996. "An Age of Genocide." *New Republic* (January 29):27–36.

Riis, Jacob A. 1957 (1890). *How the Other Half Lives.* New York: Hill & Wang.

Ringer, Benjamin B. 1967. *The Edge of Friendliness.* New York: Basic Books.

Rioux, Marcel, and Yves Martin (eds.). 1978. *French-Canadian Society.* Toronto: Macmillan of Canada.

Rist, Ray C. 1978. *Guest Workers in Germany: The Prospects for Pluralism.* New York: Praeger.

Ritterband, Paul. 1995. "Modern Times and Jewish Assimilation." Pp. 377–394 in Robert M. Seltzer and Norman J. Cohen (eds.), *The Americanization of the Jews.* New York: New York University Press.

Robertson, Ian, and Phillip Whitten. 1978. "Introduction." Pp. xi–xx in Ian Robertson and Phillip Whitten (eds.), *Race and Politics in South Africa.* New Brunswick, N.J.: Transaction.

Robinson, Eugene. 1996. "Over the Brazilian Rainbow." *Washington Post* (December 18–24):23.

Roche, John Patrick. 1982. "Suburban Ethnicity: Ethnic Attitudes and Behavior Among Italian

Americans in Two Suburban Communities." *Social Science Quarterly* 63:145–153.

Rocher, Guy. 1976. "Multiculturalism: The Doubts of a Francophone." In *Second Canadian Conference on Multiculturalism.* Ottawa: Minister of Supply and Services.

Rodal, Berel. 1991. "The Canadian Conundrum: Two Concepts of Nationhood." Pp. 156–174 in Uri Ra'anan, Maria Mesner, Keith Armes, and Kate Martin (eds.), *State and Nation in Multi-Ethnic Societies.* Manchester, England: Manchester University Press.

Rodrigues, Jose Honorio. 1967. *The Brazilians.* Trans. Ralph Edward Dimmick. Austin, Tex.: University of Texas Press.

Rodriguez, Clara. 1974. *The Ethnic Queue in the United States: The Case of Puerto Ricans.* San Francisco: R&E Research Associates.

———. 1989. *Puerto Ricans: Born in the U.S.A.* Boston: Unwin and Hyman.

Rodríguez-Vecchini, Hugo. 1994. "Foreward: Back and Forward." Pp. 29–102 in Carlos Antonio Torre, Hugo Rodríguez Vecchini, and William Burgos (eds.), *The Commuter Nation: Perspectives on Puerto Rican Migration.* Río Piedras, Puerto Rico: Editorial de la Universidad de Puerto Rico.

Rolle, Andrew. 1980. *The Italian Americans: Troubled Roots.* New York: Free Press.

Roof, Wade Clark. 1972. "Residential Segregation of Blacks and Racial Inequality in Southern Cities: Toward a Causal Model." *Social Problems* 19:393–407.

———. 1978. "'The Negro as an Immigrant Group'—A Research Note on Chicago's Racial Trends." *Ethnic and Racial Studies* 1:452–464.

———. 1979. "Race and Residence: The Shifting Basis of American Race Relations." *Annals of the American Academy of Political and Social Science* 44:1–12.

Rose, Hilary, and Steven Rose. 1978. "The IQ Myth." *Race and Class* 20:63–74.

Rose, Peter I. 1968. *The Subject Is Race: Traditional Ideologies and the Teaching of Race Relations.* New York: Oxford University Press.

———. 1977. *Strangers in Their Midst.* Merrick, N.Y.: Richwood.

———. 1985. "Asian Americans: From Pariahs to Paragons." Pp. 181–212 in Nathan Glazer (ed.), *Clamor at the Gates: The New American Immigration.* San Francisco: ICS Press.

Rose, Richard. 1971. *Governing Without Consensus: An Irish Perspective.* Boston: Beacon.

———. 1976. *Northern Ireland: Time of Choice.* Washington, D.C.: American Enterprise Institute for Public Policy Research.

———. 1991. "Northern Ireland: The Irreducible Conflict." Pp. 133–150 in Joseph V. Montville (ed.), *Conflict and Peacemaking in Multiethnic Societies.* New York: Lexington Books.

Rose, Richard, Ian McAllister, and Peter Mair. 1978. *Is There a Concurring Majority About Northern Ireland?: Studies in Public Policy.* Glasgow: University of Strathclyde.

Rosen, Bernard C. 1959. "Race, Ethnicity, and the Achievement Syndrome." *American Sociological Review* 24:47–60.

Royal Commission on Bilingualism and Biculturalism. 1969a. *Report.* Vol. 3, *The Work World.* Ottawa: Queen's Printer.

———. 1969b. *Report.* Vol. 4, *The Cultural Contribution of the Other Ethnic Groups.* Ottawa: Queen's Printer.

Rubin, Israel. 1972. *Satmar: An Island in the City.* Chicago: Quadrangle.

Rudolph, Susanne Hoeber, and Lloyd I. Rudolph. 1993. "Modern Hate." *New Republic* (March 22):24–29.

Rudwick, Elliott M. 1964. *Race Riot at East St. Louis.* Carbondale, Ill.: Southern Illinois University Press.

Rudwick, Elliott, and August Meier. 1969. "Negro Retaliatory Violence in the Twentieth Century." Pp. 406–417 in August Meier and Elliott Rudwick (eds.), *The Making of Black America: Essays in Negro Life and History.* Vol. 2. New York: Atheneum.

Runciman, W. G. 1966. *Relative Deprivation and Social Justice.* Berkeley: University of California Press.

Russell-Wood, A. J. R. 1968. "Race and Class in Brazil, 1937–1967, a Re-Assessment: A Review." *Race* 10:185–191.

Ryan, William. 1967. "Savage Discovery: The Moynihan Report." Pp. 457–466 in Lee Rainwater and William L. Yancey (eds.), *The Moynihan Report and the Politics of Controversy.* Cambridge, Mass.: MIT Press.

———. 1975. *Blaming the Victim.* Rev. ed. New York: Vintage.

Saad, Lydia. 1995. "Immigrants See U.S. As Land of Opportunity." *Gallup Poll Monthly* (July):19–33.

Sachs, Albie. 1975. "The Instruments of Domination in South Africa." Pp. 223–249 in Leonard Thompson and Jeffrey Butler (eds.), *Change in Contemporary South Africa.* Berkeley: University of California Press.

———. 1992. "'Watch Out—There's a Constitution About.'" *South Africa International* 22(April):184–189.

Sagarin, Edward. 1971. *The Other Minorities.* Waltham, Mass.: Xerox.

Said, Edward W. 1981. *Covering Islam.* New York: Pantheon.

Salholy, Eloise. 1987. "Do Colleges Set Asian Quotas?" *Newsweek* 109(February 9):6.

Samora, Julian, and Richard A. Lamanna. 1967. *Mexican-Americans in a Midwest Metropolis: A Study of East Chicago.* Advance Report 8, Mexican-American Study Project. Los Angeles: University of California, Los Angeles.

Samuda, Ronald J. 1975. *Psychological Testing of American Minorities.* New York: Harper & Row.

Sandberg, Neil. 1974. *Ethnic Identity and Assimilation: The Polish American Community.* New York: Praeger.

———. 1986. *Jewish Life in Los Angeles.* Lanham, Md.: University Press of America.

Sanders, Ronald. 1992. *Lost Tribes and Promised Lands: The Origins of American Racism.* New York: HarperPerennial.

Sanders, Thomas G. 1981. *Racial Discrimination and Black Consciousness in Brazil.* American Universities Field Staff Reports, No. 42. Hanover, N.H.: American Universities Field Staff.

———. 1982. "Brazilian Population in 1982: Growth, Migration, Race, Religion." *UFSI Reports* 1982, No. 42.

Santiago, Anne M. 1991. *The Spatial Dimensions of Ethnic and Racial Stratification.* Ann Arbor, Mich.: Population Studies Center, University of Michigan.

Sarlin, Stuart H., and Eugene D. Tate. 1976. "'All in the Family': Is Archie Funny?" *Journal of Communication* 26:61–67.

Saunders, John. 1972. "Class, Color, and Prejudice: A Brazilian Counterpoint." Pp. 141–165 in Ernest Q. Campbell (ed.), *Racial Tensions and National Identity.* Nashville, Tenn.: Vanderbilt University Press.

Savage, Michael. 1975. "Major Patterns of Group Interaction in South African Society." Pp. 280–302 in Leonard Thompson and Jeffrey Butler (eds.), *Change in Contemporary South Africa.* Berkeley: University of California Press.

Saywell, John. 1977. *The Rise of the Parti Québécois 1967–76.* Toronto: University of Toronto Press.

Schemo, Diana Jean. 1995. "Elevators Don't Lie: Intolerance in Brazil." *New York Times* (August 30):7A.

Schermerhorn, R. A. 1949. *These Our People: Minorities in American Culture.* Boston: Heath.

———. 1970. *Comparative Ethnic Relations: A Framework for Theory and Research.* New York: Random House.

Schiller, Herbert I. 1973. *The Mind Managers.* Boston: Beacon.

Schlemmer, Lawrence. 1988. "South Africa's National Party Government." Pp. 7–54 in Peter L. Berger and Bobby Godsell (eds.), *A Future South Africa: Visions, Strategies and Realities.* Cape Town: Human & Rousseau Tafelberg.

———. 1991. "Between Race, Class and Culture: Social Divisions in South Africa's Political Transition and Their Policy Implications." Pp. 175–197 in Uri Ra'anan, Maria Mesner, Keith Armes, and Kate Martin (eds.), *State and Nation in Multi-Ethnic Societies.* Manchester, England: Manchester University Press.

Schlesinger, Arthur M., Jr. 1992. *The Disuniting of America: Reflections on a Multicultural Society.* New York: Norton.

Schmidhauser, John R. 1960. *The Supreme Court: Its Politics, Personalities, and Procedures.* New York: Holt, Rinehart & Winston.

Schmitt, David E. 1994. "Resolving Conflict in Bicommunal Political Systems." Pp. 175–189 in Adrian Guelke (ed.), *New Perspectives on the Northern Ireland Conflict.* Aldershot, England: Avebury.

Schneider, Ronald M. 1996. *Brazil: Culture and Politics in a New Industrial Powerhouse.* Boulder, Colo.: Westview.

Schooler, Carmi. 1976. "Serfdom's Legacy: An Ethnic Continuum." *American Journal of Sociology* 81:1265–1286.

Schöpflin, George. 1993. "The Rise and Fall of Yugoslavia." Pp. 172–203 in John McGarry and Brendan O'Leary (eds.), *The Politics of Ethnic Conflict Regulation.* London: Routledge.

Schuman, Howard. 1974. "Are Whites Really More Liberal? Blacks Aren't Impressed." *Psychology Today* (September):82–86.

———. 1982. "Free Will and Determinism in Public Beliefs About Race." Pp. 345–350 in Norman R. Yetman and C. Hoy Steele (eds.), *Majority and Minority: The Dynamics of Race and Ethnicity in American Life.* 3d ed. Boston: Allyn & Bacon.

Schuman, Howard, Charlotte Steeh, and Lawrence Bobo. 1985. *Racial Attitudes in America: Trends and Interpretations.* Cambridge, Mass.: Harvard University Press.

Schwartz, Mildred A. 1967. *Trends in White Attitudes Toward Negroes.* Chicago: National Opinion Research Center.

Sciortino, Giuseppe. 1992. "From Guests to Hosts: Italy as a New Country of Immigration." Paper presented at the Annual Meeting of the American Sociological Association, Pittsburgh, Pa.

Scott, Nolvert P., Jr. 1975. "The Black Peoples of Canada." Pp. 143–162 in Donald G. Baker (ed.), *Politics of Race: Comparative Studies.* Westmead, England: Saxon House.

Sealey, D. Bruce, and Antoine S. Lussier. 1975. *The Métis: Canada's Forgotten People.* Winnipeg: Pemmican.

Selznick, Gertrude J., and Stephen Steinberg. 1969. *The Tenacity of Prejudice: Anti-Semitism in Contemporary America.* New York: Harper & Row.

Shaheen, Jack G. 1984. *The TV Arab.* Bowling Green, Ohio: Bowling Green State University Popular Press.

Sharot, Stephen. 1973. "The Three-Generations Thesis and the American Jews." *British Journal of Sociology* 24:151–164.

Shea, Christopher. 1995. "Under UCLA's Elaborate System Race Makes a Big Difference." *Chronicle of Higher Education* (April 18):A12–A14.

Sheane, Michael. 1977. *Ulster and Its Future After the Troubles.* Bredbury, England: Highfield Press.

Sheatsley, Paul B. 1966. "White Attitudes Toward the Negro." Pp. 303–324 in Talcott Parsons and Kenneth B. Clark (eds.), *The Negro American.* Boston: Houghton Mifflin.

Shibutani, Tamotsu, and Kian M. Kwan. 1965. *Ethnic Stratification: A Comparative Approach.* New York: Macmillan.

Shils, Edward. 1968. "Ideology." Pp. 66–76 in David L. Sills (ed.), *International Encyclopedia of the Social Sciences.* Vol. 7. New York: Macmillan.

Shimoni, Gideon. 1988. "South African Jews and the Apartheid Crisis." Pp. 3–58 in *American Jewish Year Book 1988.* New York: American Jewish Committee.

Sibley, Mulford Q. 1963. *The Quiet Battle.* Chicago: Quadrangle.

Siembieda, William J. 1975. "Suburbanization of Ethnics by Color." *Annals of the American Academy of Political and Social Science* 422:118–128.

Siggner, Andrew J. 1980. "A Socio-Demographic Profile of Indians in Canada." Pp. 31–65 in J. Rick Ponting and Roger Gibbins, *Out of Irrelevance: A Socio-Political Introduction to Indian Affairs in Canada.* Toronto: Butterworths.

———. 1986. "The Socio-Demographic Conditions of Registered Indians." *Canadian Social Trends* (Winter):2–9.

Silberman, Charles E. 1985. *A Certain People: American Jews and Their Lives Today.* New York: Summit.

Silva, Helga. 1984. "More Latins Winning Jobs at City Level." *Miami Herald* (May 18):2C.

Silva, Nelson do Valle. 1985. "Updating the Cost of Not Being White in Brazil." Pp. 42–55 in Pierre-Michel Fontaine (ed.), *Race, Class, and Power in Brazil.* Los Angeles: Center for Afro-American Studies, University of California, Los Angeles.

Simmons, Ozzie G. 1971. "The Mutual Images and Expectations of Anglo-Americans and Mexican-Americans." Pp. 62–71 in Nathaniel N. Wagner and Marsha J. Haug (eds.), *Chicanos: Social and Psychological Perspectives.* St. Louis: Mosby.

Simon, Julian L. 1991. "The Case for Greatly Increased Immigration." *The Public Interest* 102(Winter):89–103.

Simon, Rita James. 1978. *Continuity and Change: A Study of Two Ethnic Communities in Israel.* Cambridge, England: Cambridge University Press.

Simons, Marlise. 1988. "Brazil's Blacks Feel Prejudice 100 Years After Slavery's End." *New York Times* (May 14):1, 6.

Simpson, George Eaton. 1968. "Assimilation." Pp. 428–444 in David L. Sills (ed.), *International Encyclopedia of the Social Sciences.* Vol. 1. New York: Macmillan.

Simpson, George Eaton, and J. Milton Yinger. 1972. *Racial and Cultural Minorities: An Analysis of Prejudice and Discrimination.* 4th ed. New York: Harper & Row.

Sindler, Allan P. 1978. *Bakke, DeFunis, and Minority Admissions: The Quest for Equal Opportunity.* New York: Longman.

Skellington, Richard. 1996. *"Race" in Britain Today.* 2d ed. London: Sage.

Skidmore, Thomas. 1972. "Toward a Comparative Analysis of Race Relations Since Abolition in Brazil and the United States." *Journal of Latin American Studies* 4(May):1–28.

———. 1974. *Black into White: Race and Nationality in Brazilian Thought.* New York: Oxford University Press.

———. 1990. "Racial Ideas and Social Policy in Brazil, 1870–1940." Pp. 7–36 in Richard Graham (ed.), *The Idea of Race in Latin America, 1870–1940.* Austin, Tex.: University of Texas Press.

———. 1993. "Bi-racial U.S.A. vs. Multi-racial Brazil: Is the Contrast Still Valid?" *Journal of Latin American Studies* 25:373–386.

Sklare, Marshall. 1969. "The Ethnic Church and the Desire for Survival." Pp. 101–117 in Peter I. Rose (ed.), *The Ghetto and Beyond.* New York: Random House.

———. 1971. *America's Jews.* New York: Random House.

Sklare, Marshall, and Joseph Greenblum. 1967. *Jewish Identity in the Suburban Frontier: A Study of Group Survival in the Open Society.* New York: Basic Books.

Slawson, John, and Lawrence Bloomgarden. 1965. *The Unequal Treatment of Equals: The Social Club . . . Discrimination in Retreat.* New York: Institute of Human Relations, American Jewish Committee.

Smith, Anthony D. 1981. *The Ethnic Revival.* Cambridge, England: Cambridge University Press.

Smith, David J., and Gerald Chambers. 1991. *Inequality in Northern Ireland.* Oxford: Clarendon Press.

Smith, Elmer Lewis. 1958. *The Amish People.* New York: Exposition Press.

Smith, M. G. 1965. *The Plural Society in the British West Indies.* Berkeley: University of California Press.

———. 1969. "Institutional and Political Conditions of Pluralism." Pp. 27–65 in Leo Kuper and M. G. Smith (eds.), *Pluralism in Africa.* Berkeley: University of California Press.

Smith, Robin. 1987. "The Black Trade Unions: From Economics to Politics." Pp. 90–106 in Jesmond Blumenfeld (ed.), *South Africa in Crisis.* London: Croom Helm.

Smith, T. Lynn. 1963. *Brazil: People and Institutions.* Baton Rouge, La.: Louisiana State University Press.

———. 1974. *Brazilian Society.* Albuquerque, N. Mex.: University of New Mexico Press.
Smith, Tom W. 1980. *A Compendium of Trends on General Social Survey Questions* (NORC Report No. 129). Chicago: National Opinion Research Center.
———. 1990a. "Ethnic Images." *GSS Topical Report No. 19.* Chicago: National Opinion Research Center.
———. 1990b. *Jewish Attitudes Toward Blacks & Race Relations.* New York: American Jewish Committee.
———. 1991. *What Do Americans Think about Jews?* New York: American Jewish Committee.
Smooha, Sammy. 1978. *Israel: Pluralism and Conflict.* London: Routledge & Kegan Paul.
———. 1988. "Internal Divisions in Israel at Forty." *Middle East Review* 20(Summer):26–36.
Sniderman, Paul M., and Thomas Piazza. 1993. *The Scar of Race.* Cambridge, Mass.: Belknap Press.
Snipp, C. Matthew. 1986. "The Changing Political and Economic Status of the American Indians: From Captive Nations to Internal Colonies." *American Journal of Economics and Sociology* 45:145–157.
———. 1989. *American Indians: The First of This Land.* New York: Russell Sage Foundation.
Solis-Garza, Luis A. 1972. "Cesar Chavez: The Chicano Messiah?" Pp. 297–305 in Edward Simmen (ed.), *Pain and Promise: The Chicano Today.* New York: New American Library.
South African Information Service. 1979. *South Africa 1979: Official Yearbook for the Republic of South Africa.* Pretoria: Department of Foreign Affairs and Information.
South African Institute of Race Relations. 1992. *Race Relations Survey 1991/92.* Johannesburg: South African Institute of Race Relations.
———. 1993. *Race Relations Survey 1992/93.* Johannesburg: South African Institute of Race Relations.
———. 1994. "A Picture of the Population." *Fast Facts* (September):1–3.
Sowell, Thomas. 1978. "Three Black Histories." Pp. 7–64 in Thomas Sowell (ed.), *Essays and Data on American Ethnic Groups.* Urban Institute.
———. 1981. *Ethnic America.* New York: Basic Books.
———. 1990. *Preferential Policies: An International Perspective.* New York: William Morrow.
Spear, Allan H. 1967. *Black Chicago.* Chicago: University of Chicago Press.
Spicer, Edward H. 1980. "American Indians." Pp. 58–122 in Stephen Thernstrom (ed.), *Harvard Encyclopedia of American Ethnic Groups.* Cambridge, Mass.: Harvard University Press.
Stampp, Kenneth M. 1956. *The Peculiar Institution: Slavery in the Ante-Bellum South.* New York: Knopf.
Stanbury, W. T. 1975. *Success and Failure: Indians in Urban Society.* Vancouver: University of British Columbia Press.
Stanley, David T., Dean E. Mann, and Jameson W. Doig. 1967. *Men Who Govern.* Washington, D.C.: Brookings Institution.
Staples, Robert. 1981. "The Black American Family." Pp. 217–244 in Charles H. Mindeland and Robert W. Habenstein (eds.), *Ethnic Families in America.* 2d ed. New York: Elsevier.
Stark, Andrew. 1992. "English-Canadian Opposition to Quebec Nationalism." Pp. 123–158 in R. Kent Weaver (ed.), *The Collapse of Canada?* Washington, D.C.: Brookings Institution.
Statistics Canada. 1988. *Current Demographic Analysis: Income of Immigrants in Canada.* Ottawa: Minister of Supply and Services.
———. 1989a. *The Nation: Language,* Part 1. Ottawa: Minister of Supply and Services.
———. 1989b. *Dimensions: Profile of Ethnic Groups.* Ottawa: Minister of Supply and Services.
———. 1992. *Immigration and Citizenship: The Nation.* Ottawa: Ministry of Industry, Science, and Technology.
———. 1993a. *Ethnic Origin.* Ottawa: Industry, Science and Technology Canada.
———. 1993b. *Language, Tradition, Health, Lifestyle and Social Issues: 1991 Aboriginal Peoples Survey.* Ottawa: Ministry of Industry, Science, and Technology.

———. 1995. *Canada at a Glance 1995.* 12-581E. Ottawa: Communications Division.

Steeh, Charlotte, and Howard Schuman. 1992. "Young White Adults: Did Racial Attitudes Change in the 1980s?" *American Journal of Sociology* 98:340–367.

Steele, Shelby. 1991. *The Content of Our Character: A New Vision of Race in America.* New York: HarperPerennial.

Steger, Wilbur. 1973. "Economic and Social Costs of Residential Segregation." Pp. 83–113 in Marion Clawson (ed.), *Modernizing Urban Land Policy.* Baltimore, Md.: Johns Hopkins University Press.

Stein, Ben. 1985. "'Miami Vice': It's So Hip You'll Want to Kill Yourself." *Public Opinion* 8(October/November):41–43.

Steinberg, Stephen. 1989. *The Ethnic Myth: Race, Ethnicity and Class in America.* Updated and expanded ed. New York: Atheneum.

Steiner, Stan. 1970. *La Raza: The Mexican Americans.* New York: Harper & Row.

Steinfels, Peter. 1992. "Debating Intermarriage, and Jewish Survival." *New York Times* (October 18):1, 16.

Stember, Charles H. 1966. *Jews in the Mind of America.* New York: Basic Books.

Stille, Alexander. 1992. "Italy: No Blacks Need Apply." *Atlantic Monthly* 269 (February):28–38.

Stoddard, Ellwyn R. 1973. *Mexican Americans.* New York: Random House.

Stokes, Bruce. 1988. "Learning the Game." *National Journal* 21(October 22):2649–2654.

Stone, John. 1973. *Colonist or Uitlander: A Study of the British Immigrant in South Africa.* Oxford: Clarendon Press.

Stonequist, Everett V. 1937. *The Marginal Man.* New York: Charles Scribner's Sons.

Strodtbeck, Fred L. 1958. "Family Interaction, Values, and Achievement." Pp. 147–165 in Marshall Sklare (ed.), *The Jews: Social Patterns of an American Group.* Glencoe, Ill.: Free Press.

Stultz, Newell M. 1980. "Some Implications of African 'Homelands' in South Africa." Pp. 194–216 in Robert M. Price and Carl G. Rotberg (eds.), *The Apartheid Regime: Political Power and Racial Domination.* Berkeley: Institute of International Studies, University of California.

Sudetic, Chuck. 1994. "Serbs of Sarajevo Stay Loyal to Bosnia." *New York Times* (August 26):8A.

Sunahara, M. Ann. 1980. "Federal Policy and the Japanese Canadians: The Decision to Evacuate, 1942." Pp. 93–120 in K. Victor Ujimoto and Gordon Hirabayashi (eds.), *Visible Minorities and Multiculturalism: Asians in Canada.* Toronto: Butterworths.

Suro, Roberto. 1989. "Employers Are Looking Abroad for the Skilled and the Energetic." *New York Times* (July 16):IV-4.

Suzuki, Teiiti. 1969. *The Japanese Immigrant in Brazil.* Tokyo: University of Tokyo Press.

Swain, Carol M. 1995. "The Future of Black Representation." *American Prospect* (Fall):78–83.

Swinton, David H. 1992. "The Economic Status of African Americans: Limited Ownership and Persistent Inequality." Pp. 61–117 in National Urban League, *The State of Black America 1992.* New York: National Urban League.

Szabo, Joan C. 1989. "Opening Doors for Immigrants." *Nation's Business* 77(August):48–49.

Szymanski, Albert. 1976. "Social Discrimination and White Gain." *American Sociological Review* 41:403–414.

Taeuber, Karl E. 1975. "Racial Segregation: The Persisting Dilemma." *Annals of the American Academy of Political and Social Science* 422:87–96.

———. 1979. "Housing, Schools, and Incremental Segregative Effects." *Annals of the American Academy of Political and Social Science* 441:157–167.

Taeuber, Karl E., and Alma F. Taeuber. 1964. "The Negro as an Immigrant Group: Recent Trends in Racial and Ethnic Segregation in Chicago." *American Journal of Sociology* 69:374–382.

———. 1965. *Negroes in Cities.* Chicago: Aldine.

Taft, Ronald. 1963. "The Assimilation Orientation of Immigrants and Australians." *Human Relations* 16:279–293.

Tannenbaum, Frank. 1947. *Slave and Citizen: The Negro in the Americas.* New York: Knopf.

Tasker, Fredric. 1980. "Anti-Bilingualism Measure Approved in Dade County." *Miami Herald* (November 5):1.

Taylor, D. Garth. 1979. "Housing, Neighborhoods, and Race Relations: Recent Survey Evidence." *Annals of the American Academy of Political and Social Science* 441:26–40.

Taylor, D. Garth, Paul B. Sheatsley, and Andrew M. Greeley. 1978. "Attitudes Toward Racial Integration." *Scientific American* 238(June):42–49.

Taylor, Paul. 1995. "South Africa: 'A Miracle That Keeps Evolving.'" *Washington Post* (May 1–7):16.

Taylor, Ronald L. 1979. "Black Ethnicity and the Persistence of Ethnogenesis." *American Journal of Sociology* 84:1401–1423.

Taylor, William L. 1995. "Affirmative Action: The Questions to Be Asked." *Poverty and Race* (May/June):2–3.

Telles, Edward E. 1992. "Residential Segregation by Skin Color in Brazil." *American Sociological Review* 57:186–197.

tenBroek, Jacobus, Edward N. Barnhart, and Floyd W. Matson. 1968. *Prejudice, War and the Constitution.* Berkeley: University of California Press.

Thomas, Piri. 1967. *Down These Mean Streets.* New York: Knopf.

Thomas, Robert K. 1966. "Colonialism: Classic and Internal." *New University Thought* 4(Winter):37–44.

Thomas, W. I., and Florian Znaniecki. 1918. *The Polish Peasant in Europe and America.* Vol. 1. New York: Knopf.

Thompson, John Herd, and Morton Weinfeld. 1995. "Entry and Exit: Canadian Immigration Policy in Context." *Annals of the American Academy of Political and Social Science* 538:185–198.

Thompson, Leonard M. 1964. "The South African Dilemma." Pp. 178–218 in Louis Hartz, *The Founding of New Societies.* New York: Harcourt, Brace & World.

———. 1975. "Introduction." Pp. ix–xv in Leonard Thompson and Jeffrey Butler (eds.), *Change in Contemporary South Africa.* Berkeley: University of California Press.

———. 1985. *The Political Mythology of Apartheid.* New Haven, Conn.: Yale University Press.

———. 1990. *A History of South Africa.* New Haven, Conn.: Yale University Press.

Thompson, Leonard M., and Andrew Prior. 1982. *South African Politics.* New Haven, Conn.: Yale University Press.

Thomson, Dale. 1995. "Language, Identity, and the Nationalist Impulse: Quebec." *Annals of the American Academy of Political and Social Science* 538(March):69–82.

Thornton, Russell. 1987. *American Indian Holocaust and Survival: A Population History Since 1492.* Norman, Okla.: University of Oklahoma Press.

———. 1995. "North American Indians and the Demography of Contact." Pp. 213–230 in Vera Lawrence Hyatt and Rex Nettleford (eds.), *Race, Discourse, and the Origin of the Americas: A New World View.* Washington, D.C.: Smithsonian Institution Press.

Thurow, Lester. 1969. *Poverty and Discrimination.* Washington, D.C.: Brookings Institution.

Tien, Chan-Liu. 1995. "The Role of Asian Americans in Higher Education." *Migration World* 23(4):23–25.

Time. 1960. "Grosse Pointe's Gross Points." 75(April 25):25.

———. 1987. "The New Whiz Kids." 131(August 31):42–51.

———. 1995. "The Rise of English Only." (October 9):49.

Tinker, John N. 1973. "Intermarriage and Ethnic Boundaries: The Japanese American Case." *Journal of Social Issues* 29:49–66.

Tiryakian, Edward A. 1967. "Sociological Realism: Partition for South Africa." *Social Forces* 46:208–221.

Toch, Hans. 1965. *The Social Psychology of Social Movements.* Indianapolis, Ind.: Bobbs-Merrill.

Toplin, Robert Brent. 1981. *Freedom and Prejudice: The Legacy of Slavery in the United States and Brazil.* Westport, Conn.: Greenwood.

Train, Arthur. 1974 (1912). "Imported Crime." Pp. 184–188 in Wayne Moquin and Charles

Van Doren (eds.), *A Documentary History of the Italian Americans.* New York: Praeger.

Troper, Harold M. 1972. *Only Farmers Need Apply.* Toronto: Griffen House.

Trueheart, Charles, and Dennis McAuliffe, Jr. 1995. "A Way of Life at Risk." *Washington Post* (September 18–24):18–19.

Tumin, Melvin M. 1964. "Ethnic Group." P. 243 in Julius Gould and William L. Kolb (eds.), *Dictionary of the Social Sciences.* New York: Free Press.

Turner, Jonathan H., and Edna Bonacich. 1980. "Toward a Composite Theory of Middleman Minorities." *Ethnicity* 7:144–158.

Turner, Ralph H., and Lewis M. Killian. 1972. *Collective Behavior.* 2d ed. Englewood Cliffs, N.J.: Prentice-Hall.

Tuttle, William M., Jr. 1970. *Race Riot: Chicago in the Red Summer of 1919.* New York: Atheneum.

Tyler, Gus. 1975. "Introduction: A People on the Move." Pp. 1–18 in Gus Tyler (ed.), *Mexican-Americans Tomorrow.* Albuquerque, N. Mex.: University of New Mexico Press.

Ulster Year Book. 1963–65. Belfast: Her Majesty's Stationery Office.

———. 1983. Belfast: Her Majesty's Stationery Office.

UNFPA (United Nations Population Fund). 1993. *The Individual and the World: Population, Migration and Development in the 1990s.* New York: UNFPA.

USA Today. 1989. "Blacks' Views of Bias in the USA." (September 5):6A.

———. 1995a. "Affirmative Action: The Public Reaction." (March 24):2A–3A.

———. 1995b. "The New Immigrants." (June 30–July 2):1A, 2A.

U.S. Bureau of the Census. 1971. *Statistical Abstract of the United States: 1971.* Washington, D.C.: U.S. Government Printing Office.

———. 1973. *Statistical Abstract of the United States: 1973.* Washington, D.C.: U.S. Government Printing Office.

———. 1975. *Historical Statistics of the United States: Colonial Times to 1970.* Part 1. Washington, D.C.: U.S. Government Printing Office.

———. 1978. "Perspectives on American Husbands and Wives." *Current Population Reports.* Series P-23, No. 77. Washington, D.C.: U.S. Government Printing Office.

———. 1979. "The Social and Economic Status of the Black Population in the United States: An Historical Overview, 1790–1978." *Current Population Reports.* Series P-23, No. 80. Washington, D.C.: U.S. Government Printing Office.

———. 1981. "Race of the Population by States: 1980." *1980 Census of Population.* PC80-S1-3. Washington, D.C.: U.S. Government Printing Office.

———. 1982a. "Ancestry and Language in the United States, November 1979." *Current Population Reports.* Series P-23, No. 116. Washington, D.C.: U.S. Government Printing Office.

———. 1982b. *Statistical Abstract of the United States: 1982–83.* Washington, D.C.: U.S. Government Printing Office.

———. 1983. *America's Black Population: 1970 to 1982, A Statistical View.* Special Publication PIO/POP-83-1. Washington, D.C.: U.S. Government Printing Office.

———. 1988. *Statistical Abstract of the United States: 1988.* Washington, D.C.: U.S. Government Printing Office.

———. 1989a. *Statistical Abstract of the United States: 1989.* Washington, D.C.: U.S. Government Printing Office.

———. 1989b. "Population Profile of the United States: 1989." *Current Population Reports.* Series P-23, No. 159. Washington, D.C.: U.S. Government Printing Office.

———. 1989c. "Money Income and Poverty Status in the United States: 1987." *Current Population Reports.* Series P-60, No. 161. Washington, D.C.: U.S. Government Printing Office.

———. 1989d. "Voting and Registration in the Election of November 1988" (Advance Report). *Current Population Reports.* Series P-20, No. 435. Washington, D.C.: U.S. Government Printing Office.

———. 1991a. *Statistical Abstract of the United States: 1991*. Washington, D.C.: U.S. Government Printing Office.

———. 1991b. "The Hispanic Population in the United States: March 1990." *Current Population Reports.* Series P-20, No. 449. Washington, D.C.: U.S. Government Printing Office.

———. 1991c. "Population Profile of the United States 1991." *Current Population Reports.* P-23-173. Washington, D.C.: U.S. Government Printing Office.

———. 1992a. "The Black Population in the United States: March 1991." *Current Population Reports.* P20-404. Washington, D.C.: U.S. Government Printing Office.

———. 1992b. "Educational Attainment in the United States: March 1991 and 1990." *Current Population Reports.* P20-462. Washington, D.C.: U.S. Government Printing Office.

———. 1992c. *Statistical Abstract of the United States: 1992*. Washington, D.C.: U.S. Government Printing Office.

———. 1992d. "The Asian and Pacific Islander Population in the United States: March 1991 and 1990." *Current Population Reports.* P20-459. Washington, D.C.: U.S. Government Printing Office.

———. 1992e. "Marital Status and Living Arrangements: March 1992." *Current Population Reports.* P20-468. Washington, D.C.: U.S. Government Printing Office.

———. 1992f. "The Hispanic Population in the United States: March 1992." *Current Population Reports.* P20-465. Washington, D.C.: U.S. Government Printing Office.

———. 1993a. *1990 Census of Population, Asians and Pacific Islanders in the United States.* 1990 CP-3-5. Washington, D.C.: U.S. Government Printing Office.

———. 1993b. *1990 Census of Population and Housing Supplementary Reports.* Series CPH-S-1-1. Washington, D.C.: U.S. Government Printing Office.

———. 1993c. *Ancestry of the Population in the United States, 1990*. CP-3-2. Washington, D.C.: U.S. Government Printing Office.

———. 1993d. *1990 Census of Population: Social and Economic Characteristics, Metropolitan Areas.* Washington, D.C.: U.S. Government Printing Office.

———. 1994a. *Statistical Abstract of the United States: 1994*. Washington, D.C.: U.S. Government Printing Office.

———. 1994b. "The Hispanic Population in the United States: March 1993." *Current Population Reports.* Series P20-475. Washington, D.C.: U.S. Government Printing Office.

———. 1995a. "The Black Population in the United States: March 1994 and 1993." *Current Population Reports.* Series P20-480. Washington, D.C.: U.S. Government Printing Office.

———. 1995b. *Statistical Abstract of the United States: 1995*. Washington, D.C.: U.S. Government Printing Office.

———. 1996. *Population Projections of the United States by Age, Sex, Race and Hispanic Origin.* Washington, D.C.: U.S. Government Printing Office.

U.S. Commission on Civil Rights. 1970. *Mexican Americans and the Administration of Justice in the Southwest.* Washington, D.C.: U.S. Government Printing Office.

———. 1972. *The Excluded Student: Educational Practices Affecting Mexican Americans in the Southwest.* Mexican American Educational Study Report 3. Washington, D.C.: U.S. Government Printing Office.

———. 1974. *Toward Equality in Education for Mexican Americans.* Mexican American Educational Study Report 6. Washington, D.C.: U.S. Government Printing Office.

———. 1976. *Puerto Ricans in the Continental United States: An Uncertain Future.* Washington, D.C.: U.S. Government Printing Office.

———. 1977. *Window Dressing on the Set: Women and Minorities in Television.* Washington, D.C.: U.S. Government Printing Office.

———. 1979. *Window Dressing on the Set: An Update.* Washington, D.C.: U.S. Government Printing Office.

———. 1985. *Issues in Housing Discrimination: A Consultation/Hearing of the United States Commission on Civil Rights, November 12–13, 1985.*

Washington, D.C.: U.S. Government Printing Office.

———. 1986a. *The Economic Status of Americans of Southern and Eastern European Ancestry.* Washington, D.C.: Clearinghouse Publication.

———. 1986b. *Recent Activities Against Citizens and Residents of Asian Descent.* Washington, D.C.: U.S. Government Printing Office.

———. 1988. *The Economic Status of Americans of Asian Descent: An Exploratory Investigation.* Washington, D.C.: Clearinghouse Publication.

———. 1992. *Civil Rights Issues Facing Asian Americans in the 1990s.* Washington, D.C.: U.S. Government Printing Office.

U.S. Department of Education. 1991. *Indian Nations at Risk: An Educational Strategy for Action.* Washington, D.C.: U.S. Department of Education.

———. 1992. *Digest of Education Statistics.* Washington, D.C.: U.S. Government Printing Office.

U.S. Department of the Interior, Bureau of Indian Affairs. 1991. *American Indians Today.* 3d ed. Washington, D.C.: U.S. Department of the Interior, Bureau of Indian Affairs.

U.S. Department of Labor. 1983. *Employment and Earnings, October 1983.* Vol. 30, No. 10. Washington, D.C.: U.S. Government Printing Office.

———. 1989. *Monthly Labor Review* (April). Washington, D.C.: U.S. Government Printing Office.

———. 1992. *Monthly Labor Review* (July). Washington, D.C.: U.S. Government Printing Office.

U.S. Immigration and Naturalization Service. 1991. *1990 Statistical Yearbook of the Immigration and Naturalization Service.* Washington, D.C.: U.S. Government Printing Office.

———. 1994. *Statistical Yearbook of the Immigration and Naturalization Service 1993.* Washington, D.C.: U.S. Government Printing Office.

Utter, Jack. 1993. *American Indians: Answers to Today's Questions.* Lake Ann, Mich.: National Woodlands.

Valdez, Avelardo. 1983. "Recent Increases in Intermarriage by Mexican-American Males." *Social Science Quarterly* 64:136–144.

Valentine, Charles A. 1971. "The 'Culture of Poverty': Its Scientific Significance and Its Implications for Actions." Pp. 193–225 in Eleanor Burke Leacock (ed.), *The Culture of Poverty: A Critique.* New York: Simon & Schuster.

van den Berghe, Pierre L. 1967. *South Africa: A Study in Conflict.* Berkeley: University of California Press.

———. 1970. *Race and Ethnicity: Essays in Comparative Sociology.* New York: Basic Books.

———. 1976. "Ethnic Pluralism in Industrial Societies: A Special Case?" *Ethnicity* 3:242–255.

———. 1978. *Race and Racism: A Comparative Perspective.* 2d ed. New York: Wiley.

———. 1979a. "The Impossibility of a Liberal Solution in South Africa." Pp. 56–67 in Pierre van den Berghe (ed.), *The Liberal Dilemma in South Africa.* New York: St. Martin's.

———. 1979b. "Nigeria and Peru: Two Contrasting Cases in Ethnic Pluralism." *International Journal of Comparative Sociology* 20:162–174.

———. 1981. *The Ethnic Phenomenon.* New York: Elsevier.

van der Merwe, Hendrik W., and David Welsh. 1980. "Identity, Ethnicity and Nationalism as Political Forces in South Africa: The Case of Afrikaners and Coloured People." Pp. 263–274 in Jacques Dofny and Akinsola Akiwowo (eds.), *National and Ethnic Movements.* Beverly Hills, Calif.: Sage.

Vander Zanden, James W. 1983. *American Minority Relations.* 4th ed. New York: Knopf.

Van Valey, Thomas L., Wade Clark Roof, and Jerome Wilcox. 1977. "Trends in Residential Segregation: 1960–1970." *American Journal of Sociology* 82:826–844.

van zyl Slabbert, F. 1975. "Afrikaner Nationalism, White Politics, and Political Change in South Africa." Pp. 3–18 in Leonard Thompson and Jeffrey Butler (eds.), *Change in Contemporary South Africa.* Berkeley: University of California Press.

———. 1987. "Incremental Change or Revolution?" Pp. 399–409 in Jeffrey Butler, Richard Elphick, and David Welsh (eds.), *Democratic Liberalism in South Africa: Its History and Prospect.* Middletown, Conn: Wesleyan University Press.

Vecoli, Rudolph J. 1964. "Contadini in Chicago: A Critique of the Uprooted." *Journal of American History* 51:404–417.

Veltman, Calvin. 1983. *Language Shift in the United States.* Berlin: Mouton.

Verba, Sidney, and Norman H. Nie. 1972. *Participation in America.* New York: Harper & Row.

Vidal, David. 1978. "Many Blacks Shut Out of Brazil's Racial 'Paradise.'" *New York Times* (June 5):1, 10.

Vidmar, Neil, and Milton Rokeach. 1974. "Archie Bunker's Bigotry: A Study in Selective Perception and Exposure." *Journal of Communication* 24:36–47.

Volkan, Vamik, and Max Harris. 1992. "Negotiating a Peaceful Separation: A Psychopolitical Analysis of Current Relationships Between Russia and the Baltic Republics." *Mind and Human Interaction* 4(December):20–23.

Wagenheim, Kal. 1975. *A Survey of Puerto Ricans in the United States Mainland in the 1970s.* New York: Praeger.

Wagley, Charles. 1963. "Race Relations in an Amazon Community." Pp. 116–141 in Charles Wagley (ed.), *Race and Class in Rural Brazil.* 2d ed. New York: Columbia University Press.

———. 1971. *An Introduction to Brazil.* Rev. ed. New York: Columbia University Press.

Wagley, Charles, and Marvin Harris. 1958. *Minorities in the New World: Six Case Studies.* New York: Columbia University Press.

Wagner, Nathaniel N., and Marsha J. Haug. 1971. *Chicanos: Social and Psychological Perspectives.* St. Louis: Mosby.

Walker, James W. St. G. 1980. *A History of Blacks in Canada: A Study Guide for Teachers and Students.* Ottawa: Minister of Supply and Services.

Wallace, Martin. 1970. *Drums and Guns: Revolution in Ulster.* London: Geoffrey Chapman.

———. 1971. *Northern Ireland: 50 Years of Self-Government.* New York: Barnes & Noble.

Walsh, James H. 1992. "Migration and European Nationalism." *Migration World* 20(4):19–22.

Ward, Lewis B. 1965. "The Ethnics of Executive Selection." *Harvard Business Review* 43(March/April):6–39.

Warner, W. Lloyd, and James Abegglen. 1963. *Big Business Leaders in America.* New York: Atheneum.

Warner, W. Lloyd, and Leo Srole. 1945. *The Social Systems of American Ethnic Groups.* New Haven, Conn.: Yale University Press.

Warren, Christopher L., and John F. Stack, Jr. 1986. "Immigration and the Politics of Ethnicity and Class in Metropolitan Miami." Pp. 61–79 in John F. Stack, Jr. (ed.), *The Primordial Challenge: Ethnicity in the Contemporary World.* New York: Greenwood.

Washburn, Wilcomb E. 1975. *The Indian in America.* New York: Harper & Row.

Washington Post. 1995. "Middle-Class Views in Black and White." (October 16–22):8.

Waters, Mary C. 1990. *Ethnic Options: Choosing Identities in America.* Berkeley: University of California Press.

Waters, Mary C., and Karl Eschbach. 1995. "Immigration and Ethnic and Racial Inequality in the United States." *Annual Review of Sociology* 21:419–446.

Watson, James L. 1977. *Between Two Cultures: Migrants and Minorities in Britain.* Oxford: Basil Blackwell.

Wax, Murray L. 1971. *Indian Americans: Unity and Diversity.* Englewood Cliffs, N.J.: Prentice-Hall.

Waxman, Chaim I. 1981. "The Fourth Generation Grows Up: The Contemporary American Jewish Community." *Annals of the American Academy of Political and Social Science* 454:70–85.

Webster, Peggy Lovell, and Jeffrey W. Dwyer. 1988. "The Cost of Being Nonwhite in Brazil." *Sociology and Social Research* 72:136–138.

Weiss, Lowell. 1994. "Timing Is Everything." *Atlantic Monthly* (January):32–37.

Welch, Susan, John Comer, and Michael Steinman. 1973. "Political Participation Among Mexican Americans: An Exploratory Examination." *Social Science Quarterly* 53:779–813.

Welsh, David. 1975. "The Politics of White Supremacy." Pp. 51–78 in Leonard Thompson and Jeffrey Butler (eds.), *Change in Contemporary South Africa.* Berkeley: University of California Press.

Wertheimer, Jack. 1989. "Recent Trends in American Judaism." Pp. 63–162 in *American Jewish Year Book 1989.* New York: American Jewish Committee.

———. 1993. *A People Divided: Judaism in Contemporary America.* New York: Basic Books.

Westerman, Marty. 1989. "Death of the Frito Bandito." *American Demographics* 11(March):28–32.

Westie, Frank P. 1964. "Race and Ethnic Relations." Pp. 576–618 in Robert E. L. Faris (ed.), *Handbook of Modern Sociology.* Chicago: Rand McNally.

Whitaker, Mark. 1995. "Whites v. Blacks." *Newsweek* (October 16):28–35.

White, George. 1993. "Anger Flares on Both Sides of Counter." *Los Angeles Times* (June 8):H7.

White, Michael J. 1987. *American Neighborhoods and Residential Differentiation.* New York: Russell Sage Foundation.

Whitfield, Stephen J. 1995. "Movies in America as Paradigms of Accommodation." Pp. 79–94 in Robert M. Seltzer and Norman J. Cohen (eds.), *The Americanization of the Jews.* New York: New York University Press.

Whitney, Craig R. 1988. "Belfast: Strife-Wracked but No Beirut." *New York Times* (October 30):10.

Wiechers, Marinus. 1989. *South African Political Terms.* Cape Town: Tafelberg.

Wilkerson, Isabel. 1989. "Many Who Are Black Favor New Term for Who They Are." *New York Times* (January 31):1, 8.

———. 1993. "Black Mediator Serves As a Bridge to Koreans." *New York Times* (June 2):A7.

Wilkie, Mary E. 1977. "Colonials, Marginals and Immigrants: Contributions to a Theory of Ethnic Stratification." *Comparative Studies in Society and History* 19:67–95.

Willems, Emilio. 1960. "Brazil." Pp. 119–146 in International Sociological Association, *The Positive Contribution by Immigrants.* Paris: UNESCO.

Willhelm, Sidney M. 1971. *Who Needs the Negro?* Garden City, N.Y.: Doubleday.

Williams, J. Allen, Jr., Peter G. Beeson, and David P. Johnson. 1973. "Some Factors Associated with Income Among Mexican Americans." *Social Science Quarterly* 53:710–715.

Williams, Lena. 1995. "Not Just a White Man's Game." *New York Times* (November 9):C1, C6.

Williams, Robin M., Jr. 1964. *Strangers Next Door: Ethnic Relations in American Communities.* Englewood Cliffs, N.J.: Prentice-Hall.

———. 1977. *Mutual Accommodation: Ethnic Conflict and Cooperation.* Minneapolis, Minn.: University of Minnesota Press.

———. 1979. "Structure and Process in Ethnic Relations: Increased Knowledge and Unanswered Questions." Paper presented at the annual meeting of the American Sociological Association, Boston.

———. 1994. "The Sociology of Ethnic Conflicts: Comparative International Perspectives." *Annual Review of Sociology* 20:49–79.

Willie, Charles V. 1978. "The Inclining Significance of Race." *Society* 15(July/August):10–15.

Wilson, Clint C., II, and Félix Gutiérrez. 1995. *Race, Multiculturalism, and the Media: From Mass to Class Communication.* 2d ed. Thousand Oaks, Calif.: Sage.

Wilson, Kenneth L., and Alejandro Portes. 1980. "Immigrant Enclaves: An Analysis of the Labor Market Experiences of Cubans in Miami." *American Journal of Sociology* 86:295–315.

Wilson, Monica, and Leonard Thompson. 1969. *The Oxford History of South Africa.* New York: Oxford University Press.

Wilson, Reginald. 1989. "The State of Black Higher Education: Crisis and Promise." Pp.

121–135 in *The State of Black America 1989*. New York: National Urban League.

Wilson, William J. 1973. *Power, Racism, and Privilege: Race Relations in Theoretical and Sociohistorical Perspectives.* New York: Free Press.

———. 1980. *The Declining Significance of Race.* 2d ed. Chicago: University of Chicago Press.

———. 1981. "The Black Community in the 1980s: Questions of Race, Class, and Public Policy." *Annals of the American Academy of Political and Social Science* 454:26–41.

———. 1987. *The Truly Disadvantaged: The Inner City, the Underclass, and Public Policy.* Chicago: University of Chicago Press.

———. 1994. "Race-Neutral Programs and the Democratic Coalition." Pp. 159–173 in Nicolaus Mills (ed.), *Debating Affirmative Action: Race, Gender, Ethnicity, and the Politics of Inclusion.* New York: Delta.

Winks, Robin. 1971. *The Blacks in Canada.* New Haven, Conn.: Yale University Press.

Winnick, Louis. 1990. *New People in Old Neighborhoods: The Role of New Immigrants in Rejuvenating New York's Communities.* New York: Russell Sage Foundation.

Wirth, Louis. 1945. "The Problem of Minority Groups." Pp. 347–372 in Ralph Linton (ed.), *The Science of Man in the World Crisis.* New York: Columbia University Press.

———. 1956 (1928). *The Ghetto.* Chicago: University of Chicago Press.

Wolfinger, Raymond E. 1966. "Some Consequences of Ethnic Politics." Pp. 42–54 in M. Kent Jennings and L. Harmon Zeigler (eds.), *The Electoral Process.* Englewood Cliffs, N.J.: Prentice-Hall.

Wong, Francisco Raimundo. 1974. "The Political Behavior of Cuban Migrants." Ph.D. dissertation, University of Michigan.

Wong, Morrison G. 1986. "Post-1965 Asian Immigrants: Where Do They Come From, Where Are They Now, and Where Are They Going?" *Annals of the American Academy of Political and Social Science* 487:150–168.

———. 1989. "A Look at Intermarriage Among the Chinese in the United States in 1980." *Sociological Perspectives* 32:87–107.

Wood, Charles H., and Jose Alberto Magno de Carvalho. 1988. *The Demography of Inequality in Brazil.* Cambridge, England: Cambridge University Press.

Wood, Floris W. (ed.). 1990. *An American Profile—Opinions and Behavior, 1972–1989.* Detroit, Mich.: Gale Research.

Woodrum, Eric. 1981. "An Assessment of Japanese American Assimilation, Pluralism, and Subordination." *American Journal of Sociology* 87:157–169.

Woodward, C. Vann. 1974. *The Strange Career of Jim Crow.* 3d ed. New York: Oxford University Press.

Worsnop, Richard. 1991. "Asian Americans." *CQ Researcher* 1(December 3):945–968.

———. 1992. "Native Americans." *CQ Researcher* 2(May 8):385–408.

Wren, Christopher S. 1989. "South Africans Put Rising Faith in Negotiations." *New York Times* (September 4):1, 6.

———. 1991. "South Africa Scraps Law Defining People by Race." *New York Times* (June 18):A1.

Wright, Charles R. 1986. *Mass Communication: A Sociological Perspective.* 3d ed. New York: Random House.

Wright, Lawrence. 1994. "One Drop of Blood." *New Yorker* (July 25):46–50, 52–55.

Wright, Ronald. 1992. *Stolen Continents: The Americas Through Indian Eyes Since 1492.* Boston: Houghton Mifflin.

Yaffee, James. 1968. *The American Jews.* New York: Random House.

Yans-McLaughlin, Virginia. 1982. *Family and Community: Italian Immigrants in Buffalo, 1880–1930.* Urbana, Ill.: University of Illinois Press.

Yates, Steven. 1994. *Civil Wrongs: What Went Wrong with Affirmative Action.* San Francisco: ICS Press.

Yetman, Norman R. (ed.). 1991. *Majority and Minority: The Dynamics of Race and Ethnicity in American Life.* 5th ed. Boston: Allyn & Bacon.

Yinger, J. Milton. 1981. "Toward a Theory of Assimilation and Dissimilation." *Ethnic and Racial Studies* 4:249–264.

Young, Donald. 1932. *American Minority Peoples.* New York: Harper & Row.

Young, Robert A. 1995. *The Secession of Quebec and the Future of Canada.* Montreal and Kingston: McGill-Queen's University Press.

Zenner, Walter P. 1991. *Minorities in the Middle: A Cross-Cultural Analysis.* Albany, N.Y.: State University of New York Press.

Zimmerman, Ben. 1963. "Race Relations in the Arid Sertao." Pp. 82–115 in Charles Wagley (ed.), *Race and Class in Rural Brazil.* 2d ed. New York: Columbia University Press.

Zinsmeister, Karl. 1987. "Asians: Prejudice from Top and Bottom." *Public Opinion* 10(July/August):8–10, 59.

Zureik, Elia T. 1979. *The Palestinians in Israel: A Study in Internal Colonialism.* London: Routledge & Kegan Paul.

Zweigenhaft, Richard L. 1984. *Who Gets to the Top? Executive Suite Discrimination in the Eighties.* New York: American Jewish Committee.

———. 1987. "Minorities and Women of the Corporation: Will They Attain Seats of Power?" Pp. 37–62 in G. W. Domhoff and T. R. Dye (eds.), *Power Elites and Organizations.* Newbury Park, Calif.: Sage.

Zweigenhaft, Richard L., and G. William Domhoff. 1982. *Jews in the Protestant Establishment.* New York: Praeger.

INDEX

B

D

E

F

G

J

O

P

Q

R

S

T